本书英文版于2019年由世界卫生组织（World Health Organization）出版，书名为：WHO Study Group on Tobacco Product Regulation. Report on the Scientific Basis of Tobacco Product Regulation: Seventh Report of a WHO Study Group (WHO Technical Report Series, No. 1015)

© World Health Organization 2019

世界卫生组织（World Health Organization）授权中国科技出版传媒股份有限公司（科学出版社）翻译出版本书中文版。中文版的翻译质量和对原文的忠实性完全由科学出版社负责。当出现中文版与英文版不一致的情况时，应将英文版视作可靠和有约束力的版本。

中文版《烟草制品管制科学基础报告：WHO研究组第七份报告》
©中国科技出版传媒股份有限公司（科学出版社）2022

WHO技术报告系列 1015

WHO烟草制品管制研究小组

烟草制品管制科学基础报告

WHO研究组第七份报告

胡清源 侯宏卫 等 译

科学出版社
北　京

内 容 简 介

本报告介绍了 WHO 烟草制品管制研究小组第九次会议的结论和建议。讨论了新型烟草制品、传统烟草制品管制科学基础方面的优先事项，以及 WHO《烟草控制框架公约》第七次缔约方会议的要求。报告了以下主题：①加热型烟草制品的发展和管制建议；②电子烟碱传输系统中烟碱的临床药理学；③全球烟碱降低策略的研究进展；④降低卷烟烟气中有害物质暴露的监管策略；⑤烟草制品中调味剂的科学研究；⑥烟草制品中的糖对产品管制的影响；⑦燃烧型烟草制品中有害成分优先级清单的扩展研究；⑧无烟烟草制品中有害物质的分析方法；⑨水烟的发展及管制建议。研究组关于每个主题的建议在相关章节末尾列出，最后一章为总体建议。

本书会引起吸烟与健康、烟草化学和公共卫生学诸多应用领域科学家的兴趣，为客观评价烟草制品的管制和披露措施提供必要的参考。

图书在版编目（CIP）数据

烟草制品管制科学基础报告：WHO 研究组第七份报告 / WHO 烟草制品管制研究小组著；胡清源等译. —北京：科学出版社，2022.7
（WHO 技术报告系列；1015）
书名原文：WHO Study Group on Tobacco Product Regulation. Report on the Scientific Basis of Tobacco Product Regulation: Seventh Report of a WHO Study Group
ISBN 978-7-03-072634-6

Ⅰ. ①烟⋯ Ⅱ. ①W⋯ ②胡⋯ Ⅲ. ①烟草制品–科学研究–研究报告 Ⅳ. ① TS45

中国版本图书馆 CIP 数据核字（2022）第 113800 号

责任编辑：刘 冉 / 责任校对：杜子昂
责任印制：吴兆东 / 封面设计：北京图阅盛世

科学出版社 出版
北京东黄城根北街 16 号
邮政编码：100717
http://www.sciencep.com

北京中石油彩色印刷有限责任公司 印刷
科学出版社发行 各地新华书店经销

*

2022 年 7 月第 一 版 开本：720 × 1000 1/16
2022 年 7 月第一次印刷 印张：32 3/4
字数：660 000
定价：198.00 元
（如有印装质量问题，我社负责调换）

WHO Technical Report Series 1015

WHO Study Group on Tobacco Product Regulation

Report on the Scientific Basis of Tobacco Product Regulation: Seventh Report of a WHO Study Group

译者名单

胡清源　侯宏卫　陈　欢
张小涛　田永峰　付亚宁
刘　彤　韩书磊　李国宇
王红娟　田雨闪

译 者 序

2003年5月，第56届世界卫生大会*通过了《烟草控制框架公约》（FCTC），迄今已有包括我国在内的180个缔约方。根据FCTC第9条和第10条的规定，授权世界卫生组织（WHO）烟草制品管制研究小组（TobReg）对可能造成重要公众健康问题的烟草制品管制措施进行鉴别，提供科学合理的、有根据的建议，用于指导成员国进行烟草制品管制。

自2007年起，WHO陆续出版了七份烟草制品管制科学基础报告，分别是945、951、955、967、989、1001和1015。WHO烟草制品管制科学基础系列报告阐述了降低烟草制品的吸引力、致瘾性和毒性等烟草制品管制相关主题的科学依据，内容涉及烟草化学、代谢组学、毒理学、吸烟与健康等烟草制品管制的多学科交叉领域，是一系列以科学研究为依据、对烟草管制发展和决策有重大影响意义的技术报告。将其引进并翻译出版，可以为相关烟草科学研究的科技工作者提供科学性参考。希望引起吸烟与健康、烟草化学和公共卫生学等诸多应用领域科学家的兴趣，为客观评价烟草制品的管制和披露措施提供必要的参考。

第一份报告（945）由胡清源、侯宏卫、韩书磊、陈欢、刘彤、付亚宁翻译，全书由韩书磊负责统稿；

第二份报告（951）由胡清源、侯宏卫、刘彤、付亚宁、陈欢、韩书磊翻译，全书由刘彤负责统稿；

第三份报告（955）由胡清源、侯宏卫、付亚宁、陈欢、韩书磊、刘彤翻译，全书由付亚宁负责统稿；

第四份报告（967）由胡清源、侯宏卫、陈欢、刘彤、韩书磊、付亚宁翻译，全书由陈欢负责统稿；

第五份报告（989）由胡清源、侯宏卫、陈欢、刘彤、韩书磊、付亚宁翻译，全书由陈欢负责统稿；

第六份报告（1001）由胡清源、侯宏卫、韩书磊、陈欢、刘彤、付亚宁、王红娟翻译，全书由韩书磊负责统稿；

---

* 世界卫生大会 (World Health Assembly，WHA) 是世界卫生组织的最高决策机构，每年召开一次。

第七份报告（1015）由胡清源、侯宏卫、陈欢、张小涛、田永峰、付亚宁、刘彤、韩书磊、李国宇、王红娟、田雨闪翻译，全书由刘彤负责统稿。

由于译者学识水平有限，本中文版难免有错漏和不当之处，敬请读者批评指正。

2022年4月

目　　录

WHO 烟草制品管制研究小组第九次会议	xv
致谢	xix
缩略语	xxiii
概述	xxvii
参考文献	xxviii
1. 引言	1
1.1　参考文献	2
2. 加热型烟草制品	3
2.1　引言	3
2.2　加热型烟草制品科学基础	4
2.3　加热型烟草制品简史	5
2.4　新型产品	9
2.4.1　释放物	9
2.4.2　暴露生物标志物	10
2.5　消费者对加热型烟草制品的认知	14
2.6　在选定市场中的普及率	16
2.7　菲利普·莫里斯国际在美国提出的"风险降低的烟草制品"申请	16
2.8　对监管和烟草控制政策的影响	18
2.9　研究和政策建议	19
2.10　参考文献	20
3. 电子烟碱传输系统中烟碱的临床药理学	28
3.1　引言	28
3.2　ENDS 操作	28
3.3　电子烟液中的烟碱浓度	30

3.4　向 ENDS 使用者的烟碱传输 ·· 31
　　3.5　ENDS 释放物中的有害物质 ·· 33
　　　　3.5.1　烟碱释放物 ·· 33
　　　　3.5.2　非烟碱有害释放物 ·· 35
　　3.6　ENDS 在戒烟中的潜在作用 ·· 36
　　3.7　ENDS 对健康的潜在影响 ·· 48
　　　　3.7.1　与 ENDS 使用有关的行为轨迹 ··· 48
　　　　3.7.2　ENDS 和电子非烟碱传输系统的危害 ···································· 51
　　3.8　证据汇总、研究差距和由证据得出的政策问题 ·································· 53
　　3.9　参考文献 ·· 56

4. 全球烟碱降低策略：科学现状 ·· 68
　　4.1　背景 ·· 68
　　4.2　烟碱降低的个体影响 ·· 71
　　　　4.2.1　行为补偿和有害物质暴露 ·· 71
　　　　4.2.2　建立或维持烟碱成瘾的阈值 ·· 71
　　　　4.2.3　使用 VLNC 卷烟后的烟草戒断 ·· 72
　　　　4.2.4　临床试验中的不依从性 ·· 73
　　　　4.2.5　不利的健康影响和弱势人群 ·· 73
　　　　4.2.6　小结 ·· 74
　　4.3　烟碱降低对人群的影响 ·· 75
　　　　4.3.1　VLNC 卷烟对普通卷烟的替代 ·· 75
　　　　4.3.2　VLNC 卷烟对其他烟草制品的替代 ······································ 76
　　　　4.3.3　黑市 ·· 78
　　　　4.3.4　VLNC 卷烟的操纵 ·· 79
　　　　4.3.5　关于 VLNC 卷烟和烟碱降低的观念和态度 ································ 80
　　　　4.3.6　小结 ·· 80
　　4.4　烟碱降低的监管方法 ·· 81
　　　　4.4.1　烟碱降低的可行性和潜在挑战 ·· 82
　　　　4.4.2　成功实施烟碱降低政策的先决条件 ······································ 82
　　　　4.4.3　实施烟碱降低政策的策略 ·· 84

		4.4.4 烟碱降低的异议思考 ································· 84

    4.4.5 小结 ··· 85
  4.5 研究问题 ··· 85
  4.6 政策建议 ··· 86
  4.7 参考文献 ··· 87

5. 降低卷烟烟气中有害物质暴露的监管策略 ··················· 100
  5.1 引言 ··· 100
  5.2 WHO 技术报告系列 951 中所述的卷烟烟气成分监管 ······ 100
  5.3 烟草行业对 WHO 技术报告系列 951 的响应 ··············· 102
  5.4 烟气成分含量与生物标志物之间的关系 ························ 104
  5.5 产品监管策略中有害物质水平的强制性降低 ·················· 104
  5.6 建议强制性降低的有害成分和建议的限量 ····················· 105
  5.7 强制性降低有害物质水平的实施 ··································· 107
  5.8 结论与建议 ·· 107
  5.9 参考文献 ··· 108

6. 烟草制品中调味剂的科学研究 ···································· 113
  6.1 引言 ··· 113
  6.2 调味烟草和烟碱产品使用的流行病学 ···························· 114
  6.3 调味产品：认知、尝试、摄入和监管 ···························· 115
  6.4 烟草和电子烟行业的调味烟草发展简史 ························ 116
  6.5 香味感知系统 ··· 118
  6.6 香味受体：香料感知和编码新科学 ································ 119
  6.7 调味剂的毒理学效应 ··· 120
  6.8 结论 ··· 121
    6.8.1 优先研究事项建议 ··· 122
    6.8.2 政策建议 ··· 122
  6.9 参考文献 ··· 123

7. 烟草制品中的糖含量 ·· 128
  7.1 引言 ··· 128
  7.2 不同类型烟草制品中的糖 ··· 129

- 7.2.1 糖和含糖添加剂的类型 ······ 129
- 7.2.2 糖和含糖添加剂的量 ······ 129
- 7.2.3 添加和内源性糖的总水平 ······ 130
- 7.2.4 烟草品种、产品和用途的区域和文化差异 ······ 131
- 7.3 糖对烟草制品释放物的影响 ······ 131
  - 7.3.1 无烟烟草制品 ······ 131
  - 7.3.2 糖和简单混合物的热解产物 ······ 131
  - 7.3.3 卷烟和其他燃烧型烟草制品 ······ 132
- 7.4 糖对烟草制品毒性的影响 ······ 133
  - 7.4.1 无烟烟草 ······ 133
  - 7.4.2 添加和不添加糖的卷烟 ······ 133
- 7.5 糖对烟草制品致瘾性的影响 ······ 134
  - 7.5.1 pH 值和游离态烟碱 ······ 134
  - 7.5.2 烟草和烟气中药理活性化合物的形成 ······ 134
- 7.6 依赖与戒烟 ······ 135
- 7.7 糖对烟草制品吸引力的影响 ······ 135
  - 7.7.1 认知：感官特征 ······ 135
  - 7.7.2 吸烟体验和行为：香味、适口性、易吸入性、使用频率 ······ 135
  - 7.7.3 开始吸烟 ······ 136
- 7.8 不同管辖区域对糖的监管 ······ 136
- 7.9 结论 ······ 137
- 7.10 建议 ······ 138
  - 7.10.1 下一步研究内容 ······ 138
  - 7.10.2 政策 ······ 138
- 7.11 致谢 ······ 140
- 7.12 参考文献 ······ 140

8. 燃烧型烟草制品中有害成分的优先级清单更新 ······ 147

- 8.1 引言 ······ 147
- 8.2 优先级清单的编制背景 ······ 148
  - 8.2.1 有害物质优先级清单选择标准 ······ 148

　　　　8.2.2　缔约方会议关于可燃型烟草优先成分和释放物的关键决定……149
　8.3　关于有害性的新科学知识概述……153
　　　　8.3.1　醛（乙醛，丙烯醛，甲醛，巴豆醛，丙醛，丁醛）……153
　　　　8.3.2　芳香胺（3-氨基联苯，4-氨基联苯，1-氨基萘，2-氨基萘）
　　　　　　　……156
　　　　8.3.3　烃类化合物（苯，1,3-丁二烯，异戊二烯，甲苯）……156
　　　　8.3.4　多环芳烃（苯并[a]芘）……157
　　　　8.3.5　烟草特有亚硝胺……157
　　　　8.3.6　生物碱（烟碱）……158
　　　　8.3.7　苯酚（邻苯二酚，间、对和邻甲酚，苯酚，对苯二酚，间苯
　　　　　　　二酚）……159
　　　　8.3.8　其他有机化合物（丙酮，丙烯腈，吡啶，喹啉）……159
　　　　8.3.9　金属和类金属（砷，镉，铅，汞）……160
　　　　8.3.10　其他成分（氨，一氧化碳，氰化氢，氮氧化物）……161
　8.4　分析方法的可用性……161
　　　　8.4.1　WHO TobLabNet 优先有害物质分析标准方法……161
　　　　8.4.2　其他优先有害物质分析方法概述……162
　8.5　品牌间有害成分变化的最新动态……165
　8.6　清单上有害物质的未来重新评估标准……168
　8.7　新型有害物质的选择标准……168
　8.8　研究需求和监管建议……171
　　　　8.8.1　研究需求、数据差距和未来工作……171
　　　　8.8.2　监管建议和对缔约方的支持……172
　8.9　参考文献……173

9. 测量和降低无烟烟草制品中有害物质浓度的方法……185
　9.1　引言……185
　9.2　产品组成……187
　9.3　农业实践和生产过程中导致的有害化合物形成和积累……189
　9.4　产品添加剂……191
　9.5　无烟烟草制品中的有害物质及降低有害物质影响的方法……191

9.6 微生物检测	193
9.6.1 通过细胞活性快速检测活性微生物	194
9.7 用于测量无烟烟草中有害物质的分析方法概述	195
9.8 监管方法和对策	195
9.9 政策建议和摘要	198
9.9.1 免责声明	201
9.10 参考文献	201

10. 水烟抽吸：流行性、健康影响和减少使用的干预措施 ... 208

10.1 引言	208
10.2 流行性、健康影响和减少使用的有效干预措施	208
10.2.1 水烟使用的区域和全球模式	209
10.2.2 急性和慢性健康影响	211
10.2.3 文化习俗以及开始和持续使用	213
10.2.4 调味剂的影响	214
10.2.5 低烟碱产品的依赖性	214
10.2.6 干预	215
10.3 未来需要的研究	216
10.4 政策建议	217
10.4.1 水烟使用的相关政策	219
10.5 结论	219
10.6 参考文献	220

11. 总体建议 ... 227

11.1 主要建议	227
11.2 对公共卫生政策的意义	228
11.3 对 WHO 计划的影响	229

Contents

Participants in the ninth meeting of the WHO Study Group on Tobacco Product
Regulation ··· xvii
Acknowledgements ··· xxi
Abbreviations ·· xxv
Commentary ·· xxix
    References ·· xxx
1. Introduction ·· 231
    1.1 References ·· 232
2. Heated tobacco products ·· 233
    2.1 Introduction ·· 233
    2.2 The science of heated tobacco products ·· 234
    2.3 A brief history of heated tobacco products ······································· 236
    2.4 Recent products ·· 240
        2.4.1 Emissions ··· 240
        2.4.2 Biomarkers of exposure ·· 241
    2.5 Consumer perceptions of heated tobacco products ························· 246
    2.6 Uptake in selected markets where products are available ················ 248
    2.7 Application by Philip Morris International for status as a "modified risk
         tobacco product" in the USA ·· 249
    2.8 Implications for regulation and tobacco control policies ················· 250
    2.9 Recommendations for research and policy ······································· 252
    2.10 References ·· 254
3. Clinical pharmacology of nicotine in electronic nicotine delivery systems ········ 260
    3.1 Introduction ·· 260
    3.2 ENDS operations ·· 261
    3.3 Nicotine concentration in e-liquids ··· 262

3.4 Nicotine delivery to ENDS users ···264
3.5 Toxicant content of ENDS emissions ··266
　　3.5.1 Nicotine emissions ··266
　　3.5.2 Emissions of non-nicotine toxicants ·······································268
3.6 Potential role of ENDS in smoking cessation ···································269
3.7 Potential health impact of ENDS ··283
　　3.7.1 Behavioural trajectories associated with use of ENDS ················284
　　3.7.2 Harm from ENDS and electronic non-nicotine delivery systems ·····287
3.8 Summary of evidence, research gaps and policy issues derived from the evidence ··289
3.9 References ···295

4. A global nicotine reduction strategy: state of the science ·················304
4.1 Background ···304
4.2 Individual outcomes of nicotine reduction ······································308
　　4.2.1 Behavioural compensation and exposure to toxicants ···············308
　　4.2.2 Threshold for establishing or maintaining nicotine addiction ···········309
　　4.2.3 Tobacco cessation after use of VLNC cigarettes ·······················310
　　4.2.4 Non-compliance in clinical trials ··310
　　4.2.5 Adverse health effects and vulnerable populations ·····················311
　　4.2.6 Summary ··312
4.3 Population impact of nicotine reduction ···313
　　4.3.1 VLNC cigarettes as replacements for regular cigarettes ··············314
　　4.3.2 Substitution of VLNC cigarettes with alternative tobacco products ··315
　　4.3.3 The black market ··317
　　4.3.4 Manipulation of VLNC cigarettes ··318
　　4.3.5 Beliefs and attitudes regarding VLNC cigarettes and nicotine reduction ···319
　　4.3.6 Summary ··320
4.4 Regulatory approaches to nicotine reduction ···································322
　　4.4.1 Feasibility of nicotine reduction and potential challenges ············322
　　4.4.2 Prerequisites for successful implementation of a nicotine reduction policy ··323
　　4.4.3 Strategies for implementation of a nicotine reduction policy ·········325
　　4.4.4 Philosophical objections to nicotine reduction ························326

4.4.5 Summary ·········326
4.5 Research questions ·········327
4.6 Policy recommendations ·········328
4.7 References ·········329

5. A regulatory strategy for reducing exposure to toxicants in cigarette smoke·····340

5.1 Introduction ·········340
5.2 Regulation of cigarette smoke constituents described in WHO Technical Report Series No. 951 ·········340
5.3 Industry response to WHO Technical Report Series No. 951 ·········343
5.4 Relation between smoke constituent levels and biomarkers ·········345
5.5 Use of mandated lowering of toxicant levels in a product regulatory strategy ·········346
5.6 Toxicants recommended for mandated lowering and recommended limits ·········347
5.7 Implementation of mandated lower levels of toxicants ·········348
5.8 Conclusions and recommendations ·········349
5.9 References ·········350

6. The science of flavour in tobacco products ·········354

6.1 Introduction ·········354
6.2 Epidemiology of use of flavoured tobacco and nicotine products ·········355
6.3 Flavoured products: perceptions, experimentation, uptake and regulation ·········357
6.4 A brief history of the development of flavoured tobacco by the tobacco and e-cigarette industry ·········358
6.5 Sensory systems that contribute to flavour ·········360
6.6 Flavour receptors: a new science of flavour sensing and coding ·········362
6.7 Toxicological effects of flavours ·········363
6.8 Conclusions ·········365
 6.8.1 Recommended priorities for research ·········366
 6.8.2 Recommended policies ·········367
6.9 References ·········367

7. Sugar content of tobacco products ·········372

7.1 Introduction ·········372
7.2 Sugars in different types of tobacco product ·········373

7.2.1 Types of sugars and sugar-containing additives ·················374
7.2.2 Amounts of sugars and sugar-containing additives ···············374
7.2.3 Total levels of added and endogenous sugars ··················375
7.2.4 Regional and cultural differences in tobacco varieties, products and use ··················376
7.3 Effects of sugars on levels of emissions from tobacco products ··················376
   7.3.1 Smokeless tobacco products ··················376
   7.3.2 Pyrolysis products of sugars and simple mixtures ··················377
   7.3.3 Cigarettes and other combusted tobacco products ··················377
7.4 Effects of sugars on the toxicity of tobacco products ··················378
   7.4.1 Smokeless tobacco ··················378
   7.4.2 Cigarettes with and without added sugar ··················379
7.5 Effects of sugar on the addictiveness of tobacco products ··················379
   7.5.1 pH and free nicotine ··················380
   7.5.2 Formation of pharmacologically active compounds in tobacco and smoke ··················380
7.6 Dependence and quitting ··················381
7.7 Effects of sugars on the attractiveness of tobacco products ··················381
   7.7.1 Perception: sensory characteristics ··················381
   7.7.2 Smoking experience and behaviour: flavour, palatability, ease of inhalation, frequency of use ··················382
   7.7.3 Initiation ··················382
7.8 Regulation of sugars according to jurisdiction ··················382
7.9 Conclusions ··················384
7.10 Recommendations ··················384
   7.10.1 Further research ··················384
   7.10.2 Policy ··················385
7.11 Acknowledgements ··················387
7.12 References ··················387

8. Updated priority list of toxicants in combusted tobacco products ··················393
8.1 Introduction ··················393
8.2 Background of preparation of the priority list ··················394
   8.2.1 Criteria for selection of toxicants for the priority list ··················395
   8.2.2 Key decisions of the Conference of the Parties on priority contents and emissions of combustible tobacco ··················397

8.3 Overview of new scientific knowledge on toxicity ··400
    8.3.1 Aldehydes (acetaldehyde, acrolein, formaldehyde, crotonaldehyde, propionaldehyde, butyraldehyde) ··400
    8.3.2 Aromatic amines (3-aminobiphenyl, 4-aminobiphenyl, 1-aminonaphthalene, 2-aminonaphthalene) ··404
    8.3.3 Hydrocarbons (benzene, 1,3-butadiene, isoprene, toluene) ···············404
    8.3.4 Polycyclic aromatic hydrocarbons (benzo[a]pyrene) ························405
    8.3.5 Tobacco-specific N-nitrosamines ··406
    8.3.6 Alkaloids (nicotine) ··407
    8.3.7 Phenols (catechol, m-, p- and o-cresols, phenol, hydroquinone, resorcinol) ··408
    8.3.8 Other organic compounds (acetone, acrylonitrile, pyridine, quinolone) ··409
    8.3.9 Metals and metalloids (arsenic, cadmium, lead, mercury) ···············409
    8.3.10 Other constituents (ammonia, carbon monoxide, hydrogen cyanide, nitrogen oxides) ··411
8.4 Availability of analytical methods ··411
    8.4.1 Standardized WHO TobLabNet methods for the analysis of priority toxicants ··411
    8.4.2 Overview of methods for the remaining priority toxicants ···············412
8.5 Update on variations in toxicants among brands ··416
8.6 Criteria for future re-evaluation of toxicants on the list ··418
8.7 Criteria for selection of new toxicants ··419
8.8 Research needs and regulatory recommendations ··422
    8.8.1 Research needs, data gaps and future work ································422
    8.8.2 Regulatory recommendations and support to Parties ·······················423
8.9 References ··425

9. Approaches to measuring and reducing toxicant concentrations in smokeless tobacco products ··434
9.1 Introduction ··434
9.2 Product composition ··437
9.3 Agricultural practices and manufacturing processes that result in the formation and accumulation of harmful compounds ································439
9.4 Product additives ··441

9.5 Harmful agents in smokeless tobacco products and methods to reduce their effects ·················441

9.6 Detection of microorganisms ·················445

 9.6.1 Rapid detection of live microorganisms by cell viability ·················445

9.7 Overview of analytical methods used to measure toxicants in smokeless tobacco ·················446

9.8 Regulatory approaches and responses ·················447

9.9 Policy recommendations and summary ·················449

 9.9.1 Disclaimer ·················452

9.10 References ·················453

10. Waterpipe tobacco smoking: prevalence, health effects and interventions to reduce use ·················459

10.1 Introduction ·················459

10.2 Prevalence, health effects and effective interventions to reduce use ·················460

 10.2.1 Regional and global patterns of waterpipe tobacco use ·················460

 10.2.2 Acute and chronic health effects ·················462

 10.2.3 Cultural practices and initiation and maintenance of use ·················464

 10.2.4 Influence of flavourings ·················466

 10.2.5 Dependence liability of low-nicotine products ·················467

 10.2.6 Interventions ·················467

10.3 Future research ·················470

10.4 Policy recommendations ·················471

 10.4.1 Policies relevant to waterpipe tobacco use ·················473

10.5 Conclusions ·················474

10.6 References ·················475

11. Overall recommendations ·················481

11.1 Main recommendations ·················482

11.2 Significance for public health policies ·················483

11.3 Implications for the Organization's programmes ·················483

WHO 烟草制品管制研究小组第九次会议

美国明尼苏达州明尼阿波利斯，2017年12月5~7日

参加者①

D. L. Ashley博士，美国公共卫生部退役海军少将；美国佐治亚州立大学（亚特兰大）人口健康科学系研究教授

O. A. Ayo-Yusuf教授，Sefako Makgatho卫生科学大学（南非比勒陀利亚）副校长

A. R. Boobis教授，英国伦敦帝国理工学院医学系药理学与治疗学中心毒理学教授

M. V. Djordjevic博士，美国国家癌症研究所（贝塞斯达）癌症控制与人口科学部行为研究计划烟草控制研究处项目负责人

S. K. Hammond博士，美国加利福尼亚大学伯克利分校公共卫生学院环境卫生学教授

D. K. Hatsukami博士，美国明尼苏达大学（明尼阿波利斯）精神病学教授

A. Opperhuizen博士，荷兰乌得勒支风险评估与研究办公室主任

G. Zaatari博士，WHO 烟草制品管制研究小组主席；贝鲁特美国大学（黎巴嫩贝鲁特）病理学与实验医学系教授兼主任

WHO烟草实验室网络（TobLabNet）主席
Nuan Ping Cheah 博士，新加坡卫生科学局药物、化妆品和卷烟测试实验室主任

发言人
Stephen S. Hecht博士，美国明尼苏达大学（明尼阿波利斯）教授
Suchitra Krishnan-Sarin博士，美国耶鲁医学院（纽黑文）精神病学教授
Mohammed Jawad博士（在读），英国伦敦帝国理工学院公共卫生政策评估系

---

① 未能参会：Mike Daube 教授，澳大利亚科廷大学（西澳大利亚州珀斯）健康学荣誉退休教授；P. Gupta 博士，Healis Sekhsaria 公共卫生研究所（印度孟买）所长

Richard O'Connor博士，美国Roswell Park综合癌症中心（纽约布法罗）健康行为系肿瘤学教授

Armando Peruga博士，智利Desarrollo大学（圣地亚哥）医学院流行病学和卫生政策中心研究员

Patricia Richter博士，美国疾病控制与预防中心（亚特兰大）烟草和挥发物科副主任

Irina Stepanov博士，美国明尼苏达大学（明尼阿波利斯）环境健康科学部和共济会癌症中心副教授

Reinskje Talhout博士，荷兰国家公共卫生与环境研究所（RIVM）（比尔特霍芬）健康防护中心高级科学顾问

Geoff Wayne先生，美国波特兰独立顾问

观察员

Katja Bromen博士，欧洲委员会（比利时布鲁塞尔）烟草控制小组政策官员，健康和食品安全、人类源性物质和烟草控制司总干事

Denis Chonière先生，加拿大管制物质和烟草管理局（安大略省渥太华）烟草制品管理办公室主任

WHO FCTC秘书处

Carmen Audera-Lopez博士，世界卫生组织（瑞士日内瓦）技术官员

秘书处（世界卫生组织非传染性疾病预防司无烟草行动组）

Sarah Emami女士，高级顾问

Ranti Fayokun博士，科学家

Vinayak Prasad博士，项目理事

Moira Sy女士，初级顾问

Participants in the ninth meeting of the WHO Study Group on Tobacco Product Regulation

Minneapolis, Minnesota, United States of America, 5–7 December 2017

Members[1]

Dr D.L. Ashley, Rear-Admiral (retired), Public Health Service; Research Professor, Department of Population Health Sciences, Georgia State University, Atlanta (GA), United States

Professor O.A. Ayo-Yusuf, Deputy Vice-Chancellor for Research, Sefako Makgatho Health Sciences University, Pretoria, South Africa

Professor A.R. Boobis, Professor of Toxicology, Centre for Pharmacology and Therapeutics, Department of Medicine, Imperial College London, London, England

Dr M.V. Djordjevic, Program Director and Project Officer, Tobacco Control Research Branch, Behavioral Research Program, Division of Cancer Control and Population Sciences, National Cancer Institute, Bethesda (MD), United States

Dr S.K. Hammond, Professor of Environmental Health Sciences, School of Public Health, University of California at Berkeley, Berkeley (CA), United States

Dr D.K. Hatsukami, Professor of Psychiatry, University of Minnesota, Minneapolis (MN), United States

Dr A. Opperhuizen, Director, Office for Risk Assessment and Research, Utrecht, Netherlands

Dr G. Zaatari (Chair), Professor and Chairman, Department of Pathology and Laboratory Medicine, American University of Beirut, Beirut, Lebanon

Chair of WHO Tobacco Laboratory Network (TobLabNet)

Dr Nuan Ping Cheah, Director of Pharmaceutical, Cosmetics and Cigarette Testing Laboratory, Health Sciences Authority, Singapore

Presenters

Dr Stephen S. Hecht, Professor, University of Minnesota, Minneapolis (MN), United States

Dr Suchitra Krishnan-Sarin, Professor of Psychiatry, Yale School of Medicine, New Haven (CT), United States

Dr Mohammed Jawad, PhD Candidate, Public Health Policy Evaluation Unit, Imperial College London, London, England

---

1 Unable to attend: Professor Mike Daube, Emeritus Professor, Faculty of Health Sciences, Curtin University, Perth, Western Australia, Australia; Dr P. Gupta, Director, Healis Sekhsaria Institute for Public Health, Mumbai, India.

Dr Richard O'Connor, Professor of Oncology, Department of Health Behavior, Roswell Park Comprehensive Cancer Center, Buffalo (NY), United States

Dr Armando Peruga, Researcher, Centre of Epidemiology and Health Policies, School of Medicine, Clínica Alemana, Universidad del Desarrollo, Santiago, Chile

Dr Patricia Richter, Deputy Chief, Tobacco and Volatiles Branch, Centers for Disease Control and Prevention, Atlanta (GA), United States

Dr Irina Stepanov, Associate Professor, Division of Environmental Health Sciences and the Masonic Cancer Center, University of Minnesota, Minneapolis (MN), United States

Dr Reinskje Talhout, Senior Scientific Adviser, Centre for Health Protection, National Institute for Public Health and the Environment (RIVM), Bilthoven, Netherlands

Mr Geoff Wayne, independent consultant, Portland (OR), United States

Observers

Dr Katja Bromen, Policy Officer, Tobacco Control Team, European Commission, Directorate-General for Health and Food Safety, Substances of Human Origin and Tobacco Control, Brussels, Belgium

Mr Denis Chonière, Director, Tobacco Products Regulatory Office, Controlled Substances and Tobacco Directorate, Health Canada, Ottawa, Ontario, Canada

Secretariat of the WHO Framework Convention on Tobacco Control

Dr Carmen Audera-Lopez, Technical Officer, Geneva, Switzerland

Secretariat (Tobacco Free Initiative, Prevention of Noncommunicable Diseases, WHO, Geneva, Switzerland)

Ms Sarah Emami, Senior Consultant

Dr Ranti Fayokun, Scientist

Dr Vinayak Prasad, Programme Manager

Ms Moira Sy, Junior Consultant

致　　谢

世界卫生组织烟草制品管制研究小组（TobReg）对提供本报告基础背景文件的作者表示感谢。本报告在Vinayak Prasad博士和世界卫生组织非传染性疾病预防司前司长Douglas Bettcher博士的监督和支持下，由Ranti Fayokun博士和Sarah Emami女士多方协调得以完成。

TobReg感谢世界卫生组织《烟草控制框架公约》（WHO FCTC）第9条和第10条两个工作组的主要贡献，Katie Bromen博士（欧洲委员会）和Denis Chonière先生（加拿大）帮助确保了WHO和TobReg充分响应缔约方会议的有关要求。

TobReg感谢美国明尼苏达大学共济会癌症中心不遗余力地主持会议，并感谢Dorothy Hatsukami博士邀请专家确保会议成功。

TobReg感谢世界卫生组织《烟草控制框架公约》秘书处协助起草缔约方会议的要求，作为背景文件的基础。

Acknowledgements

The WHO Study Group on Tobacco Product Regulation (TobReg) expresses its gratitude to the authors of the background papers used as the basis for this report. Production of the report was coordinated by Dr Ranti Fayokun and Ms Sarah Emami, with the supervision and support of Dr Vinayak Prasad and Dr Douglas Bettcher, former Director of the WHO Department for Prevention of Noncommunicable Diseases.

TobReg acknowledges two of the key facilitators of the Working Group on Articles 9 and 10 of the WHO Framework Convention on Tobacco Control (WHO FCTC), Dr Katja Bromen (European Commission) and Mr Denis Chonière (Canada), who helped ensure that WHO and TobReg adequately responded to relevant requests of the COP.

TobReg also expresses its gratitude to the Masonic Cancer Center at the University of Minnesota, USA, for generously hosting the meeting and to Dr Dorothy Hatsukami for facilitating the invitation and making the meeting a success.

TobReg thanks the Convention Secretariat of the WHO FCTC for facilitating drafting of the requests of the COP, which served as a basis for some of the background papers.

缩　略　语

CI	置信区间
CO	一氧化碳
COP	缔约方会议
EHCSS	电加热吸烟系统
EN&NNDS	电子烟碱和非烟碱传输系统
ENDS	电子烟碱传输系统
ENNDS	电子非烟碱传输系统
FDA	美国食品药品监督管理局
FTC	（美国）联邦贸易委员会
GC	气相色谱
GRAS	一般认为安全
HCI	加拿大深度抽吸模式
HPHC	有害和潜在有害成分
HTP	加热型烟草制品
IARC	国际癌症研究机构
ISO	国际标准化组织
LC	液相色谱
MRTP	风险降低的烟草制品
MS	质谱
MS/MS	串联质谱
NNAL	4-(甲基亚硝基氨基)-1-(3-吡啶基)-1-丁醇
NNK	4-(甲基亚硝基氨基)-1-(3-吡啶基)-1-丁酮
NNN	N'-亚硝基降烟碱
NRT	烟碱替代疗法
PAH	多环芳烃
PMI	菲利普·莫里斯国际
PREP	潜在降低暴露的产品
qPCR	定量聚合酶链反应
SOP	标准操作规程
THP	烟草加热产品
THS	烟草加热系统
TobLabNet	烟草实验室网络
TobReg	烟草制品管制研究小组
TRP	瞬时受体电位
TSNA	烟草特有亚硝胺
VLNC	极低烟碱含量
VOC	挥发性有机化合物
WHO FCTC	世界卫生组织《烟草控制框架公约》

Abbreviations

CI	confidence interval
CO	carbon monoxide
COP	Conference of the Parties of the WHO Framework Convention on Tobacco Control
EHCSS	electrically heated cigarette smoking system
EN&NNDS	electronic nicotine and non-nicotine delivery systems
ENDS	electronic nicotine delivery systems
ENNDS	electronic non-nicotine delivery systems
FDA	United States Food and Drug Administration
FTC	Federal Trade Commission (USA)
GC	gas chromatography
GRAS	generally recognized as safe
HCI	Health Canada intense
HPHC	harmful and potentially harmful constituent
HTP	heated tobacco product
IARC	International Agency for Research on Cancer
ISO	International Organization for Standardization
LC	liquid chromatography
MRTP	modified risk tobacco product
MS	mass spectrometry
MS/MS	tandem mass spectrometry
NNAL	4-(methylnitrosamino)-1-(3-pyridyl)-1-butanol
NNK	4-(methylnitrosamino)-1-(3-pyridyl)-1-butanone
NNN	N´-nitrosonornicotine
NRT	nicotine replacement therapy
PAH	polycyclic aromatic hydrocarbon
PMI	Philip Morris International
PREP	potentially reduced exposure product
qPCR	quantitative polymerase chain reaction
SOP	standard operating procedure
THP	tobacco heating product
THS	tobacco heating system
TobLabNet	Tobacco Laboratory Network (WHO)
TobReg	WHO Study Group on Tobacco Product Regulation
TRP	transient receptor potential
TSNA	tobacco-specific N´-nitrosamine
VLNC	very low-nicotine content
VOC	volatile organic compound
WHO FCTC	WHO Framework Convention on Tobacco Control

概　　述

减少烟草使用相关危害的科学方法和监管方法引起了人们的极大关注。这在一定程度上推动了新产品的开发，例如被宣传为危害较小或有助于戒烟的电子烟碱传输系统（ENDS）和加热型烟草制品（HTP）。这种关注可以追溯到几十年前，早在2003年世界卫生组织烟草制品管制研究小组（TobReg）成立之前。世界卫生组织在早期的报告中已广泛讨论了产品管制问题。然而，在过去的十年中，市场得到了广泛的发展，产品和行业活动、行业报告的范围不断扩大，并且提供了更多的研究，国家数据和出版物也越来越多。

本报告主要基于为2017年12月的TobReg会议准备的背景文件，涵盖了与烟草制品监管相关的关键问题。在过去的一年中，与第4~10章相关的出版物数量有所增加，随着ENDS和HTP市场的发展，科学文献以及国家监测计划或调查的数据也明显增加。例如，对2003~2018年间发布的电子烟（ENDS的一个子集）相关出版物进行的文献计量分析表明，从2017年至2018年增长了近24%[1]。本报告基于委托时的最佳证据，根据特定的职权范围编写。世界卫生组织意识到关于电子烟碱传输系统和市场发展的出版物的数量增加，本报告并未涉及包括加拿大、欧洲和美国等的年轻人使用这些产品的最新趋势以及限制使用的监管方法。ENDS部分仅限于这些产品中烟碱的临床药理学，其依据是截至2017年12月的可用证据，并对2018年3月至2018年12月期间的文献进行了审查后进行了一些内容补充。仅包括被认为与该主题相关的出版物。

世界卫生组织在2014年和2016年，分别就与电子烟碱传输系统有关的健康风险、这些产品在帮助吸烟者戒烟和减少其对烟碱依赖方面的功效及对烟草控制的干扰以及世界卫生组织《烟草控制框架公约》的实施，向《烟草控制框架公约》缔约方会议第六次会议（FCTC / COP / 6/10 Rev.1）和第七次会议（FCTC / COP / 7/11）提交了两份报告。这些报告促成缔约方会议达成两项决定，一是请各缔约方广泛追求四个监管目标，根据各国法律和公共卫生目的，禁止或限制这些产品的生产、进口、分销、展示、销售和使用；二是出于对人体健康的高度保护，酌情考虑将电子烟碱传输系统作为烟草制品、医药产品、消费品或其他类别加以禁止或限制。

世界卫生组织及其技术小组（包括TobReg）正在监测市场发展、研究、出版物、综述、行业活动以及关于电子烟碱传输系统（包括电子烟）和其他新型产品的动态，并审查全球监测机制，以确保对这些产品进行有效的监测和评估。世界卫生组织将对不断增多的出版物、新产品、新变种和口味的证据进行全面审查，届时将更新2016年报告。审查还将包括有关这些产品的影响、市场趋势、对青少年使用这些产品的关注以及主要烟草公司的参与，尤其是与世界卫生组织《烟草控制框架公约》第5.3条有关的证据。

  本报告提供了截至2017年12月的最佳证据，并在审查后于2018年3~12月进行了补充。应将其与指导WHO成员国制定有关烟草制品政策和法规的早期证据结合起来阅读。

参 考 文 献

[1] Briganti M, Delnevo CD, Brown L, Hastings SE, Steinberg MB. Bibliometric analysis of electronic cigarette publications: 2003–2018. Int J Environ Res Public Health. 2019; 16(3): 320.

Commentary

Scientific and regulatory approaches to reducing the harm associated with tobacco use have attracted much interest. This has fuelled in part the development of newer products, such as electronic nicotine delivery systems (ENDS) and heated tobacco products (HTPs), which have been promoted as less harmful or as aids to smoking cessation. The interest dates back several decades, well before the establishment of the WHO Study Group on Tobacco Product Regulation (TobReg) in 2003. The topic of product regulation has been addressed extensively in earlier WHO reports. During the past decade however, the market has developed extensively, with an ever-increasing range of products and industry activity and reports, as well as more research, country data and publications being made available.

This technical report, based largely on background papers prepared for a meeting of TobReg in December 2017, covers pertinent and critical issues in tobacco product regulation. While there was some increase in the number of publications relevant to sections 4–10 in the past year, there were marked increases in the number of scientific publications, data from national surveillance programmes/surveys and market developments with respect to ENDS and HTPs. For example, a bibliometric analysis of publications on electronic cigarettes (a subset of ENDS) published between 2003 and 2018 indicated an increase of nearly 24% between 2017 and 2018 (1). The reports in this publication were written according to specific terms of reference and are based on the best evidence available at the time they were commissioned. WHO is aware of the unprecedented increase in the number of publications on ENDS and on market developments, recent trends in the use of these products by young people in some countries, including Canada, some countries in Europe and the United States of America, and regulatory approaches to curtailing use, which are not addressed in this report. The section on ENDS is limited to the clinical pharmacology of nicotine in these products and is based on evidence available up to December 2017, with some additions after review between March and December 2018. Only publications considered to be relevant to the topic were included.

WHO submitted two reports to the sixth (FCTC/COP/6/10 Rev. 1) and seventh (FCTC/COP/7/11) sessions of the Conference of the Parties to the WHO Framework Convention on Tobacco Control (WHO FCTC), in 2014 and 2016, respectively, on the health risks associated with ENDS, the efficacy of these products in helping smokers to quit smoking and reduce their nicotine dependence, interference with tobacco control and implementation of the WHO FCTC. These reports led to two decisions by the Conference of the Parties, in which Parties were invited to broadly pursue four regulatory objectives to prohibit or restrict the manufacture, importation, distribution, presentation, sale and use of these products, as appropriate to their national laws and public health objectives and to consider prohibiting or regulating ENDS, including as tobacco products, medicinal products, consumer products, or other categories, as appropriate, taking into account a high level of protection for human health.

WHO and its technical groups, including TobReg, are monitoring market developments, research, publications, commentaries, industry activities and debate on ENDS (including e-cigarettes) and other novel products and are reviewing global surveillance mechanisms to ensure effective monitoring and evaluation of these products. WHO will address the increasing numbers of publications and new products, variants and flavours in its next comprehensive review of the evidence, which will update the 2016 report. The review will also include the mounting evidence on the effects of these products, marketing trends, concern about their use by young people and the engagement of the major tobacco companies, especially in relation to Article 5.3 of the WHO FCTC.

This report provides the best evidence available up to December 2017, with some additions after review between March and December 2018. It should be read in relation to the earlier evidence that guided the development of policies and regulations by WHO Member States on tobacco products up to that time.

Reference

1. Briganti M, Delnevo CD, Brown L, Hastings SE, Steinberg MB. Bibliometric analysis of electronic cigarette publications: 2003–2018. Int J Environ Res Public Health. 2019;16(3):320.

1. 引　　言

有效的烟草制品管制是全面烟草控制计划的重要组成部分，包括强制性对成分和释放物进行监管、披露测试结果相关信息、设定适当限值以及对产品包装和标签制定标准。世界卫生组织《烟草控制框架公约》（WHO FCTC）[1]的第9、10和11条以及实施第9条和第10条[2]的部分指南中涵盖了烟草制品管制的内容。世界卫生组织的其他资料，包括《烟草制品管制：基本手册》[3]和《烟草制品管制：实验室检测能力建设》[4]，在这方面为成员国提供了支持。

世界卫生组织总干事于2003年正式组建了世界卫生组织烟草制品管制研究小组（TobReg），以填补烟草制品管制空白。其任务是向WHO总干事提供有关烟草制品管制的政策建议。TobReg由产品监管、烟草依赖治疗、毒理学以及烟草制品成分和释放物实验室分析等领域的国际科学专家组成。这些专家来自WHO六大地区的成员国。作为WHO的正式实体，TobReg通过总干事向WHO执行委员会提交技术报告，以提请成员国注意WHO在烟草制品管制方面的工作。这些技术报告基于TobReg讨论、评估和审查的未发表的背景资料。

TobReg第九次会议于2017年12月5~7日在美国明尼阿波利斯举行，由美国明尼苏达大学共济会癌症中心主办。与会者讨论了FCTC/COP 7（4），FCTC/COP 7（9）文件概述的世界卫生组织《烟草控制框架公约》缔约方会议第七次会议提出的烟碱和新型产品以及烟草制品管制的优先事项，并提出了以下要求：

- 继续监测和审查新型烟草制品（如加热型烟草制品）的市场发展和使用情况；
- 收集有关无烟烟草制品中导致其有害性、致瘾性和吸引力的化学物质的含量水平和释放物，测量这些化学物质的分析方法以及市售产品中含量水平的科学信息，并确定减少无烟烟草制品中有害物质的技术方法；
- 促进以下研究以预防吸食水烟和促进戒烟：文化相关干预措施的研究，流行病学研究，急性和慢性健康风险研究，文化习俗研究，开始吸烟和维持使用研究，调味剂对引发、维持使用和增加使用的影响研究，低烟碱烟草制品成瘾的风险研究，基于信息技术和通信等概念的有效政策研究。

为响应这些要求，世界卫生组织委托编写了以下背景文件：
1. 加热型烟草制品（第2章）；

2. 电子烟碱传输系统中烟碱的临床药理学（第3章）；

3. 全球烟碱降低策略：科学现状（第4章）；

4. 降低卷烟烟气中有害物质暴露的监管策略（第5章）；

5. 烟草制品中调味剂的科学研究（第6章）；

6. 烟草制品中的糖含量（第7章）；

7. 燃烧型烟草制品中有害成分的优先级清单更新（第8章）；

8. 测量和降低无烟烟草制品中有害物质浓度的方法（第9章）；

9. 水烟抽吸：流行性、健康影响和减少使用的干预措施（第10章）。

1.1 参 考 文 献

[1] WHO Framework Convention on Tobacco Control. Geneva: World Health Organization; 2003 (http://www.who.int/fctc/en/, accessed 14 May 2019).

[2] Partial guidelines on implementation of Articles 9 and 10. Geneva: World Health Organization; 2012 (https://www.who.int/fctc/guidelines/Guideliness_Articles_9_10_rev_240613.pdf, accessed 14 January 2019).

[3] Tobacco product regulation: basic handbook. Geneva: World Health Organization; 2018 (https://www.who.int/tobacco/publications/prod_regulation/basic-handbook/en/, accessed 14 January 2019).

[4] Tobacco product regulation: building laboratory testing capacity. Geneva: World Health Organization; 2018 (https://www.who.int/tobacco/publications/prod_regulation/building-laboratory-testing-capacity/en/, accessed 14 January 2019).

2. 加热型烟草制品

Dr Richard J. O'Connor, Professor of Oncology, Department of Health Behavior, Roswell Park Comprehensive Cancer Center, Buffalo, New York, USA

如今，烟草公司继续向关注健康的吸烟者制造其新产品安全的假象。

——2006年世界无烟日

2.1 引　　言

加热型烟草制品（HTP），有时也称为"加热不燃烧烟草制品"，是烟草行业创造的一个术语，是一种新型的"潜在降低暴露的产品"（PREP）或"风险降低的烟草制品"（MRTP）。这个概念是在20世纪80年代由美国烟草公司菲利普·莫里斯（Philip Morris）和雷诺（RJ Reynolds）提出的，并分别上市了第一代这类产品——雅阁（Accord）和派米雷（Premier）。从那时起，这些概念上相似的产品一直在不断发展，现在已经占领了重要的市场份额。电子烟的引入、积极的市场营销和日益普及，可能促进了这类产品的成功，其部分原因是改变了人们对于传统卷烟和使用烟碱传输装置的社会规范和观念。20世纪90年代和21世纪有大量关于加热型烟草制品的文献发表，也有新型产品的文献发表，尽管其中大部分来自烟草行业。然而，很少有关于新型产品在市场上的流行性和替代率方面的研究，因为这些产品中有许多是试销，而不是广泛上市销售。尽管如此，实验室和现场研究仍可以提供有关这种替代可能性的信息。

本综述基于截至2017年10月可获得的有关加热型烟草制品的文献，包括其历史、设计及烟碱和有害物质向使用者的传输和营销（包括在线广告和销售），加热型烟草制品技术，制造商关于降低有害性、危害、风险和暴露的声明，与传统卷烟的比较，消费者对这些替代产品的看法，以及这些产品对监管和市场政策的影响。审查的重点是烟草公司销售的加热型烟草制品，但也涵盖可用作烟草的"干药草"雾化器（通常是大麻）、手持式或便携式产品。通常用于抽吸大麻的台式"雾化器"以及水烟或麻醉型产品被排除在外。审查主要基于已发表的文献，必要时还基于新闻报道和新闻稿、股东报告、科学报告和互联网文章。

PubMed 搜索关键词：heat not burn; heat-not-burn; heated tobacco; tobacco heating; Accord; Eclipse; Heatbar; Premier; THS [tobacco heating system]; vaporizer; PREP; MRTP; THP [tobacco heating product]; iQOS; glo。

2.2 加热型烟草制品科学基础

加热型烟草制品（HTP）基于以下原理：与吸烟相关的大多数危害性物质是由燃烧过程引起的。在传统卷烟中，燃烧锥的温度可以达到900℃，并且沿着烟杆的温度中位值为600℃。这可能会导致燃烧、热解、热合成和无数其他反应，从而导致超过7000种化合物被确定为卷烟烟气的成分[1]。多环芳烃（PAH）、杂环芳胺和一些挥发性有机化合物（VOC）（例如苯、1,3-丁二烯、丙烯醛、甲苯）主要是燃烧形成的。烟草特有亚硝胺（TSNA），例如N'-亚硝基降烟碱（NNN）和4-(甲基亚硝基氨基)-1-(3-吡啶基)-1-丁酮（NNK）存在于烤烟中，并在典型的卷烟温度下以近线性方式部分转移到烟气中。在卷烟燃烧过程中会形成一些TSNA。烟草中存在的有毒金属（例如镉）也可能在典型的烟草燃烧温度下转移到卷烟烟气中。然而，燃烧烟草对于"挥发"烟碱尽管是有效的，但不是必需的，从毒理学风险和消费者可接受性的角度来看，不燃烧而以可吸入形式从烟草中释放烟碱的替代方法是可取的。摄入烟碱的一种方法是将烟草加热到使烟碱挥发但不会燃烧植物材料的温度，同时保持类似于吸烟行为的视觉效果。原则上，不通过燃烧而是通过挥发烟碱的方式将产生一种不太复杂且毒性成分较少的气溶胶。

烟碱不能有效地以气体形态传输。为了将烟碱传输到使用者的肺部，必须将烟碱雾化以悬浮在烟气颗粒物中。在过去的几十年中，是通过四种主要方法实现的。首先是具有嵌入式热源的类似卷烟的装置雾化烟碱，这是Premier和Eclipse以及菲利普·莫里斯国际（PMI）的"Platform 2"产品的基本原理。第二种方法是使用外部热源，将特殊设计的卷烟中的烟碱雾化。这是Accord、Heatbar、iQOS和glo的基本设计。在PMI的HTP中使用的烟草显然不是典型的烟丝，而是薄片（一种重组烟草），其中包括5%~30%（质量分数）多元醇、乙二醇酯和脂肪酸等雾化剂，如甘油、赤藓糖醇、1,3-丁二醇、四甘醇、三甘醇、柠檬酸三乙酯、碳酸丙烯酯、月桂酸乙酯、三乙酸甘油酯、内消旋赤藓糖醇、二乙酸甘油酯混合物、辛二酸二乙酯、柠檬酸三乙酯、苯甲酸苄酯、乙酸苄基苯酯、香草酸乙酯、三丁酸甘油酯、乙酸月桂酯、月桂酸、肉豆蔻酸和丙二醇[2]。这类组合物作为气溶胶形成基质与加热系统一起使用是有利的。这种新型卷烟（长45 mm，直径7 mm）包含约320 mg的烟草材料，比传统卷烟（约700 mg）要少得多。在iQOS中，烟草通过插入加热棒中的加热片进行加热，热量通过卷烟[3]上部的滤嘴进行散发。然后，气溶胶通过中空的醋酸纤维管和聚合物膜过滤后进入口腔。产品的设计温度不超过350℃，大概可以抽吸14口或6 min[3]。英美烟草公司将其glo产品描述为一个加热管，该加热管由两个单独控制的腔组成，通过设备上的按钮激活，在30~40 s预热时间内达到工作温度（240℃）[4]。插入加热腔中的长82 mm、直径5 mm的烟支中包含大约260 mg的薄片和14.5%的甘油（雾化剂）。烟支上的通风孔被描述为"……提供适当的吸力并有助于……蒸汽凝结和冷凝……"所必需的[4]。烟支由烟杆、管状冷却段和烟嘴组成。

第三种方法是使用加热的密封室直接雾化烟叶中的烟碱——这是个性化干雾化器（例如Pax）的基本原理；但是，目前尚不清楚在这种装置中使用烟草的流行情况。第四种方法是使用电子烟碱传输系统（ENDS）从少量烟草中提取香味成分（更多信息请参见第3章）。英美烟草公司的iFuse似乎是一种ENDS-烟草制品的混合产品，气溶胶通过烟草时提取香味物质，然后被使用者吸入[5]。当蒸汽通过烟草加热腔时（从35℃到32℃），蒸汽损失了少量热量，表明烟草被加热了。但是，由于电子烟液中烟碱的含量为1.86 mg/mL，每台装置烟碱的传输量为20~40 μg/口，因此很难估算装置中烟草对传输烟碱的贡献。据报道，在吸烟机抽吸条件下，有害物质的释放量与不带烟草的ENDS几乎相同，意味着烟草的贡献极小[5]。日本烟草国际公司Ploom TECH的运作方式与此类似，只是类似ENDS的成分中似乎不含烟碱。

2.3　加热型烟草制品简史

历史的视角对于理解当前加热型烟草制品的状况是很重要的。表2.1列出了已引入市场进行测试的HTP。图2.1中按照时间线对各种HTP的上市（和退市）进行了介绍。这一市场领域中的活动是集中的，2006~2008年和2015~2017年是特别活跃的时期。

表2.1　按制造商排序的加热型烟草制品

公司	商品名称*	简要描述	目前情况
Philip Morris USA/Altria	Accord	THS由小支卷烟和外部加热装置组成。有烟草和薄荷风味。于1998年在美国弗吉尼亚州里士满以及日本的测试市场中推出。广告的重点是减少二手烟，而不是减少潜在的健康风险	2006年退市，不再出售
Philip Morris International	Heatbar	THS由小支卷烟（加热棒）、外部加热装置与抽吸检测装置组成。共有四个混合型口味。设备有多种颜色。2006年在瑞士推出	2008年退市，不再出售

续表

公司	商品名称*	简要描述	目前情况
Philip Morris International	iQOS	THS由小支卷烟（HEETS）和外部加热装置组成。使用万宝路（Marlboro）品牌。2015年在日本推出	目前在近40个国家销售，包括加拿大、意大利、日本和英国等。正在拓展其他市场。于2017年5月向FDA提交MRTP申请，并于2018年1月由烟草制品科学咨询委员会审查
Philip Morris International	THS 2.2	THS在一系列科学期刊上发表的论文中都有描述	作为iQOS进行销售
Philip Morris International	Platform 2	压缩碳热源（概念上类似于Eclipse）	目前还不清楚市场上是否有这种产品。在几个PMI的演示中商标为TEEPS
Japan Tobacco International	Ploom	Ploom于2007年成立独立公司。2011年与日本烟草国际公司签订了营销和商业化协议，后者于2015年直接购买了整套设备的专利和设计。目前有Mevius卷烟品牌名称。Ploom的前公司更名为Pax，主要生产干草雾化器	目前在日本和瑞士销售
RJ Reynolds / Reynolds American	Premier	类似于卷烟的装置，用于加热和雾化烟草中的香味物质。烟杆由装有烟草颗粒的铝胶囊制成，顶端带有碳以加热烟草。于1988年在美国圣路易斯、菲尼克斯和图森市进行测试销售。被作为药物传输装置向FDA投诉	由于销售不佳，于1989年退市。目前不在市场上销售
RJ Reynolds / Reynolds American	Eclipse	卷烟状装置。烟杆由薄片组成，用碳加热和挥发烟碱。在1996~2000年间进行了多种版本的试销，并开始了更广泛的销售，附带广告和健康声明。被美国佛蒙特州提起诉讼	2007年退出美国市场；此后仍保持有限的销售。宣布于2017年10月向FDA提交"实质等同"申请，以将改进的版本引入美国市场
RJ Reynolds / Reynolds American	Revo	2014~2015年，类似Eclipse的产品在美国威斯康星州试销	2015年退市
British American Tobacco	glo	glo™将专有的Kent Neostiks™加热到大约240℃，以提供与卷烟相似的令人满意的味道，同时减少约90%的有害成分。glo™还提供了许多附加功能，包括不燃烧或没有烟灰，减少手、头发、衣服和周围环境的异味。Neostiks有三种口味	在加拿大、日本、韩国、俄罗斯和瑞士销售
British American Tobacco	iFuse	电子烟和HTP的混合体。烟弹中装有加香的电子烟液，腔体中装有烟支。加热元件使电子烟液雾化，在被使用者吸入之前经过烟草加热腔。于2015年推出	在罗马尼亚进行试销
干药草雾化器			
Pax	Pax 2	矩形装置。用于加热植物材料的"电磁炉"，"Kapton薄膜柔性加热器"。不能与液体、蜡或浓缩物一起使用。容量1.6 mL。有四种颜色	可以在线购买，也可以从授权零售商处购买
Pax	Pax 3	矩形装置。用于加热植物材料的"电磁炉"。也可与特殊插入物的浓缩物一起使用。当装置检测到吸嘴上的抽吸动作时，雾化器即会启动。电池为3500 mAh。容量0.17~0.35 g。有四种颜色	可以在线购买，也可以从授权零售商处购买

续表

公司	商品名称*	简要描述	目前情况
V2	Pro Series 3	圆柱形装置。三合一雾化器,适用于电子烟液、散烟或蜡的不同烟弹和烟嘴。热传导加热腔。预设温度160~180℃。电池容量650 mAh。有三种颜色	在线购买
V2	Pro Series 7	矩形装置。三合一雾化器,适用于电子烟液、散烟或蜡的不同烟弹和烟嘴。更大的容量,更多的自定义选项。高温热传导加热腔有三个温度(200℃、215℃、225℃)。1800 mAh电池;3.7~4.7 V。三种颜色	在线购买
Vapor Fi	Orbit	圆柱形装置。用于干燥植物材料的1.7 mL容量腔体。温度范围182~216℃;2200 mAh电池。两种颜色(黑色、红色)	在线购买
Vapor Fi	Atom	矩形装置。温度范围182℃、210℃、240℃。将植物材料装进外加热腔中。3000 mAh电池;3.3~4.2 V。运动感应。外观上类似于Pax设备	在线购买
Atmos	Jump	圆柱形装置。最高温度200℃;1300 mAh电池;电压3.4 V。有四种颜色	在线购买

FDA:美国食品药品监督管理局;HTP:加热型烟草制品;MRTP:风险降低的烟草制品;PMI:菲利普·莫里斯国际;THS:烟草加热系统

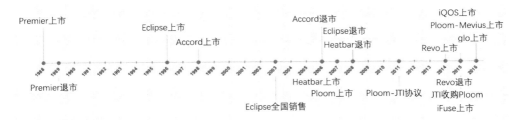

图2.1 加热型烟草制品推出时间表(1988~2016年)

Eclipse

Eclipse是Premier(派米雷)产品的延续,于20世纪80年代末推出,1996~2007年可在美国的各种测试市场中见到,并被进行了广泛的研究[6-13]。该款类似于卷烟的产品在顶端嵌入了一个碳加热源,以加热重组烟草和甘油,产生包含烟碱的气溶胶。要使用它,吸烟者会像抽吸传统卷烟一样点燃进行加热和抽吸。在两个单独的为期5天的试验中,Breland及其同事[6,7]发现,Eclipse降低了吸烟者烟碱和NNK暴露量,但增加了对一氧化碳(CO)的暴露量,为了获得相同的烟碱量需要更深度的抽吸方式。Lee等[10]在实验室中发现了类似的Eclipse抽吸结果。Fagerstrom及其同事[9]指出,吸烟者在使用Eclipse的两周内,CO暴露量从21.0 ppm(parts per million,百万分之一)增加到了33.0 ppm。在随后的一项8周研究[8]中,选择了10名参与者继续使用Eclipse，发现血浆中烟碱含量与基准没有差异，但是使用Eclipse的吸烟者呼出气体中CO水平高于吸卷烟（分别为32.5 ppm与22.5 ppm）。Stewart及其同事[13]报告称，使用Eclipse持续4周的一小组吸烟者显示出较少的肺泡上皮损伤，但碳氧血红蛋白水平和其他氧化应激指标升高。Pauly及其同事[14]确定了由于玻璃纤维污染而导致的Eclipse潜在的健康风险。他们指出，在处理包装或产品本身时，碳加热元件周围的绝缘材料可能会磨损和变松，并使吸烟者接触这些纤维。先前的研究表明，在114个人的肺部样本中[15]证明有87%吸入了纤维和塑料，表明这不仅仅是理论上的担忧。随后的消费者调查[16]显示，他们认为这是健康风险。雷诺公司（RJ Reynolds）强烈反对这些观察结果，并发表了几项研究来反驳，认为纤维太大而无法吸入，变松的纤维的数量没有报道得那么多[17-20]。

Eclipse于2003年在全国范围内首次发布时就提出了健康声明，理由是它可能降低患癌、支气管炎和肺气肿的风险[21]。该制造商的广告声称，Eclipse减少了烟气有害成分的释放量，并暗示，除戒烟外，Eclipse是吸烟者的"下一个最佳选择"。2005年，美国佛蒙特州总检察长提起诉讼，并在2010年认定该广告违反了消费者保护法和总和解协议。随后在2013年，雷诺公司被判赔偿800万美元。2015年，将Eclipse重新包装并命名为"Revo"在美国威斯康星州进行了短暂的测试销售。2017年11月，雷诺公司宣布，他们将在2018年推出Eclipse的改进版本；截至2019年1月尚未启动。

Accord

Accord于1998年首先在美国销售，随后在日本上市。该产品是iQOS的前身，具有外部电池供电的加热装置，专门设计的"卷烟"被插入其中用于抽吸。吸烟激活了装置中的加热棒，并产生气溶胶。Accord被推销为"更干净的"卷烟，不含烟灰或二手烟；没有提出健康声明。一项早期研究的回顾表明，与传统卷烟相比，使用Accord可降低一氧化碳的摄入和减缓心动过速[22,23]。然而，这与不良的戒断作用抑制相关[22,23]。在一项同时使用Accord和抽吸卷烟的研究中，在使用Accord产品6周后抑制了抽吸卷烟和CO暴露剂量依赖；也就是说，使用Accord的频率越高，参与者吸卷烟的次数就越少。此外，当参与者减少每天的卷烟数量时，他们的吸烟强度并没有增加，并且Accord烟中的一氧化碳含量较低[24]。对参与者后续采访的分析表明，研究参与者认为Accord是"更安全的卷烟"。Roethig等[25]研究的第二代电加热卷烟显示，与传统卷烟相比，某些暴露生物标志物（例如烟碱、1-羟基芘）减少了43%~85%。

2.4 新型产品

2.4.1 释放物

表2.2总结了已发表的有关某些有害和潜在有害成分（HPHC）[26]在吸烟机抽吸条件下的释放物的文献，以下三种HTP是发表数据最多的：Eclips（ISO抽吸模式）、电加热吸烟系统（EHCSS，ISO抽吸模式）和THS 2.2/iQOS9[加拿大深度抽吸(HCI)模式]，这些数据是由制造商公布的。PMI实验表明，THS 2.2气溶胶不含固体碳衍生颗粒，这与其不燃烧的说法一致[27]。总的来说，其有害和潜在有害成分含量低于参比卷烟（1R6F，该参比卷烟已被证明了许多受关注释放物的水平[28]）。在行业内[29]和行业外[30,31]进行的研究在很大程度上重复了这些发现，特别是Eclipse和EHCSS中的烟碱含量低于参比卷烟。英美烟草公司在2017年"烟碱和烟草研究学会"会议上展示了两张关于其HTP的海报[5,32]，并于2017年末在杂志的增刊中发表了8篇研究[4,33-40]。该产品被称为THP 1.0，在商业市场上被称为glo。它将烟草（细支卷烟中包含有14.5%甘油的薄片）加热到最高250℃[32]。热成像仪分析表明，该产品加热到100℃释放出水分，然后到240℃释放出甘油，然后在350℃分解[32]。在改良的HCI抽吸模式下进行抽吸的数据显示，TobReg优先级有害物质的释放量远低于参比卷烟（表2.2）[41]。另一张海报[42]给出了对室内空气影响的详细信息，并声称THP 1.0有害成分的释放量低于传统卷烟。在英美烟草公司的iFuse与其他HTP（在本报告中称为cTHP，描述符合iQOS）的比较中[43]，cTHP释放出更高水平的乙醛（每10次抽吸125 μg vs. 35.9 μg），而iFuse释放出更多的甲醛（每10次抽吸< 4.18 μg vs. 38.7 μg）。这两种产品的乙醛释放量都大大低于3R4F参比卷烟（>-88%），但iFuse的甲醛释放物水平与传统卷烟（-8%）相当。Bekki等[31]的结果表明，iQOS中烟碱和亚硝胺从烟丝到气溶胶的转移率与传统卷烟相当，甚至略高于传统卷烟。

表2.2 参比卷烟1R6F、Eclipse、电加热吸烟系统（EHCSS）、烟草加热系统（THS）和加热型烟草制品（THP）释放的有害成分

组分	1R6F（μg/支）ISO[28]	1R6F（μg/支）HCI[28]	Eclipse（μg/支）ISO[21]	EHCSS（μg/支）ISO[44]	THS 2.2（μg/支）HCI[45]	THP 1.0（μg/支）HCI[32]
乙醛	522	1552	84.2	179	219	111
丙烯醛	43	154	11.5	27.3	11.3	2.22
丙烯腈	7.0	24	0.44	0.258	<0.032	
4-氨基联苯	1.2	2.3	0.058	<0.051	<0.005	
2-萘酚	8	14	0.123	0.046	<0.012	
氨	9.0	30	4.37	14.2	4.01	

续表

组分	1R6F（μg/支）ISO[28]	1R6F（μg/支）HCI[28]	Eclipse（μg/支）ISO[21]	EHCSS（μg/支）ISO[44]	THS 2.2（μg/支）HCI[45]	THP 1.0（μg/支）HCI[32]
苯	33	88		0.363	0.649	<0.056
苯并[a]芘	6.8	15	1.2	<0.19	<1.0	<0.35
CO（mg）	10.1	28.0	7.5	0.465	0.531	<0.223
甲醛	27	104		12.9	5.53	3.29
异戊二烯	320	881		34.3	2.35	<0.135
烟碱	0.721	1.90	0.18	0.313	1.32	0.462
NNK（ng）	71	187	31.8	6.18	6.7	6.61
NNN（ng）	85	212	26	19.8	17.2	24.7
甲苯	53	150		1.48	2.59	<0.204
焦油（mg）	8.58	29.1	3.2	3.1	10.3	13.6

EHCSS：电加热吸烟系统；HCI：加拿大深度抽吸模式；ISO：国际标准化组织抽吸模式；NNK：4-(甲基亚硝基氨基)-1-(3-吡啶基)-1-丁酮；NNN：N'-亚硝基降烟碱；THP：加热型烟草制品；THS：烟草加热系统

2.4.2 暴露生物标志物

2008年，菲利普·莫里斯美国公司和PMI发表了一系列关于其电加热吸烟系统（EHCSS）的文章，包括毒理学、释放物（也包括二手烟）以及短期和长期临床暴露的研究以及随机试验[44,46-48]。随后在2012年发表了一系列论文[49-56]。2016年，PMI发表了另一系列研究，有关THS 2.2[3,45,57-59]，报告了一系列产品评估的结果。表2.3和表2.4列出了短期暴露后（约1周，一般为常住人口）的生物标志物数据[47,49,51-54,58,59]。在一些人群的研究中，除烟碱以外，其他关键有害物质的水平均低于持续吸烟者的水平。由于这些研究的参与者通常只限于研究期间，因此，这些发现可能无法推广到实际使用条件（例如，产品使用方式，与卷烟或其他燃烧或非燃烧产品一起使用）。

表2.3A 在电加热吸烟系统的短期研究结束时，使用传统卷烟、加热型烟草制品以及可获得的戒烟情况下的生物标志物水平（日本和美国）

生物标志物	HPHC	日本（6天）[53] 卷烟	EHCSS	戒烟	美国（8天）[47] 卷烟	EHCSS	戒烟	日本（8天）[52] EHCSS-K6a	EHCSS-K3a	戒烟
烟碱当量	烟碱（mg/24h）	7.2	3.4	0.6	16.1	8.2	0.2	9.6	5.1	3.9	0.2
烟碱	烟碱	NR	NR	NR	NR	NR	NR	19	10.1	8.1	0.5
可替宁	烟碱（ng/mL）	150.5	79.4	11.1	NR	NR	NR	117.9	111.2	79.5	7.4
NNAL	NNK（ng/24h）	188	95	80	601.3	169.1	117.8	316	100	102	83
COHb	一氧化碳（%）[6]	5.1	2.1	1.9	6.4	2.1	1.1	5.4	2.1	2.2	1.7
MHBMA	1,3-丁二烯（μg/24h）	1.5	0.6	0.4	4.8	1.5	1.5	1.5	0.7	0.7	0.4
3-HPMA	丙烯醛（mg/24h）	1.2	0.8	0.5	1859	1056.7	302.3	1.4	1.1	1	0.5

续表

生物标志物	HPHC	日本（6天）[53]			美国（8天）[47]			日本（8天）[52]			
		卷烟	EHCSS	戒烟	卷烟	EHCSS	戒烟	卷烟	EHCSS-K6[a]	EHCSS-K3[a]	戒烟
S-PMA	苯（μg/24h）	2.3	0.4	0.5	5.3	1.1	0.4	2.2	0.6	0.5	0.3
1-OHP	苯并[a]芘（ng/24h）	106.5	38.4	37.4	149.6	58.1	41.1	135.3	56	59	42
3-羟基苯并[a]芘	苯并[a]芘	NR	NR	NR	NR	NR	NR	NR	NR	NR	NR
4-氨基联苯	4-氨基联苯（ng/24h）	9.4	4.4	4.7	12.6	3.1	2	15	4.2	3.9	4.6
1-萘胺	1-萘胺	NR	NR	NR	NR	NR	NR	NR	NR	NR	NR
2-萘胺	2-萘胺（ng/24h）	4.6	4.4	4.7	20	4.2	2.2	27.2	6.9	6.1	5.9
邻氨基甲苯	甲苯（ng/24h）	66.2	33.6	33.8	87.6	58.5	40.9	103.9	31.1	27.7	33.8
HMPMA	巴豆醛（mg/24h）	0.5	0.2	0.1	2320.7	750.4	299	1.3	0.6	0.6	0.5

COHb：碳氧血红蛋白；EHCSS：电加热吸烟系统；HMPMA：羟甲基丙基巯基尿酸；HPHC：有害和潜在有害成分；3-HPMA：3-羟丙基巯基尿酸；MHBMA：单羟基丁烯基巯基尿酸；NNAL：4-(甲基亚硝基氨基)-1-(3-吡啶基)-1-丁醇；NNK：4-(甲基亚硝基氨基)-1-(3-吡啶基)-1-丁酮；NR：未报告；1-OHP：1-羟基芘；S-PMA：S-苯基巯基尿酸

a. K6和K3是测试系统的两个版本

表2.3B　在电加热吸烟系统的短期研究结束时，使用传统卷烟、加热型烟草制品以及可获得的戒烟情况下的生物标志物水平（英国和韩国）

生物标志物	HPHC	韩国（8天）[51]		英国（8天）[54]			
		EHCSS-K3[a]	戒烟	卷烟	EHCSS-K6[a]	EHCSS-K3[a]	戒烟
烟碱当量	烟碱（mg/24h）	4.2	0.1	13.1	8.6	6.5	0
烟碱	烟碱	8.6	0.1	NR	NR	NR	NR
可替宁	烟碱（ng/mL）	85.8	0.1	234.3	158.3	125	0
NNAL	NNK（ng/24h）	80.4	45.1	293.6	100.6	104.3	59.2
COHb	一氧化碳(%)	0.9	0.5	5.7	1.4	1.9	0.5
MHBMA	1,3-丁二烯（μg/24h）	0.6	0.3	5.1	1.5	2.6	0.3
3-HPMA	丙烯醛（mg/24h）	2.4	1.7	1.8	1.3	1.2	0.5
S-PMA	苯（μg/24h）	1.6	1.4	6.2	0.9	1.3	0.2
1-OHP	苯并[a]芘（ng/24h）	143.6	127	181.6	71.9	73.1	75.1
3-羟基苯并[a]芘	苯并[a]芘	NR	NR	NR	NR	NR	NR
4-氨基联苯	4-氨基联苯（ng/24h）	5.5	5.5	NR	NR	NR	NR
1-萘胺	1-萘胺	NR	NR	NR	NR	NR	NR
2-萘胺	2-萘胺（ng/24h）	5.4	5.5	NR	NR	NR	NR
邻氨基甲苯	甲苯（ng/24h）	29.1	30.2	135	58	49.3	47.6
HMPMA	巴豆醛（mg/24h）	2.0	1.7	5.2	2.6	2.6	1.7

COHb：碳氧血红蛋白；EHCSS：电加热吸烟系统；HMPMA：羟甲基丙基巯基尿酸；HPHC：有害和潜在有害成分；3-HPMA：3-羟丙基巯基尿酸；MHBMA：单羟基丁烯基巯基尿酸；NNAL：4-(甲基亚硝基氨基)-1-(3-吡啶基)-1-丁醇；NNK：4-(甲基亚硝基氨基)-1-(3-吡啶基)-1-丁酮；NR：未报告；1-OHP：1-羟基芘；S-PMA：S-苯基巯基尿酸

a. K6和K3是测试系统的两个版本

表2.4 在发表的有关 THS 2.2的短期研究结束时，使用传统卷烟、加热型烟草制品以及可获得的戒烟情况下的生物标志物水平

生物标志物	HPHC	日本[58,60]			波兰[59]		
		THS	卷烟	戒烟	THS	卷烟	戒烟
烟碱当量	烟碱（mg/g肌酐）	5.44	5.52	0.15	10.6	9.76	0.14
烟碱	烟碱	19.13	21.34	0.1	20.74	19.01	0.1
可替宁	烟碱	161	164.3	2.96	239.99	219.73	2.05
NNAL	NNK（pg/mg肌酐）	37.77	76.55	28.63	49.65	107.04	41.51
COHb	一氧化碳（%）	2.39	5.14	2.37	1.06	4.51	0.99
MHBMA	1,3-丁二烯（pg/mg肌酐）	107.39	450.19	92.18	192.93	2399.4	163.17
3-HPMA	丙烯醛（ng/mg肌酐）	311.08	599.67	199.04	402.26	931.01	245.69
邻甲苯胺	甲苯（pg/mg肌酐）	143.77	850.02	126.34	164.45	2922.81	143.7
1-羟基	苯并[a]芘（pg/mg肌酐）	73.02	149.62	62.99	81.22	182.85	85.13
3-OH-苯并[a]芘	苯并[a]芘（fg/mg肌酐）	29.52	96.42	24.47	37.07	130.29	33.64
4-氨基联苯	4-氨基联苯（pg/mg肌酐）	1.53	8.57	1.49	1.9	12.58	1.6
1-萘胺	1-萘胺（pg/mg肌酐）	2.47	57.08	2.45	3.3	89.37	2.56
N-硝基异烟碱	N-硝基异烟碱（pg/mg肌酐）	1.31	4.64	0.18	1.55	5.99	0.16
2-萘胺	2-萘胺（pg/mg肌酐）	2.33	13.38	2.27	2.96	25.32	2.52
CEMA	丙烯腈（ng/mg肌酐）	10.61	54.19	9.04	13.18	99.48	12.6
HEMA	环氧乙烷（pg/mg肌酐）	997.76	2099.41	806.29	1342.4	4504	1248.27
HMPMA	巴豆醛（pg/mg肌酐）	59.51	157.83	47.84	86.65	376.78	63.25
S-BMA	甲苯（ng/mg肌酐）	2098.09	2354.17	2192.86	NR	NR	NR

CEMA：2-氯乙基甲基丙烯酸酯；COHb：碳氧血红蛋白；HEMA：甲基丙烯酸羟乙酯；HMPMA：羟甲基丙基硫基尿酸；3-HPMA：3-羟丙基硫基尿酸；MHBMA：单羟基丁烯基硫基尿酸；NNAL：4-(甲基硝基氨基)-1-(3-吡啶基)-1-丁酮；NNK：4-(甲基亚硝基氨基)-1-(3-吡啶基)-1-丁酮；NNN：N'-亚硝基降烟碱；NR：未报告；1-OHP：1-羟基芘；S-BMA：S-苄基硫基尿酸；S-PMA：S-苯基硫基尿酸；THS：烟草加热系统

  PMI还发布了EHCSS（12周）和THS 2.2（90天）的长期临床研究结果。在美国有60名参与者使用EHCSS 12周后，4-(甲基亚硝基氨基)-1-(3-吡啶基)-1-丁醇（NNAL，一种NNK的暴露生物标志物）的水平下降了63%（$P<0.0001$），碳氧血红蛋白（CO暴露生物标志物）降低了23%（$P<0.0001$），S-苯基硫基尿酸（苯暴露生物标志物）降低了49%（$P<0.0001$）[48]（文章没有提供每次平均估计值）。日本76位参与者使用THS 2.2 90天后，NNAL的水平下降了73%，碳氧血红蛋白的水平下降了42%，S-苯基硫基尿酸的水平下降了23%[61]。此外还测量了许多其他暴露生物标志物（表2.5），并对参与者进行了90天的随访，进行是否继续吸烟和戒烟的跟踪调查。使用THS比传统卷烟减少了某些成分的暴露，但是与戒烟期间相比，包括亚硝胺和丙烯醛在内的许多成分的暴露量仍然较高。

表2.5　日本90天随机分配THS 2.2、持续吸烟或戒烟的参与者在研究结束时的生物标志物水平

生物标志物	HPHC	THS	卷烟	戒烟	与卷烟的差异（%）	与戒烟的差异（%）
烟碱当量	烟碱（mg/g肌酐）	7	6	0	8	1751
NNAL	NNK（pg/mg肌酐）	23	95	14	−76	67
NNN	NNN（pg/mg肌酐）	1个	4	0	−67	438
COHb	CO（%）	3	6	3	−48	−2
MHBMA	1,3-丁二烯（pg/mg肌酐）	142	785	137	−82	4
3-HPMA	丙烯醛（ng/mg肌酐）	386	696	276	−44	40
S-PMA	苯（pg/mg肌酐）	146	1157	144	−87	1
1-OHP	苯并[a]芘（pg/mg肌酐）	85	167	88	−49	−3
3-羟基苯[a]芘	苯并[a]芘（fg/mg肌酐）	30	87	29	−65	4
4-氨基联苯	4-氨基联苯（pg/mg肌酐）	2	10	2	−78	−12
1-萘胺	1-萘胺（pg/mg肌酐）	4	55	4	−94	−16
2-萘胺	2-萘胺（pg/mg肌酐）	2	149	3	−98	−11
邻甲苯胺	甲苯（pg/mg肌酐）	68	126	78	−46	−12
CEMA	丙烯腈（ng/mg肌酐）	8	84	8	−91	−6
HEMA	环氧乙烷（pg/mg肌酐）	1742	3739	1633	−53	7
HMPMA	巴豆醛（ng/mg肌酐）	154	299	159	−48	−3

CEMA：2-氯乙基甲基丙烯酸酯；CO：一氧化碳；COHb：碳氧血红蛋白；HEMA：甲基丙烯酸羟乙酯；HMPMA：羟甲基丙基巯基尿酸；HPHC：有害和潜在有害成分；3-HPMA：3-羟丙基巯基尿酸；MHBMA：单羟基丁烯基巯基尿酸；NNAL：4-(甲基亚硝基氨基)-1-(3-吡啶基)-1-丁醇；NNK：4-(甲基亚硝基氨基)-1-(3-吡啶基)-1-丁酮；NNN：N'-亚硝基降烟碱；1-OHP：1-羟基芘；S-PMA：S-苯基巯基尿酸

使用补偿公式

$$1-[(\ln 标志物1-\ln 标志物0)]/(\ln 产物1-\ln 产物0)$$

比较吸烟机测量的释放物的变化（EHCSS的ISO释放量、THS的HCI释放量，每个都与具有特定HPHC认证值的1R6F参比卷烟进行比较）与暴露生物标志物水平，很明显，存在潜在的实质性补偿，包括THS对烟碱的补偿几乎是全部补偿（表2.6）。获得更高水平的烟碱的能力可能部分解释了上一代HTP和iQOS在使用方面的差异。

表2.6　相对于参比卷烟，PMI加热型烟草制品中某些有害和潜在有害成分（HPHC）的潜在补偿

HPHC	ECHSS（日本）（%）	EHCSS（美国）（%）	THS（日本）（%）	THS（波兰）（%）
丙烯醛	79	67	75	68
丙烯腈	NR	NR	64	55
4-氨基联苯	79	62	55	50
2-萘酚	99	67	69	62
苯	66	71	64	41
苯并[a]芘	77	78	74	70

续表

HPHC	ECHSS（日本）(%)	EHCSS（美国）(%)	THS（日本）(%)	THS（波兰）(%)
一氧化碳（mg）	78	73	81	63
烟碱	58	63	96	123
NNK（ng）	80	63	79	77
甲苯	85	91	84	79

ECHSS：电加热吸烟系统；NNK：4-(甲基亚硝基氨基)-1-(3-吡啶基)-1-丁酮；NR：未报告；THS：烟草加热系统

尽管已经公布了关于THP 1.0产品暴露的研究方案，但尚未发现有关该研究的生物标志物报告[62]。英美烟草公司发表了一项关于人类使用glo产品以确定使用模式的研究[38]。其中3个研究组将3种产品（glo-薄荷醇，glo-烟草，glo + iQOS）带回家中长达14天，进行了4次实验室拜访，而第4组仅在实验室中使用glo。这些措施包括抽吸轮廓、口腔水平暴露和口腔插入深度。将产品带回家的参与者完成了产品使用的每日日记。

总体而言，glo产品的抽吸容量约为60 mL，平均每次抽吸10~12口，持续时间为1.8~2.0 s，每次间隔为8 s。单次抽吸容量和总体积明显高于参比卷烟，但与iQOS产品相当。参与者报告在家中4天期间每天使用12~15支卷烟，每天约8~12支glo或iQOS。与普通卷烟相比，使用glo的每种产品其口腔中释放物的暴露量明显要低，尤其是对于烟碱而言。平均口部插入深度为7.7 mm，该公司以此为依据而忽略HCI抽吸方案中的通气孔封闭的争论[38]。

迄今为止，仅在一项已发表的研究中检查了消费者使用个人雾化器来加热烟草的情况[63]，其中将Pax（装有1 g自卷卷烟烟草）与吸烟者自己的卷烟品牌进行了比较，以及一支e-GO型电子烟，15名参与者参加了一个固定抽吸模式的实验室实验。与吸烟者自己的品牌相比，雾化器增加的血浆烟碱含量较低（14.3 ng/mL vs. 24.4 ng/mL），与电子烟相当，并没有增加CO的暴露。没有其他生物标志物的报道。

2.5　消费者对加热型烟草制品的认知

人群调查和烟草行业营销研究表明，消费者对声称或暗示具有降低健康风险的产品有强烈的需求[64-70]。大学生们在看到每种产品的广告后，对Eclipse的评价没有万宝路那么正面，但也没有万宝路那么负面，尽管没有人尝试过该产品[71]。Shiffman及其同事在两项研究中显示[67,69]，吸烟者认为Eclipse减少了危害，而使用Eclipse的兴趣与戒烟意愿降低相关。Hamilton及其同事[64]向吸烟者展示了几种PREP和传统烟草制品的广告，发现尽管广告中没有任何健康声明，但PREP仍然被认为具有较小的健康风险,并传达了有关健康和安全的正面信息。来自日本的数据[72]表明,极少数的消费者知道并对HTP感兴趣[72]。尽管接受调查的8240名受访者中有48%知道电子烟和/或HTP,但任何产品的实际使用率都非常低。6.6%的人使用了其中的某个产品,绝大多数使用了电子烟。这些在日本没有合法销售,但可以由个人携带入境供自己使用。只有0.51%的受访者曾经使用过Ploom,0.55%的受访者曾经使用过iQOS(在HTP的使用者中分别为7.8%和8.4%)。后来的一项研究[73]显示,2016~2017年间,iQOS的使用率急剧上升,从2016年的0.6%上升到2017年的3.6%。

新产品失败的一个常见原因是消费者因口感差而拒绝[74,75]。感官特性是卷烟设计的重要组成部分[76,77]。对Eclipse和Accord的主观影响的临床研究发现产品相对于自己的品牌被认为是负面的,满意度普遍较低,厌恶度较高[6,7,10,22,23]。参加焦点小组会议的吸烟者表示,他们极不喜欢曾尝试过的PREP卷烟,包括Eclipse和Accord,几乎所有人都报告说他们不会向其他吸烟者推荐该产品[78,79]。Caraballo和同事们[78]发现,尝试过Eclipse的吸烟者普遍不喜欢该产品,因此不会推荐给其他吸烟者。多数声称不喜欢Eclipse卷烟的吸烟者认为它们太温和,并表示其提供的烟碱量不足以满足其渴望。许多人报告说他们不喜欢这种味道。Hughes及其同事[80]报告说,尝试过Eclipse的吸烟者通常不喜欢它,尽管他们认为它比传统卷烟更安全。对Pax雾化器的研究[63]表明,该雾化器确实能部分抑制戒断症状,但与自己品牌卷烟相比,其满意度和美味程度明显低。

PMI在其THS设备上发布的研究包括对产品的主观反应数据,但EHCSS和上一代产品的研究报告中未提供这些数据。这些数据通常对于理解吸烟者为何使用某种产品及其替代品的有效性非常重要[81,82]。这项研究是使用改良的卷烟评估调查表[83]和有关吸烟渴望来衡量渴望的简短调查表[84]进行的。卷烟评估问卷包括12个项目(7个等级的量表),分为五个问卷:吸烟满意度、反感、渴望减少、呼吸道感觉和心理奖赏的享受[83]。吸烟欲望问卷由10个项目组成,提供7个等级的单一分值[84]。

在日本的THS实验室研究过程中,THS的平均满意度得分下降的幅度大于传统卷烟[平均值=−0.69;95%置信区间(CI),−0.34和−1.04][58]。这些产品评估问卷中的其他得分与卷烟得分没有变化或没有实质性差异。THS和传统卷烟的吸烟欲望调查问卷得分均未发现差异,两者均低于戒烟(如预期)。在波兰的一项类似研究中,观察到的THS和卷烟在主观效果上的差异相差很大,并且在满意度上具有统计学意义(平均值=−1.26;95%CI,−0.85,−1.68),渴望减少(平均值=−1.12);95%CI,−0.66,−1.58),感觉(平均值=−1.00;95%CI,−0.64,−1.36)和奖励(平均值=−0.72;95%CI,−0.39,−1.06)。THS和卷烟之间的渴望分数没有显著差异,并且两者均显著低于戒烟者。这些研究表明,波兰吸烟者对THS的看法不及日本吸烟者。这可能会对从一个市场到另一个市场的观察结果的概括产生影响。THS 2.1的早期研究[85]显示了类似的结果，THS的满意度平均得分在第5天比传统卷烟低1.4点（$P<0.001$）。在奖励、感觉和渴望分量表上也发现了显著差异，在所有情况下，THS的评分均低于卷烟。在日本进行的长期使用研究[61]表明，随着持续使用时间超过90天，满意度的差异逐渐消失。这些数据表明，吸烟欲望调查问卷上的THS分数随着时间的推移而增加，戒烟分数也一样（以美国明尼苏达州烟碱戒烟量表[86]为衡量标准）。这可能表明人们对该产品作为长期替代品有些不满意。

2.6 在选定市场中的普及率

在意大利和日本，已经进行了3项有关人群HTP使用率的研究。在意大利，有1%的从不吸烟者，0.8%的既往吸烟者和3%的当前吸烟者尝试过iQOS[87]。Tabuchi及其同事[73]的数据显示，2015年之后，iQOS的当前使用率（从0.3%升至3.6%）和Ploom的使用率（从0.3%升至1.2%）呈上升趋势，2017年glo的使用率相当（0.8%）。使用HTP的预测人群是当前吸烟（戒烟意愿影响更大）、生活在更贫困的地区以及看过电视宣传iQOS的。双重使用很普遍（72%）。与传统卷烟不同的是，在日本教育程度与HTP使用之间似乎没有反比关系[88]。

HTP在日本获得市场份额的一个重要原因是不允许出售含烟碱的ENDS。因此，在吸烟率相对较高且烟草控制法律相对较弱的国家/地区，HTP可能会填补市场的差距。iQOS已在欧盟（允许销售ENDS）和加拿大（直到2018年不允许销售）的多个市场中进行销售。在2017年向纽约消费者分析集团提交的报告中[89]，PMI报告称，在2016年第四季度，iQOS的市场份额在日本已达到4.9%，但在瑞士（1.7%）、葡萄牙（0.7%）、罗马尼亚（0.6%）等市场营销重点领域的市场份额大幅下降，意大利（0.4%）和俄罗斯（0.3%）该产品至少已售出一年。根据其使用者小组的说法，向iQOS的转换率很高，从瑞士的54%到日本的72%不等。

2.7 菲利普·莫里斯国际在美国提出的"风险降低的烟草制品"申请

2017年5月，PMI向美国食品药品监督管理局（FDA）提交其THS/iQOS的MRTP申请，以供审查。正在考虑的三个健康声明是：

- ■ "从卷烟完全转换为iQOS可以减少与烟草有关疾病的风险。"
- ■ "与继续吸烟相比，完全转换为iQOS带来的危害风险较小。"
- ■ "从卷烟完全转换到iQOS，可大大减少人体暴露于有害和潜在有害化学物质的机会。"

根据《烟草控制法案》，完整申请的修订版本[②]在FDA网站（https://www.FDA.gov/tobaccoproducts/labelingandadvertising/ucm54628 htm）供公众审查和给出意见。有关该设备的许多技术信息已被删除。下文讨论了在公开可用部分中引用的但未在上面进行讨论的发现（即未发表的研究）。

在线材料包括2013~2014年在美国进行的另一项THS 2.2人体试验（NCT01989156）。在随机分配的160名参与者中，有88名完成了为期90天的完整研究。与在日本和波兰进行的研究一样，与吸烟5天和90天后相比，暴露的关键生物标志物明显减少。然而，在这项研究中，在日本戒烟的依从性大大低于非临床阶段，表明日本的经验不一定能推广到美国。

MRTP文件还报告了一系列研究，以测试在美国市场上针对iQOS提出的特定要求。描述了一系列定性和定量研究，最后进行了三个"评估阶段"研究，研究了在各种情况下（包装、小册子、直接邮寄）对健康声明和风险感知的理解。PMI开发了用于本研究的多因素风险感知量表。研究使用了大约2500名成年吸烟者（有和无戒烟意愿），2500名既往吸烟者和2500名非吸烟者的3个样本进行。iQOS产品被评为介于卷烟和戒烟之间的中等风险，可与电子烟的风险等级相比。尽管有证据表明大约25%的参与者推断出伤害减少了（这可能是错误的）。这些说法改变了人们对iQOS的风险认知，或导致吸烟者尝试iQOS的意愿增加，这一点并不明显。

在德国、意大利、日本、韩国、瑞士和美国报道了有关产品转换的一系列观察性研究。在前五个国家中，对2089名每日吸烟者进行了"整体报价测试"。iQOS可以免费使用4周，并且产品使用情况记录在电子日记中。在为期4周的试用期结束时，改用为THS的比例从德国的10%到韩国的37%不等，双重使用率从日本的32%到韩国的39%不等。这项在美国进行的研究中包括1106名当前每日吸烟者，他们在1周的基准后可以免费使用iQOS 4周。产品使用情况（卷烟和Heat-sticks烟支）记录在电子日记中。在研究结束时，大约15%的受试者已改用THS，按照研究中的定义（Heatsticks烟支使用量超过总消费烟支数的70%），而22%是双重使用者（30% ~ 70%是Heatsticks烟支）。

有限的上市后数据主要来自日本，借鉴了iQOS采购登记册。报告称，2016年1~7月，单独使用iQOS的比例（占总消费量的95%以上）从52%增加到65%。对日本两组iQOS购买者群体（2015年9月和2016年5月）的Markov过渡模型表明，转换为iQOS的吸烟者不太可能回归单独使用卷烟（尽管执行摘要中提供了很少的模型细节）。

[②] 对申请进行修订，以删除根据美国法律被视为商业机密的信息

在2018年1月FDA烟草制品科学咨询委员会会议上③，PMI提供了支持iQOS的证据，包括本报告中的许多数据。FDA对提交的文件进行了初步评估，委员会对证据基础的某些部分表示关注，特别是"完全转换"的定义以及对消费者理解这些声明的研究的局限性。委员会不支持风险改良的声明，支持暴露改良的声明，并对声明措辞在风险传播方面的无效表示关注。委员会向FDA提出的建议不具有约束力。

2.8 对监管和烟草控制政策的影响

很难预测非传统烟草制品的使用情况。烟草行业对上一代PREP的实验证据分析表明，烟草行业试图将焦点从生产"更安全"产品的责任转移到吸烟者未能采用他们所生产的"更安全"产品上[75]。

尽管如此，该市场领域似乎正在增长而不是萎缩，并且在至少一个国家中占有重要的市场份额，PMI声称日本的iQOS销售过100万个设备[90]。现有的人群研究（均由制造商进行）表明，iQOS比传统卷烟释放的有害物质更少，并且可以作为传统卷烟的有效的短期替代品，这是通过烟碱的释放和在卷烟评估问卷所测得的主观效果来评估的。一项为期90天的研究表明，结果相似。但是，必须记住，这些研究的参与者进行了补偿抽吸，因此结果可能无法反映实际的使用方式，包括同时使用多种产品。2017年12月，食品、消费品和环境中化学物质的有害性、致癌性和致突变性委员会评估了英国市场上的两种HTP（iQOS和iFuse）并得出结论[91]：

"虽然吸烟者转用'加热不燃烧'烟草制品可能降低了风险，但仍存在残留风险，对于吸烟者来说完全戒烟会更有益。这应成为最大限度地减少吸烟风险长期策略的一部分。"

"产品逐渐变化"可能是另一个重要的考虑因素。PMI在提交纽约消费产业分析师协会的报告中[89]，描述2014~2017年在日本测试的原型产品试用版对设备的某些修改，包括更好的外观、刀片自动清洁、更好的使用者界面、更快的充电速度、蓝牙连接以及随附的移动应用程序。使用颜色来增加设备的吸引力。因此，产品在推出后可能是"移动目标"。这些做法与烟草行业的做法没有什么不同，烟草行业会随着时间和市场的不同[92-94]对卷烟产品进行细微调整，例如2017年的万宝路卷烟不一定与2010年的万宝路卷烟相同，在法国销售的万宝路卷烟也不一定与在美国销售的万宝路卷烟相同。此外，来自研究的产品数据可能无法完全反映当前消费者可获得的产品。因此，支持产品的研究不一定是针对最终上市的产

---

③ 参见 https://www.fda.gov/downloads/AdvisoryCommittees/CommitteesMeetingMaterials/TobaccoProductsScientificAdvisoryCommittee/UCM599235.pdf.

品，而要使用一个原型甚至一系列不同的原型。尽管这种做法本身并不有害，但应确定所研究产品和市场产品之间在设计、功能或外观上的差异，包括它们对消费者使用的影响。FDA要求报告对现有产品的变更，以进行销售授权，可能对其进行监控，并对影响公共卫生的变更进行更严格的审查和要求（例如，HPHC递送的变化，实质性的设计变更）。欧盟成员国根据《欧盟烟草制品指令》也有类似规定。其他国家的监管机构应考虑采取类似的对产品变更进行通报和授权的要求。

随着非传统烟草制品的引入，侧流烟气释放物和二手烟暴露进一步受到关注。一些烟草制造商声称某些HTP可使侧流烟气的暴露量降至最低[42,46,95,96]，而一些研究表明该水平很高[97,98]。释放物可能部分取决于产品的设计。侧流烟气释放物对健康和监管有影响，需要进一步研究以确定HTP的环境暴露特征。例如，根据健康声明的措辞，无烟立法可能会也可能不会涉及HTP。但是，预防原则将支持在法规中涵盖此类释放物。

2.9 研究和政策建议

需要独立的科学证据来验证企业科学家关于减少暴露和降低风险的声明。截至2017年10月，经同行评审发表的文献主要是由PMI对其THS 2.2产品进行的研究，重点关注了HPHC中的部分物质。在这些研究中没有发现关于金属（如镉）暴露生物标志物。金属（类金属）作为致癌物和强化剂（镉、镍、钴、砷），以及作为本身有毒的物质（铅、铜）都令人担忧。尽管大多数研究表明，与吸烟相比，HPHC暴露减少，但研究并未解决THS产生新的有害物质暴露的可能性，无论是加热系统本身还是烟草中使用的添加剂。在向FDA烟草制品科学咨询委员会提交的MRTP报告中，当被这些问题所困扰时，PMI承认iQOS气溶胶中的50种成分的含量高于传统卷烟烟气中的含量，其中3种是iQOS独有的；在iQOS中，约有750种成分的含量等于或低于常规卷烟烟气中的含量，并且>4000种是卷烟烟气中特有的。在50种成分中，有4种被认定为毒理学关注物质（缩水甘油、糠醇、3-氯-1,2-丙二醇和糠醛）。

尽管已经发布了一项针对随机试验的研究方案，但关于英美烟草公司HTP的公开数据很少[62]，这表明该公司正在这一方面进行研究。尚未找到有关日本烟草国际公司的Ploom产品的公开研究报告，同样缺乏暴露有关个人雾化器的HPHC数据。有关用于烟草的个人雾化器的使用和销售的数据非常难以找到，因此无法与该细分市场进行比较。使用这种设备来抽吸大麻的现象似乎正在增加[99-101]。

对于近期和长期研究重点的一些建议是：

■ 使用经过验证的工具（示例[102]）监测产品可用性、销售和市场营销；

- 监测产品使用情况，包括完全转换、双重（或多重）使用、与电子烟碱传输系统相比的使用情况以及非烟草使用者的主动使用情况，尤其关注那些本来不会吸烟的低风险年轻人；
- 验证已报告的产品中39种优先级有害物质和/或FDA HPHC清单的成分和释放物；
- 评估未被普遍接受的清单（例如霍夫曼分析物、HPHC）所涵盖的加热型烟草制品产生的潜在新的有害物质；
- 评估气溶胶粒径分布；
- 评估设备的功能和安全性（例如电池）；
- 产品的跨市场比较，例如，所有市场上的iQOS是否相同，以及产品的特性、成分和释放物及其随时间的变化情况是否不同；
- 根据典型的使用方式（包括双重使用或同时使用加热不燃烧型产品和卷烟）对使用者进行独立的临床和生物标志物分析；
- 烟草使用者和非使用者对产品的公众认知（意识、使用意图、风险）；
- 有关谁在购买产品、购买原因和使用方式的信息；
- 研究吸烟者向HTP和雾化产品（ENDS）的转化率，以确定这些产品是否以吸烟者可以接受的方式阻止吸烟；
- 对产品引进和使用的潜在人口层面影响进行建模（例如SimSmoke、微观模拟）；
- 调查营销策略对使用行为的影响，包括这些产品是作为补充产品还是替代产品销售；
- 热处理对产品中除烟草以外的成分（例如调味剂）的影响；
- 研究暴露于二手烟释放物（包括对儿童和孕妇的影响）及其对环境空气质量的影响。

2.10 参 考 文 献

[1] Rodgman A, Perfetti TA. The chemical components of tobacco and tobacco smoke. 2nd ed. Boca Raton (FL): CRC Press; 2013.

[2] Batista RN. Reinforced web of reconstituted tobacco. Google Patents WO2015193031A1; 2015 (https://patents.google.com/patent/WO2015193031A1/en, accessed 1 October 2018).

[3] Smith MR, Clark B, Ludicke F, Schaller P, Vanscheeuwijck P, Hoeing J et al. Evaluation of the tobacco heating system 2.2. Part 1: Description of the system and the scientific assessment program. Regul Toxicol Pharmacol. 2016;81(Suppl 2):S17–26.

[4] Eaton D, Jakaj B, Forster M, Nicol J, Mavropoulou E, Scott K et al. Assessment of tobacco heating product THP1.0. Part 2: Product design, operation and thermophysical characterisation. Regul Toxicol Pharmacol. 2018;93:4–13.

[5] Scott JK, Poynton S, Margham J, Forster M, Eaton D, Davis P et al. Controlled aerosol release to heat tobacco: product operation and aerosol chemistry assessment. Poster. Society for Research on Nicotine and Tobacco; 2–5 March 2016; Chicago (IL); 2016 (https://www.researchgate.net/publication/298793405_Controlled_aerosol_release_to_heat_tobacco_product_operation_and_aerosol_chemistry_assessment, accessed October 2018).

[6] Breland AB, Buchhalter AR, Evans SE, Eissenberg T. Evaluating acute effects of potential reduced-exposure products for smokers: clinical laboratory methodology. Nicotine Tob Res. 2002;4(Suppl 2): S131–40.

[7] Breland AB, Kleykamp BA, Eissenberg T. Clinical laboratory evaluation of potential reduced exposure products for smokers. Nicotine Tob Res. 2006;8(6):727–38.

[8] Fagerstrom KO, Hughes JR, Callas PW. Long-term effects of the Eclipse cigarette substitute and the nicotine inhaler in smokers not interested in quitting. Nicotine Tob Res. 2002; 4(Suppl 2): S141–5.

[9] Fagerstrom KO, Hughes JR, Rasmussen T, Callas PW. Randomised trial investigating effect of a novel nicotine delivery device (Eclipse) and a nicotine oral inhaler on smoking behaviour, nicotine and carbon monoxide exposure, and motivation to quit. Tob Control. 2000;9(3):327–33.

[10] Lee EM, Malson JL, Moolchan ET, Pickworth WB. Quantitative comparisons between a nicotine delivery device (Eclipse) and conventional cigarette smoking. Nicotine Tob Res. 2004;6(1):95–102.

[11] Rennard SI, Umino T, Millatmal T, Daughton DM, Manouilova LS, Ullrich FA et al. Evaluation of subclinical respiratory tract inflammation in heavy smokers who switch to a cigarette-like nicotine delivery device that primarily heats tobacco. Nicotine Tob Res. 2002;4(4):467–76.

[12] Stapleton JA, Russell MA, Sutherland G, Feyerabend C. Nicotine availability from Eclipse tobacco-heating cigarette. Psychopharmacology. 1998;139(3):288–90.

[13] Stewart JC, Hyde RW, Boscia J, Chow MY, O'Mara RE, Perillo I et al. Changes in markers of epithelial permeability and inflammation in chronic smokers switching to a nonburning tobacco device (Eclipse). Nicotine Tob Res. 2006;8(6):773–83.

[14] Pauly JL, Lee HJ, Hurley EL, Cummings KM, Lesses JD, Streck RJ. Glass fiber contamination of cigarette filters: an additional health risk to the smoker? Cancer Epidemiol Biomarkers Prev. 1998;7(11):967–79.

[15] Pauly JL, Stegmeier SJ, Allaart HA, Cheney RJ, Zhang PJ, Mayer AG et al. Inhaled cellulosic and plastic fibers found in human lung tissue. Cancer Epidemiol Biomarkers Prev. 1998;7(5):419–28.

[16] Cummings KM, Hastrup JL, Swedrock T, Hyland A, Perla J, Pauly JL. Consumer perception of risk associated with filters contaminated with glass fibers. Cancer Epidemiol Biomarkers Prev. 2000;9(9):977–9.

[17] Higuchi MA, Ayres PH, Swauger JE, Morgan WT, Mosberg AT. Quantitative analysis of potential transfer of continuous glass filament from eclipse prototype 9-014 cigarettes. Inhal Toxicol.Heated tobacco products 2000;12(11):1055–70.

[18] Higuchi MA, Ayres PH, Swauger JE, Morgan WT, Mosberg AT. Analysis of potential transfer of continuous glass filament from Eclipse cigarettes. Inhal Toxicol. 2000;12(7):617–40.

[19] Swauger JE, Foy JW. Safety assessment of continuous glass filaments used in Eclipse. Inhal Toxicol. 2000;12(11):1071–84.

[20] Higuchi MA, Ayres PH, Swauger JE, Deal PA, Guy T, Morton M et al. Quantification of

[21] Slade J, Connolly GN, Lymperis D. Eclipse: does it live up to its health claims? Tob Control. 2002;11(Suppl 2):ii64–70.

[22] Buchhalter AR, Eissenberg T. Preliminary evaluation of a novel smoking system: effects on subjective and physiological measures and on smoking behavior. Nicotine Tob Res. 2000;2(1):39–43.

[23] Buchhalter AR, Schrinel L, Eissenberg T. Withdrawal-suppressing effects of a novel smoking system: comparison with own brand, not own brand, and de-nicotinized cigarettes. Nicotine Tob Res. 2001;3(2):111–8.

[24] Hughes JR, Keely JP. The effect of a novel smoking system – Accord – on ongoing smoking and toxin exposure. Nicotine Tob Res. 2004;6(6):1021–7.

[25] Roethig HJ, Zedler BK, Kinser RD, Feng S, Nelson BL, Liang Q. Short-term clinical exposure evaluation of a second-generation electrically heated cigarette smoking system. J Clin Pharmacol. 2007;47(4):518–30.

[26] United States Food and Drug Administration. Harmful and potentially harmful constituents in tobacco products and tobacco smoke; established list. Fed Reg. 2012:20034–7.

[27] Pratte P, Cosandey S, Goujon-Ginglinger C. Investigation of solid particles in the mainstream aerosol of the Tobacco Heating System THS2.2 and mainstream smoke of a 3R4F reference cigarette. Hum Exp Toxicol. 2017;36(11):1115–20.

[28] Center for Tobacco Reference Products. Certificate of Analysis 1R6F Certified Reference Cigarette. Vol. 2016-002CTRP. Lexington (KY): University of Kentucky; 2016.

[29] Li X, Luo Y, Jiang X, Zhang H, Zhu F, Hu S et al. Chemical analysis and simulated pyrolysis of tobacco heating system 2.2 compared to conventional cigarettes. Nicotine Tob Res. 2018. doi:10.1093/ntr/nty005.

[30] Farsalinos KE, Yannovits N, Sarri T, Voudris V, Poulas K. Nicotine delivery to the aerosol of a heat-not-burn tobacco product: comparison with a tobacco cigarette and e-cigarettes. Nicotine Tob Res. 2018;27:e76–8.

[31] Bekki K, Inaba Y, Uchuyama S, Kunugita N. Comparison of chemicals in mainstream smoke in heat-not-burn tobacco and combustion cigarettes. J Univ Occup Environ Health Japan. 2017;39(3):201–7.

[32] Jakaj B, Eaton D, Forster M, Nicol T, Liu C, McAdam K et al. Characterizing key thermophysical processes in a novel tobacco heating product THP1.0(T). Poster. Society for Research on Nicotine and Tobacco; 7–11 March 2017, Florence, Italy; 2017 (http://www.bat-science.com/groupms/sites/BAT_9GVJXS.nsf/vwPagesWebLive/DOAKAK4V/$FILE/SRNT%202017%20poster%2036-Liu%20et%20al_JMcA%20KM.pdf?openelement, accessed 1 October 2018).

[33] Forster M, Fiebelkorn S, Yurteri C, Mariner D, Liu C, Wright C et al. Assessment of novel tobacco heating product THP1.0. Part 3: Comprehensive chemical characterisation of harmful and potentially harmful aerosol emissions. Regul Toxicol Pharmacol. 2018;93:14–33.

[34] Murphy J, Liu C, McAdam K, Gaça M, Prasad K, Camacko O et al. Assessment of tobacco heating product THP1.0. Part 9: The placement of a range of next-generation products on an emissions continuum relative to cigarettes via pre-clinical assessment studies. Regul Toxicol

Pharmacol. 2018;93:92–104.

[35] Taylor M, Thorne D, Carr T, Breheny D, Walker P, Proctor C et al. Assessment of novel tobacco heating product THP1.0. Part 6: A comparative in vitro study using contemporary screening approaches. Regul Toxicol Pharmacol. 2018;93:62–70.

[36] Thorne D, Breheny D, Proctor C, Gaca M. Assessment of novel tobacco heating product THP1.0. Part 7: Comparative in vitro toxicological evaluation. Regul Toxicol Pharmacol. 2018;93:71–83.

[37] Forster M, McAughey J, Prasad K, Mavropoulou E, Proctor C. Assessment of tobacco heating product THP1.0. Part 4: Characterisation of indoor air quality and odour. Regul Toxicol Pharmacol. 2018;93:34–51.

[38] Gee J, Prasad K, Slayford S, Gray A, Nother K, Cunningham A et al. Assessment of tobacco heating product THP1.0. Part 8: Study to determine puffing topography, mouth level exposure and consumption among Japanese users. Regul Toxicol Pharmacol. 2018:93:84–91.

[39] Jaunky T, Adamson J, Santopietro S, Terry A, Thorne D, Breheny D et al. Assessment of tobacco heating product THP1.0. Part 5: In vitro dosimetric and cytotoxic assessment. Regul Toxicol Pharmacol. 2018;93:52–61.

[40] Proctor C. Assessment of tobacco heating product THP1.0. Part 1: Series introduction. Regul Toxicol Pharmacol. 2018;93:1–3.

[41] WHO Study Group on Tobacco Product Regulation. Report on the scientific basis of tobacco product regulation. Seventh report of a WHO study group. Geneva: World Health Organization; 2015.

[42] Forster M, McAughey J, Liu C, McAdam K, Murphy J, Proctor C. Characterization of a novel tobacco heating product THP1.0(T): Indoor air quality. In: Society for Research on Nicotine and Tobacco, Florence, Italy, 7–11 March 2017.

[43] Breheny D, Adamson J, Azzopardi D, Baxter A, Bishop E, Carr T et al. A novel hybrid tobacco product that delivers a tobacco flavour note with vapour aerosol (Part 2): In vitro biological assessment and comparison with different tobacco-heating products. Food Chem Toxicol. 2017;106(Pt A):533–46.

[44] Werley MS, Freelin SA, Wrenn SE, Gerstenberg B, Roemer E, Schramke H et al. Smoke chemistry, in vitro and in vivo toxicology evaluations of the electrically heated cigarette smoking system series K. Regul Toxicol Pharmacol. 2008;52(2):122–39.

[45] Schaller JP, Keller D, Poget L, Pratte P, Kaelin E, McHugh D et al. Evaluation of the tobacco heating system 2.2. Part 2: Chemical composition, genotoxicity, cytotoxicity, and physical properties of the aerosol. Regul Toxicol Pharmacol. 2016;81(Suppl 2):S27–47.

[46] Frost-Pineda K, Zedler BK, Liang Q, Roethig HJ. Environmental tobacco smoke (ETS) evaluation of a third-generation electrically heated cigarette smoking system (EHCSS). Regul Toxicol Pharmacol. 2008;52(2):118–21.

[47] Frost-Pineda K, Zedler BK, Oliveri D, Feng S, Liang Q, Roethig HJ. Short-term clinical exposure evaluation of a third-generation electrically heated cigarette smoking system (EHCSS) in adult smokers. Regul Toxicol Pharmacol. 2008;52(2):104–10.

[48] Frost-Pineda K, Zedler BK, Oliveri D, Liang Q, Feng S, Roethig HJ. 12-week clinical exposure evaluation of a third-generation electrically heated cigarette smoking system (EHCSS) in adult

[49] Martin Leroy C, Jarus-Dziedzic K, Ancerewicz J, Lindner D, Kulesza A, Magnette J. Reduced exposure evaluation of an electrically heated cigarette smoking system. Part 7: A one-month, randomized, ambulatory, controlled clinical study in Poland. Regul Toxicol Pharmacol. 2012;64(2 Suppl):S74–84.

[50] Schorp MK, Tricker AR, Dempsey R. Reduced exposure evaluation of an electrically heated cigarette smoking system. Part 1: Non-clinical and clinical insights. Regul Toxicol Pharmacol. 2012;64(2 Suppl):S1–10.

[51] Tricker AR, Jang IJ, Martin Leroy C, Lindner D, Dempsey R. Reduced exposure evaluation of Heated tobacco products an electrically heated cigarette smoking system. Part 4: Eight-day randomized clinical trial in Korea. Regul Toxicol Pharmacol. 2012;64(2 Suppl):S45–53.

[52] Tricker AR, Kanada S, Takada K, Leroy CM, Lindner D, Schorp MK et al. Reduced exposure evaluation of an electrically heated cigarette smoking system. Part 5: 8-day randomized clinical trial in Japan. Regul Toxicol Pharmacol. 2012;64(2 Suppl):S54–63.

[53] Tricker AR, Kanada S, Takada K, Martin Leroy C, Lindner D, Schorp MK et al. Reduced exposure evaluation of an electrically heated cigarette smoking system. Part 6: 6-day randomized clinical trial of a menthol cigarette in Japan. Regul Toxicol Pharmacol. 2012;64(2 Suppl):S64–73.

[54] Tricker AR, Stewart AJ, Leroy CM, Lindner D, Schorp MK, Dempsey R. Reduced exposure evaluation of an electrically heated cigarette smoking system. Part 3: Eight-day randomized clinical trial in the UK. Regul Toxicol Pharmacol. 2012;64(2 Suppl):S35–44.

[55] Urban HJ, Tricker AR, Leyden DE, Forte N, Zenzen V, Feuersenger A et al. Reduced exposure evaluation of an electrically heated cigarette smoking system. Part 8: Nicotine bridging – estimating smoke constituent exposure by their relationships to both nicotine levels in mainstream cigarette smoke and in smokers. Regul Toxicol Pharmacol. 2012;64(2 Suppl):S85–97.

[56] Zenzen V, Diekmann J, Gerstenberg B, Weber S, Wittke S, Schorp MK. Reduced exposure evaluation of an electrically heated cigarette smoking system. Part 2: Smoke chemistry and in vitro toxicological evaluation using smoking regimens reflecting human puffing behavior. Regul Toxicol Pharmacol. 2012;64(2 Suppl):S11–34.

[57] Dayan AD. Investigating a toxic risk (self-inflicted) the example of conventional and advanced studies of a novel tobacco heating system. Regul Toxicol Pharmacol. 2016;81 Suppl 2:S15–6.

[58] Haziza C, de La Bourdonnaye G, Merlet S, Benzimra M, Ancerewicz J, Donelli A et al. Assessment of the reduction in levels of exposure to harmful and potentially harmful constituents in Japanese subjects using a novel tobacco heating system compared with conventional cigarettes and smoking abstinence: a randomized controlled study in confinement. Regul Toxicol Pharmacol. 2016;81:489–99.

[59] Haziza C, de La Bourdonnaye G, Skiada D, Ancerewicz J, Baker G, Picavet P et al. Evaluation of the tobacco heating system 2.2. Part 8: 5-day randomized reduced exposure clinical study in Poland. Regul Toxicol Pharmacol. 2016;81(Suppl 2):S139–50.

[60] Haziza C, de La Bourdonnaye G, Skiada D, Ancerewicz J, Baker G, Picavet P et al. Biomarker of exposure level data set in smokers switching from conventional cigarettes to tobacco heating system 2.2, continuing smoking or abstaining from smoking for 5 days. Data Brief. 2017;10:283–

[61] Lüdicke F, Picavet P, Baker G, Haziza C, Poux V, Lama N et al. Effects of switching to the mentholtobacco heating system 2.2, smoking abstinence, or continued cigarette smoking on clinically relevant risk markers: a randomized, controlled, open-Label, multicenter study in sequential confinement and ambulatory settings (Part 2). Nicotine Tob Res. 2018;20(2):173–82.

[62] Gale N, McEwan M, Eldridge AC, Sherwood N, Bowen E, McDermott S et al. A randomised, controlled, two-centre open-label study in healthy Japanese subjects to evaluate the effect on biomarkers of exposure of switching from a conventional cigarette to a tobacco heating product. BMC Public Health. 2017;17(1):673.

[63] Lopez AA, Hiler M, Maloney S, Eissenberg T, Breland AB. Expanding clinical laboratory tobacco product evaluation methods to loose-leaf tobacco vaporizers. Drug Alcohol Depend. 2016;169:33–40.

[64] Hamilton WL, Norton G, Ouellette TK, Rhodes WM, Kling R, Connolly GN. Smokers' responses to advertisements for regular and light cigarettes and potential reduced-exposure tobacco products. Nicotine Tob Res. 2004;6(Suppl 3):S353–62.

[65] O'Connor RJ, Hyland A, Giovino GA, Fong GT, Cummings KM. Smoker awareness of and beliefs about supposedly less-harmful tobacco products. Am J Prev Med. 2005;29(2):85–90.

[66] Parascandola M, Augustson E, O'Connell ME, Marcus S. Consumer awareness and attitudes related to new potential reduced-exposure tobacco product brands. Nicotine Tob Res. 2009;11(7):886–95.

[67] Shiffman S, Pillitteri JL, Burton SL, Di Marino ME. Smoker and ex-smoker reactions to cigarettes claiming reduced risk. Tob Control. 2004;13(1):78–84.

[68] Hund LM, Farrelly MC, Allen JA, Chou RH, St Claire AW, Vallone DM et al. Findings and implications from a national study on potential reduced exposure products (PREPs). Nicotine Tob Res. 2006;8(6):791–7.

[69] Shiffman S, Jarvis MJ, Pillitteri JL, Di Marino ME, Gitchell JG, Kemper KE. UK smokers' and ex-smokers' reactions to cigarettes promising reduced risk. Addiction. 2007;102(1):156–60.

[70] Levy DT, Cummings KM, Villanti AC, Niaura R, Abrams DB, Fong GT et al. A framework for evaluating the public health impact of e-cigarettes and other vaporized nicotine products. Addiction. 2017;112(1):8–17.

[71] O'Connor RJ, Ashare RL, Fix BV, Hawk LW, Cummings KM, Schmidt WC. College students' expectancies for light cigarettes and potential reduced exposure products. Am J Health Behav. 2007;31(4):402–10.

[72] Tabuchi T, Kiyohara K, Hoshino T, Bekki K, Inaba Y, Kunugita N. Awareness and use of electronic cigarettes and heat-not-burn tobacco products in Japan. Addiction. 2016;111(4):706–13.

[73] Tabuchi T, Gallus S, Shinozaki T, Nakaya T, Kunugita N, Colwell B. Heat-not-burn tobacco product use in Japan: its prevalence, predictors and perceived symptoms from exposure to secondhand heat-not-burn tobacco aerosol. Tob Control. 2018;27(e):e25–33.

[74] Fairchild A, Colgrove J. Out of the ashes: the life, death, and rebirth of the "safer" cigarette in the United States. Am J Public Health. 2004;94(2):192–204.

[75] Wayne GF. Potential reduced exposure products (PREPs) in industry trial testimony. Tob

[76] Carpenter CM, Wayne GF, Connolly GN. The role of sensory perception in the development and targeting of tobacco products. Addiction. 2007;102(1):136–47.

[77] Cook BL, Wayne GF, Keithly L, Connolly G. One size does not fit all: how the tobacco industry has altered cigarette design to target consumer groups with specific psychological and psychosocial needs. Addiction. 2003;98(11):1547–61.

[78] Caraballo RS, Pederson LL, Gupta N. New tobacco products: do smokers like them? Tob Control. 2006;15(1):39–44.

[79] O'Hegarty M, Richter P, Pederson LL. What do adult smokers think about ads and promotional materials for PREPs? Am J Health Behav. 2007;31(5):526–34.

[80] Hughes JR, Keely JP, Callas PW. Ever users versus never users of a "less risky" cigarette. Psychol Addict Behav. 2005;19(4):439–42.

[81] Hanson K, O'Connor R, Hatsukami D. Measures for assessing subjective effects of potential reduced-exposure products. Cancer Epidemiol Biomarkers Prev. 2009;18(12):3209–24.

[82] Rees VW, Kreslake JM, Cummings KM, O'Connor RJ, Hatsukami DK, Parascanola M et al. Assessing consumer responses to potential reduced-exposure tobacco products: a review of tobacco industry and independent research methods. Cancer Epidemiol Biomarkers Prev. 2009;18(12):3225–40.

[83] Cappelleri JC, Bushmakin AG, Baker CL, Merikle E, Olufade AO, Gilbert DG. Confirmatory factor analyses and reliability of the modified cigarette evaluation questionnaire. Addict Behav. 2007;32(5):912–23.

[84] Cox LS, Tiffany ST, Christen AG. Evaluation of the brief questionnaire of smoking urges (QSUbrief) in laboratory and clinical settings. Nicotine Tob Res. 2001;3(1):7–16.

[85] Lüdicke F, Baker G, Magnette J, Picavet P, Weitkunat R. Reduced exposure to harmful and potentially harmful smoke constituents with the tobacco heating system 2.1. Nicotine Tob Heated tobacco products Res. 2017;19(2):168–75.

[86] Hughes JR, Hatsukami D. Signs and symptoms of tobacco withdrawal. Arch Gen Psychiatry. 1986;43(3):289–94. 87.

[87] Liu X, Lugo A, Spizzichino L, Tabuchi T, Pacifici R, Gallus S. Heat-not-burn tobacco products: concerns from the Italian experience. Tob Control. 2018. doi: 10.1136/tobaccocontrol-2017-054054.

[88] Miyazaki Y, Tabuchi T. Educational gradients in the use of electronic cigarettes and heat-notburn tobacco products in Japan. PLoS One. 2018;13(1):e0191008.

[89] Calentzopoulos A, Olczak J, King M. Consumer Analyst Group of New York (CAGNY) Conference. New York: Philip Morris International; 2017.

[90] Camilleri LC, Calantzopoulos A. Annual meeting of shareholders. Philip Morris International; 2016.

[91] Statement on the toxicological evaluation of novel heat-not-burn tobacco products (press release). London: Food Standards Agency; 2017 (https://cot.food.gov.uk/sites/default/files/heat_not_burn_tobacco_statement.pdf, accessed 1 October 2018).

[92] King B, Borland R, Abdul-Salaam S, Polzin G, Ashley D, Watson C et al. Divergence between strength indicators in packaging and cigarette engineering: a case study of Marlboro varieties in

[93] Australia and the USA. Tob Control. 2010;19(5):398–402.
Wayne GF, Connolly GN. Regulatory assessment of brand changes in the commercial tobacco product market. Tob Control. 2009;18(4):302–9.

[94] Wu W, Zhang L, Jain RB, Ashley DL, Watson CH. Determination of carcinogenic tobacco-specific nitrosamines in mainstream smoke from US-brand and non-US-brand cigarettes from 14 countries. Nicotine Tob Res. 2005;7(3):443–51.

[95] Mitova MI, Campelos PB, Goujon-Ginglinger CG, Maeder S, Mottier N, Rouget EG et al. Comparison of the impact of the tobacco heating system 2.2 and a cigarette on indoor air quality. Regul Toxicol Pharmacol. 2016;80:91–101.

[96] Tricker AR, Schorp MK, Urban HJ, Leyden D, Hagedorn HW, Engl J et al. Comparison of environmental tobacco smoke (ETS) concentrations generated by an electrically heated cigarette smoking system and a conventional cigarette. Inhal Toxicol. 2009;21(1):62–77.

[97] Protano C, Manigrasso M, Avino P, Sernia S, Vitali M. Second-hand smoke exposure generated by new electronic devices (IQOS(R) and e-cigs) and traditional cigarettes: submicron particle behaviour in human respiratory system. Ann Ig. 2016;28(2):109–12.

[98] O'Connell G, Wilkinson P, Burseg KMM, Stotesbury SJ, Pritchard JD. Heated tobacco products create side-stream emissions: implications for regulation. J Environ Anal Chem. 2015;2(5):163.

[99] McDonald EA, Popova L, Ling PM. Traversing the triangulum: the intersection of tobacco, legalized marijuana and electronic vaporisers in Denver, Colorado. Tob Control. 2016;25(Suppl 1):i96–102.

[100] Lee DC, Crosier BS, Borodovsky JT, Sargent JD, Budney AJ. Online survey characterizing vaporizer use among cannabis users. Drug Alcohol Depend. 2016;159:227–33.

[101] Morean ME, Kong G, Camenga DR, Cavallo DA, Krishnan-Sarin S. High school students' use of electronic cigarettes to vaporize cannabis. Pediatrics. 2015;136(4):611–6.

[102] Henriksen L, Ribisl KM, Rogers T, Moreland-Russell S, Barker DM, Esquivel NS et al. Standardized tobacco assessment for retail settings (STARS): dissemination and implementation research. Tob Control. 2016;25(Suppl 1):i67–74.

3. 电子烟碱传输系统中烟碱的临床药理学

Armando Peruga, Centre of Epidemiology and Health Policies, School of Medicine, Clínica Alemana, University of Desarrollo, Chile

Thomas Eissenberg, Center for the Study of Tobacco Products, Department of Psychology, Virginia Commonwealth University, USA

3.1 引　　言

电子烟碱传输系统（ENDS）是一类由多个部分组成的产品，其中电动线圈用于加热包含烟碱、溶剂（例如丙二醇、植物甘油）以及常用香料的液体基质或电子烟液。使用者吸入所产生的气溶胶，其中包含浓度可变的烟碱[1]。烟碱是一种产生依赖性的中枢神经系统刺激剂。在许多国家，包括在欧盟和美国这两个最大的市场中，ENDS作为普通消费品或烟草制品被管制[2]。

理想情况下，出于公共卫生考虑，向公众销售的含有烟碱等对中枢神经系统起作用的药物的ENDS产品，应该几乎没有滥用或依赖的可能性。这种观点是正确的，除非需要某种程度的滥用能力来维持依从性，并支持替代具有更大滥用能力和危害的物质。由于ENDS声称在戒烟和减少吸烟中可能发挥作用，因此属于这一类。

本背景文件的目的是对2018年3~12月的文献进行综述，在对ENDS的烟碱含量和烟碱传递进行回顾后补充了一些内容，并探讨影响烟碱和非烟碱有害物质释放的因素。此外，我们回顾了ENDS在戒烟中的潜在作用以及对人群健康的潜在影响。我们还找出一些相关的研究差距，并提出政策建议。

3.2　ENDS操作

了解ENDS的操作方式很有用。图3.1是常见的ENDS装置的示意图。加热线圈连接到封闭在加热芯中的电源（通常为电池，图中未显示），该加热芯又被含烟碱的电子烟液包围。当电流通过时，线圈加热，从而蒸发加热芯中的一些电子烟液。当使用者从ENDS的烟嘴吸入空气时，蒸汽被带走并再次冷凝形成气溶胶，使用者将其吸入。

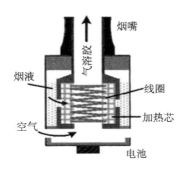

图3.1　ENDS操作示意图
资料来源：Alan Shihadeh博士,贝鲁特美国大学（黎巴嫩）

几个因素会影响气溶胶携带的烟碱量，包括流经电子烟碱传输系统（ENDS）的电能，使用者的吸入行为（或"抽吸曲线"）以及电子烟液中烟碱的量[3]。电功率（W）是电池电压（V）和线圈电阻（Ω）的函数，$W = V^2/\Omega$。早期ENDS型号的功率≤10 W，但目前市场上销售的设备的功率≥250 W[4,5]。通常，通过使用低电阻（例如<1 Ω）的线圈,向线圈或组合施加变化的电压来获得更高的功率。

抽吸曲线的变量包括抽吸次数、持续时间和抽吸体积以及抽吸间隔。使用者抽吸的曲线非常个性化。但是，有经验的ENDS使用者通常比没有ENDS使用经验的吸烟者抽吸更长的时间[6-9]（图3.2）。

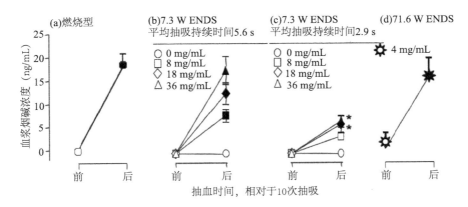

图3.2　使用燃烧型卷烟和电子烟之后的血浆烟碱浓度
(a) $N = 32$[8]; (b) $N = 33$[8]; (c) $N = 31$[8]; (d) $N = 11$[4]（抽吸曲线不可用）
资料来源：改编自先前发表的资料[1]，增加了抽吸持续时间数据并更新了(d)组数据

3.3 电子烟液中的烟碱浓度

ENDS中使用的含烟碱的电子烟液装在预装的烟弹或续液瓶中,具体取决于所用设备的类型。市售电子烟液中的烟碱浓度可以达到36 mg/mL或更高[1],使用者可以在销售点从多种浓度中进行选择;一些制造商提供与电子烟液相关的标签信息。但是,对于制造商在多大程度上准确地告知消费者电子烟液代表样品中的烟碱浓度尚无全面的研究。现有研究根据便利性样本给出了部分图片。在烟碱含量上具有清晰标签信息的电子烟液的比例是未知的。一些研究表明此类信息并非始终可用[10,11]或在制造商标签上可解释[12]。但是,烟碱的浓度通常以总体积的百分比或mg/mL的形式进行标注。表3.1列出了据称含有烟碱的电子烟液中烟碱的浓度并与制造商标注的浓度进行比较。

表3.1 电子烟液中烟碱浓度测量值与标注值的比较

第一作者和参考文献编号	电子烟液容器的类型	样品数		
		分析样品数量	>标注浓度 ± 10%	>标注浓度 ± 25%
Beauval[13]	续液瓶	2	0	0
Buettner-Schmidt[14]	续液瓶	70	36	NA
Cameron[15]	预填充烟弹和续液瓶	21	13	7
Cheah[10]	烟弹	8[a]	8[b]	7[b]
Davis[16]	续液瓶	81	36	21
El-Hellani[17]	预填充烟弹	4	4	4
Etter[18]	续液瓶	35	4	0
Etter[19]	续液瓶	34	10	3
Farsalinos[20]	续液瓶	21	9	0
Goniewicz[21]	续液瓶	62	25	7
Kim[22]	续液瓶	13	7	2
Kirschner[23]	续液瓶	6	6	4
Kosmider[24]	续液瓶	9	2	0
Lisko[25]	续液瓶	29	15	7
Pagano[26]	预填充烟弹	4	3	2
Peace[27]	续液瓶	27	16	7
Rahman[28]	续液瓶	69	65	53
Raymond[29]	续液瓶	35	22	22
Trehy[30]	预填充烟弹	22	22	19
Trehy[30]	续液瓶	17	8	6

NA:不可用
a. 分析的品牌数量;未提供分析的样品数量
b. 至少有一个样品的烟碱浓度高于标准的品牌数

大多数研究表明，烟碱浓度低于制造商报告的浓度。除一项研究以外，所有研究均表明，一些样品中烟碱浓度低于或高于产品标注烟碱浓度的10%，符合美国制造商协会推荐的质量标准[31]。53%样本的中位值，烟碱浓度在标注上误报了至少10%，26%样本的中位值，烟碱浓度被误报了至少25%。

我们只知道3项关于相同品牌和型号、不同批次的电子烟液的烟碱浓度一致性研究。一个批次[19]的中位值差异为0.5%，另外两个批次的中位值差异分别为15%[16]和16%[32]。

其他研究表明，某些标注不含烟碱的产品确实具有可测量的烟碱含量。表3.2列出了对电子烟液中烟碱浓度进行分析的研究，并与标注的烟碱差异情况进行了比较。几乎一半的研究报告一些称不含烟碱的电子烟液中存在少量烟碱。此外，在据称没有烟碱的电子烟液样本中约有5%的样品烟碱浓度很高。

表3.2 测量标注为零烟碱的电子烟液中烟碱的浓度

第一作者和参考文献编号	分析样品数量	烟碱含量 > 0.1 mg/mL 样品数	烟碱含量 > 10 mg/mL 样品数	> 0.1 mg/mL烟碱含量的样品浓度
Beauval[13]	2	0	0	—
Cheah[10]	2	0	0	—
Davis[16]	10	0	0	—
Goniewicz[21]	28	3	0	0.8~0.9
Kim[22]	20	0	0	—
Lisko[25]	5	0	0	—
Omaiye[33]	125	17	2	0.4~20.4
Raymond[29]	35	6	6	5.7~23.9
Trehy[30]	8	2	2	12.9~24.8/烟弹
Trehy[30]	5	2	2	12~21
Westenberger[34]	5	0	0	—

3.4 向 ENDS 使用者的烟碱传输

ENDS的烟碱传输特性可能是该产品能在多大程度上替代卷烟的一个重要决定因素。图3.2展示了电子烟中烟碱浓度、使用行为和设备功率对ENDS相对于卷烟的烟碱传输特性的影响。图3.2(a)[9]显示了吸烟者以30 s的间隔抽吸10口卷烟的烟碱递送曲线。图3.2(b)显示了使用者以30 s的间隔抽吸10口，平均抽吸时间为5.6 s，装有0、8 mg/mL、18 mg/mL或36 mg/mL烟碱电子烟液的7.3 W ENDS的烟碱递送曲线。显然，电子烟液中烟碱的浓度会影响烟碱向使用者血液的传输。当7.3 W的ENDS与36 mg/mL烟碱电子烟液配对时，使用者抽吸时间为5.6 s、抽吸间隔为30 s时，ENDS超过了卷烟的烟碱传输特性[8]。

抽吸持续时间也是ENDS烟碱传输效率的一个影响因素。图3.2(c)[8]中使用的装置和电子烟液烟碱浓度与图3.2(b)中相同，但研究参与者采用了较短的抽吸时间（平均2.9 s）。当抽吸持续时间较短，且装置和电子烟液特性均保持不变时，烟碱的释放量就会减少。图3.2(d)显示了当使用者以30 s的抽吸间隔抽吸10口时，功率更高的ENDS设备（平均功率71.6 W）的烟碱递送曲线[4]。当这些更高功率的设备与4 mg/mL烟碱液体配对使用时，它们的烟碱释放曲线与卷烟近似。

总体而言，至少在某些情况下，这些数据表明，某些电子烟碱传输系统可以与卷烟相同的速率向静脉血液传输相同剂量的烟碱。不幸的是，很少有研究比较ENDS和卷烟将烟碱传输到动脉血中的能力，而这是中枢神经系统暴露于药物的重要指标[35]。在迄今唯一的此类比较中，从7.3 W ENDS中用36 mg/mL烟碱烟液进行10口抽吸（抽吸间隔30 s）导致的平均动脉烟碱浓度（最大12 ng/mL）低于10口（最大浓度为27 ng/mL）卷烟抽吸（间隔30 s），但达到峰值浓度的时间没有差异[36]。然而，样本很少（ENDS为4个，卷烟为3个），并且未测量抽吸时间。在这项研究的控制条件下，正电子发射断层扫描成像显示该ENDS有效地将烟碱传递至中枢神经系统。

尽管在某些条件下用于生成图3.2数据的ENDS可以像卷烟一样有效地传输烟碱，但许多ENDS不能[6,9,37-41]。ENDS烟碱递送的这种异质性与烟碱替代产品相反，后者虽然能够以较低的速度实现较低的血浆浓度，但可以更可靠地递送烟碱。例如，如图3.3(a)[42]所示，烟碱口香糖可能需要≥30 mm才能达到血浆峰值浓度，而

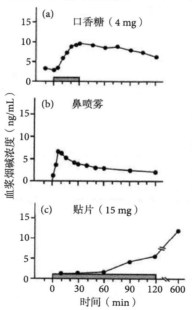

图3.3 几种形式的单剂量烟碱给药之前、期间和之后的血浆烟碱浓度
灰色条表示产品使用持续时间
资料来源：参考文献[42]，经麻省医学会许可转载

图3.3(c)显示烟碱贴片可能需要超过2 h[43,44]；其他治疗产品（例如烟碱含片）也在此时间范围内提供烟碱[43]。据推测，尽管没有对这种推测进行过实验性检验，但将烟碱像卷烟一样有效地传输到血液和大脑中的烟碱替代品，更可能替代卷烟，因为临床试验中使用的ENDS并不能有效地传输烟碱[45]。

3.5　ENDS释放物中的有害物质

ENDS的有害物质释放物取决于多种因素，包括设备构造、设备功率、烟液成分和使用行为。下文将回顾有关ENDS有害物质释放量的文献，从烟碱开始，然后转向非烟碱有害物质（有关文献的综述，请参见Breland等[1]及美国卫生和公共服务部[46]）。

3.5.1　烟碱释放物

ENDS产生的烟碱的"释放量"是在特定的抽吸方案下ENDS产生的气溶胶中烟碱的含量（以mg计）。理解从ENDS获得烟碱的量被认为对于理解ENDS使用者的烟碱药代动力学很重要。一篇文献综述[47]确定了7篇关于烟碱释放量的研究[30,34,48-52]；此后，有关这个问题的其他几项研究也已发表[3,33,53,54]。

在这些研究中烟碱的释放量变化很大，这取决于使用的ENDS类型、电子烟液的烟碱浓度和用于获得气溶胶的抽吸方式。一些方法学问题使研究的可比性变得复杂化，包括以下事实：ISO抽吸模式下抽吸ENDS方法无法激活某些ENDS模型。尽管在这些研究中，ENDS的烟碱释放量与吸烟机抽吸卷烟的烟碱释放量没有完全可比性，但它们通常远低于卷烟释放的烟碱含量[47]。然而文献有限，有两个重要原因。首先，烟碱的释放量不能捕获烟碱的释放速率，这不仅是烟碱量的测量，也是对使用者获得烟碱的速率的测量。烟碱的释放速率几乎可以肯定地与烟碱的传递速率有关，烟碱的传递速率可能是含烟碱产品替代卷烟能力的关键因素，因为卷烟中的烟碱在血液中迅速达到峰值水平并进入大脑[55]。其次，电子烟碱传输系统及电子烟液的异质性很大，以致对特定电子烟碱传输系统的研究结果可能无法推广到另一个电子烟碱传输系统。

为了解决第一个问题，人们越来越关注测量烟碱"通量"，即电子烟碱传输系统释放烟碱的速率[56,57]。可以测量烟碱通量（通常以μg/s为单位），并且可以对电子烟碱传输系统和卷烟进行比较。那些模仿卷烟烟碱通量的电子烟，比不模仿卷烟烟碱通量的电子烟更可能很好地替代卷烟。为了解决第二个问题，已经开发了基于物理学的数学模型来预测电子烟碱传输系统的烟碱通量[58]，甚至是对那些尚未制造的ENDS。该模型考虑了在电流开始流动后线圈加热所需的时间以及在抽吸之间线圈冷却的时间。它还解释了可以从线圈中带走热量的各种方式：通过空气,通过电子烟液蒸发时的潜热,通过金属焊料传导到设备主体,并辐射到周围环境。模型的输入是加热器线圈的长度、直径、电阻和热容,电子烟液的组成和热力学性质(包括烟碱浓度),抽吸速率和持续时间,以及抽吸间隔和周围的空气温度。在模型测试中,作者将其预测结果与实际烟碱通量测量结果进行了比较,其中包括功率、抽吸曲线、ENDS类型(罐或雾化器)和烟液成分变化的100种情况。数学上预测的烟碱通量与测量值高度相关($r=0.85$,$P<0.0001$)[58]。此外,该模型准确地预测了烟碱通量对装置功率和烟碱浓度、电子烟液中丙二醇和植物甘油的比例以及使用者的抽吸时间依赖性(图3.4)。图3.4显示,设备的电功率越高,达到给定通量所需的电子烟碱浓度就越低。卷烟通量为100 μg/s,这些线表示的ENDS烟碱通量相当于卷烟的两倍、一倍和一半。给定ENDS功率与烟液烟碱浓度之间的关系如图3.4所示,通过将功率较高的ENDS与较高浓度的烟液配对,可以获得比卷烟大得多的烟碱通量。该图未显示某些功率超过100 W的ENDS的烟碱通量[4,5]。

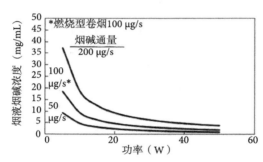

图3.4 ENDS功率与电子烟液中烟碱浓度的关系及其对烟碱通量的影响
资料来源:参考文献[58],经贝鲁特美国大学(黎巴嫩)Alan Shihadeh博士许可转载

关于ENDS烟碱释放物的另一个重要问题是电子烟液和气溶胶中烟碱以生物利用度较高的游离态烟碱与生物利用度较低的质子化形态的相对含量[17]。电子烟中烟碱释放物的一些研究报告了烟碱的释放量,但没有确定所使用的方法是对总烟碱还是仅对其中一种烟碱形态进行了测量[38,58],因此难以与所报告的结果进行比较或评估烟碱对使用者的影响。在对此问题的评估中,在19种商用电子烟液中的游离态烟碱变化很大(10%~90%),重要的是,在气溶胶中也存在差异[17,59],这可能是另一个影响ENDS烟碱向使用者传输的因素。因此,除了测量烟碱通量外,还应确定气溶胶中烟碱的形态。总体而言,就烟碱向使用者的传输而言,来自电子烟碱传输系统的烟碱释放量存在很大差异,可以通过仔细考虑影响烟碱的许多因素来解释和预测,特别是电子烟碱传输系统的功率、电子烟液成分和使用行为。

3.5.2　非烟碱有害释放物

ENDS气溶胶中的非烟碱有害物质存在于烟液中，或在加热烟液时形成。加热前存在于烟液中的成分包括丙二醇和植物甘油，特意添加的调味剂和其他化合物以及非人为添加的污染物，它们合在一起占大多数电子烟液含量的80%~97%[60]。雾化的丙二醇是一种呼吸道刺激物[61-64]，当以大剂量静脉注射给药时，可能会导致致命的乳酸性酸中毒[65]。临床前研究还表明，大剂量植物甘油可能具有毒性[66,67]。每天、长期吸入雾化的丙二醇和/或植物甘油对健康的影响尚不清楚。电子烟液中使用的调味剂通常是添加到食品中的化合物，在加热和雾化后它们对人肺的影响尚不清楚[68]。在电子烟液和气溶胶中至少发现了三种调味剂引起了人们的健康担忧：二乙酰（黄油味）可导致闭塞性细支气管炎[69]；苯甲醛（水果味）具有细胞毒性和遗传毒性[70]；肉桂醛（肉桂味）也具有细胞毒性和遗传毒性[71]，可引起肺细胞炎症反应[72]。污染物包括二甘醇、乙二醇和乙醇[73,74]。即使实施了严格的质量控制以确保无污染的电子烟液，长期、每天、频繁地吸入雾化的丙二醇和植物甘油以及许多化学调味剂（这些调味剂通常混合在一种烟液中）所产生的不确定影响对ENDS使用者构成了潜在的健康威胁。

烟液加热时形成的非烟碱有害物质包括金属、挥发性醛、呋喃和苯。在一项11个"第一代"ENDS品牌（形状像卷烟的一次性ENDS）的研究中，三个品牌每5 min抽吸4.3 s，两个系列共抽吸60口，并分析所产生的气溶胶中的元素，包括金属[75]。结果显示品牌之间存在很大差异，但大多数品牌产生的气溶胶中发现了许多金属，"在某些情况下，其浓度明显高于传统卷烟"。作者得出的结论是ENDS气溶胶中的大多数元素和金属可能源自雾化器中的组件，例如加热丝、焊点、加热芯和护套。这些结果表明了ENDS的结构如何影响气溶胶的非烟碱有害物质的情况。

在对具有1.5 Ω加热元件和可变电压电池（3.3~5.0 V）的新一代ENDS的研究中，比较了多种电子烟液（均为6 mg/mL烟碱）产生的气溶胶的醛含量（4 s每口，共抽吸10口，91 mL/口）[76]。系统将功率从9.1 W调节到16.6 W，ENDS气溶胶中均存在乙醛、丙酮、丙烯醛和甲醛，并且随着抽吸量的增加，醛的产生成比例地增加，并且当功率> 11.7 W时，醛的产量急剧增加。ENDS气溶胶中存在醛已得到充分证明[77-79]，设备功率对形成醛所起到的作用是：将ENDS功率从4.1 W增加到8.8 W，大约导致挥发性醛的释放量增加3倍[80-83]。也有人认为调味剂会在加热过程中形成非烟碱有害物质[84-87]。例如，在电子烟液中加热甜味剂可能会使使用者接触呋喃（一种有毒的化合物）。在一项研究[88]中，使用了VaporFi铂金罐ENDS（2.3 Ω）在各种条件下产生气溶胶，包括功率（4.2 W和10.8 W），抽吸持续时间（4 s和8 s）和甜味剂（山梨糖醇、葡萄糖和蔗糖），某些呋喃的单位抽吸释放量可与卷烟报告的值相当，而且设备功率也是一个影响因素：将功率从4.3 W增加到10.8 W，呋喃释放量增加了两倍。关于苯，将ENDS功率从6 W增加到13 W，可将这种致癌物的释放量提高100倍[89]，尽管该水平仍远低于卷烟烟气中的水平。易挥发的醛、呋喃和苯都是通过电子烟液（例如丙二醇、植物甘油、甜味剂）成分的热解而形成的，再加上设备功率的增加会增加ENDS气溶胶中这些有害物质的含量，表明大功率电子烟碱传输系统是一个特殊的公共卫生问题。迄今为止，大多数关于ENDS气溶胶有害物质特性的研究仅限于功率为25 W或更低的设备[80,83,88,90]，此处报告的许多数据可能与某些地方常见的高功率设备无关[4,5]。

3.6　ENDS 在戒烟中的潜在作用

共检索到了6篇叙述性综述[91-96]和6篇系统性综述[97]，其中5篇是荟萃分析[98-103]，论述了电子烟碱和非烟碱传输系统（EN&NNDS）在减少吸烟和戒烟中的作用。两项荟萃分析[100,102]涵盖了截至2016年1月的可用研究。

对证据质量的所有5项系统性审查[97,98,100,102,103]都得出结论，现有研究证据的确定性是低或极低的，主要是由于纳入的横断面研究和队列研究存在局限性，且许多已发表文章缺乏细节。鉴于这些局限性，El Dib等[102]和Malas等[97]得出结论，不能从这些综述得出可靠的推论，并且证据仍然没有定论。同样，对系统评价的回顾得出的结论是："总体而言，仅有有限的证据表明电子烟可能是促进戒烟的有效辅助手段"[104]。然而，其他系统性综述得出了不同的结论。虽然Kalkhoran和Glantz[101]确定"电子烟与吸烟者的戒烟率显著降低有相关性"，Hartmann-Boyce等[100]和Rahman等[98]结论是使用电子烟与戒烟和减少吸烟有关。Khoudigian等[103]只包括随机临床试验。结论中的显著差异是由于选择符合条件的研究的标准不同以及在进行审查时研究的可用性。表3.3总结了每次审查中使用的研究。

结论的差异并非来自随机临床试验提供的证据。对少数现有试验的荟萃分析表明，与安慰剂相比，使用ENDS可以使戒烟的可能性增加两倍。两项荟萃分析[98,99]估计愿意戒烟的风险比为2.29（95%CI, 1.05, 4.96），一项荟萃分析[102]评估为2.03（95%CI, 0.94, 4.38），另一项[103]估计为2.02（95%CI, 0.97, 4.22）。差异是由于分析的两项随机临床试验和对缺失数据处理导致的权重的轻微变化。更具体而言，不同的结论产生来自所审查的纵向和横断面研究提出的相互矛盾的证据。下面，我们集中于纵向研究的证据，因为在横截面研究中很难解释可能的关联方向。

最近一次的系统性回顾包括已发布的7项新的纵向研究，涉及EN&NNDS使用者和非使用者之间戒烟的差异[105-110]，包括对前一个随访时间的更新[111]。表3.4根据样本属性、参与者使用的EN&NNDS产品的特性、ENDS使用的典型化方式、烟碱依赖和戒断的标准以及结果的摘要纵向总结了研究的结果。总结了7项研究，在系统性评价中发现EN&NNDS使用与戒烟之间存在统计学上显著的正相关或负相关。还总结了未包括在综述中的7个纵向研究，共16项研究。为了选择最佳的纵向研究来评估证据，我们认为应该至少在以下三种条件下测量使用电子烟碱传输系统和戒烟之间的相关性，以得出有效的结果：

- 标准1　应该知道所使用的电子烟液是否包含烟碱以及所用设备的类型（功率）。理想情况下，应根据测试的传递烟碱能力对设备进行分类，但这在人群研究中可能会被证明是困难的，如果没有对参与者使用的设备进行实验室测试。那么，很难评估这种关联是否与ENDS作为烟碱替代辅助手段的潜在作用有关。我们知道某些ENDS装置在某些情况下可以传输类似卷烟的烟碱含量[4,8]；但是，在美国[112]和许多其他国家，由于低功率和其他因素而使用不能传输烟碱的ENNDS或ENDS仍然很普遍。
- 标准2　分析必须区分使用ENDS戒烟的人和不使用ENDS戒烟的人。许多人出于除戒烟以外的其他原因使用电子烟，包括减少吸烟[113]，在不允许吸烟的情况下在室内使用或出于娱乐目的[114,115]。如果像预期的那样，对戒烟的实际影响不同甚至相反，那么将为了戒烟和不是为了戒烟的ENDS使用者进行合并可能会偏向零关联。
- 标准3　电子烟碱传输系统的使用方式必须准确，以便区分固定使用和暂时、不稳定使用情况，以评估电子烟碱传输系统对人群健康的影响[116,117]。由于电子烟碱传输系统的使用是一种相对较新的人群行为，因此许多人可能会短暂地试用EN&NNDS，不会采用固定的使用方式。将ENDS的"曾经使用"与"从不使用"进行比较，例如，可以将一生中仅使用过一次ENDS的人归类为使用者，尽管认定为吸烟者的标准是人们一生中至少抽过100支烟。将试验者与稳定的使用者混为一谈可能会导致上一段所述的偏差。

考虑到这些标准，我们发现在检查的14项研究中：

- 只有两个描述了所用设备的类型（标准1）；
- 12项研究没有限制或分析使用EN&NNDS的原因，尽管两项研究包括对某些变量的调整或分析，这些变量可以用作使用EN&NNDS的代表（标准2）；
- 7项研究仅比较了曾经使用过和从未使用过EN&NNDS的使用者，3项使用了当前使用的粗略测量，6项使用了更详尽的频率测量（标准3）。

7个纵向研究至少满足3个标准之一；没有1项符合所有3个标准。来自7项研究的综合证据表明，他们的样本由不同的亚组组成，这些亚组在使用EN&NNDS戒烟方面受到了不同的影响。因此，可以假设某些吸烟者可以通过频繁或密集使用某些类型的电子烟（ENDS）来成功戒烟，而其他吸烟者则没有差别，甚至妨碍戒烟。这些研究的结果见表3.5。

表3.3 比较电子烟碱和非烟碱传输系统作为戒烟辅助手段有效性的研究

可供审查的研究	Franck[91] (2003.09)	Harrell[92] (2013.12)	Rahman[98] (2014.05)	McRobbie[99] (2014.08)	Lam[93] (2015.03)	Ioakeimidis[94] (2015.06)	Kalkhoran[101] (2015.07)	Hartmann-Boyce[100] (2016.01)	El Dib[102] (2016.01)	Malas[97] (2016.02)	Khoudigian[103] (2016.05)
队列研究											
Polosa,2011	√										
Adkison,2013		√									
Caponnetto,2013b										√	
Ely,2013				√				√			
Van Staden,2013			√								
Vickerman,2013		√									
Borderud,2014				√		√	√	√			
Choi,2014							√				
Etter,2014			√	√							
Farsalinos,2014				√			√	√			
Grana,2014				√						√	
Nides,2014						√		√			
Pearson,2014	√			√			√			√	
Polosa,2014			√				√	√			
Prochaska,2014							√				
Wagener,2014									√		
Al-Delaimy,2015				√			√				
Biener,2015							√		√		
Brose,2015							√		√		
Harrington,2015						√			√		
Hitchman,2015							√		√		
Manzoli,2015								√		√	
McRobbie,2015								√			
Oncken,2015											
Pacifici,2015								√			

续表

可供审查的研究	Franck[91] (2003.09)	Harrell[92] (2013.12)	Rahman[98] (2014.05)	McRobbie[99] (2014.08)	Lam[93] (2015.03)	Ioakeimidis[94] (2015.06)	Kalkhoran[101] (2015.07)	Hartmann-Boyce[100] (2016.01)	El Dib[102] (2016.01)	Malas[97] (2016.02)	Khoudigian[103] (2016.05)
Pavlov, 2015							✓				
Polosa 2015							✓	✓			
Shi, 2015							✓				
Sutfin, 2015											
横断面研究											
Siegel, 2011		✓								✓	
Popova, 2013		✓								✓	
Dawkins, 2013[37]	✓							✓			
Goniewicz, 2013[32]		✓									
Pokhrel, 2013											
Brown, 2014			✓				✓			✓	
Christensen, 2014							✓			✓	
McQueen, 2015							✓				
Tackett, 2015				✓							
对照组随机对照试验											
Bullen, 2010	✓										✓
Bullen, 2013[45]	✓	✓		✓	✓	✓		✓	✓	✓	✓
Caponnetto, 2013a	✓	✓		✓	✓	✓		✓	✓	✓	✓
Caponnetto, 2014				✓				✓	✓	✓	
Adriaens, 2014					✓				✓		
无对照组随机对照试验											
Hajek, 2015							✓	✓			
未知											
Humair, 2014				✓							

表3.4 纵向研究的特点表明，在现有的系统性综述中，使用电子烟碱和非烟碱传输系统与戒烟之间存在显著的正相关或负相关，但最新研究未包括在这些评论中

参考文献	收集数据年份	样品					产品		EN&NNDS使用					烟碱依赖	戒烟标准	结果
		国家和基准	年龄范围(岁)	处于基准水平人数	随访	随访保留率	EN&NNDS类型	烟液烟碱	频率	数量	持续时间	原因				EN&NNDS使用与戒烟的关系
[118]	2010	美国普通人群中具有全国代表性的吸烟者样本	≥18	5255	1年	63%	NM	NM	曾使用 vs. 从未使用	NM	NM	NM		M	SR随访30天	考虑到最后一次尝试吸烟时药物的使用、年龄、性别、种族、受教育程度、每日吸烟、曾经吸烟，曾使用过电子烟的吸烟者比没有使用过的吸烟者戒烟成功率低
[119]	2011~2012	美国吸烟者戒烟热线[a]	≥18	2758	7个月	35%	NM	NM	曾使用 vs. 使用≥1个月 vs. 使用<1个月	NM	NM	NNU		M NU	SR随访30天	未使用过ENDS, 31.3%；使用少于1个月, 16.6%；使用超过1个月, 21.7% ($P<0.001$，未针对NRT进行调整)
[120]	2011~2012	美国目前在基准水平的吸烟者	18~59	1000	1年	23%	NM	NM	EN&NNDS使用者分类：曾使用、可能使用、不会使用	NM	NM	NM		M	SR随访1个月	在基准水平检查时是否在随访期间曾经在随访期间戒烟断的可能性显著低于那些在基准未使用过电子烟的自述从未使用过电子烟的吸烟者（AOR=0.41；95% CI=0.186, 0.93），这些吸烟者经过致瘾性（早晨第一次吸烟的时间）、年龄、性别、受教育程度、种族、戒烟愿望和吸烟状况的调整。"使用过"只包括随访时回答一致的受访者

续表

参考文献	收集数据年份	样品				产品		EN&NNDS使用				烟碱依赖	戒烟标准	结果 EN&NNDS使用与戒烟的关系
		国家和基准	年龄范围(岁)	处于基准水平人数	随访	随访保留率	EN&NNDS类型	烟液烟碱	频率	数量	持续时间	原因			
[121]	2011~2012	美国两个大都市地区代表性吸烟者样本	18~65	1374	24年	51%	NM	NM	是否使用过电子烟；如果使用过，过去30天中有多少天使用。当前是每天使用，某些天使用，还是根本不使用。如果经常使用，曾经"相当经常地"使用包括：密集使用，每日使用1个月以上；间断性使用，非每日使用1个月以上；未使用或者最多使用一两次	M		SR随访1个月	在性别、年龄、种族、受教育水平和基准吸烟水平进行调整后，烟者戒烟的可能性是从未使用过电子烟戒烟者的6倍仅使用过一至两次的人的6倍以上。尽管没有统计学意义，但非使用者或实验者退出的可能性是非使用者或实验者的3倍	
[122]	2012	美国，参加NRT的吸烟者从一个免费目公开的互联网戒烟计划中招募，随机分为社交网络整合或没有社交网络x2(获得免费的NRT，未获得)	30~52	3408	3个月	62%	NM	NM	在随访期间曾使用与从未使用	NM	NM	NM	M	SR随访1个月	在调整了（按基准衡量）性别、年龄、种族/民族、受教育程度、父母、治疗分配、过去一年的尝试戒烟次分、过去一年对戒烟次数、每天吸烟数，自我戒烟效率和改变阶段后，使用电子烟戒烟者的吸烟率比不使用电子烟的吸烟率低。在过去3个月中对其他戒烟方法的使用以及尝试戒烟的次数进行了进一步调整，相关性不显著

续表

参考文献	收集数据年份	样品		处于基准水平人数	随访	随访保留率	产品		方式 EN&NNDS使用				烟碱依赖	戒烟标准	结果	
		国家和基准	年龄范围(岁)				EN&NNDS类型	烟液烟碱	频率	数量	持续时间	原因				EN&NNDS使用与戒烟次数的关系
[123]	2012~2013	目前正在接受戒烟治疗的美国癌症患者	NR	781	1年	53%	NM	NM	选择：基准为过去30天<1次抽吸，过去7天内使用过vs.常规随访时未使用	NM	NM	NM	M	SR随访1天	在校正烟碱依赖、戒烟次数和癌症诊断后，当前吸烟者和非吸烟者一样有可能吸烟。意向治疗分析表明，非电子烟使用者戒烟的可能性是电子烟使用者的两倍	
[124]	2012~2013	英国，吸烟者	≥18	1759	1年	44%	卷烟	NM	每天，少于每天但每周至少1次，少于每月至少1次，完全不使用	NM	NM	NM[b]	NM[c]	SR "我在过去一年里完全戒烟了"	与随访时非电子烟使用者比较：非日常使用类似卷烟的使用者戒烟的可能性较小，日常使用类似烟弹的使用者戒烟的可能性较小（P <0.001），日常使用电子烟壶的可能性更大（P=0.0012）。每天吸烟的人更容易戒烟（P=0.4），根据年龄、性别、收入、受教育程度、戒烟动机和吸烟欲望进行了调整	
[111]	2013~2015	意大利，随机抽样6个月以上的每日吸烟者，6个月以上的EN&NNDS使用者	30~75	932	2年	69%	NM	NM	EN&NNDS使用次数：每月≥50次	NM	每月持续使用EN&NNDS的数量	NM	使用一氧化碳测试随访30天的子样本	双重使用并没有提高随访时戒烟的机会，但减少了吸烟		
[110]	2014	加拿大，参加戒烟计划、可享受免费的NRT和行为咨询的吸烟者	≥16	6526	6个月	—	NM	NM	随访前3个月的EN&NNDS使用情况	NM	NM	M NU	M	SR普遍戒断7天	在6个月随访中，使用EN&NNDS与戒烟较少有关(AOR=0.50；0.39，0.64)。根据吸烟的严重程度和戒烟的信心进行了调整	

续表

参考文献	收集数据年份	样品			产品		方式 EN&NNDS使用				烟碱依赖	戒烟标准	结果 EN&NNDS使用与戒烟的关系
		国家和基准水平	年龄范围(岁)	处于基准水平人数	随访	随访保留率	EN&NNDS类型	烟液烟碱	频率	数量	持续时间	原因			
[106]	2014~2015		18~25	190	6个月	99%	NM	NM		NM	NM	NM	M	6个月随访7天,唾液可替宁验证	在调整了性别、年龄、烟友、家庭成员,基准依赖和烟碱水平之后,EN&NNDS使用率(AOR, 0.56; 95%CI, 0.24, 1.35)比非烟碱使用者低得多。他们比不吸毒的人更频繁地尝试戒烟
[105]	2014~2015		15~85	2057	6个月	69.00%	NM	NM	电子烟的使用:过去30天定期使用。不使用,很少或从未使用过	NM	NM	NM	NM	6个月随访7天	基准双重使用者与吸烟者相比,在调整了年龄、性别、6个月内戒烟的意图后,随访后戒烟的可能性较小(12.5% vs. 9.5%, $P=0.18$, AOR, 1.2; 95%CI, 0.8, 1.9)。在基准和吸烟量较高的情况下,过去30天中至少戒烟24小时,他们更有可能将吸烟量减少一半,并尝试戒烟
[107]	2014		≥18	5672	NR	NR	NR	NR	过去30天内使用EN&NNDS的天数,分类为1~5天、6~20天、30天	NM	随访前1个月	NM	NM	随访自述30天完全戒断	日常EN&NNDS使用者的戒烟率非日常使用者的戒烟率无差异。非日常使用者的戒烟率低于日常使用者。结果在年龄、性别、教育程度、烟草使用、戒烟计划类型和咨询的调整与NRT配合使用上进行了调整

续表

参考文献	收集数据年份	样品 国家和基准	样品 年龄范围(岁)	样品 处于基准水平人数	随访	随访保留率	产品 EN&NNDS类型	产品 烟液烟碱	方式 EN&NNDS使用 频率	方式 EN&NNDS使用 数量	方式 EN&NNDS使用 持续时间	方式 EN&NNDS使用 原因	烟碱依赖	戒烟标准	结果 EN&NNDS使用与戒烟的关系
[108]	2013~2015		≥25	5124	1年	83%	报告使用的烟弹、烟芯和罐装产品	NM	过去30天内使用EN&NNDS的天数	NM	NM	NM	第一支烟时间	随访时自述30天完全戒断	每天使用EN&NNDS的使用者在使用非烟弹、可填充罐装时，戒烟的可能性是非使用者的8倍。参与实验不服用任何其他戒烟产品的人戒烟倾向于每日使用者的一半。非每日使用的人戒烟倾向低于非日使用者，但两者之间并不显著。较少在统计学上并不再关联出现的戒烟现象和不可填充罐装烟弹的使用者中
[109]	2012~2015	美国，已接受住院咨询和计划出院后戒烟的每日吸烟者，鼓励他们使用NRT	≥18	对照=677，干预=680	6个月	对照=75%，干预=77%	NM	NM	过去7天和30天内使用EN&NNDS的天数	NM	NM	所有参加者计划戒烟	每天抽一支烟的时间和第一支烟的时间	可靠宁验证戒烟7日点流行率	在随访前7天，非EN&NNDS使用者中有主动机戒烟但实际上戒烟的比例略高于吸烟者，但差异无统计学意义
[125]	2014~2016	美国，1年以上每日吸烟≥5支且未寻求戒烟治疗的吸烟者	≥18	对照=22，组1=25，组2=21	3个月	对照=73%，组1=76%，组2=71%	卷烟	组1=3周提供16 mg/mL烟碱；组2=24 mg/mL烟碱；对照=不提供ENDS，不提供也不排除购买ENDS	过去一周内使用EN&NNDS的天数	每周使用EN&NNDS的天数	每天抽几次烟	从基准到随访		共同验证的随访7日戒断	24 mg/mL烟碱电子烟液使用中9.5%戒烟，16 mg/mL对照组有4.0%戒烟，差异无统计学意义。没有对ENDS的使用频率进行分析。戒烟动机是相互影响的，在不同的小组中有不同的变化

续表

参考文献	收集数据年份	样品				产品			方式 EN&NNDS使用				结果		EN&NNDS使用与戒烟的关系
		国家和基准	年龄范围(岁)	处于基准水平人数	随访	随访保留率	EN&NNDS类型	烟液烟碱	频率	数量	持续时间	原因	烟碱依赖	戒烟标准	
[126]	2015~2016	意大利，10年以上每日吸烟≥10支且有强烈戒烟动机但未戒烟成功或使用NRT的吸烟者	≥55	每组70			3.3~4.2 V工作电压	组1=END+支持安慰剂，END+支持控制组=无EN&NNDS						EN&NNDS使用者的戒烟率高于对照组，而EN&NNDS使用者的戒烟率无差异。作者指出，最终使用者的气溶胶中含有的烟碱非常少（0.1 mg/次）。	

AOR：调整后的优势比；CI：置信区间；CO：一氧化碳；EN&NNDS：电子烟烟碱和非烟碱传输系统；M：测量；NM：未测量；NRT：烟碱替代疗法；NU：不用于分析；SR：自述；U：用于分析

a. 作者没有报告打电话的人在登记时是否有戒烟。一些参与者参加了多次通话，作者没有说明他们是否是"新"来电者。b. 测量了戒烟的欲望。c. 测量了吸烟冲动的强度

表3.5　关于使用EN&NNDS戒烟的有效性的7个纵向研究的结果

参考文献	按照标准的结果				使用设备类型	戒烟用途
	使用频率					
Berry[108]	日常使用EN&NNDS的吸烟者戒烟率高于非日常使用EN&NNDS的吸烟者，特别是使用可充液烟弹装置的吸烟者戒烟率高于非使用可充液烟弹装置的吸烟者					
	使用EN&NNDS	AOR[a]	95%CI	P		
	非使用者	参考				
	试验者	0.5	0.26，1.00	0.05		
	当前使用者（非每天）	0.5	0.17，1.47	0.21		
	当前使用者（每天）	7.9	4.45，13.95	<0.001		
	EN&NNDS按设备类型使用					
	非使用者					
	每日使用者					
	非烟弹	10.1	54，10.1	<0.001		
	可再装	9.1	4.9，17.0	<0.001		
	罐装	10.2	5.4，19.4	<0.001		
	非罐装	3.6	1.04，12.2	0.04		
	实验性或非日常性使用者					
	烟弹	0.3	0.1，0.9	0.03		
	不可填充	0.25	0.1，0.9	0.04		
	非罐装	0.34	0.1，0.8	0.02		
	a. 根据年龄、性别、种族、地区、收入、吸烟频率、吸烟强度、烟碱依赖性和以前尝试戒烟进行调整					
Hitchman[124]	EN&NNDS使用	AOR[a]	95%CI	P	尽管衡量了戒烟的动机，但没有收集到有关于ENNDS可以被专门用来戒烟的信息	
	非使用者	参考				
	非每日仿真烟使用者	0.35	0.20，0.60	0.002		
	每日仿真烟使用者	0.74	0.39，1.42	0.36		
	非每日雾化烟使用者	0.7	0.29，1.68	0.42		
	日常雾化烟使用者	2.7	1.48，4.89	0.001		
	a. 根据性别、年龄、受教育程度、收入、戒烟动机、强烈的吸烟欲望进行调整					
Biener[121]	每天使用过EN&NNDS的吸烟者比那些未使用过EN&NNDS的使用者更容易戒烟。然而，非日常使用者比不使用的人戒烟的可能性更小，尽管该关联在统计上不显著				EN&NNDS用于戒烟的动机由两个参数测量得出：1年内戒烟的期望值和6个月内戒烟的计划。非每日使用EN&NNDS的吸烟者1年内期望继续吸烟的可能性比非使用者高出6倍。每日使用者与非使用者对继续吸烟的期望相似。不使用EN&NNDS的吸烟者比使用EN&NNDS的吸烟者有戒烟计划的可能性更低，但差异无统计学意义	
	EN&NNDS使用	AOR[a]	95%CI			
	非使用者	参考				
	当前使用者（非每天）	0.31	0.04，2.8			
	当前使用者（每天）	6.07	1.15，24.4			
	a. 根据性别、年龄、种族、受教育程度和吸烟的基准水平进行调整					
Manzoli[111]	定期使用ENDS或ENNDS的吸烟者（每周抽吸≥50）与不使用EN&NNDS的吸烟者的戒烟率相似（AOR=1.25；95%CI，0.85，1.84）。根据年龄、性别、体重指数、婚姻状况、受教育程度、职业、饮酒、高血压、高胆固醇症、糖尿病、自述健康状况、吸烟年限（电子烟使用者以前吸烟）、每日吸烟的卷烟数量（或仅使用电子烟每天抽吸量）调整基准					

续表

参考文献	按照标准的结果			使用设备类型	戒烟用途
	使用频率				
Subialka Nowariak[107]	EN&NNDS的日常和非日常使用者的戒烟行为不同于非使用者。虽然每日使用者和非使用者的戒烟率没有差异，但非每日使用者的戒烟率低于非使用者				
	过去30天内使用EN&NNDS	AOR	95% CI	P	
	非使用者	参考			
	不经常使用（1~5天）	0.35	0.2, 0.59	<0.001	
	中等程度（6~29天）	0.5	0.32, 0.8	0.004	
	每日使用（30天）	1.16	0.71, 1.7	0.453	
	根据年龄、性别、受教育程度、烟草种类、保健专业人员的建议、国家方案和药物使用情况进行调整				
Rigotti[109]	在过去7天内曾使用过安非他明的人士中，戒烟率为13.7%，而从未使用过安非他明的人士中，戒烟率为17.9%。风险差异为－4.3%(95% CI，－13.6,5.1)，无统计学意义。在ENDS使用者中，戒烟率略低，但在统计上没有显著降低				
Zawertailo[110]				在6个月的随访中，未使用EN&NNDS戒烟的EN&NNDS使用者（粗略测量为过去3个月内至少使用过一次）的戒烟率与非使用者相似（37.6% vs. 42.0%；P=0.43），而自述使用EN&NNDS帮助戒烟的EN&NNDS使用者的戒烟率显著低于非使用者	

AOR：调整后的优势比；CI：置信区间；ENDS：电子烟碱传输系统；ENDS：电子非烟碱传输系统；EN&NNDS：电子烟碱和非烟碱传输系统

Giovenco等对2010年以来戒烟的吸烟者进行了一项横断面研究[127]。根据美国2014~2015年全国健康调查的报道，这一假说得到了一定的支持：与从未使用过ENDS的人相比，每天使用ENDS的人戒烟的比例是原来的3倍[128]。有趣的是，Giovenco等研究发现，每天使用ENDS的使用者与曾经使用ENDS的使用者相反，他们的戒烟率分别是未使用ENDS的人的2.6和1.5倍。在不同的亚组中，戒烟成功或失败可能受到以下因素的影响：

■ 鼓励使用EN&NNDS，包括戒烟；
■ 使用电子烟碱传输系统的数量、频率和持续时间的模式；
■ 使用的技术，包括设备和电子烟液的类型；
■ 吸烟者的类型，包括对烟碱的依赖程度以及以前成功和不成功戒烟尝试的历史；
■ 电子烟碱传输系统和烟草使用的监管环境[129-133]。

进一步支持某些吸烟者可以通过使用电子烟碱传输系统（ENDS）成功戒烟的事实包括电子烟碱传输系统可能是卷烟的经济替代品[134-136]，以及在两个主要的EN&NNDS市场上吸烟率下降的趋势没有逆转。美国成年人当前的吸烟率从2005年的20.9%下降到2015年的15.1%，下降了27.7%（$P<0.05$）[137]，包括2014~2015年显著下降1.7个百分点，这与2014~2015年的戒烟率显著上升相吻合，作者将其部分归因于EN&NNDS的使用。针对政策环境可能影响戒烟尝试的其他变化对结果进行了调整，例如增加税收和美国疾病控制与预防中心的"前吸烟者提示"媒体宣传活动。

在英国，2016年成年（≥18岁）吸烟者的比例为15.8%，是自2010年度人口调查开始以来的最低纪录[138]。同时，在英格兰使用EN&NNDS的增加与戒烟尝试的成功增加相关[139]。

这些数据本身并不能证明人们使用EN&NNDS是一种有效的戒烟辅助手段。它们的确表明，使用EN&NNDS至少并没有改变英国吸烟率下降的趋势。

3.7　ENDS对健康的潜在影响

由于某些电子烟碱传输系统可以帮助某些吸烟者戒烟，它们对人群的潜在健康益处是什么？使用电子烟碱传输系统对人群健康的总体影响主要取决于两个因素：一个是电子烟碱传输系统帮助预防吸烟的能力；另一个是与诸如吸烟之类的既定替代方法相比，使用电子烟碱传输系统的相对风险[140]。

3.7.1　与ENDS使用有关的行为轨迹

如果电子烟碱传输系统（ENDS）预防吸烟，则它们不会诱使不吸烟者吸烟，而是诱使吸烟者戒烟，并且在理想情况下应戒烟。换句话说，无论一个人的最初身份是什么，从不吸烟者、当前或既往吸烟者，与使用电子烟碱传输系统相关的行为路径或轨迹都必须远离吸烟，并最终摆脱对烟碱的依赖。图3.5显示了从不吸烟者、当前吸烟者或既往吸烟者的初始状态到四个可能的最终状态之一的27种可能路径：吸烟者、ENDS使用者、双重使用者或双重非使用者。图3.5中的轨迹网表示两种烟碱产品之间的行为路径。实际上，药品、烟草或消费品两种以上产品之间的竞争，可能会使情况变得复杂。

因此，了解使用ENDS对人群健康影响的第一步是估计处于每个初始状态的人随着时间的推移在四种最终状态中的一种状态结束的概率。概率是语境敏感的，因此无法在EN&NNDS和烟草的不同文化和法规环境之间转移。估计概率很复杂，尤其是当缺乏表征它们的经验证据时。讨论的重点是两个最相关的轨迹组合，看在这些组合中，EN&NNDS可以对健康起到促进作用，还是不利于健康。一个是吸烟者戒烟的轨迹（图3.5中的蓝线），另一个是非吸烟者吸烟的轨迹（图3.5中的红线）。

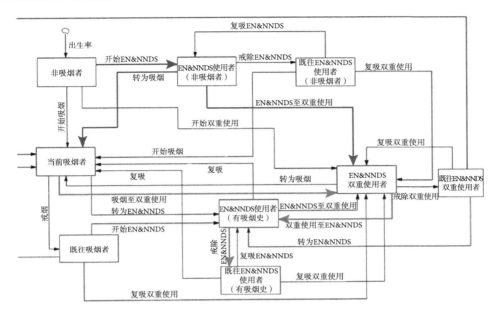

图3.5　与ENDS使用相关的轨迹网
EN&NNDS，电子烟碱和非烟碱传输系统
资料来源：修改自参考文献[141]

吸烟者戒烟的轨迹

我们在上面讨论了EN&NNDS在戒烟中作用的证据。与关于ENNDS是支持还是妨碍戒烟的两极化讨论相反，我们得出结论，EN&NNDS使用对戒烟的影响可能取决于个人使用和吸烟的方式、态度和行为、技术和监管环境。ENDS对于戒烟的总体有用性可能取决于使用ENDS对其产生影响的亚组的优势。例如，Giovenco等[127]显示ENDS每日使用者戒烟的频率是从未使用者的3.2倍；但是，每日使用者仅占样本的5.1%。非每日使用者和曾经尝试过ENDS的使用者分别占样本的9.8%和33.1%，比不使用ENDS的使用者戒烟的频率低2.6和1.5倍。总体而言，使用EN&NNDS的总样本的调整戒断率为26.5%，不使用EN&NNDS的则为28.2%（表3.6）。考虑到EN&NNDS非每日使用者和既往使用者在人群中占主导地位，在每日使用者中，妨碍戒烟要比促进戒烟更为显著。

非吸烟者吸烟的轨迹

从不吸烟的年轻人使用ENDS后尝试吸烟的可能性更大。美国3项纵向研究的荟萃分析[142-145]表明，在基线检查时使用过一次ENDS的年轻人比从未使用过ENDS的年轻人更容易尝试吸烟，约为2倍。

表3.6 理论上对使用和不使用电子烟碱和非烟碱传输系统（EN&NNDS）的吸烟者戒断率的影响

EN&NNDS使用者类型	EN&NNDS使用率（%）	对EN&NNDS使用的贡献率（%）	因使用EN&NNDS而调整[a]的戒断率（%）	不使用[b]EN&NNDS的戒断率（%）
每日使用者	5.1	52.2	4.6	1.4
非每日使用者	9.8	12.1	1.1	2.8
既往使用者	33.1	20.2	6.3	9.3
非使用者	51.9	28.2	14.7	14.7
总计	100	—	26.5	28.2

a. 每日和非每日使用者、既往使用者和非使用者的调整戒断率分别为3.18、0.38、0.67和1
b. 如果整个人群都是非使用者，戒烟率为28.2%

最近的荟萃分析[146]包括之前提到的3项研究和其他6项研究[147]得出的结论是，曾经使用ENDS的年轻人随后开始吸烟的可能性比从未使用ENDS的使用者高约3.5倍。作者还报道说，在过去30天内使用ENDS可使在未来30天内至少吸烟一次的机会增加4倍。英国的两项纵向研究[148,149]表明，ENDS的实验使用和随后的尝试吸烟之间存在相似的关联。但是，到目前为止可用的数据还不能证明这种明显的关联是有因果关系的，或者主要是由于使用ENDS。

由于多种原因，这种关联很难理解[150,151]。在大多数纵向研究中，对这些产品的使用在一生中或过去30天内至少测量到一次。这些回忆期涵盖了年轻人形成时期的各种行为，包括更频繁地尝试使用电子烟碱传输系统和吸烟，这是试探性的和易变的，并且不那么普遍。可以假设，与偶尔吸一口烟等易变性、暂时性的ENDS使用相比，固定的ENDA使用模式更能确定未来吸烟的可能性。

此外，对于这种关联有三种理论解释。第一个是"共同责任猜想"。根据该理论，ENDS的使用和吸烟是彼此独立的，因为它们是潜在的高风险行为倾向的结果。因此，已经提出，很大一部分尝试ENDS然后吸烟的年轻人不管是否存在ENDS都会尝试吸烟。ENDS在吸烟之前使用而非之后的事实是由于多种因素，包括ENDS的新颖性。第二种理论是"重新规范化"假说，据此，ENDS的使用在年轻人中广泛且频繁，其使用的方式和举止使他们联想到吸烟。ENDS使用与吸烟之间的相似性有助于在社会学习框架内从一种产品到另一种产品的轨迹。第三种理论是"催化剂"理论，包括开始电子烟碱传输系统使用的六个假设：风味、健康、价格、榜样、隐蔽性和接受性。提出了另外三个假设来解释向吸烟的过渡：成瘾、可及性和经验[152]。证明这些理论中的任何一个都将面临关键的方法学挑战[153]。在一些纵向研究中，对测量常见的易感性特征的变量进行了调整；但是，残留的混杂因素总是使ENDS使用与吸烟之间的联系更加混乱，没有人能够毫无疑问地证明哪种假设或组合最能解释从不使用烟碱到ENDS使用再到后来的吸烟的过渡。

一些"从不吸烟者"尝试ENDS并最终吸烟的事实必须与这一事实相一致，即在这两个ENDS市场最为突出的国家，目前年轻人中吸烟的流行率继续下降。一项综述[142]显示，在美国等一些国家[154]，每月至少一次使用ENDS（也可能是EN&NNDS）的流行率迅速增加，而在英国等其他国家，不吸烟者的使用率则稳定在极低的水平。

3.7.2 ENDS和电子非烟碱传输系统的危害

尽管EN&NNDS可能会通过吸烟和戒烟的轨迹来引导人群，但使用ENDS对健康的总体影响取决于与其使用相关的健康风险。使用EN&NNDS的长期健康影响仍是未知的，要确定这种影响需要调查许多具有良好特征的使用者的健康状况，并且追踪这些使用者多年。同时，关于EN&NNDS毒性的结论主要基于化学和毒理学研究的经验证据，而在较小程度上还基于临床研究。对这些研究的综述使许多作者得出或多或少的结论是，EN&NNDS并非无害，但通常比卷烟危险性低[155-160]。尤其是与吸烟相关的疾病导致的死亡。根据健康状况的类型来指定和表征EN&NNDS使用的健康风险。

癌症风险

EN&NNDS设备功率设置、烟液配方和使用的理想组合产生的包含致癌化学物质的气溶胶效力不足卷烟烟气致癌力的1%，比药用烟碱吸入器的致死力高两个数量级。但是，如图3.6所示，一些产品和环境会大大增加EN&NNDS气溶胶的致癌风险，有时接近卷烟烟气的致癌风险[161]。当使用者向雾化器线圈[76]施加过多的功率时，似乎会形成具有较高致癌性的气溶胶。有人认为，这种情况仅在"干烧"条件下发生[162]——EN&NNDS使用者容易发现的情况。但是，没有经验证据表明仅仅是由于干烧情况引起的，或者，如果是的话，这种情况发生的频率。

心血管风险

关于使用EN&NNDS所引起的心血管疾病风险是否低至其致癌潜力存在争议。有人认为ENDS气溶胶的主要心血管风险是烟碱的毒性所致，对于健康使用者而言，烟碱的短期心血管风险似乎较低[163]。在2015~2017年进行的临床和细胞培养研究中，研究了使用电子烟碱传输系统和心血管疾病风险指标（包括心率、血压和迷走神经张力）之间的关系，血小板聚集和黏附，主动脉僵硬和内皮功能，表达抗氧化防御和免疫系统功能的基因，以及氧化应激指数。所审查的6项研究显示出的显著不良心血管影响中，3项研究发现ENDS对生理心血管风险指标的影响小于卷烟，而其他3项研究发现ENDS与卷烟具有相同的作用。一些研究表明，尽管添加烟碱可以增强这些不良心血管效应，但它们与烟碱无关[164]。

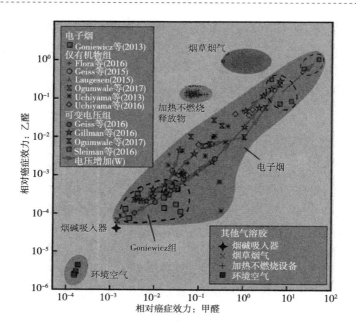

图3.6　电子烟碱和非烟碱传输系统产生的气溶胶中以及烟草烟气、加热不燃烧设备、烟碱吸入器和环境空气中的甲醛和乙醛的致癌性

资料来源：经BMJ Publishing Group Ltd.许可，引自参考文献[161]。

肺部危险

尽管EN&NNDS气溶胶的毒性可能比卷烟烟气低，并且死亡率比卷烟低，但长期戒烟者和双重吸烟者对肺部毒性的降低仍然是未知的。对此主题综述的作者得出的结论是，EN&NNDS引起的炎症可能差异性地影响肺癌和慢性阻塞性肺疾病的风险[165]。因此，最新的经验证据表明EN&NNDS气溶胶比卷烟烟气的有害性更小，但没有经验数据来量化暴露于EN&NNDS气溶胶和烟草烟气的相对风险。

已经做了一些努力来模拟EN&NNDS的潜在人群影响[166-168]；但是，结果仅与输入模型的数据一样好。鉴于数据很少，目前尚不清楚在最坏情况和最佳情况下计算电子烟碱传输系统的益处时应包括哪些因素[169,170]，尤其是对于电子烟碱传输系统在帮助人们戒烟方面的功效以及其相对于卷烟的安全性等变量。

使用电子烟碱传输系统对人群健康的影响的量化非常复杂，因为必须考虑许多变量。现有证据表明，电子烟碱传输系统可能对人群健康产生积极影响，特别是在制定了适当的电子烟碱传输系统法规以最大限度地发挥其益处和降低其风险的情况下。

3.8 证据汇总、研究差距和由证据得出的政策问题

ENDS是由很多种类组成的产品，烟碱和非烟碱有害物质的特性各不相同，取决于其结构、功率、电子烟液成分、烟碱浓度和使用行为等因素。烟碱的传输量范围从零到超过相当抽吸数量的卷烟剂量。与大多数类型NRT相比，来自ENDS的烟碱到达使用者血液的速度更快，而且至少在某些ENDS中，烟碱的浓度更高。ENDS对于某些吸烟者在某些情况下戒烟可能有效，而对其他吸烟者在不同情况下的戒烟效果可能相反。ENDS对戒烟有利还是有害，似乎取决于技术、最终使用者的动机和消费行为、寻求ENDS使用的吸烟者类型，以及ENDS使用和烟草使用的监管环境。

很难将证据转化为EN&NNDS在戒烟中的潜在作用。证据不允许建议支持或反对将ENDS和ENNDS用作戒烟辅助剂的一揽子政策。然而，它指出了政策制定者需要监管的四个领域。

ENDS监管中烟碱通量的概念：在烟碱替代方案中希望最大限度发挥ENDS技术潜力的监管者，应将烟碱释放速率（即烟碱通量）作为决定的主要因素。实际上，不应孤立地监管影响烟碱通量的因素。尽管此类标准也应基于临床评估（即对ENDS使用者及不使用者的影响），但是ENDS烟碱通量可以对监管所用的产品标准进行数学建模。

烟碱通量与有害物质特征之间的关系：上面的推论是，获得不同烟碱通量的条件可能会影响有害物质特征，因为某些增加烟碱通量的相同因素（例如功率）也会增加气溶胶中某些有害物质的浓度，例如醛类物质。因此，监管机构要考虑制造商和政府应如何告知使用者产生足够的烟碱通量和相关的有害物质之间的平衡。

电子烟液中烟碱的浓度：尽管有一些关于烟碱浓度标注的行业指南，但许多电子烟液上的标签并未标明浓度，或者不提供准确的信息，使ENDS使用者无法了解电子烟液中烟碱浓度的准确信息，不能获得控制烟碱自我给药的重要信息。

ENDS设备和电子烟液的标签和质量控制：所有电子烟液上的标签应清晰显示每个容器中烟碱的总量、游离烟碱与质子化烟碱的比例及浓度（mg/mL）；另外，应指示电子烟液烟碱浓度不高于某个浓度值，例如0.1 mg/mL。必须使用质量控制来确保标签信息的准确性和生产标准的一致性。

尽管本报告未讨论该主题，但是有确凿证据表明，除了通过雾化吸入以外，接触电子烟液中烟碱还可能危害健康，有时甚至致命[171]。为了避免意外接触烟碱，监管者应该考虑要求所有电子烟液容器都有儿童防护装置。

对本报告中描述的有关电子烟碱传输系统的问题制定适当政策和法规需要对制造商的披露要求，以及有效的、有组织的、系统的国家监督。要求制造商提供

的关键公开数据包括市售设备的电压、电阻和功率以及电子烟液成分。此外，应进行监测，以确定消费者对电子烟碱传输系统的使用行为，例如哪些人使用电子烟碱传输系统、使用的目的、使用什么产品、如何使用以及使用频率。

表3.7总结了通过电子烟碱传输系统传输烟碱，其对戒烟的影响以及对人群健康影响的前瞻性证据。该表还列出了每个证据要素在研究和政策问题上的差距。

表3.7 证据、研究差距和政策问题摘要

主题	证据	研究差距	政策和监管问题
有效的烟碱传输	**技术** 在烟碱释放的水平和速度方面，可以生产与卷烟类似的烟碱通量。烟碱通量受下列因素影响： - 施加于给定电阻的线圈上的电压：功率越高，气溶胶中的烟碱浓度越高； - 烟液浓度：烟液浓度越高，给定功率值的气溶胶浓度越高； - 使用者的吸烟行为：吸烟时间越长，烟碱含量越高	ENDS在设备技术和电子烟液组件方面差别很大。虽然控制烟碱传递的ENDS技术特性是已知的，但还需要进一步的研究才能将这些发现推广到所有ENDS，而不仅仅是单个产品。 需要进行程序性和跨学科研究，结合气溶胶研究、分析化学和临床实验室方法完成预测烟碱通量的数学模型，并开始尽可能描述预测非烟碱有毒物质的类似模型。 烟碱通量与戒烟之间的关系必须在人群水平上进行表征。描述烟碱通量和潜在滥用可能有助于解释和防止非烟草使用者通过ENDS开始吸烟	在ENDS监管中烟碱通量的概念是决定烟碱ENDS替代卷烟烟碱能力的主要因素，监管机构在努力最大限度地发挥ENDS技术的烟碱替代潜力时应考虑这一因素。 仅对影响烟碱通量的一个因素（如烟液浓度）进行管制也可能导致ENDS使用者改变其他因素（如增加功率）。这种变化可能增加也可能不增加烟碱流量，但可能增加健康风险（例如，在较高功率值下增加挥发性醛的释放）。 那些不能以与卷烟相似的速度和浓度释放烟碱的ENDS，至少应该有不能帮助戒烟的警示
	ENDS释放物的毒性特征与烟碱通量的关系 增加烟碱通量的一些因素，如功率，也会增加气溶胶中某些有害物质的浓度，如醛类	在各种条件下，需要进行进一步研究，以确定ENDS释放的毒性特征与影响烟碱通量的因素之间的关联，例如： - 烟液组成：溶剂类型和调味物质。许多用于ENDS烟液的调味料是可食用的，在加热和吸入后没有经过安全测试。 - 设备结构：包括一般设计、使用的金属、烟芯材料	有害物质特性信息：监管机构可考虑如何让使用者了解影响在产生足够的烟碱通量和相关的有害物质释放之间取得平衡的因素
	烟液中烟碱的浓度 尽管有一些关于烟液烟碱浓度标签的行业指南，但许多烟液都没有此类标签，很难解释，或者通常无法提供准确的信息。 剥夺ENDS使用者有关烟液中烟碱浓度的准确信息会使他们无法获得决定是否自行服用烟碱的充分信息	一个主要的研究空白是如何对消费者有用的方式传达这些信息，并增加使用ENDS来提高戒烟率的机会。这同样适用于ENDS功率。一些ENDS使用可能没有意识到，他们使用的ENDS不能有效地输送烟碱，无论它含有多大浓度的烟碱	烟液体中烟碱浓度的范围烟液中烟碱的最低和最高含量以及浓度的监管应确保以下两者之间的平衡： - 确保足够的浓度达到烟碱替代的足够通量； - 意外接触含有烟碱的电子烟液的风险。 每个烟液容器的烟碱总量应足够小以避免在包装安全措施失效时发生致命或严重意外烟碱中毒的风险。

续表

主题	证据	研究差距	政策和监管问题
有效的烟碱传输			烟碱浓度应足够高以提供足够的烟碱通量，并在线圈加热过度以获得更多烟碱气溶胶时限制有毒物质的产生。 ENDS设备和烟液的标签和质量控制 消费者必须拥有关于产品设计特性和烟液成分的准确、可靠信息，包括： - 所列成分包括所有烟液成分（即无污染物）； - 所有烟液容器都贴有标签以提供有关烟碱含量、游离态烟碱与质子化烟碱的比率以及浓度的准确信息。 应禁止已知具有严重风险的烟液污染物（如二甘醇、双乙酰）。 必须采用质量控制以确保标签信息的准确性和生产标准的实施
ENDS作为戒烟辅助手段的有效性	在某些情况下，ENDS对某些吸烟者的戒烟是有效的。在不同的情况下，它可能对其他吸烟者产生相反的影响 影响ENDS作为戒烟辅助手段的可能性的条件包括： - ENDS装置和烟液的适当组合以类似于卷烟的水平和速度输送烟碱； - 以最少的数量、频率和持续时间使用ENDS戒烟； - 吸烟者的类型，包括烟碱依赖程度和以往成功和不成功戒烟的历史； - ENDS和烟草使用的监管环境	需要更好地了解ENDS使用可能促进或不利于戒烟的情况，包括： - 有助于戒烟的ENDS特征：烟碱释放特征与戒烟密切相关；是否需要加调味物质，如果需要，哪些种类、多少量的调味物质可以最大限度地戒烟；哪些使用行为和使用频率与戒烟和避免再次吸烟最密切相关； - ENDS可能有效戒烟的吸烟者亚群和ENDS无效戒烟的吸烟者亚群； - 保持卷烟和ENDS双重使用的ENDS的特征，以及如何对其进行操纵以鼓励戒烟； - 如何帮助长期ENDS使用者停止ENDS使用，如果这是他们希望的结果	监管者应该意识到ENDS使用可能对戒烟产生相反的影响。然而，如果没有进一步的研究，就不可能提出政策建议，以最大限度地发挥其帮助戒烟的潜力，并最大限度地减少其对戒烟的有害影响

续表

主题	证据	研究差距	政策和监管问题
ENDS潜在的健康影响	**ENDS总体健康影响** - ENDS具有引导吸烟者远离吸烟的能力，同时从不劝阻不吸烟者吸烟，避免前吸烟者复发； - 与不使用ENDS或不吸烟者使用所说的替代品相关的健康风险	上文描述了ENDS作为戒烟辅助工具的有效性方面的研究差距。在以下情况下，需要进一步研究ENDS使用与不使用和吸烟之间的长期健康后果：吸烟者（包括成人、儿童和孕妇）以及不吸烟者	针对本报告所描述的ENDS问题制定适当的政策和法规需要有效的、有组织的国家和全球对ENDS市场类型及其使用情况的监测。更具体地说，需要提供有关谁使用它们、目的是什么的信息（包括传输其他滥用物质）和使用的产品（包括电压、电阻、功率，烟液成分、使用频率的测量）
		在某些情况下，体外（例如细胞制剂）或体内（例如动物模型）方法可能适用于解决问题或指导后续的临床研究。对于人群研究，与ENDS使用相关的疾病不一定是与吸烟相关的疾病。因此，必须特别注意可能与ENDS使用有关的疾病状态和疾病生物标志物，首先要关注已知ENDS释放物，包括丙二醇、植物甘油、香料、甜味剂和这些烟液成分在加热时产生的副产物	

3.9 参考文献

[1] Breland A, Soule E, Lopez A, Ramôa C, El-Hellani A, Eissenberg T et al. Electronic cigarettes: what are they and what do they do? Ann N Y Acad Sci. 2016;1394:5–30.

[2] Kennedy RD, Awopegba A, De León E, Cohen JE. Global approaches to regulating electronic cigarettes. Tob Control 2016;26:440–5.

[3] Talih S, Balhas Z, Eissenberg T, Salman R, Karaoghlanian N, El Hellani H et al. Effects of user puff topography, device voltage, and liquid nicotine concentration on electronic cigarette nicotine yield: measurements and model predictions. Nicotine Tob Res. 2014;17:150–7.

[4] Wagener T, Floyd E, Stepanov I, Driskill LM, Frank SG, Meier E et al. Have combustible cigarettes met their match? The nicotine delivery profiles and harmful constituent exposures of secondgeneration and third-generation electronic cigarette users. Tob Control. 2016;26:e23–8.

[5] Rudy A, Leventhal A, Goldenson NI, Eissenberg T. Assessing electronic cigarette effects and regulatory impact: challenges with user self-reported device power. Drug Alcohol Depend. 2017;179:337–40.

[6] Farsalinos K, Spyrou A, Stefopoulos C, Tsimopoulou K, Kourkoveli P, Tsiapris D et al. Nicotine absorption from electronic cigarette use: comparison between experienced consumers (vapers) and naïve users (smokers). Sci Rep. 2015;5:11629.

[7] Hua M, Yip H, Talbot P. Mining data on usage of electronic nicotine delivery systems (ENDS) from YouTube videos. Tob Control. 2011;22:103–6.

[8] Hiler M, Breland A, Spindle T, Maloney S, Lipato T, Karaoghlanian N et al. Electronic cigarette user plasma nicotine concentration, puff topography, heart rate, and subjective effects: the influence of liquid nicotine concentration and user experience. Exp Clin Psychopharmacol. 2017;25(5):380–92.

[9] Vansickel A, Cobb C, Weaver M, Eissenberg T. A clinical laboratory model for evaluating the acute effects of electronic "cigarettes": nicotine delivery profile and cardiovascular and subjective effects. Cancer Epidemiol Biomarkers Prev. 2010;19:1945–53.

[10] Cheah N, Chong N, Tan J, Morsed FA, Yee SK. Electronic nicotine delivery systems: regulatory and safety challenges: Singapore perspective. Tob Control. 2012;23:119–25.

[11] Chaudhry IW, Leigh NJ, Smith DM, O'Connor RJ, Goniewicz ML. Labeling information on electronic nicotine delivery systems. Tob Regul Sci. 2017;3:3–9.

[12] Buonocore F, Marques Gomes ACN, Nabhani-Gebara S, Barton SJ, Calabrese G. Labelling of electronic cigarettes: regulations and current practice. Tob Control. 2016;26:46–52.

[13] Beauval N, Antherieu S, Soyez M, Gengler N, Grova N, Howsam M et al. Chemical evaluation of electronic cigarettes: multicomponent analysis of liquid refills and their corresponding aerosols. J Anal Toxicol. 2017;41:670–8.

[14] Buettner-Schmidt K, Miller DR, Balasubramanian N. Electronic cigarette refill liquids: childresistant packaging, nicotine content, and sales to minors. J Pediatric Nurs. 2016;31:373–9.

[15] Cameron JM, Howell DN, White JR, Andrenyak DM, Layton ME, Roll JM. Variable and potentially fatal amounts of nicotine in e-cigarette nicotine solutions: Table 1. Tob Control 2013;23:77–8.

[16] Davis B, Dang M, Kim J, Talbot P. Nicotine concentrations in electronic cigarette refill and doit-yourself fluids. Nicotine Tob Res. 2014;17:134–41.

[17] El-Hellani A, El-Hage R, Baalbaki R, Salman R, Talih S, Shihadeh A et al. Free-base and protonated nicotine in electronic cigarette liquids and aerosols. Chem Res Toxicol. 2015;28:1532–7.

[18] Etter J, Zäther E, Svensson S. Analysis of refill liquids for electronic cigarettes. Addiction. 2013;108:1671–9.

[19] Etter J, Bugey A. E-cigarette liquids: constancy of content across batches and accuracy of labeling. Addictive Behav. 2017;73:137–43.

[20] Farsalinos KE, Gillman IG, Melvin MS, Paolantonio AR, Gardow WJ, Humphries KE et al. Nicotine Clinical pharmacology of nicotine in electronic nicotine delivery systems levels and presence of selected tobacco-derived toxins in tobacco flavoured electronic cigarette refill liquids. Int J Environ Res Public Health. 2015;12:3439–52.

[21] Goniewicz ML, Gupta R, Lee YH, Reinhardt S, Kim S, Kim B et al. Nicotine levels in electronic cigarette refill solutions: a comparative analysis of products from the US, Korea, and Poland. Int J Drug Policy. 2015;26:583–8.

[22] Kim S, Goniewicz M, Yu S, Kim B, Gupta R. Variations in label information and nicotine levels in electronic cigarette refill liquids in South Korea: regulation challenges. Int J Environ Res Public Health. 2015;12:4859–68.

[23] Kirschner RI, Gerona R, Jacobitz KL. Nicotine content of liquid for electronic cigarettes. Clin Toxicol. 2013;51:684–4.

[24] Kosmider L, Sobczak A, Szołtysek-Bołdys I, Prokopowicz A, Skórka A, Abdulafeez O et al. Assessment of nicotine concentration in electronic nicotine delivery system (ENDS) liquids and precision of dosing to aerosol. Przegl Lek. 2015;72:500–4.

[25] Lisko JG, Tran H, Stanfill SB, Blount BC, Watson CH. Chemical composition and evaluation of nicotine, tobacco alkaloids, pH, and selected flavors in e-cigarette cartridges and refill solutions. Nicotine Tob Res. 2015;17:1270–8.

[26] Pagano T, DiFrancesco AG, Smith SB, George J, Wink G, Rahman I et al. Determination of nicotine content and delivery in disposable electronic cigarettes available in the United States by gas chromatography–mass spectrometry. Nicotine Tob Res. 2015;18:700–7.

[27] Peace MR, Baird TR, Smith N, Wolf CE, Poklis JL, Poklis A. Concentration of nicotine and glycols in 27 electronic cigarette formulations. J Anal Toxicol. 2016;40:403–7.

[28] Rahman A, Nik Mohamed MH, Mahmood S. Nicotine estimations in electronic cigarette eliquids among Malaysian marketed samples. Anal Chem Lett. 2018;8:54–62.

[29] Raymond BH, Collette-Merrill K, Harrison RG, Jarvis S, Rasmussen RJ. The nicotine content of a sample of e-cigarette liquid manufactured in the United States. J Addict Med. 2018;12:127–31.

[30] Trehy ML, Ye W, Hadwiger ME, Moore TW, Allgire JF, Woodruff JT et al. Analysis of electronic cigarette cartridges, refill solutions, and smoke for nicotine and nicotine related impurities. J Liquid Chromatogr Relat Technol. 2011;34:1442–58.

[31] E-liquid manufacturing standards 2017 version 2.3.2. Mooresville (NC): American E-liquid Manufacturing Standards Association; 2017 (http://www.aemsa.org/standards/, accessed 19 September 2017).

[32] Goniewicz ML, Hajek P, McRobbie H. Nicotine content of electronic cigarettes, its release in vapour and its consistency across batches: regulatory implications. Addiction. 2013;109:500–7.

[33] Omaiye E, Cordova I, Davis B, Talbot P. Counterfeit electronic cigarette products with mislabeled nicotine concentrations. Tob Regul Sci. 2017;3:347–57.

[34] Westenberger B. Evaluation of e-cigarettes. St Louis (MO): Center for Drug Evaluation and Research, Office of Pharmaceutical Science, Office of Testing and Research; 2009 (https://www.fda.gov/AboutFDA/CentersOffices/OfficeofMedicalProductsandTobacco/CDER/ucm088761.htm, accessed 1 October 2018).

[35] Henningfield J, London E, Benowitz N. Arterial–venous differences in plasma concentrations of nicotine after cigarette smoking. JAMA. 1990;263:2049–50.

[36] Baldassarri SR, Hillmer AT, Anderson JM, Jatlow P, Nabulsi N, Labaree D et al. Use of electronic cigarettes leads to significant beta2-nicotinic acetylcholine receptor occupancy: evidence from a PET imaging study. Nicotine Tob Res. 2018;20(4):425–33.

[37] Dawkins L, Corcoran O. Acute electronic cigarette use: nicotine delivery and subjective effects in regular users. Psychopharmacology. 2013;231:401–7.

[38] Hajek P, Goniewicz ML, Phillips A, Myers Smith K, West O, McRobbie H. Nicotine intake from electronic cigarettes on initial use and after 4 weeks of regular use. Nicotine Tob Res. 2014;17:175–9.

[39] Nides MA, Leischow SJ, Bhatter M, Simmons M. Nicotine blood levels and short-term smoking reduction with an electronic nicotine delivery system. Am J Health Behav. 2014;38:265–74.

[40] St Helen G, Havel C, Dempsey DA, Jacob P 3rd, Benowitz NL. Nicotine delivery, retention and

pharmacokinetics from various electronic cigarettes. Addiction. 2015;111:535-44.
[41] Yan XS, D'Ruiz C. Effects of using electronic cigarettes on nicotine delivery and cardiovascular function in comparison with regular cigarettes. Regul Toxicol Pharmacol 2015;71:24-34.
[42] Henningfield JE. Nicotine medications for smoking cessation. N Engl J Med. 1995;333:1196-203.
[43] Choi JH, Dresler CM, Norton MR, Strahs KR. Pharmacokinetics of a nicotine polacrilex lozenge. Nicotine Tob Res. 2003;5:635-44.
[44] Evans SE, Blank M, Sams C, Weaver MF, Eissenberg T. Transdermal nicotine-induced tobacco abstinence symptom suppression: nicotine dose and smokers' gender. Exp Clin Psychopharmacol. 2006;14:121-35.
[45] Bullen C, Howe C, Laugesen M, McRobbie H, Parag V, Williman J et al. Electronic cigarettes for smoking cessation: a randomised controlled trial. Lancet 2013;382:1629-37.
[46] E-Cigarette use among youth and young adults. A report of the Surgeon General. Atlanta (GA): Department of Health and Human Services, National Center for Chronic Disease Prevention and Health Promotion, Office on Smoking and Health; 2016.
[47] 47. Schroeder MJ, Hoffman AC. Electronic cigarettes and nicotine clinical pharmacology. Tob Control 2014;23:ii30-5.
[48] Cobb NK, Byron MJ, Abrams DB, Shields PG. Novel nicotine delivery systems and public health: the rise of the "e-cigarette". Am J Public Health. 2010;100:2340-2.
[49] McAuley TR, Hopke PK, Zhao J, Babaian S. Comparison of the effects of e-cigarette vapor and cigarette smoke on indoor air quality. Inhal Toxicol. 2012;24:850-7.
[50] Pellegrino RM, Tinghino B, Mangiaracina G, Marani A, Vitali M, Protano C et al. Electronic cigarettes: an evaluation of exposure to chemicals and fine particulate matter (PM). Ann Ig. 2012;24:79-88.
[51] Goniewicz ML, Kuma T, Gawron M, Knysak J, Kosmider L. Nicotine levels in electronic cigarettes. Nicotine Tob Res. 2012;15:158-66.
[52] Farsalinos KE, Romagna G, Tsiapras D, Kyrzopoulos S, Voudris V. Evaluation of electronic cigarette use (vaping) topography and estimation of liquid consumption: implications for research protocol standards definition and for public health authorities' regulation. Int J Environ Res Public Health 2013;10:2500-14.
[53] Laugesen M. Nicotine and toxicant yield ratings of electronic cigarette brands in New Zealand. NZ Med J. 2015;128:77-82.
[54] Tayyarah R, Long GA. Comparison of select analytes in aerosol from e-cigarettes with smoke from conventional cigarettes and with ambient air. Regul Toxicol Pharmacol. 2014;70:704-10.
[55] Henningfield JE, Keenan RM. Nicotine delivery kinetics and abuse liability. J Consult Clin Psychol 1993;61:743-50.
[56] Shihadeh A, Eissenberg T. Electronic cigarette effectiveness and abuse liability: predicting and regulating nicotine flux. Nicotine Tob Res. 2014;17:158-62.
[57] Eissenberg T, Shihadeh A. Nicotine flux: a potentially important tool for regulating electronic cigarettes. Nicotine Tob Res. 2014;17:165-7.
[58] Talih S, Balhas Z, Salman R, El-Hage R, Karaoghlanian N, El-Hellani A et al. Transport phenomena governing nicotine emissions from electronic cigarettes: model formulation and

[58] experimental investigation. Aerosol Sci Technol. 2016;51:1–11.

[59] Stepanov I, Fujioka N. Bringing attention to e-cigarette pH as an important element for research and regulation. Tob Control. 2014;24:413–4.

[60] Han S, Chen H, Zhang X, Liu T, Fu Y. Levels of selected groups of compounds in refill solutions for electronic cigarettes. Nicotine Tob Res. 2015;18:708–14.

[61] Burr GA, Van Gilder TJ, Trout BD, Wilcox TC, Driscoll R. Health hazard evaluation report HETA Clinical pharmacology of nicotine in electronic nicotine delivery systems 90-0355-2449, Actors' Equity Association/The League of American Theatres and Producers, Inc. Washington (DC): National Institute for Occupational Safety and Health, Hazard Evaluations and Technical Assistance Branch Division of Surveillance, Hazard Evaluations, and Field Studies; 1994 (http://www.cdc.gov/niosh/hhe/reports/pdfs/1990-0355-2449.pdf, accessed 1 October 2018).

[62] Moline JM, Golden AL, Highland JH, Wilmarth KR, Kao AS. Health effects evaluation of theatrical smoke, haze, and pyrotechnics. Report to Equity-League Pension and Health Trust Funds, 2000.

[63] Varughese S, Teschke K, Brauer M, Chow Y, van Netten C, Kennedy SM. Effects of theatrical smokes and fogs on respiratory health in the entertainment industry. Am J Ind Med. 2005;47:411–8.

[64] Wieslander G. Experimental exposure to propylene glycol mist in aviation emergency training: acute ocular and respiratory effects. Occup Environ Med. 2001;58:649–55.

[65] Neale BW, Mesler EL, Young M, Rebuck JA, Weise WJ. Propylene glycol-induced lactic acidosis in a patient with normal renal function: a proposed mechanism and monitoring recommendations. Ann Pharmacother. 2005;39:1732–6.

[66] Deichmann W. Glycerol:–effects upon rabbits and rats–. Am Ind Hyg Assoc Q 1941;2:5–6.

[67] Hine CH, Anderson HH, Moon HO, Dunlop MK, Morse MS. Comparative toxicity of synthetic and natural glycerin. Arch Ind Hyg Occup Med. 1953;7:282–91.

[68] Safety assessment and regulatory authority to use flavors: focus on electronic nicotine delivery systems and flavored tobacco products. Washington (DC): Federal Emergency Management Agency; 2013 (https://www.femaflavor.org/safety-assessment-and-regulatory-authority-useflavors-focus-electronic-nicotine-delivery-systems, accessed 18 Oct 2017).

[69] Farsalinos KE, Kistler KA, Gillman G, Voudris V. Evaluation of electronic cigarette liquids and aerosol for the presence of selected inhalation toxins. Nicotine Tob Res. 2014;17:168–74.

[70] Kosmider L, Sobczak A, Prokopowicz A, Kurek J, Zaciera M, Knysak J et al. Cherry-flavoured electronic cigarettes expose users to the inhalation irritant, benzaldehyde. Thorax. 2016;71:376–7.

[71] Behar RZ, Luo W, Lin SC, Wang Y, Valle J, Pankow JE et al. Distribution, quantification and toxicity of cinnamaldehyde in electronic cigarette refill fluids and aerosols. Tob Control 2016;25:ii94–102.

[72] Gerloff J, Sundar IK, Freter R, Sekera ER, Friedman AE, Robinson R et al. Inflammatory response and barrier dysfunction by different e-cigarette flavoring chemicals identified by gas chromatography–mass spectrometry in e-liquids and e-vapors on human lung epithelial cells and fibroblasts. Appl In Vitro Toxicol. 2017;3:28–40.

[73] Hutzler C, Paschke M, Kruschinski S, Henkler F, Hahn J, Luch A. Chemical hazards present in liquids and vapors of electronic cigarettes. Arch Toxicol. 2014;88:1295–308.

[74] Varlet V, Farsalinos K, Augsburger M, Thomas A, Etter JF. Toxicity assessment of refill liquids for electronic cigarettes. Int J Environ Res Public Health. 2015;12:4796–815.

[75] Williams M, Bozhilov K, Ghai S, Talbot P. Elements including metals in the atomizer and aerosol of disposable electronic cigarettes and electronic hookahs. PLoS One. 2017;12:e0175430.

[76] Ogunwale MA, Li M, Ramakrishnam Raju MV, Chen Y, Nantz MH, Conklin DJ et al. Aldehyde detection in electronic cigarette aerosols. ACS Omega 2017;2:1207–14.

[77] Herrington JS, Myers C. Electronic cigarette solutions and resultant aerosol profiles. J Chromatogr A 2015;1418:192–9.

[78] Talih S, Balhas Z, Salman R, Karaoghlanian N, Shihadeh A. "Direct dripping": a high-temperature, high-formaldehyde emission electronic cigarette use method. Nicotine Tob Res. 2015;18:453–9.

[79] Bekki K, Uchiyama S, Ohta K, Inaba Y, Nakagome H, Kunugita N. Carbonyl compounds generated from electronic cigarettes. Int J Environ Res Public Health. 2014;11:11192–200.

[80] Sleiman M, Logue JM, Montesinos VN, Russell ML, Litter MI, Gundel LA et al. Emissions from electronic cigarettes: key parameters affecting the release of harmful chemicals. Environ Sci Technol. 2016;50:9644–51.

[81] Wang P, Chen W, Liao J, Matsuo T, Ito K, Fowles J et al. A device-independent evaluation of carbonyl emissions from heated electronic cigarette solvents. PLoS One. 2017;12:e0169811.

[82] Havel CM, Benowitz NL, Jacob P 3rd, St Helen G. An electronic cigarette vaping machine for the characterization of aerosol delivery and composition. Nicotine Tob Res. 2017;19(10):1224–31.

[83] Geiss O, Bianchi I, Barrero-Moreno J. Correlation of volatile carbonyl yields emitted by e-cigarettes with the temperature of the heating coil and the perceived sensorial quality of the generated vapours. Int J Hyg Environ Health. 2016;219:268–77.

[84] Khlystov A, Samburova V. Flavoring compounds dominate toxic aldehyde production during e-cigarette vaping. Environ Sci Technol. 2016;50:13080–5.

[85] Farsalinos K, Gillman G, Kistler K, Yannovits N. Comment on "Flavoring compounds dominate toxic aldehyde production during e-cigarette vaping". Environ Sci Technol. 2017;51:2491–2.

[86] Khlystov A, Samburova V. Response to comment on "Flavoring compounds dominate toxic aldehyde production during e-cigarette vaping". Environ Sci Technol. 2017;51:2493–4.

[87] Tierney P, Karpinski C, Brown J, Luo W, Pankow JF. Flavour chemicals in electronic cigarette fluids. Tob Control. 2015;25:e10–5.

[88] Soussy S, EL-Hellani A, Baalbaki R, Salman R, Shihadeh A, Saliba NA. Detection of 5-hydroxymethylfurfural and furfural in the aerosol of electronic cigarettes. Tob Control. 2016;25:ii88–93.

[89] Pankow JF, Kim K, McWhirter KJ, Luo W, Escobedo JO, Strongin RM et al. Benzene formation in electronic cigarettes. PLoS One. 2017;12:e0173055.

[90] Gillman I, Kistler K, Stewart E, Paolantonio AR. Effect of variable power levels on the yield of total aerosol mass and formation of aldehydes in e-cigarette aerosols. Regul Toxicol Pharmacol. 2016;75:58–65.

[91] Franck C, Budlovsky T, Windle S, Filion KB, Eisenberg MJ. Electronic cigarettes in North America: history, use, and implications for smoking cessation. Circulation. 2014;129:1945–52.

[92] Harrell PT, Simmons VN, Correa JB, Padhya TA, Brandon TH. Electronic nicotine delivery systems ("e-cigarettes"): review of safety and smoking cessation efficacy. Otolaryngol Head Neck Surg. 2014;151:381–93.

[93] Lam C, West A. Are electronic nicotine delivery systems an effective smoking cessation tool? Can J Resp Ther. 2015;51:93–8.

[94] Ioakeimidis N, Vlachopoulos C, Tousoulis D. Efficacy and safety of electronic cigarettes for smoking cessation: a critical approach. Hell J Cardiol. 2016;57:1–6.

[95] Waghel RC, Battise DM, Ducker ML. Effectiveness of electronic cigarettes as a tool for smoking cessation or reduction. J Pharm Technol. 2014;31:8–12.

[96] Orr KK, Asal NJ. Efficacy of electronic cigarettes for smoking cessation. Ann Pharmacother 2014;48:1502–6.

[97] Malas M, van der Tempel J, Schwartz R, Mimichiello A, Lightfoot C, Noormohamed A et al. Electronic cigarettes for smoking cessation: a systematic review. Nicotine Tob Res. 2016;18:1926–36.

[98] Rahman MA, Hann N, Wilson A, Mnatzganian G, Worrell-Carter L. E-cigarettes and smoking cessation: evidence from a systematic review and meta-analysis. PLoS One. 2015;10:e0122544.

[99] McRobbie H, Bullen C, Hartmann-Boyce J, Hajek P. Electronic cigarettes for smoking cessation and reduction. Cochrane Database Syst Rev. 2014;(12):CD010216.

[100] Hartmann-Boyce J, McRobbie H, Bullen C, Begh R, Stead LF, Hajek P. Electronic cigarettes for smoking cessation. Cochrane Database Syst Rev. 2016;9:CD010216.

[101] Kalkhoran S, Glantz SA. E-cigarettes and smoking cessation in real-world and clinical settings: a systematic review and meta-analysis. Lancet Resp Med. 2016;4:116–28.

[102] El Dib R, Suzumura EA, Akl EA, Gomaa H, Agarwal A, Chang Y et al. Electronic nicotine delivery Clinical pharmacology of nicotine in electronic nicotine delivery systems systems and/or electronic non-nicotine delivery systems for tobacco smoking cessation or reduction: a systematic review and meta-analysis. BMJ Open. 2017;7:e012680.

[103] Khoudigian S, Devji T, Lytvyn L, Campbell K, Hopkins R, O'Reilly D. The efficacy and short-term effects of electronic cigarettes as a method for smoking cessation: a systematic review and a meta-analysis. Int J Public Health. 2016;61:257–67.

[104] Stratton K, Kwan LY, Eaton DL, editors. Smoking cessation among adults. Chapter 17. In: Public health consequences of e-cigarettes. Washington (DC): National Academies Press;2018;541–88 (http://nationalacademies.org/hmd/Reports/2018/public-health-consequences-of-e-cigarettes.aspx, accessed 20 April 2018).

[105] Pasquereau A, Guignard R, Andler R, Nguyen-Thanh V. Electronic cigarettes, quit attempts and smoking cessation: a 6-month follow-up. Addiction. 2017;112:1620–8.

[106] Wang M, Li W, Wu Y, Lam TH, Chan SS. Electronic cigarette use is not associated with quitting of conventional cigarettes in youth smokers. Pediatric Res. 2017;82:14–8.

[107] Subialka Nowariak EN, Lien RK, Boyle RG, Amato MS, Beebe LA. E-cigarette use among treatment-seeking smokers: moderation of abstinence by use frequency. Addict Behav 2018;77:137–42.

[108] Berry KM, Reynolds LM, Collins JM, Siegel MB, Fetterman JL, Hamburg NM et al. E-cigarette initiation and associated changes in smoking cessation and reduction: the Population

Assessment of Tobacco and Health Study, 2013–2015. Tob Control 2018; doi:10.1136/tobaccocontrol-2017-054108.

[109] Rigotti NA, Chang Y, Tindle HA, Kalkhoran SM, Levy DE, Regan S et al. Association of e-cigarette use with smoking cessation among smokers who plan to quit after a hospitalization. Ann Internal Med. 2018;168(9):613–20.

[110] Zawertailo L, Pavlov D, Ivanova A, Ng G, Baliunas D, Selby P. Concurrent e-cigarette use during tobacco dependence treatment in primary care settings: association with smoking cessation at three and six months. Nicotine Tob Res. 2016;19:183–9.

[111] Manzoli L, Flacco ME, Ferrante M, La Vecchia C, Siliquini R, Ricciardi W et al. Cohort study of electronic cigarette use: effectiveness and safety at 24 months. Tob Control 2017;26:284–92.

[112] Barrington-Trimis JL, Gibson LA, Halpern-Felsher B, Harrell MB, Kong G, Krishnan-Sarin S et al. Type of e-cigarette device used among adolescents and young adults: findings from a pooled analysis of 8 studies of 2,166 vapers. Nicotine Tob Res. 2018;20(2):271–4.

[113] Hummel K, Hoving C, Nagelhout GE, de Vries H, van den Putte B, Candel MJJM et al. Prevalence and reasons for use of electronic cigarettes among smokers: findings from the International Tobacco Control (ITC) Netherlands Survey. Int J Drug Policy. 2014;26(6):601–8.

[114] Attitudes of Europeans towards tobacco and electronic cigarettes (Special Eurobarometer 458). Brussels: European Union; 2017.

[115] Ayers JW, Leas EC, Allem JP, Benton A, Dredze M, Althouse BM et al. Why do people use electronic nicotine delivery systems (electronic cigarettes)? A content analysis of Twitter, 2012–2015. PLoS One. 2017;12(3):e0170702.

[116] Pearson JL, Hitchman SC, Brose LS, Bauld L, Glasser AM, Vilanti AC et al. Recommended core items to assess e-cigarette use in population-based surveys. Tob Control. 2018;27(3):341–6.

[117] Amato MS, Boyle RG, Levy D. How to define e-cigarette prevalence? Finding clues in the use frequency distribution. Tob Control. 2015;25:e24–9.

[118] Shi Y, Pierce J, White M, Vijayaraghavan M, Compton W, Conway K et al. E-cigarette use and smoking reduction or cessation in the 2010/2011 TUS-CPS longitudinal cohort. BMC Public Health. 2016;16(1):1105.

[119] Vickerman K, Carpenter K, Altman T, Nash CM, Zbikowski SM. Use of electronic cigarettes among state tobacco cessation quitline callers. Nicotine Tobacco Res. 2013;15:1787–791.

[120] Al-Delaimy W, Myers M, Leas E, Strong DR, Hofstetter CR. E-Cigarette use in the past and quitting behavior in the future: a population-based study. Am J Public Health. 2015;105:1213–9.

[121] Biener L, Hargraves JL. A longitudinal study of electronic cigarette use among a populationbased sample of adult smokers: association with smoking cessation and motivation to quit. Nicotine Tob Res. 2014;17:127–33.

[122] Pearson J, Stanton C, Cha S, Niaura RS, Luta G, Graham AL. E-Cigarettes and smoking cessation: insights and cautions from a secondary analysis of data from a study of online treatmentseeking smokers. Nicotine Tob Res. 2014;17:1219–27.

[123] Borderud S, Li Y, Burkhalter J, Sheffer CE, Ostroff JS. Electronic cigarette use among patients with cancer: Characteristics of electronic cigarette users and their smoking cessation outcomes. Cancer. 2014;120:3527–35.

[124] Hitchman SC, Brose LS, Brown J, Robson D, McNeill A. Associations between e-cigarette type,

[125] Carpenter M, Heckman B, Wahlquist A, Wagener TL, Goniewicz ML, Gray KM et al. A naturalistic, randomized pilot trial of e-cigarettes: uptake, exposure, and behavioral effects. Cancer Epidemiol Biomarkers Prev. 2017;26:1795–803.

[126] Masiero M, Lucchiari C, Mazzocco K, Veronesi G, Maisonneuve P, Jemos C et al. E-Cigarettes may support smokers with high smoking-related risk awareness to stop smoking in the short run: preliminary results by randomized controlled trial. Nicotine Tob Res. 2018. doi:10.1093/ntr/nty175.

[127] Giovenco DP, Delnevo CD. Prevalence of population smoking cessation by electronic cigarette use status in a national sample of recent smokers. Addict Behav. 2017;76:129–34.

[128] Zhu S, Zhuang Y, Wong S, Cummins SE, Tedeschi GJ. E-cigarette use and associated changes in population smoking cessation: evidence from US current population surveys. BMJ. 2017;358:j3262.

[129] Cho HJ, Dutra LM, Glantz SA. Differences in adolescent e-cigarette and cigarette prevalence in two policy environments: South Korea and the United States. Nicotine Tob Res. 2018;20(8):949–53.

[130] Benmarhnia T, Leas E, Hendrickson E, Trinidad DR, Strong DR, Pierce JP. The potential influence of regulatory environment for e-cigarettes on the effectiveness of e-cigarettes for smoking cessation: different reasons to temper the conclusions from inadequate data. Nicotine Tob Res. 2018;20(5):659.

[131] Yong HH, Hitchman S, Cummings K, Borland R, Gravely SML, McNeill A et al. Does the regulatory environment for e-cigarettes influence the effectiveness of e-cigarettes for smoking cessation? Longitudinal findings from the ITC four country survey. Nicotine Tob Res. 2017;19(11):1268–76.

[132] Yong HH, Borland R, Balmford J, McNeill A, Hitchman S, Driezen P et al. Trends in e-cigarette awareness, trial, and use under the different regulatory environments of Australia and the United Kingdom. Nicotine Tob Res. 2015;17:1203–11.

[133] Yong HH, Borland R, Hitchman SCB, Cummings KM, Gravely SML, McNeill AD et al. Response to letter to the Editor by Benmarhnia T, Leas E, Hendrickson E, Trinidad D, Strong D, Pierce J. The potential influence of regulatory environment for e-cigarettes on the effectiveness of ecigarettes for smoking cessation: different reasons to temper the conclusions from inadequate data. Nicotine Tob Res. 2017;20(5):660–1.

[134] Grace RC, Kivell BM, Laugesen M. Estimating cross-price elasticity of e-cigarettes using a simulated demand procedure. Nicotine Tob Res. 2014;17:592–8.

[135] Johnson MW, Johnson PS, Rass O, Pacek LR. Behavioral economic substitutability of e-cigarettes, tobacco cigarettes, and nicotine gum. J Psychopharmacol. 2017;31:851–60.

[136] Snider SE, Cummings KM, Bickel WK. Behavioral economic substitution between conventional cigarettes and e-cigarettes differs as a function of the frequency of e-cigarette use. Drug AlClinical pharmacology of nicotine in electronic nicotine delivery systems cohol Depend. 2017;177:14–22.

[137] Jamal A, King BA, Neff LJ, Whitmill J, Babb SD, Graffunder CM. Current cigarette smoking

among adults – United States, 2005–2015. Morbid Mortal Wkly Rep. 2016;65:1205–11.

[138] Adult smoking habits in the UK: 2016. Cigarette smoking among adults including the proportion of people who smoke, their demographic breakdowns, changes over time, and e-cigarettes. Lonson: Office for National Statistics; 2017 (https://www.ons.gov.uk/peoplepopulationandcommunity/healthandsocialcare/healthandlifeexpectancies/bulletins/adultsmokinghabitsingreatbritain/2016#toc, accessed October 2018).

[139] Beard E, West R, Michie S, Brown J. Association between electronic cigarette use and changes in quit attempts, success of quit attempts, use of smoking cessation pharmacotherapy, and use of stop smoking services in England: time series analysis of population trends. BMJ 2016;354:i4645.

[140] McRobbie H. Modelling the population health effects of e-cigarettes use: current data can help guide future policy decisions. Nicotine Tob Res. 2016;19:131–2.

[141] Hill A, Camacho OM. A system dynamics modelling approach to assess the impact of launching a new nicotine product on population health outcomes. Regul Toxicol Pharmacol 2017;86:265–78.

[142] Yoong S, Tzelepis F, Wiggers JH, Oldmeadow C, Chai K, Paul CL et al. Prevalence of smoking-proxy electronic inhaling system (SEIS) use and its association with tobacco initiation in youths: a systematic review. Geneva: World Health Organization; 2015 (http://who.int/tobacco/industry/product_regulation/BackgroundPapersENDS2_4November.pdf?ua=1, accessed 3 October 2017).

[143] Leventhal AM, Strong DR, Kirkpatrick MG, Unger JB, Sussman S, Riggs NR et al. Association of electronic cigarette use with initiation of combustible tobacco product smoking in early adolescence. JAMA. 2015;314:700.

[144] Primack B, Soneji S, Stoolmiller M, Fine MJ, Sargent JD. Progression to traditional cigarette smoking after electronic cigarette use among US adolescents and young adults. JAMA Pediatrics. 2015;169:1018–23.

[145] Wills T, Knight R, Sargent J, Gibbons FX, Pagano I, Williams RJ. Longitudinal study of e-cigarette use and onset of cigarette smoking among high school students in Hawaii. Tob Control 2016;26:34–9.

[146] Soneji S, Barrington-Trimis J, Wills T, Leventhal AM, Unger JB, Gibson LA et al. Association between initial use of e-cigarettes and subsequent cigarette smoking among adolescents and young adults. JAMA Pediatrics. 2017;171:788.

[147] Miech R, Patrick M, O'Malley P et al. E-cigarette use as a predictor of cigarette smoking: results from a 1-year follow-up of a national sample of 12th grade students. Tobacco Control 2017. doi:10.1136/tobaccocontrol-2016-053291.

[148] Best C, Haseen F, Currie D, Ozakinci G, MacKintosh AM, Stead M et al. Relationship between trying an electronic cigarette and subsequent cigarette experimentation in Scottish adolescents: a cohort study. Tob Control. 2017. doi:10.1136/tobaccocontrol-2017-053691.

[149] Conner M, Grogan S, Simms-Ellis R, Flett K, Sykes-Muskett B, Cowap L et al. Do electronic cigarettes increase cigarette smoking in UK adolescents? Evidence from a 12-month prospective study. Tob Control. 2017. doi:10.1136/tobaccocontrol-2016-053539.

[150] Kozlowski LT, Warner KE. Adolescents and e-cigarettes: Objects of concern may appear larger

than they are. Drug Alcohol Depend. 2017;174:209–14.

[151] Gartner CE. E-cigarettes and youth smoking: be alert but not alarmed. Tob Control 2017. doi:10.1136/tobaccocontrol-2017-054002.

[152] Schneider S, Diehl K. Vaping as a catalyst for smoking? An initial model on the initiation of electronic cigarette use and the transition to tobacco smoking among adolescents. Nicotine Tob Res 2015;18:647–53.

[153] Lee PN. Appropriate and inappropriate methods for investigating the "gateway" hypothesis, with a review of the evidence linking prior snus use to later cigarette smoking. Harm Reduction J. 2015;12:8.

[154] Jamal A, Gentzke A, Hu SS, Cullen KA, Apelberg BJ, Homa DM et al. Tobacco use among middle and high school students – United States, 2011–2016. Morb Mortal Wkly Rep. 2017;66:597–603.

[155] Dinakar C, O'Connor GT. The health effects of electronic cigarettes. N Engl J Med. 2016;375:1372–81.

[156] Hua M, Talbot P. Potential health effects of electronic cigarettes: a systematic review of case reports. Prev Med Rep. 2016;4:169–78.

[157] Farsalinos KE, Polosa R. Safety evaluation and risk assessment of electronic cigarettes as tobacco cigarette substitutes: a systematic review. Ther Adv Drug Saf. 2014;5:67–86.

[158] Burstyn I. Peering through the mist: systematic review of what the chemistry of contaminants in electronic cigarettes tells us about health risks. BMC Public Health. 2014;14:18.

[159] Pisinger C. A systematic review of health effects of electronic cigarettes. Geneva: World Health Organization; 2017 (http://www.who.int/tobacco/industry/product_regulation/BackgroundPapersENDS3_4November-.pdf, accessed 30 September 2017).

[160] Glasser AM, Collins L, Pearson JL, Abudayyeh H, Niaura RS, Abrams DB et al. Overview of electronic nicotine delivery systems: a systematic review. Am J Prev Med. 2017;52:e33–66.

[161] Stephens WE. Comparing the cancer potencies of emissions from vapourised nicotine products including e-cigarettes with those of tobacco smoke. Tob Control. 2017. doi:10.1136/tobaccocontrol-2017-053808.

[162] Farsalinos KE, Voudris V, Poulas K. E-cigarettes generate high levels of aldehydes only in "dry puff" conditions. Addiction. 2015;110:1352–6.

[163] Benowitz NL, Burbank AD. Cardiovascular toxicity of nicotine: implications for electronic cigarette use. Trends Cardiovasc Med. 2016;26:515–23.

[164] Shields PG, Berman M, Brasky TM, Freudenheim JL, Mathe E, McElroy JP et al. A review of pulmonary toxicity of electronic cigarettes in the context of smoking: a focus on inflammation. Cancer Epidemiol Biomarkers Prev. 2017;26:1175–91.

[165] Levy DT, Borland R, Villanti AC, Niaura R, Yuan Z, Zhang Y et al. The application of a decision-theoretic model to estimate the public health impact of vaporized nicotine product initiation in the United States. Nicotine Tob Res 2016;19:149–59.

[166] Kalkhoran S, Glantz SA. Modeling the health effects of expanding e-cigarette sales in the United States and United Kingdom. JAMA Internal Med. 2015;175:1671.

[167] Levy D, Borland R, Lindblom E, Goniewicz ML, Meza R, Holford TR et al. Potential deaths averted in USA by replacing cigarettes with e-cigarettes. Tob Control 2017. doi:10.1136/

tobaccocontrol-2017-053759.

[168] Soneji S, Sung H, Primack B, Pierce JP, Sargent J. Problematic assessment of the impact of vaporized nicotine product initiation in the United States. Nicotine Tob Res 2016;19:264–5.

[169] Glantz SA. Need for examination of broader range of risks when predicting the effects of new tobacco products. Nicotine Tob Res. 2016;19:266–7.

[170] Levy DT, Borland R, Fong GT, Vilanti AC, Niaura R, Meza R et al. Developing consistent and transparent models of e-cigarette use: reply to Glantz and Soneji et al. Nicotine Tob Res. 2016;19:268–70.

[171] Stratton K, Kwan LY, Eaton DL, editors. Injuries and poisonings. Chapter 14. In: Public health consequences of e-cigarettes. Washington (DC): National Academies Press;2018;541–88 (http://nationalacademies.org/hmd/Reports/2018/public-health-consequences-of-e-cigarettes.aspx, accessed 20 April 2018).

4. 全球烟碱降低策略：科学现状

Geoffrey Ferris Wayne, Portland, OR, USA

Eric Donny, Department of Physiology and Pharmacology, Wake Forest School of Medicine, Winston-Salem (NC), USA

Kurt M. Ribisl, Health Behavior, Gillings School of Global Public Health, University of North Carolina School of Medicine, Chapel Hill (NC), USA

4.1 背 景

2015年，TobReg发布了一份咨询性说明以支持将卷烟中烟碱含量降低至低于发展或维持成瘾所必需水平的政策[1]。烟碱降低的政策可能会最大限度地减少从尝试卷烟使用到依赖的发展、减少使用时间、促进戒烟、降低成瘾吸烟者的吸烟率以及鼓励那些想要吸烟的人过渡到危害较小的产品或其他来源的烟碱，从而对公众健康产生重大影响[1,2]。世界上大约有10亿人吸烟，并且全球卷烟和其他燃烧型烟草制品（包括雪茄和自制卷烟）的消费量继续增加[3]。世界卫生组织《烟草控制框架公约》第9条授权缔约方对包括烟碱在内的烟草制品的成分和释放物进行管制[4,5]。有明显的证据表明将卷烟中烟碱含量降低到非常低的水平可以减少对卷烟的依赖[1,6,7]。

WHO的咨询说明描述了烟碱降低政策的潜在健康结果，同时注意到许多研究项目仍在进行中，包括极低烟碱含量（VLNC）卷烟的临床试验。该咨询说明确定了一些有关可能的健康结果的开放性问题，包括：

- ■ VLNC卷烟在未吸烟的青少年、成年人和非依赖吸烟者中的使用和效果；
- ■ VLNC卷烟在弱势人群（例如中度或重度精神疾病患者和孕妇）中的使用和效果；
- ■ VLNC卷烟与其他形式的烟碱或其他药物的使用和效果；
- ■ 长期使用VLNC卷烟的影响。

自发布咨询说明以来的关键讨论集中在上述开放性问题[8-10]，难以得出有关公开市场上限制性临床试验结果的结论[8,9,11]以及潜在的影响，例如认为VLNC产品的毒性比普通卷烟低[11-13]。该咨询说明未提供有关如何制定烟碱降低政策的详细建议，突出了实施烟碱降低政策的实际挑战[9,14]。

本章更新了关于烟碱降低和使用VLNC产品的快速发展的科学，并解决了有关2015年咨询说明发布以后的重要讨论中提出的烟碱降低政策的问题。自从2013年12月在TobReg第七次会议上提交有关烟碱降低的背景文件并开始讨论烟碱降低以来，已经发表了100多项相关研究。根据缔约方大会第七次会议向世界卫生组织《烟草控制框架公约》提交的FCTC COP 7(14)号决定，公约秘书处和世界卫生组织召集了许多学科的专家召开面对面会议讨论减少烟草致瘾性的措施（2018年5月15~16日，德国柏林）。截至2017年9月，在撰写本章时（2017年9月19日，参见表4.1）各种临床试验仍在进行中，下面讨论这些试验的范围和相关的初步结果。此外，还讨论了FDA推行烟碱降低策略的计划[15]。

表4.1 截至2017年9月19日，正在进行或仍在分析中的极低烟碱含量（VLNC）卷烟的临床试验

首席研究员	临床试验编号	目的	关键特点	样本量	预计完成日期
Cassidy	NCT02587312	确定降低卷烟中烟碱含量如何影响青少年吸烟行为	青少年	90	2019年5月
Cinciripini	NCT02964182	评估卷烟和电子烟中不同烟碱含量的影响	结合电子烟	480	2020年11月
Colby	NCT03194256	确定卷烟中烟碱含量以及电子烟液中烟碱浓度和风味如何影响青少年吸烟者对这些产品的反应和使用	基于实验室；青少年电子烟替代	120	2022年2月
Donny	NCT02301325	评估有/没有经皮给药烟碱的低烟碱卷烟的影响	经皮给药烟碱	240	2017年4月
Donny	NCT03185546	评估VLNC卷烟、具有不同烟碱含量的电子烟和电子烟调味剂对吸烟的影响	使用各种烟碱含量和口味的电子烟	480	2021年9月
Drobes	NCT02796391	确定卷烟中烟碱含量逐渐减少与立即减少以及有针对性的行为治疗相结合对戒烟和中间结果的影响	戒烟	220	2021年4月
Foulds, Evins	NCT01928758	评估逐渐减少卷烟中烟碱对伴有情绪和/或焦虑症的吸烟者的吸烟行为、毒素暴露和精神症状的影响	情感障碍	280	2018年10月
Ganz	NCT01964807	测试可提供广泛浓度的烟碱、微粒和其他心血管毒素的产品以确定与烟草相关的成分如何对心血管疾病产生不良影响	基于实验室；心血管作用	90	2018年9月
Hatsukami	NCT02139930	比较两种减少卷烟中烟碱含量的方法：立即减少还是逐步减少	逐步与立即	1250	2017年3月
Hatsukami	NCT03272685	确定VLNC卷烟在复杂的烟草和烟碱产品市场中的作用	获得各种替代产品	700	2022年12月
Higgins, Heil	NCT02250534	确定弱势女性群体长期暴露于各种烟碱含量卷烟的影响	弱势女性群体	282	2019年10月
Higgins, Sigmon	NCT02250664	确定在阿片类药物滥用者中长时间接触各种烟碱含量卷烟的影响	阿片类药物滥用者	282	2019年10月

续表

首席研究员	临床试验编号	目的	关键特点	样本量	预计完成日期
Klemperer	NCT03060083	通过将吸烟者随机分为改用VLNC卷烟或减少每天吸烟的数量来研究两种策略。所有吸烟者都使用烟碱贴片来帮助他们降低烟碱摄入量	比较烟碱降低与减少卷烟	74	2017年11月
Koffarnus, Bickel	NCT02951143	评估吸烟者如何购买和消费烟碱降低的卷烟	基于实验室;成本	232	2020年7月
Kollins, Mcclernon	NCT02599571	研究注意力不足过动症患者卷烟中不同烟碱水平的影响	注意力缺陷多动障碍	350	2020年6月
Mcclernon	NCT02989038	在18~25岁的低频率、非依赖性吸烟者对抽吸不同烟碱含量卷烟的自给药反应和选择	年轻的成年吸烟者	90	2019年4月
Muscat, Horn	NCT01928719	确定逐步降低卷烟中烟碱含量使社会经济地位低的吸烟者是减少还是消除对烟碱的依赖性	社会经济地位低	400	2018年10月
Oncken	NCT02048852	确定降低卷烟烟碱或薄荷醇含量或两者同时降低对育龄妇女的影响	薄荷醇;育龄妇女	320	2018年12月
Oncken, Dornelas	NCT02592772	观察降低卷烟中烟碱或薄荷醇含量或两者同时降低对男性的影响	薄荷醇;男人性;性别差异	57	2018年12月
Peters	NCT02990455	确定是或不是当前酒精使用障碍的吸烟者是否通过增加酒精摄入量或烟气暴露来应对烟碱降低的卷烟	饮酒	90	2019年11月
Richie	NCT02415270	确定改用烟碱或活性氧或氮物质含量低的烟草制品对吸烟行为以及暴露于烟草烟气和氧化应激的生物标志物的短期影响	暴露与危害生物标志物	70	2017年5月
Rohsenow	NCT01989507	确定VLNC卷烟对当前或过去患有物质使用障碍的吸烟者的影响	物质滥用	250	2019年5月
Rose	NCT02870218	评估吸烟史较短的重度、长期吸烟者和轻度吸烟者（10年内每天吸烟量<10支）使用含0.4 mg、1.0 mg、2.5 mg、5.6 mg或16.9 mg烟碱的卷烟的情况。在为期12周的研究中，参与者可以免费使用含烟碱的电子烟（15 mg/mL）	量效关系;重度和轻度吸烟者;使用电子烟	320	2019年6月
Shiffman	NCT02228824	调查不同烟碱水平的卷烟对非每日吸烟者的影响	非每日吸烟者	312	2017年7月
Strasser	NCT01898507	检查在具有不同烟碱代谢的吸烟者组中吸低烟碱卷烟的影响	烟碱代谢	210	2018年7月
Tidey	NCT02019459	检查将卷烟中烟碱含量降低至非成瘾水平是否可以减少精神分裂症吸烟者吸烟	精神分裂症	80	2018年8月
Tidey	NCT02232737	确定长期暴露于患有各种情感障碍的人的烟碱含量不同的卷烟中的影响	情感障碍	282	2019年10月

资料来源：来自clinical trials.gov和个人通信（EC Donny）的数据。2017年9月19日或之前在clinical trials.gov上发现的被标识为搜索词"低烟碱"和"降低烟碱"的研究，并被分为尚未招募、招募、应邀招募和非主动招募。少于50名参与者的研究或没有解决烟碱降低相对于普通品牌或正常烟碱含量对照影响的设计被省略。该表的先前版本由Donny等发布[16]。按预计完成日期列出

4.2 烟碱降低的个体影响

烟碱降低对个体影响的大多数估计来自随机临床试验，在这些试验中，参与者通常在双盲条件下长期抽吸普通卷烟或VLNC卷烟（参见表4.1）。实验室评估可以相对迅速地评估滥用倾向和补偿性吸烟等问题，这些评估为烟碱降低提供了有用的补充信息。在这两种情况下所用卷烟的烟丝中烟碱含量较低，通常是采用基因工程的结果（例如Quest品牌，美国国家药物滥用研究所提供的Spectrum卷烟）。尽管可以通过其他方法来烟碱降低（例如，20世纪90年代Philip Morris的Next卷烟，采用烟碱提取法），由于成本、口味和可得性等实际问题，对用这些方法制造的卷烟进行的临床研究相对较少。此外，据我们所知，尚未有评估烟碱降低对其他形式的燃烧型烟草（例如自卷烟）影响的研究。尽管很重要，迄今为止的研究结果无法预测现实世界中烟碱降低政策的全部效果。因此，上市后监测（国际烟草控制项目：http://www.itcproject.org/）将是降低烟碱管制行动的重要组成部分。

4.2.1 行为补偿和有害物质暴露

VLNC卷烟释放出的有害物质通常与普通卷烟相似[17]；因此，它们对健康的影响在很大程度上取决于吸烟行为的变化。随机临床试验证实了较早的建议[1,18,19]，即烟碱降低可减少每天抽吸的卷烟数量，而这与普通品牌卷烟和具有正常烟碱含量的参比卷烟的数量[6,20]④有关。在使用6周后，参与者通常报告相比对照组少吸烟25%~40%[6,19-21]。

人们普遍担心，吸烟者会通过更强烈的抽吸来弥补烟碱的缺乏，但这一观点并没有得到足够的证据支持。当烟丝的烟碱含量仅适度降低时[20,22]1，并且通过通风降低卷烟的烟碱释放量但烟丝烟碱保持不变（通常称为"轻型"卷烟）时，通常会观察到补偿性抽吸[23]。补偿性抽吸似乎是由每克烟草≤2.4 mg烟碱的卷烟（如10~20 mg/g的典型范围相比）引起的[6,18,19,22]，这可能是因为烟气的有害成分限制了极端的补偿抽吸行为，在烟碱含量大幅下降的情况下更温和的补偿抽吸行为对维持烟碱暴露是无效的。相反，许多研究发现吸烟会减少有害物质暴露。例如，参加者转换到VLNC卷烟呼出气体中的CO、每支烟的总抽吸容量、口腔烟气成分暴露量[24]和有害成分暴露生物标志物相似或略低[6,18,19,21,22,25-27]。

4.2.2 建立或维持烟碱成瘾的阈值

2015年世界卫生组织烟草制品管制研究小组咨询说明中提出的一个重要问

---

④ EC Donny，未发表的观察。

题是建立或维持烟碱成瘾的阈值。尚不清楚对不同的成瘾程度和不同的吸烟人群是否有一个单一的阈值；但是，数据表明烟碱应减少到不超过2.4 mg烟碱/g烟草，将烟碱含量减少到0.4 mg烟碱/g烟草可能会有更多好处。尽管大多数吸烟者无法区分2.4 mg烟碱/g烟草和0.4 mg烟碱/g烟草的VLNC卷烟，但有些人可以，这表明将烟碱含量降低到2.4 mg烟碱/g烟草以下可能会影响更多的吸烟者[28]。此外，Donny等[6]发现，尽管在≤2.4 mg烟碱/g烟草水平每天吸烟和戒烟后自述的渴望的数量显著减少，但烟碱依赖的综合措施仅在1.3 mg烟碱/g烟草（一项测试）和0.4 mg烟碱/g烟草时显著降低（多项测试），并且在随访期间戒烟尝试的次数仅在0.4 mg烟碱/g烟草时显著增加。有趣的是，对大鼠的研究还表明，需要减少85%~90%或更多（相对于大多数大鼠中增强依赖性的剂量而言）才能可靠地减少静脉滴注自给药[29]。

降低烟碱含量至≤0.4 mg烟碱/g烟草是否会限制青少年吸烟者成瘾尚不清楚。尽管一些研究表明烟碱对青春期大鼠具有更大的强化作用[30,31]，但其他研究表明，动物学会烟碱自我给药所需的阈值与维持行为所需的剂量相似或更高[32-34]。后者表明，基于成年吸烟者证据的产品标准也有望限制未接触烟草的年轻人的吸烟。此外，Donny等[6]的数据的二次分析表明，与25岁以上的成年人相比，烟碱降低会使18~24岁吸烟者的滥用指数（吸烟满意度、心理奖励和呼吸道感觉的享受）更快地下降[35]。青少年吸烟者对使用不同烟碱含量的单支卷烟的反应研究[36]发现减低烟碱的卷烟对主观的积极影响较小，对戒断或暴露无明显影响。截至2017年9月，正在对青少年和年轻的成年吸烟者进行其他研究（见表4.1）。但是，青少年（和其他人群）对烟碱更敏感的可能是考虑烟碱浓度远低于2.4 mg烟碱/g烟草的产品标准的原因之一。

4.2.3 使用VLNC卷烟后的烟草戒断

减少每天吸烟量、烟碱暴露和烟碱依赖性是否会增加戒烟尚不清楚。对打算戒烟的参与者的研究表明，使用VLNC卷烟可以提高戒烟率[19,37,38]，而对目前不愿意戒烟的参与者进行的研究发现吸烟几周甚至几个月后仍会持续吸烟[6,22,26,39]。但是，大多数研究并非旨在解决戒烟问题，也不足以确定戒烟结果，只是在相对较短的时间内对行为进行评估，不包括为参与者提供烟碱的替代来源，使用具有正常烟碱含量的市售卷烟。在这方面有两项研究很重要。在一项针对1250名不愿戒烟的吸烟者的10点临床试验中，被随机分配接受VLNC卷烟20周的参与者自述的戒烟天数（10.99）比分配给烟碱含量正常的吸烟者的戒烟天数多（3.1；P <0.0001），并且在试验结束时更有可能戒断（生化确认）[20]。在一次探索性试验中，Hatsukami等[40]发现VLNC卷烟增加了对烟碱替代产品的使用，包括ENDS、雪茄和NRT，使用的不燃烧产品越多，戒烟天数就越多。几位研究人员已经讨论了烟碱

降低和其他产品的潜在交互作用[41,42]，这是FDA宣布的主题[15]（下文将进一步讨论），也是2017年9月进行的多项临床试验的目标（见表4.1）。

4.2.4 临床试验中的不依从性

很明显自2015年世界卫生组织TobReg咨询说明以来，尽管有相反的指示并提供了免费的研究卷烟，但大多数VLNC卷烟研究的参与者仍在使用非研究卷烟。例如，在Donny等[6]的研究中，有75%~80%的参与者随机分配到VLNC研究卷烟，其中至少有部分不符合要求[43]。年轻且严重依赖的吸烟者报告说对VLNC卷烟不满意，更有可能是不依从。大多数吸烟者明显减少了对烟碱的接触，这表明他们已用VLNC卷烟代替了许多日常卷烟，但仍使用一些常规烟碱卷烟。最常报告的不依从情况是每天的第一支卷烟，这突出了不依从是由于烟碱依赖[5]。为了解决这一局限性，已使用了根据依从性概率对数据进行加权的统计方法。结果表明与未加权数据相比，对每天卷烟数量和依赖性的影响类似[44][6]。其他有关激励依从性的研究导致不依从的现象有所减少（40%~50%）（基于尿液生物标志物），并重复了VLNC卷烟对吸烟行为和依赖性的影响，除了轻度、短暂戒断症状外，几乎没有负面影响的证据[1,2]。然而，不依从可能会减弱烟碱降低的积极影响（例如戒断[45]）和消极影响（例如戒断症状）。尽管提供了免费的VLNC卷烟和依从激励措施，但仍存在不依从的情况，一些吸烟者可能寻求烟碱的其他来源，如黑市、囤积、产品篡改或者使用非卷烟产品。

4.2.5 不利的健康影响和弱势人群

对烟碱降低的潜在非预期后果分析表明，对参与者的健康影响很小。使用VLNC卷烟呼出气CO不变或减少[6,18,19,21][7]。VLNC卷烟的使用可能会暂时增加戒断症状[19,46][1]和注意力缺乏[47,48]相关的非严重不良事件的发生，尽管在第一周左右观察到的缺乏情况很少[8]。抽吸常规卷烟会抑制体重增加，增加了烟碱降低对肥胖吸烟者产生独特不良影响的可能性[49,50]。值得注意的是使用普通卷烟可能会模糊意外后果。因此，在临床试验中VLNC卷烟使用与体重增加之间的关联因参与者的不依从而模糊[49]。

VLNC卷烟的大多数临床试验都涉及相对健康的日常吸烟者，通常排除了患有严重精神疾病或不是每天吸烟的人。烟碱降低的正面和/或负面影响可能因亚

---

⑤ EC Donny, 未发表的观察。
⑥ DK Hatsukami, 个人交流。
⑦ DK Hatsukami, 个人交流。
⑧ Ribisl KM, Hatsukami D, Johnson, Huang J, Williams R, Donny E.Strategies to prevent illicit trade for very low nicotine cigarettes（未公布信息）。

群而异，截至2017年9月，针对可能处于危险之中的亚群，一些临床试验正在进行或尚未发表（表4.1）。迄今为止，关于吸烟对不同亚群潜在影响的最佳数据来自先前试验的次要分析（通常是引用资料6）和实验室研究。Bandiera等[51]报道在Benowitz逐步烟碱降低的试验中，由24个高度依赖的吸烟者组成的相对较小的亚组中每天抽烟的数量增加了[52]；但是，Donny等[6]在突然减少的试验中对高度依赖的吸烟者进行的分析表明，在随机分配给VLNC卷烟后，依赖性对每天吸烟的卷烟数量或其他可能的补偿性吸烟措施没有影响。此外，由于烟碱降低，高度依赖吸烟者的依赖性下降最大。如上所述，研究通常排除非每日吸烟者；然而，一项针对非每日吸烟者的完整试验显示，由于减少了烟碱，每天抽吸的卷烟数量显著减少（＞50%）[53]。

有酒精[54]或大麻使用史[55]的受试者吸烟减少的情况与无此类使用历史的吸烟者相似，当随机分配给VLNC卷烟时，没有增加酒精或大麻使用，尽管烟碱戒断可能会增加一些人喝酒的动机[45,56]。Tidey等[57]发现，在基线检查时有抑郁症状并不能减轻烟碱降低对吸烟量或烟碱依赖程度的影响，并且将基线时具有严重抑郁症状的吸烟者随机分配给VLNC卷烟会导致抑郁水平低于在试验结束前的对照组。这些发现与广泛研究烟碱降低对患有情感障碍的吸烟者的潜在影响的结果一致[58]。一项针对三个弱势人群（患有情感障碍、对阿片类药物依赖和社会经济情况弱势的育龄女性）吸烟者的集中实验室研究发现，以剂量依赖的方式降低烟碱水平在这些人群中同样会降低卷烟的强化作用[59]。截至2017年9月，正在进行的其他试验可能会提供更多信息，说明在弱势人群中长期使用VLNC卷烟可能会带来意想不到的后果；但是，鉴于吸烟对这些人群不成比例的危害，必须权衡任何负面影响和潜在利益[60-64]。

4.2.6 小结

- 在临床试验中，使用VLNC卷烟代替普通卷烟可以减少每天吸烟的数量，并减少有害物质的暴露。
- 适度降低烟丝中烟碱含量但不使用VLNC卷烟则发现补偿性抽吸。
- 在烟碱含量≤2.4 mg/g的情况下，吸烟量减少。在0.4 mg/g的水平上，可靠地观察到了较少的烟碱依赖性和更多的戒烟尝试，尽管这种观察必须在长期研究中得到证实。将烟碱降低至≤0.4 mg/g可能有益于目前和潜在吸烟者。
- 大多数研究并非旨在解决戒烟问题。但是，大多数证据表明，即使在不打算戒烟的吸烟者中使用VLNC卷烟也可能会增加戒烟。
- 尽管数据有限，但减少卷烟中烟碱含量可能会减少青少年吸烟。
- VLNC卷烟的使用增加了替代烟碱产品（如ENDS、雪茄、小雪茄和NRT）

代替常规卷烟的使用。值得考虑如何将烟碱降低应用于除普通卷烟之外的烟碱产品。
- 临床试验中值得注意的不依从现象表明，烟碱降低使一些吸烟者可能会寻求其他烟碱来源。
- 没有发现使用VLNC卷烟代替普通卷烟会对健康造成重大不良影响。
- 烟碱含量降低同样会降低普通卷烟在一些潜在的弱势人群中的强化作用，这些人群包括患有情感障碍的吸烟者、使用酒精和其他物质的人群以及处于弱势社会经济情况的育龄女性。

4.3　烟碱降低对人群的影响

虽然临床试验提供了VLNC卷烟对个体行为和健康的影响，当使用其代替普通卷烟时，估计烟碱降低政策对公众健康的影响需要在市场条件下向更广泛的人群预测这些发现[9]，因为这可能与临床试验有很大不同，并且会受到行业促销活动的影响[9,11]。例如，观察性研究表明，与临床试验预测的相比使用NRT后卷烟消费量的减少较小，可能是因为在实际情况下使用的产品不足[65]。在临床环境中，对药物的依从性可能比在真实情况下更好，而在VLNC卷烟的临床试验中，情况可能相反，因为在试验之外常规烟碱烟草制品商业可用性[44]。

在缺乏基于人群的烟碱降低证据的情况下，世界卫生组织的咨询说明得出结论，来自临床研究和实验动物研究的数据强有力地表明降低了人群水平的风险[1,2]。但是，关于非法销售的潜在作用、烟草公司或吸烟者操纵产品的可能性、烟碱降低政策对健康观念的影响、其他高毒燃烧型烟草制品（例如雪茄烟或自卷烟）的潜在使用作为VLNC卷烟的替代品或与VLNC卷烟组合使用等问题仍然存在，但缺乏阻止使用VLNC卷烟的政策措施和烟草业市场营销的影响[1]，烟草行业营销促进了VLNC卷烟的使用，以此作为应对其他（常规）产品进一步限制的计划的一部分。在本章，我们回顾了当前有关烟碱降低可能对人群造成的反应和健康影响的证据。

4.3.1　VLNC卷烟对普通卷烟的替代

普通卷烟作为行为强化剂的功效支持了普通卷烟的流行[66]。虽然系统暴露于烟碱的情况相似，但是普通卷烟产生的烟碱被吸收的速度比其他含烟碱的产品快（尽管电子烟等替代产品使用的技术发展迅速）[66,67]。卷烟还提供了与烟碱释放相关的丰富的行为和感官刺激网络，进一步促进了吸烟的强烈条件奖赏效用[1,68,69]和VLNC卷烟强化行为的能力[1,70,71]。吸烟者消费普通卷烟时对价格上涨的敏感性相应降低[72-75]，表明其滥用倾向要大于其他形式的烟草或烟碱产品，包括口含烟[76,77]、小雪茄[75,78]、大雪茄[75]、散装烟草[75,79]、烟碱口香糖[77,80]和电子烟[80-83]。

大量的研究表明，常规烟碱卷烟的滥用风险很高，在促进当前使用者戒烟、阻止潜在使用者吸烟并减少持续使用者的烟草使用[72,74,84,85]方面增加了成本。低收入吸烟者对卷烟价格的敏感性最大[86,87]，并受到包括烟草控制和公众教育运动在内的环境因素的影响[88]。高值或"优质"卷烟品牌的吸烟者戒烟的可能性较小，这表明营销和品牌-消费者关系的重要性[89]。风味和可接受性等主观特征也在产品使用中起作用[90-92]，而感知的负面特征（例如口含烟或烟碱口香糖）超过了潜在的强化作用，导致对某些烟草替代品的需求有限，无论价格如何[77]。

VLNC卷烟至少可以部分替代普通卷烟[1,93-95]，实验室研究表明它比烟碱口香糖强化作用更强[96]。使用VLNC卷烟代替普通卷烟会减少卷烟需求[6,94]。由于普通卷烟的价格断点（不再维持购买的点）高于VLNC卷烟，如果VLNC卷烟替代普通卷烟，吸烟者将在相对较低的价格戒烟，并且一部分吸烟者对VLNC卷烟没有需求，无论价格如何[95]。这与多达一半使用VLNC卷烟的参与者的自述一致，如果只有这些卷烟可供购买，他们将在1年内停止吸烟[94]。薄荷卷烟的吸烟者报告说有类似的戒烟意图，因为他们可能会面临薄荷卷烟禁令[97]。对成本的这种敏感性表明，烟碱降低政策可以提高其他烟草控制方法的有效性，例如提价、无烟环境和媒体宣传。相反，如果减少卷烟的烟碱含量，则可以使用市场营销和其他社会方法来支持持续的需求。

如临床试验（第4.2.4节）所述，许多VLNC卷烟的吸烟者在研究用烟支数量之外补充了他们的卷烟消费量。在新西兰的一个小型预试验中，基于烟碱含量的巨大价格差异促使吸烟者减少了普通卷烟的数量，但并不能使其完全停止使用普通卷烟，而是用VLNC卷烟代替减少的数量[93]。这表明，在既有VLNC卷烟又有普通卷烟的市场中，即使价格存在明显差异，大多数经常吸烟的吸烟者也不大可能只使用VLNC卷烟[94,95]。

截至2017年9月，针对青少年的临床试验正在进行，但关于年轻人和未吸烟者的数据很少。同时，价格和其他滥用倾向研究的意义在于，与普通烟碱卷烟相比，VLNC卷烟不太可能鼓励从尝试使用到常规使用的过渡。烟碱降低的可能结果（尽管尚未证实）是可以阻止部分（可能是很大比例）儿童和青少年长期使用毒性最大形式的烟碱产品（即燃烧型产品）。相反，促销活动可以直接或间接地鼓励年轻人购买VLNC产品，让他们相信从健康角度来看，VLNC产品是可以接受的。

4.3.2　VLNC卷烟对其他烟草制品的替代

烟草制品在价格、可获得性、市场营销、社会可接受性和被认知的风险，以及烟碱的传递、感官和非药理学特征方面差异很大。这些环境因素和产品因素之间的平衡会导致烟草制品使用中的人群发生重大变化。例如，在许多国家进行的研究发现，由于价格和较低的风险认知等原因[98-100]，卷烟从生产型向自卷型转变，即使自卷产品在主观衡量指标上较差[79-100]。同时使用几种烟草制品越来越普遍[101,102]，包括与烟草同时使用电子烟和其他电子烟碱传输系统，例如HTP[103,104]。

尽管其他燃烧型烟草制品的价格比卷烟的价格更敏感，但大多数都是普通卷烟的有力替代品[75]。当卷烟价格升至小雪茄时一些吸烟者仍维持了烟草消费[78,105]，而对其他替代品（ENDS、口含烟、大雪茄、自卷烟）的兴趣也在增加[106]。当展示这些产品及其包装的图像时，大部分当前吸烟者无法准确地区分普通卷烟、小雪茄、细支雪茄、自卷烟和大雪茄，这表明在相似的产品中有替代的可能性[107]。烟草业在制造和/或销售中可能会夸大这种看法，其可能会模仿或强调燃烧型产品或替代品（例如，电子烟碱传输系统和加热不燃烧型产品）在外观或功能上的相似性。

替代品的出现可以确定多个产品之间的相互作用。例如，在2016年的一项行为研究中，一旦细支雪茄这个选择被移除，口含烟成为卷烟的重要替代品，参与者在可以购买细支雪茄时不太可能购买其他产品来破坏研究方案[108]。关于无烟烟草（湿鼻烟、口含烟、嚼烟）和卷烟使用者的替代模式的调查结果是混杂的[75,92,109-112]：吸烟者,特别是美国的吸烟者,对无烟产品的接受程度不佳[90,92,113-115]。在瑞典，尽管降低口含烟税促使许多男性吸烟者改用口含烟，但是对女性吸烟者没有达到同样的程度，这说明了社会和人口因素的重要作用[114,116]。

ENDS是普通卷烟的潜在替代品[75,108,117-120]，尽管ENDS是由各种类型的产品组成的（请参阅第3章），但至少有一些产品比NRT更有效地减少了常规卷烟的使用[119,121-123]。部分原因是主观和感觉运动特性是相似的[69]。据报道，尽管ENDS被认为不如普通卷烟令人满意[42]，但它比NRT更好地被接受，提供更多的满意度并可以更好地减少吸烟者的渴望、负面影响和压力[122,124]。ENDS仍然对价格高度敏感[81,118]，在45个国家/地区发现其价格比燃烧型产品贵[125]。与仅抽吸卷烟的人相比，同时使用卷烟和电子烟的使用者在短期内能更成功减少卷烟消费量[126]。人们普遍认为ENDS带来的风险低于普通卷烟，尽管有些人误认为ENDS与卷烟相比具有相等或更大的健康风险[127]。

电子烟碱传输系统的成年人使用者主要是当前吸烟者和既往吸烟者[104,128-137]。对年轻人的研究结果喜忧参半，一些研究表明不吸烟者有很高的兴趣和/或尝试使用[133,138,139]，另一些研究表明当前吸烟者同时使用ENDS和其他烟草制品[130-133,140-145]。一些国家报告了年轻人中ENDS的使用急剧增加。2018年，美国高中生的当前使用率超过20%[146]。ENDS使用与吸烟之间的联系引起了人们的关注，即ENDS可能会成为使用其他烟草制品，特别是普通卷烟的途径，尤其是在青少年和年轻人中[137,147-154]，尽管混杂因素使研究结果难以解释[155-159]。但是，最近在美国，人们对"青少年使用电子烟的流行"[160]以及电子烟可能鼓励年轻人抽吸常规卷烟产品的担忧进一步受到关注[161]。只要普通烟碱卷烟普遍存在,那么ENDS就不太可能在那些想要或需要戒烟的人中取代卷烟。但是,燃烧型烟草制品和其他烟草制品中烟碱含量的降低可能会增加对ENDS的使用[40]。

4.3.3 黑市

如果合法的燃烧型烟草制品中的烟碱含量很低,则正常烟碱含量的卷烟可能会被非法交易。由于尚无权限限制使用常规烟碱卷烟,因此非法贸易的信息来自其他已被禁止的或已加税的产品,消费者一直在寻找价格便宜、税率较低的产品。

非法贸易可以在零售点、街头小贩或通过互联网进行。在某些区域禁止某些烟草制品(例如,加拿大的薄荷醇和其他风味卷烟以及美国的丁香和其他风味卷烟)并未形成一个重要的黑市,并且大多数非法贸易都是为了避免卷烟消费税。美国国家研究委员会估计,2007年非法销售占全球卷烟市场总量的8.5%~21%,占84个国家/地区卷烟消费量的11.6%[162]。非法贸易的数量取决于各种因素,其中包括价格(或税收)差异的幅度,获得产品的难易程度和税收管理的效率⑨。在具有强大税收管理功能的司法管辖区中,非法贸易不多。这些管辖区的做法包括强大的"跟踪和追溯"系统,以跟踪产品从制造商到分销商和零售商的过程,以及强大的海关控制和严格的税收。尽管在许多国家使用热敏标签税票,但在美国加利福尼亚和马萨诸塞州使用加密税票,因为更耐用,并且可以包含有关供应链的编码信息[163]。许多国家没有强大的可最大限度地减少非法贸易的追踪系统,应建立这些系统,以尽量减少正常烟碱含量卷烟的非法贸易。

减少非法贸易的另一种手段是执行强有力的政策,包括罚款结构、检查可能性以及为监督和执法提供足够的人员。为了减少产生黑市的可能性,政策可以涉及制造、销售、购买、使用和/或持有正常烟碱含量的违禁产品。应禁止制造商制造生产用于国内销售的常规烟碱含量卷烟,应禁止进口商进口此类产品。对于许多其他非法产品,应对产品的制造商和销售商(例如零售商和互联网供应商)施加最严厉的惩罚和最严格的执法,而不是对拥有这些产品的吸烟者。

就像税务机构通过参观制造工厂和零售商来检查合规性一样,还应制定一项检查方案,包括对采购进行检查,以确保具有常规烟碱的产品不会在零售供应链中流通。还需要制定限制互联网供应商销售的政策。在美国,对未成年人的高价卷烟销售和违反税收政策导致对付款方式(例如Visa、Mastercard)和运输方式(UPS、FedEx)的新限制,从而大大减少了互联网卷烟供应商的数量和这些网站访问量[164]。应该制定类似的政策来限制向低烟碱产品国家的销售。例如,澳大

---

⑨ Ribisl KM, Hatsukami D, Johnson M, Huang J, Williams R, Donny E. Strategies to prevent illicit trade for very low nicotine cigarettes(未公布信息)。

利亚的烟草消费税是世界上最高的，但已经成功封锁了几乎所有销售普通卷烟的网站。此外，执法机构应进行例行测试购买，以确保互联网销售商不在受其限制的国家/地区销售常规烟碱产品。

减少非法贸易还可以通过广泛提供吸引成年消费者的替代产品。如前所述，这些产品可能包括含有烟碱的ENDS或口含烟。如果像某些国家/地区那样禁止或限制这些产品，那么黑市上对卷烟的需求可能会更大。

4.3.4 VLNC卷烟的操纵

VLNC卷烟的研究已经通过实验卷烟的方式开展，这些实验卷烟的化学和物理特性与商业卷烟的化学和物理特性足够相似，被认为可以用于行为研究[17]。实验产品不一定模仿复杂的产品设计，也不受制于用于增加商业卷烟和其他烟草制品吸引力的营销策略[89,165]。在公开市场上，品牌塑造可能会改变对VLNC卷烟的期望或看法，从而产生后续的行为影响[89]。一些公司可能会使用大量促销手段，既使用VLNC卷烟来转移人们对减少吸烟的循证措施的注意力，又以此作为维持当前产品使用的手段。还可以操纵卷烟构造的物理或化学参数来改变VLNC卷烟的配方[1]。

大量研究都探讨了烟草中非烟碱成分的行为影响，以确定它们是否增加了烟碱的滥用或交互作用。一些实验动物自身给药的研究表明，比单独使用烟碱具有更大的强化作用[166-169]，而另一些研究则没有发现其他作用或协同作用[170-172]。特定的烟草化合物的作用及其与烟碱的相互作用仍然知之甚少[1,7,173]。已证明除烟草之外，降烟碱、去甲哈尔满和乙醛也可进行自我给药[1,174-176]，但是在剂量远高于卷烟烟气的情况下[7]。单胺抑制剂可增加大鼠低剂量烟碱的自我给药，如果进一步降低烟碱的话强化作用就很小[177]。薄荷醇似乎可以增强烟碱的强化机制，从而即使在烟碱水平降低的情况下也可以促进烟碱的消费和抽吸[178,179]，这在VLNC卷烟中并未得到证实。也进行了关于烟碱类似物的开发及其作为替代品潜力的研究[180]。在VLNC卷烟中低剂量烟碱的背景下，这些成分和其他非烟碱成分可能在行为强化的决定因素中发挥更大的作用[171,181]。

烟碱降低政策的另一种可能的反应是吸烟者操纵产品[2]。在马来西亚，禁止在ENDS商店中销售含烟碱的电子烟液不仅导致黑市供应含烟碱的电子烟液，而且导致大量的自制含烟碱的电子烟液[182]。尚不知道在VLNC卷烟中添加烟碱或其他直接由消费者操纵的产品是否会产生吸引人的可接受的替代方法。尚未进行其他类型的产品操作包括改变质子化烟碱或气溶胶颗粒的尺寸分布的研究，这些决定了烟碱和其他成分的沉积和吸收[1]。

4.3.5 关于VLNC卷烟和烟碱降低的观念和态度

许多吸烟者对烟碱的观念不准确。许多研究中的大多数参与者将烟碱误认为是癌症和与吸烟相关发病率的主要原因[1,12,183]。关于使用VLNC卷烟的观念反映了这些误解，因为吸烟者认为它们比普通卷烟的危害要小得多[184,185]，与基于主观等级的预期相比，潜在地增强了对这些卷烟的兴趣和使用范围[185,186]。尽管烟碱是致癌物烟草特有亚硝胺（TSNA）的前体，烟碱降低可能会降低卷烟烟气中TSNA的含量，但没有证据表明VLNC卷烟的有害性更小。健康信息传播策略是必要的，教育吸烟者产品的相对有害性和危害。不准确的风险信念并不局限于VLNC卷烟，因为相当大比例的吸烟者无法准确评估与一系列烟草产品类别相关的风险[187,188]。针对烟碱的政策可能会进一步混淆烟碱的相对危害，这可能导致人们对NRT、ENDS或其他替代烟碱输送系统的负面看法，而不是对有害性更大的烟草制品。明确传达烟碱降低政策的目的是减少对最有害（燃烧型）烟草制品的依赖，因此对烟碱在成瘾中作用的教育至关重要。FDA[15]的公告确定了烟草制品的风险序列，是健康信息的一个示例，该信息对使用烟碱的作用与危害很敏感。

对烟碱含量的期望会影响吸烟者对常规卷烟和VLNC卷烟的主观反应[189,190]，包括通过功能磁共振成像[191,192]测得的脑岛对渴望和学习的反应。对VLNC卷烟的感官反应，特别是味道和强度等级，在观念与每日卷烟消费之间适度关联，具有较高的主观评级（在强度、味道、粗糙和柔和程度上），以及减少危害的观念支持消费量增加[185]。期望也会影响品牌选择和吸烟行为；例如，对ENDS使用者的积极期望与戒烟的可能性更大，但有意戒除ENDS的可能性较小[193]。

2015年世界卫生组织咨询说明报告了公众对烟碱降低的大力支持[1,194]，并且在新西兰和美国的基于人群的调查中证明了这种支持[186,195]，这可能反映出人们对与烟碱有关的疾病风险的看法[186]。两项调查都发现大约80%的人支持，主要是在吸烟者中，在不吸烟者和各种少数族裔人群中也是如此[186,195]。相比于监管烟碱，对薄荷醇禁令得到的支持要少得多[179,186]。试验调查与人口调查的不同之处在于提供了期望的经验基础。在一项对新西兰吸烟者首次使用VLNC卷烟的访谈研究中，吸烟者对强制性烟碱降低的兴趣降低，但支持其以比普通卷烟便宜得多的价格进行销售[196]。然而，在美国的一项试验中，吸烟者使用VLNC卷烟达6周，其实验卷烟中烟碱含量低或非常低，得出的结论是吸烟者支持烟碱含量降低的人数是反对烟碱含量降低的两倍[197]。

4.3.6 小结

■ VLNC卷烟是普通卷烟的部分但不完全的替代品。如果当前的高烟碱卷烟不再合法销售，则当前吸烟者的一部分可能会放弃吸烟，而不是改用

VLNC卷烟。大多数继续吸烟的吸烟者不太可能只使用VLNC卷烟。
- 各国应建立适当的机制以确保强有力的税收管理，并考虑采取其他执法措施来限制非法贸易，限制向需要低烟碱含量卷烟的国家销售普通烟碱含量的卷烟，并限制销售普通烟碱含量的卷烟的网站。
- 截至2017年9月，有关青少年对低烟碱产品的反应的研究正在进行中。预计烟碱的减少将阻止大量尝试使用燃烧型烟草制品的儿童和青少年成为这些产品的长期使用者。
- 燃烧型烟草制品的替代品很常见。烟碱降低政策可能必须应用于所有燃烧型烟草制品，而不是仅应用于普通卷烟，以确保将人群的使用从最具毒性和令人上瘾的产品中转移出来。
- 降低已燃烧烟草制品的烟碱含量可能会增加ENDS和HTP的替代，而ENDS和HTP则由制造商作为降低风险的产品和常规卷烟更安全的替代品在市场上销售和推广。有必要进行的独立研究，以证实企业研究和声明、适当的健康信息传播、对这些产品有效的监测和监管。
- 强有力的政策和执法体系可以帮助减少非法卷烟的供应。尽管证据有限，但减少需求的政策（例如治疗服务、替代非燃烧产品）也可能减少非法贸易。
- 吸烟者认为VLNC卷烟的危害性不如普通卷烟，这可能支持对VLNC卷烟的兴趣和使用率高于主观评分的预期。健康信息传播可以解决对风险的误解。
- 制造商或吸烟者可以控制卷烟结构的物理或化学参数，以改变VLNC卷烟的配方和增强效果，而且效果可能难以预料。
- 基于人群的调查和试验均报告了对烟碱降低的支持。
- 总的来说，研究表明，作为一项全面措施的一部分，并采取适当的保障措施，烟碱降低的政策可能是减少人群吸烟率的有效方法。

4.4 烟碱降低的监管方法

烟碱降低政策大部分关注的不是基于使用VLNC卷烟的个人影响的证据，而是基于成功实施该政策的困难[14]和实施的潜在意外后果，例如在黑市上持续供应有毒的烟碱和/或未能充分支持毒性较小的烟碱形态来替代卷烟[8,10,11,13]。这些实际挑战很难在监管行动之前进行建模[11,14]。一种建议的解决方案是监测一个或多个早期采用烟碱降低的社区的实施效果[11,198]。新西兰被认为是烟碱降低的潜在理想地点，部分原因是公众的大力支持[194]，持续讨论烟碱降低的政策[95,198,199]和限制非法销售可能的地理情况[198]。2017年7月，FDA宣布烟碱降低是一项旨在减少吸烟对健康的影响的多年计划的一部分[15]。在本节中，我们将考虑烟碱降低的可行性、各种监管方法、实施中的潜在挑战以及反对烟碱降低的更广泛的论据。

4.4.1 烟碱降低的可行性和潜在挑战

VLNC卷烟的商业化生产可以满足每克烟草0.4 mg烟碱的产品标准，而不会显著增加产品的有害性，这一点已得到明确证明[1,17,198]。外观和烟碱含量近似于普通卷烟的其他燃烧型烟草制品（例如小雪茄和自卷烟）可能是VLNC卷烟的替代品，这些烟草制品与美国的普通卷烟差不多，降低烟碱的产品标准应扩大到所有燃烧型烟草制品。如第4.3.4节所述，制造商为响应烟碱降低政策而对VLNC产品进行的更改可能会导致产品中含有较高药理学相关化合物，从而可能影响其性能。必须对产品进行监控以确保成功实施[1,93]，并验证非烟碱化合物（例如单胺氧化酶抑制剂和烟碱类似物）不会改变VLNC卷烟的强化阈值[177]。

正如世界卫生组织咨询说明[1]和随后的讨论[8,10]所指出的，如果没有常规卷烟吸烟者，尤其是烟碱依赖吸烟者的支持，烟碱降低的政策就不可能成功。虽然在人口调查中已发现了公众的广泛接受度（第4.3.5节），但有些吸烟者会考虑权衡和不可接受该项政策[200]。告知公众烟碱降低是一项长期战略的一部分，以逐步消除对有害性最大的烟草制品的依赖，并支持提供有害性较小的替代品，包括获得戒烟药物和行为治疗，将有必要维持对吸烟者的支持，并在政策制定前为吸烟者从普通卷烟转抽其他产品提供时间[1,198]。

鉴于烟碱是所有成功烟草制品的核心特征，因此烟碱降低的政策和法律挑战将是严峻的，烟草行业为破坏和错误使用这种政策所做的努力也将如此[9,14]。应该考虑的一种策略是禁止制造和销售含有常规烟碱的卷烟，而不将拥有或使用常规烟碱卷烟定为犯罪。这将使合规的负担从个人身上转移出来，持续的消费者需求可以通过教育、治疗服务和其他法律选择来缓解。尽管正在考虑烟碱降低的国家将认识到税收的潜在威胁，但人们有望在医疗保健方面节省大量费用。归根结底，净经济影响将针对每个国家进行评估，因为这将是政策可行性的一个重要决定因素，即使对公共卫生影响是明显的。

4.4.2 成功实施烟碱降低政策的先决条件

WHO咨询说明[1]列出了成功实施烟碱降低政策所必需的一些前提条件：
- 全面管制所有含烟碱和烟草的产品；
- 其他全面的烟草控制（增加税收、禁止吸烟、图形警示标签、素包装）；
- 继续向一般人群和卫生专业人员传达吸烟的健康风险；
- 提供可行的治疗方法和烟碱替代形式，以减少吸烟依赖者的戒断症状；
- 市场监管和产品检测能力。

认为烟碱是卷烟的主要有害成分的不正确观念可能会降低当前吸烟者戒烟的紧迫感[12,183]。尽管这种意想不到的后果不太可能超过烟碱降低的潜在好处，但应通过明确的信息传播加以解决[185,198]。需要进一步的研究和持续评估，以确保在不否认烟碱有害的情况下传达适当的信息（例如，说明人群之间的差异）[201]。还必须认识到，在政府对吸烟危害的基于证据的信息传播资金不足的环境中，很可能没有有效的传播，大部分的传播仍然来自烟草利益集团及其同盟。归根结底，主观影响而非卷烟有害性认知可能是预测长期吸烟行为的最佳方法[184]。

据推测，将减少危害的原则与烟草控制的其他要素结合起来的监管环境将为烟碱降低的策略提供更有利的条件[1,10,11,198]。这可能需要容易获得的、消费者可接受的非燃烧形式的烟碱，这些形式的烟碱可用于管理戒断症状并为吸烟者提供常规卷烟的可替代品，表明减少燃烧型烟草中的烟碱并不意味着禁止烟碱[41]。这需要在现实世界中进一步考虑，在现实世界中，人们已经对某些烟碱替代产品的使用和推广表示担忧。

虽然烟碱降低的方法可以大大改善公众健康，但也为烟草公司直接或间接误导消费者提供了很大的空间。例如，他们可能使用低烟碱产品来推广类似品牌的产品，以暗示新产品和老产品的安全性，并以政策顾问的身份出现。鉴于烟草行业长期以来滥用卫生政策举措进行宣传和公关，任何颁布的条例都应排除烟草公司直接或间接进行不当或误导性宣传，并具有强大的监督和执行能力。此外，法规还应确保有关VLNC卷烟或其他烟草制品健康方面的信息和建议来自卫生机构，而不是烟草公司或其代理商。建议在所有相关缔约方的承诺下，找到一种管理信息传播战略和促进VLNC的国际方法。

世界卫生组织咨询说明[1]提出了烟草或烟碱其他形式带来的潜在健康风险问题，本书其他部分也对此进行了批判性讨论（请参阅第4.3.2节）。有关促进ENDS和其他非燃烧型烟草制品替代品的许多担忧与它们支持卷烟的入门效应或使用的可能性有关；但是，在低潜在致瘾性卷烟中，这些担忧变得无关紧要[41]。同样，同时使用两个或更多产品（可能支持继续使用VLNC卷烟）也将成为问题，因为相比而言，这些卷烟不那么受欢迎，而替代产品会更令人满意[41]。然而，对其他产品的使用和健康影响（包括有害性较低的替代品）以及几种产品同时使用的方式进行评估和监控仍然是烟碱降低政策的关键组成部分[1,2,9]。限制ENDS和其他烟碱产品的市场营销以及可用性，以限制不吸烟者的试验性使用，尤其是对年轻人[10,41]。卫生机构应向公众清楚准确地传达替代产品（包括ENDS和HTP）的潜在健康风险，并应注意不要鼓励人们对替代产品上瘾。

烟碱降低策略应成为全面烟草控制战略的一部分，该战略应与其他基于证据的政策相辅相成，例如增加税收、持续和有足够资金的公共教育运动、限制公共场所吸烟、根据需要进一步管制产品、禁止烟草行业营销和公关、低成本NRT的可获得性和行为戒断治疗（例如电话咨询专线）以及针对弱势人群的特殊计划[95]。FDA在其降低烟碱的公告中报告的监管考虑是[15]继续对燃烧型卷烟进行监管，以及与燃烧型产品有害性和吸引力有关的特性的监管；ENDS和其他卷烟替代品（例如烟碱产品）的用途，可用性和对健康的影响，以及对健康声明和其他形式的风险信息传播的监管。

4.4.3　实施烟碱降低政策的策略

世界卫生组织咨询说明[1]解决了逐步降低烟碱含量与立即降低烟碱含量的问题，并指出立即降低是可取的，因为它不太可能造成意外后果，例如依赖性吸烟者短期内增加补偿性吸烟[93]。一项针对10个地点的双盲研究表明，立即降低烟碱含量而非逐步降低可在20周的干预中降低每天吸烟量和暴露生物标志物，第20周时，在经历了立即降低的组中，渴求和依赖的测量值显著降低[20]。尽管这些数据表明了立即降低的潜在益处，但值得注意的是，与其他报告一致[202]，当立即降低时有更多的参与者退出研究，这表明立即降低可能使吸烟者更加难以适应。尤其是在没有其他烟碱来源或戒烟治疗的情况下。由于吸烟者将继续获得常规卷烟，直到用完为止，因此即使在立即降低政策中，逐步引入VLNC卷烟是实际可行的结果[40]。对依赖常规卷烟的吸烟者面临立即减少吸烟量的潜在不适的担忧强调了对现成的替代形式的烟碱和/或药物的需要[40,198]。

人们对烟碱降低的反应部分反映了环境差异。例如，在瑞典降低烟碱可能会导致使用口含烟，而在欧盟大多数其他国家中，口含烟被禁止使用[11]。因此，烟碱降低的策略必须对环境差异敏感，包括其他法规和人口偏好。一些国家（例如巴西、新加坡、泰国、乌拉圭和委内瑞拉）已经限制或禁止了ENDS的销售和进口。这样的禁令并不一定会妨碍燃烧型烟草制品实行烟碱降低政策；但是，这可能会带来其他挑战。必须注意确保烟碱的替代来源（如药用烟碱）的可用性，获得烟草依赖治疗的能力以及应对潜在的非法市场所必需的能力或基础设施。

4.4.4　烟碱降低的异议思考

关于烟碱降低的一些讨论将该政策视为事实上禁止吸烟的政策[2,10,13,14]。在这种观点下，烟碱降低与其他烟草控制措施（如素包装、图形警示或禁止调味剂）有很大不同，因为它从根本上改变了产品的主要功能，即传递烟碱[14]。尽管如此，即使没有其他产品，吸烟者仍会使用VLNC卷烟，尽管随着时间的流逝，他们可能不太可能坚持使用[196]。此外，与禁止策略不同，吸烟者仍可以使用烟碱，但危害性可能较小。其他形式的烟碱的可获得性也将最大限度地减少依赖性吸烟者通过非法手段寻找烟碱的努力[198]。

烟碱降低的第二个反对意见是，从根本上说，它是"自上而下"或家长式的手段以降低烟草的健康风险[9,10]。将卷烟销售限制在VLNC卷烟中会干扰个人选择。但是，有人认为，对ENDS和烟碱的其他替代来源的支持可以使吸烟者远离卷烟，而无需实施干预式法规[11,13]。避免烟碱降低政策的好处包括消除政策上的法律或政治挑战、潜在的黑市和吸烟者的反对[9]。替代产品（包括ENDS和HTP）的长期健康后果仍然存在疑问，其替代产品的用途、功能和释放物量仍在迅速发展。最终，可能需要采取某种政策组合来降低燃烧型烟草制品相对于替代产品的吸引力。这可能包括增加相对风险的信息传播、增加低风险产品的可用性、增加低风险产品和高风险产品或烟碱产品标准之间的价格差异，从而减少对毒性最高产品的依赖。

4.4.5 小结

- 目前对生产具有主观特性与普通卷烟相似但依赖潜力降低的商用VLNC卷烟的可行性进行了确认。然而，由于烟草的燃烧和吸入，这些卷烟的有害性仍然很高。
- 临床试验中使用的VLNC卷烟的物理特性和构成相对一致；在VLNC卷烟的商业市场中，烟碱以外的产品特性可能会有很大差异，例如具有药理作用的非烟碱化合物的含量，这些差异可能需要仔细监控，或者可能需要施加限制条件来限制产品变化。
- 作为长期战略（包括全面的烟草控制计划）的一部分，公共教育运动应宣传降低烟碱含量以消除对有害性最高的烟草制品的成瘾。如果在没有实质性政策干预或教育运动的情况下，VLNC卷烟与普通卷烟同时出售，则其使用量可能仍然很低。
- 世界卫生组织咨询说明中确定的烟碱降低策略成功的前提仍然有效，即全面管制所有含烟草和烟碱的产品、监测和产品测试的能力、承诺继续进行公共教育、提供经国家相关部门批准的经过试验和测试的治疗形式。应当禁止烟草制品制造商和替代产品制造商进行不当或误导性的促销。
- 与逐渐烟碱降低含量相比，立即降低将具有明显的优势，但可能需要采取综合措施以减少意外后果。
- 降低普通卷烟烟碱含量的目的不是禁止其销售和使用，而是降低其致瘾性，以便人们可以选择是否继续使用有害性高的产品。

4.5 研究问题

- 迄今为止，研究仅集中在降低烟碱的卷烟上。未来的研究应解决减少其他燃烧型烟草制品（例如自卷烟、小雪茄、细支雪茄、水烟）中烟碱的潜在影响，以及减少未燃烧型烟草制品中烟碱的利弊。

- 应继续评估VLNC卷烟对弱势人群（包括育龄妇女）的正面和负面影响。研究青少年和非烟草使用者对VLNC卷烟可能的摄入和使用将是特别有用的信息；这主要通过在实施产品标准后的监测来评估。
- 应进一步研究依赖和不依赖型吸烟者以及刚开始使用烟草者同时使用VLNC卷烟、ENDS和其他烟草和烟碱产品的情况。
- 迄今为止进行的研究并未表明使用VLNC卷烟会增加酒精的使用或对其他物质（如大麻）的依赖，但有必要继续进行评估。
- 迄今为止，尚未开发或测试改变物理或化学特性以替代或增强烟碱药理作用的VLNC产品。
- 研究特定的公共传播策略的影响将有助于使政策的利益最大化，最大限度地减少卷烟致瘾性变化对其有害性影响的误解，并向依赖烟碱的人告知适当的替代方案。
- 应该对烟草行业为了继续推广现有产品并破坏以证据为基础的烟草控制政策而滥用烟碱降低策略的做法进行研究。还需要研究如何在国内和国际上规范这类活动。
- 几乎没有数据可用来估计非法销售或产品篡改的程度以及可能减轻烟碱降低影响的相关因素。

4.6 政策建议

- 烟碱降低的政策具有相当大的前景，可以作为减少最有害烟草制品使用并鼓励戒烟的手段。此类政策应成为综合措施的一部分，并且与旨在支持戒烟和减少吸烟的其他相关政策同时使用可能最有效。
- 燃烧型烟草制品的替代品很常见，使用和外观与普通卷烟相似的高烟碱型燃烧产品很可能会减轻烟碱降低对公众健康的潜在好处。因此，可能必须将烟碱降低政策应用于所有燃烧型产品，而不是仅仅应用于常规卷烟，以支持人们从最有害和致瘾性最大的产品中转移其他产品。
- 考虑到许多临床和行为研究表明减少对烟碱的暴露并减少了烟草消费，因此没有足够的证据表明不吸烟的青少年或不吸烟的成年人会采用并持续使用VLNC卷烟。为了最大限度地减少这些人群的意外后果，必须清楚传达VLNC卷烟和其他形式的烟碱产品的健康风险，包括VLNC卷烟并不比普通卷烟更安全的事实。
- 正如2015年世界卫生组织咨询说明中指出的那样，不应将烟碱降低的政策视为取代全面的烟草控制政策，其中应包括增加税收、有足够资金的持续的公共教育、禁止吸烟、图形警示标签、素包装和其他形式的产品监管。对所有含烟草和烟碱的产品采取一致的政策至关重要，烟草依赖治疗途径也是如此。
- 进行持续的人群监测、产品监测和测试以及执行产品标准的能力对于支持烟碱降低的政策至关重要。监测还应包括企业宣传和其他活动以及产品的供应和销售。

4.7 参 考 文 献

[1] WHO Study Group on Tobacco Product Regulation. Advisory note: global nicotine reduction strategy (WHO Technical Report Series No. 989). Geneva: World Health Organization; 2015 (http://apps.who.int/iris/bitstream/10665/189651/1/9789241509329_eng.pdf, accessed 15 May 2019).

[2] Hatsukami DK, Zaatari G, Donny E. The case for the WHO advisory note, global nicotine reduction strategy. Tob Control. 2017;26(e1):e29–30.

[3] Eriksen M, Mackay J, Schluger N, Gomeshtapeh FI, Drope J, editors. The tobacco atlas, 5th edition, Atlanta (GA), American Cancer Society; 2015.

[4] WHO Framework Convention on Tobacco Control. Geneva: World Health Organization; 2003 (http://www.who.int/fctc/en/, accessed 15 May 2019).

[5] Report on the scientific basis of tobacco product regulation: fourth report of a WHO study group (WHO Technical Report Series No. 967). Geneva: World Health Organization; 2012(http://www.who.int/tobacco/publications/prod_regulation/trs_967/en/index.html, accessed 15 May 2019).

[6] Donny EC, Denlinger RL, Tidey JW, Koopmeiners JS, Benowitz NL, Vandrey RG et al. Randomized trial of reduced-nicotine standards for cigarettes. N Engl J Med. 2015;373(14):1340–9.

[7] Smith TT, Rupprecht LE, Denlinger-Apte RL, Weeks JJ, Panas RS, Donny EC et al. Animal research A global nicotine reduction strategy: state of the science on nicotine reduction: current evidence and research gaps. Nicotine Tob Res. 2017;19(9):1005–15.

[8] Kozlowski LT. Commentary on Nardone et al. (2016): satisfaction, dissatisfaction and complicating the nicotine-reduction strategy with more nicotine. Addiction. 2016;111(12):2217–8.

[9] Kozlowski LT. Cigarette prohibition and the need for more prior testing of the WHO TobReg's global nicotine-reduction strategy. Tob Control. 2017;26(e1):e31–4.

[10] Borland R. Paying more attention to the "elephant in the room". Tob Control. 2017;26(e1):e35–6.

[11] Kozlowski LT. Let actual markets help assess the worth of optional very-low-nicotine cigarettes before deciding on mandatory regulations. Addiction. 2017;112(1):3–5.

[12] Pacek LR, Rass O, Johnson MW. Knowledge about nicotine among HIV-positive smokers: implications for tobacco regulatory science policy. Addict Behav. 2017;65:81–6.

[13] Kozlowski LT, Abrams DB. Obsolete tobacco control themes can be hazardous to public health: the need for updating views on absolute product risks and harm reduction. BMC Public Health. 2016;16:432.

[14] Kozlowski LT. Prospects for a nicotine-reduction strategy in the cigarette endgame: alternative tobacco harm reduction scenarios. Int J Drug Policy. 2015;26(6):543–7.

[15] Gottlieb S, Zeller M. A nicotine-focused framework for public health. N Engl J Med. 2017;377(12):1111–4.

[16] Donny EC, Walker N, Hatsukami DK, Bullen C. Reducing the nicotine content of combusted tobacco products sold in New Zealand. Tob Control. 2017;26:e37–42.

[17] Richter P, Steven PR, Bravo R, Lisko JG, Damian M, Gonzalez-Jimenez N et al. Characterization of Spectrum variable nicotine research cigarettes. Tob Regul Sci. 2016;2(2):94–105.

[18] Donny EC, Houtsmuller E, Stitzer ML. Smoking in the absence of nicotine: behavioral, subjective and physiological effects over 11 days. Addiction. 2007;102(2):324–34.

[19] Hatsukami DK, Kotlyar M, Hertsgaard LA, Zhang Y, Carmella SG, Jensen JA et al. Reduced nicotine content cigarettes: effects on toxicant exposure, dependence and cessation. Addiction. 2010;105(2):343–55.

[20] Hatsukami DK, Luo X, Jensen JA, al'Absi M, Allen SS, Carmella SG et al. Effect of immediate vs gradual reduction in nicotine content of cigarettes on biomarkers of smoke exposure: a randomized clinical trial. JAMA. 2018;320(9):880–91.

[21] Hatsukami DK, Hertsgaard LA, Vogel RI, Jensen JA, Murphy SE, Hecht SS et al. Reduced nicotine content cigarettes and nicotine patch. Cancer Epidemiol Biomarkers Prev. 2013;22(6):1015–24.

[22] Hammond D, O'Connor RJ. Reduced nicotine cigarettes: smoking behavior and biomarkers of exposure among smokers not intending to quit. Cancer Epidemiol Biomarkers Prev. 2014;23(10):2032–40.

[23] Kozlowski LT, O'Connor RJ. Cigarette filter ventilation is a defective design because of misleading taste, bigger puffs, and blocked vents. Tob Control. 2002;11(Suppl 1):I40–50.

[24] Ding YS, Ward J, Hammond D, Watson CH. Mouth-level intake of benzo[a]pyrene from reduced nicotine cigarettes. Int J Environ Res Public Health. 2014;11(11):11898–914.

[25] Benowitz NL, Jacob P 3rd, Herrera B. Nicotine intake and dose response when smoking reduced-nicotine content cigarettes. Clin Pharmacol Ther. 2006;80(6):703–14.

[26] Benowitz NL, Nardone N, Dains KM, Hall SM, Stewart S, Dempsey D et al. Effect of reducing the nicotine content of cigarettes on cigarette smoking behavior and tobacco smoke toxicant exposure: 2-year follow up. Addiction. 2015;110(10):1667–75.

[27] Mercincavage M, Souprountchouk V, Tang KZ, Dumont RL, Wileyto EP, Carmella SG et al. A randomized controlled trial of progressively reduced nicotine content cigarettes on smoking behaviors, biomarkers of exposure, and subjective ratings. Cancer Epidemiol Biomarkers Prev. 2016;25(7):1125–33.

[28] Perkins KA, Kunkle N, Karelitz JL, Perkins KA, Kunkle N, Karelitz JL. Preliminary test of cigarette nicotine discrimination threshold in non-dependent versus dependent smokers. Drug Alcohol Depend. 2017;175:36–41.

[29] Smith TT, Levin ME, Schassburger RL, Buffalari DM, Sved AF, Donny EC. Gradual and immediate nicotine reduction result in similar low-dose nicotine self-administration. Nicotine Tob Res. 2013;15(11):1918–25.

[30] Chen H, Matta SG, Sharp BM. Acquisition of nicotine self-administration in adolescent rats given prolonged access to the drug. Neuropsychopharmacology. 2007;32(3):700–9.

[31] Levin ED, Lawrence SS, Petro A, Horton K, Rezvani AH, Seidler FJ et al. Adolescent vs.

adultonset nicotine self-administration in male rats: duration of effect and differential nicotinic receptor correlates. Neurotoxicol Teratol. 2007;29(4):458–65.

[32] Donny EC, Hatsukami DK, Benowitz NL, Sved AF, Tidey JW, Cassidy RN. Reduced nicotine product standards for combustible tobacco: building an empirical basis for effective regulation. Prev Med. 2014;68:17–22.

[33] Smith TT, Schassburger RL, Buffalari DM, Sved AF, Donny EC. Low-dose nicotine self-administration is reduced in adult male rats naive to high doses of nicotine: implications for nicotine product standards. Exp Clin Psychopharmacol. 2014;22(5):453–9.

[34] Schassburger RL, Pitzer EM, Smith TT, Rupprecht LE, Thiels E, Donny EC et al. Adolescent rats self-administer less nicotine than adults at low doses. Nicotine Tob Res. 2016;18(9):1861–8.

[35] Cassidy RN, Tidey JW, Cao Q, Colby SM, McClernon FJ, Koopmeiners JS et al. Age moderates smokers' subjective response to very low nicotine content cigarettes: evidence from a randomized controlled trial. Nicotine Tob Res. 2018; doi:10.1093/ntr/nty079.

[36] Jackson KM, Cioe PA, Krishnan-Sarin S, Hatsukami D. Adolescent smokers' response to reducing the nicotine content of cigarettes: acute effects on withdrawal symptoms and subjective evaluations. Drug Alcohol Depend. 2018;188:153–60.

[37] Walker N, Howe C, Bullen C, Grigg M, Glover M, McRobbie H et al. The combined effect of very low nicotine content cigarettes, used as an adjunct to usual Quitline care (nicotine replacement therapy and behavioural support), on smoking cessation: a randomized controlled trial. Addiction. 2012;107(10):1857–67.

[38] McRobbie H, Przulj D, Smith KM, Cornwall D. Complementing the standard multicomponent treatment for smokers with denicotinized cigarettes: a randomized trial. Nicotine Tob Res. 2016;18(5):1134–41.

[39] Donny EC, Jones M. Prolonged exposure to denicotinized cigarettes with or without transdermal nicotine. Drug Alcohol Depend. 2009;104(1–2):23–33.

[40] Hatsukami DK, Luo X, Dick L, Kangkum M, Allen SS, Murphy SE et al. Reduced nicotine content cigarettes and use of alternative nicotine products: exploratory trial. Addiction. 2017;112(1):156–67.

[41] Benowitz NL, Donny EC, Hatsukami DK. Reduced nicotine content cigarettes, e-cigarettes and the cigarette end game. Addiction. 2017;112(1):6–7.

[42] Pechacek TF, Nayak P, Gregory KR, Weaver SR, Eriksen MP. The potential that electronic nicotine delivery systems can be a disruptive technology: results from a national survey. Nicotine Tob Res. 2016;18(10):1989–97.

[43] Nardone N, Donny EC, Hatsukami DK, Koopmeiners JS, Murphy SE, Strasser AA et al. Estimations and predictors of non-compliance in switchers to reduced nicotine content cigarettes. Addiction. 2016;111(12):2208–16.

[44] Boatman JA, Vock DM, Koopmeiners JS, Donny EC. Estimating causal effects from a randomized clinical trial when noncompliance is measured with error. Biostatistics. 2018;19(1):103–18.

[45] Dermody SS, Donny EC, Hertsgaard LA, Hatsukami DK. Greater reductions in nicotine exposure while smoking very low nicotine content cigarettes predict smoking cessation. Tob Control. 2015;24(6):536–9.

[46] Dermody SS, McClernon FJ, Benowitz N, Luo X, Tidey JW, Smith TT et al. Effects of reduced nicotine content cigarettes on individual withdrawal symptoms over time and during abstinence. A global nicotine reduction strategy: state of the science Exp Clin Psychopharmacol. 2018;26(3):223–32.

[47] Faulkner P, Ghahremani DG, Tyndale RF, Cox CM, Kazanjian AS, Paterson N et al. Reducednicotine cigarettes in young smokers: impact of nicotine metabolism on nicotine dose effects. Neuropsychopharmacology. 2017;42(8):1610–8.

[48] Keith DR, Kurti AN, Davis DR, Zvorsky IA, Higgins ST. A review of the effects of very low nicotine content cigarettes on behavioral and cognitive performance. Prev Med. 2017;104:100–16.

[49] Rupprecht LE, Koopmeiners JS, Dermody SS, Oliver JA, al'Absi M, Benowitz NL et al. Reducing nicotine exposure results in weight gain in smokers randomised to very low nicotine content cigarettes. Tob Control. 2017;26(e1):e43–8.

[50] Rupprecht LE, Donny EC, Sved AF. Obese smokers as a potential subpopulation of risk in tobacco reduction policy. Yale J Biol Med. 2015;88(3):289–94.

[51] Bandiera FC, Ross KC, Taghavi S, Delucchi K, Tyndale RF, Benowitz NL. Nicotine dependence, nicotine metabolism, and the extent of compensation in response to reduced nicotine content cigarettes. Nicotine Tob Res. 2015;17(9):1167–72.

[52] Benowitz NL, Dains KM, Hall SM, Stewart S, Wilson M, Dempsey D et al. Smoking behavior and exposure to tobacco toxicants during 6 months of smoking progressively reduced nicotine content cigarettes. Cancer Epidemiol Biomarkers Prev. 2012;21(5):761–9.

[53] Shiffman S, Kurland BF, Scholl SM, Mao JM. Nondaily smokers' changes in cigarette consumption with very low-nicotine-content cigarettes: a randomized double-blind clinical trial. JAMA Psychiatry. 2018;75(10):995–1002.

[54] Dermody SS, Tidey JW, Denlinger RL, Pacek LR, al'Absi M, Drobes DJ. The impact of smoking very low nicotine content cigarettes on alcohol use. Alcohol Clin Exp Res. 2016;40(3):606–15.

[55] Pacek LR, Vandrey R, Dermody SS, Denlinger-Apte RL, Lemieux A, Tidey JW et al. Evaluation of a reduced nicotine product standard: moderating effects of and impact on cannabis use. Drug Alcohol Depend. 2016;167:228–32.

[56] Cohn A, Ehlke S, Cobb CO. Relationship of nicotine deprivation and indices of alcohol use behavior to implicit alcohol and cigarette approach cognitions in smokers. Addict Behav. 2017;67:58–65.

[57] Tidey JW, Pacek LR, Koopmeiners JS, Vandrey R, Nardone N, Drobes DJ et al. Effects of 6-week use of reduced-nicotine content cigarettes in smokers with and without elevated depressive symptoms. Nicotine Tob Res. 2017;19(1):59–67.

[58] Gaalema DE, Miller ME, Tidey JW. Predicted impact of nicotine reduction on smokers with affective disorders. Tob Regul Sci. 2015;1(2):154–65.

[59] Higgins ST, Heil SH, Sigmon SC, Tidey JW, Gaalema DE, Stitzer ML et al. Response to varying the nicotine content of cigarettes in vulnerable populations: an initial experimental examination of acute effects. Psychopharmacology. 2017;234(1):89–98.

[60] Lawrence D, Mitrou F, Zubrick SR. Smoking and mental illness: results from population surveys in Australia and the United States. BMC Public Health. 2009;9:285.

[61] Lasser K1, Boyd JW, Woolhandler S, Himmelstein DU, McCormick D, Bor DH. Smoking and mental illness: a population-based prevalence study. JAMA. 2000;284(20):2606–10.

[62] Smith PH, Mazure CM, McKee SA. Smoking and mental illness in the US population. Tob Control. 2014;23(e2):e147–53.

[63] Adler NE, Glymour MM, Fielding J. Addressing social determinants of health and health inequalities. JAMA. 2016;316(16):1641–2.

[64] Das S, Prochaska JJ. Innovative approaches to support smoking cessation for individuals with mental illness and co-occurring substance use disorders. Expert Rev Respir Med. 2017;11(10):841–50.

[65] Beard E, Bruguera C, McNeill A, Brown J, West R. Association of amount and duration of NRT use in smokers with cigarette consumption and motivation to stop smoking: a national survey of smokers in England. Addict Behav. 2015;40:33–8.

[66] Benowitz NL, Hukkanen J, Jacob P III. Nicotine chemistry, metabolism, kinetics and biomarkers. Handb Exp Pharmacol. 2009;192:29–60.

[67] Digard H, Proctor C, Kulasekaran A, Malmqvist U, Richter A. Determination of nicotine absorption from multiple tobacco products and nicotine gum. Nicotine Tob Res. 2013;15(1):255–61.

[68] Caggiula AR, Donny EC, Palmatier MI, Liu X, Chaudhri N, Sved AF. The role of nicotine in smoking: a dual-reinforcement model. Nebr Symp Motiv. 2009;55:91–109.

[69] Van Heel M, Van Gucht D, Vanbrabant K, Baeyens F. The importance of conditioned stimuli in cigarette and e-cigarette craving reduction by e-cigarettes. Int J Environ Res Public Health. 2017;14(2):193.

[70] Rose JE. Nicotine and nonnicotine factors in cigarette addiction. Psychopharmacology. 2006;184(3–4):274–85.

[71] Perkins K, Sayette M, Conklin C, Caggiula A. Placebo effects of tobacco smoking and other nicotine intake. Nicotine Tob Res. 2003;5(5):695–709.

[72] Chaloupka FJ, Warner KE. The economics of smoking. In: Culyer AJ, Newhouse JP, editors. Handbook of health economics. Oxford: Elsevier Science; 2000:1539–627.

[73] MacKillop J, Murphy JG, Ray LA, Eisenberg DT, Lisman SA, Lum JK et al. Further validation of a cigarette purchase task for assessing the relative reinforcing efficacy of nicotine in college smokers. Exp Clin Psychopharmacol. 2008;16(1):57–65.

[74] Guindon GE, Paraje GR, Chaloupka FJ. The impact of prices and taxes on the use of tobacco products in Latin America and the Caribbean. Am J Public Health. 2015;105(3):e9–19.

[75] Zheng Y, Zhen C, Dench D, Nonnemaker JM. US demand for tobacco products in a system framework. Health Econ. 2017;26(8):1067–86.

[76] O'Connor RJ, June KM, Bansal-Travers M, Rousu MC, Thrasher JF, Hyland A et al. Estimating demand for alternatives to cigarettes with online purchase tasks. Am J Health Behav. 2014;38(1):103–13.

[77] Stein JS, Wilson AG, Koffarnus MN, Judd MC, Bickel WK. Naturalistic assessment of demand for cigarettes, snus, and nicotine gum. Psychopharmacology. 2017;234(2):245–54.

[78] Gammon DG, Loomis BR, Dench DL, King BA, Fulmer EB, Rogers T. Effect of price changes in little cigars and cigarettes on little cigar sales: USA, Q4 2011–Q4 2013. Tob Control.

2016;25(5):538–44.

[79] Tait P, Rutherford P, Saunders C. Do consumers of manufactured cigarettes respond differently to price changes compared with their roll-your-own counterparts? Evidence from New Zealand. Tob Control. 2015;24(3):285–9.

[80] Etter JF, Eissenberg T. Dependence levels in users of electronic cigarettes, nicotine gums and tobacco cigarettes. Drug Alcohol Depend. 2015;147:68–75.

[81] Huang J, Tauras J, Chaloupka FJ. The impact of price and tobacco control policies on the demand for electronic nicotine delivery systems. Tob Control. 2014;23(Suppl 3):iii41–7.

[82] McPherson S, Howell D, Lewis J, Barbosa-Leiker C, Bertotti Metoyer P, Roll J. Self-reported smoking effects and comparative value between cigarettes and high dose e-cigarettes in nicotine-dependent cigarette smokers. Behav Pharmacol. 2016;27(2–3 Spec Issue):301–7.

[83] Liu G, Wasserman E, Kong L, Foulds J. A comparison of nicotine dependence among exclusive e-cigarette and cigarette users in the PATH study. Prev Med. 2017;104:86–91.

[84] National Cancer Institute, World Health Organization. The economics of tobacco and tobacco control (NCI Tobacco Control Monograph 21). Bethesda (MD): Department of Health and Human Services; Geneva: World Health Organization; 2016.

[85] Chaloupka FJ, Yurekli A, Fong GT. Tobacco taxes as a tobacco control strategy. Tob Control. 2012;21(2):172–80.

[86] Choi SE. Are lower income smokers more price sensitive? The evidence from Korean cigarette tax increases. Tob Control. 2016;25(2):141–6.

[87] Koffarnus MN, Wilson AG, Bickel WK. Effects of experimental income on demand for potentially A global nicotine reduction strategy: state of the science real cigarettes. Nicotine Tob Res. 2015;17(3):292–8.

[88] Yang T, Peng S, Yu L, Jiang S, Stroub WB, Cottrell RR et al. Chinese smokers' behavioral response toward cigarette price: individual and regional correlates. Tob Induc Dis. 2016;14:13.

[89] Lewis M, Wang Y, Cahn Z, Berg CJ. An exploratory analysis of cigarette price premium, market share and consumer loyalty in relation to continued consumption versus cessation in a national US panel. BMJ Open. 2015;5(11):e008796.

[90] Biener L, Roman AM, McInerney SA, Bolcic-Jankovic D, Hatsukami DK, Loukas A et al. Snus use and rejection in the USA. Tob Control. 2016;25(4):386–92.

[91] Nonnemaker J, Kim AE, Lee YO, MacMonegle A. Quantifying how smokers value attributes of electronic cigarettes. Tob Control. 2016;25(e1):e37–43.

[92] Meier E, Burris JL, Wahlquist A, Garrett-Mayer E, Gray KM, Alberg AJ et al. Perceptions of snus among US adult smokers given free product. Nicotine Tob Res. 2017;20(1):22–9.

[93] Walker N, Fraser T, Howe C, Laugesen M, Truman P, Parag V et al. Abrupt nicotine reduction as an endgame policy: a randomised trial. Tob Control. 2015;24(e4):e251–7.

[94] Smith TT, Cassidy RN, Tidey JW, Luo X, Le CT, Hatsukami DK et al. Impact of smoking reduced nicotine content cigarettes on sensitivity to cigarette price: further results from a multi-site clinical trial. Addiction. 2017;112(2):349–59.

[95] Tucker MR, Laugesen M, Grace RC. Estimating demand and cross-price elasticity for very low nicotine content (VLNC) cigarettes using a simulated demand task. Nicotine Tob Res. 2018;20(4):528.

[96] Johnson MW, Bickel WK, Kirshenbaum AP. Substitutes for tobacco smoking: a behavioral economic analysis of nicotine gum, denicotinized cigarettes, and nicotine-containing cigarettes. Drug Alcohol Depend. 2004;74(3):253–64.

[97] O'Connor RJ, Bansal-Travers M, Carter LP, Cummings KM. What would menthol smokers do if menthol in cigarettes were banned? Behavioral intentions and simulated demand. Addiction. 2012;107(7):1330–8.

[98] Sureda X, Fu M, Martínez-Sánchez JM, Martínez C, Ballbé M, Pérez-Ortuño R et al. Manufactured and roll-your-own cigarettes: a changing pattern of smoking in Barcelona, Spain. Environ Res. 2017;155:167–74.

[99] Brown AK, Nagelhout GE, van den Putte B, Willemsen MC, Mons U, Guignard R et al. Trends and socioeconomic differences in roll-your-own tobacco use: findings from the ITC Europe surveys. Tob Control. 2015;24(Suppl 3):iii11–6.

[100] Ayo-Yusuf OA, Olutola BG. "Roll-your-own" cigarette smoking in South Africa between 2007 and 2010. BMC Public Health. 2013;13:597.

[101] Agaku IT, Filippidis FT, Vardavas CI, Odukoya OO, Awopegba AJ, Ayo-Yusuf OA et al. Poly-tobacco use among adults in 44 countries during 2008–2012: evidence for an integrative and comprehensive approach in tobacco control. Drug Alcohol Depend. 2014;139:60–70.

[102] Lee YO, Hebert CJ, Nonnemaker JM, Kim AE. Multiple tobacco product use among adults in the United States: cigarettes, cigars, electronic cigarettes, hookah, smokeless tobacco, and snus. Prev Med. 2014;62:14–9.

[103] Gravely S, Fong GT, Cummings KM, Yan M, Quah AC, Borland R et al. Awareness, trial, and current use of electronic cigarettes in 10 countries: findings from the ITC project. Int J Environ Res Public Health. 2014;11(11):11691–704.

[104] Palipudi KM, Mbulo L, Morton J, Mbulo L, Bunnell R, Blutcher-Nelson G et al. Awareness and current use of electronic cigarettes in Indonesia, Malaysia, Qatar, and Greece: findings from 2011–2013 global adult tobacco surveys. Nicotine Tob Res. 2016;18(4):501–7.

[105] Delnevo CD, Giovenco DP, Miller Lo EJ. Changes in the mass-merchandise cigar market since the Tobacco Control Act. Tob Regul Sci. 2017;3(2 Suppl 1):S8–16.

[106] Jo CL, Williams RS, Ribisl KM. Tobacco products sold by Internet vendors following restrictions on flavors and light descriptors. Nicotine Tob Res. 2015;17(3):344–9.

[107] Casseus M, Garmon J, Hrywna M, Delnevo CD. Cigarette smokers' classification of tobacco products. Tob Control. 2016;25(6):628–30.

[108] Quisenberry AJ, Koffarnus MN, Hatz LE, Epstein LH, Bickel WK. The experimental tobacco marketplace I: Substitutability as a function of the price of conventional cigarettes. Nicotine Tob Res. 2016;18(7):1642–8.

[109] Dave D, Saffer H. Demand for smokeless tobacco: role of advertising. J Health Econ. 2013;32(4):682–97.

[110] Adams S, Cotti C, Fuhrmann D. Smokeless tobacco use following smoking bans in bars. South Econ J. 2013;80(1):162–80.

[111] Tam J, Day HR, Rostron BL, Apelberg BJ. A systematic review of transitions between cigarette and smokeless tobacco product use in the United States. BMC Public Health. 2015;15:258.

[112] Lund KE, Vedøy TF, Bauld L. Do never smokers make up an increasing share of snus users

as cigarette smoking declines? Changes in smoking status among male snus users in Norway 2003–15. Addiction. 2017;112(2):340–8.

[113] Hatsukami DK, Severson H, Anderson A, Vogel RI, Jensen J, Broadbent B et al. Randomised clinical trial of snus versus medicinal nicotine among smokers interested in product switching. Tob Control. 2016;25(3):267–74.

[114] Allen A, Vogel RI, Meier E, Anderson A, Jensen J, Severson HH et al. Gender differences in snus versus nicotine gum for cigarette avoidance among a sample of US smokers. Drug Alcohol Depend. 2016;168:8–12.

[115] Berman ML, Bickel WK, Harris AC, LeSage MG, O'Connor RJ, Stepanov I et al. Consortium on methods evaluating tobacco: research tools to inform FDA regulation of snus. Nicotine Tob Res. 2018;20(11):1292–300.

[116] Chaloupka FJ, Sweanor D, Warner KE. Differential taxes for differential risks – toward reduced harm from nicotine-yielding products. N Engl J Med. 2015;373(7):594–7.

[117] Grace RC, Kivell BM, Laugesen M. Estimating cross-price elasticity of e-cigarettes using a simulated demand procedure. Nicotine Tob Res. 2015;17(5):592–8.

[118] Stoklosa M, Drope J, Chaloupka FJ. Prices and e-cigarette demand: evidence from the European Union. Nicotine Tob Res. 2016;18(10):1973–80.

[119] Johnson MW, Johnson PS, Rass O, Pacek LR. Behavioral economic substitutability of e-cigarettes, tobacco cigarettes, and nicotine gum. J Psychopharmacol. 201731(7);851–60.

[120] Snider SE, Cummings KM, Bickel WK. Behavioral economic substitution between conventional cigarettes and e-cigarettes differs as a function of the frequency of e-cigarette use. Drug Alcohol Depend. 2017;177:14–22.

[121] Beard E, West R, Michie S, Brown J. Association between electronic cigarette use and changes in quit attempts, success of quit attempts, use of smoking cessation pharmacotherapy, and use of stop smoking services in England: time series analysis of population trends. BMJ. 2016;354:i4645.

[122] O'Brien B, Knight-West O, Walker N, Parag V, Bullen C. E-cigarettes versus NRT for smoking reduction or cessation in people with mental illness: secondary analysis of data from the ASCEND trial. Tob Induc Dis. 2015;13(1):5.

[123] Brown J, Beard E, Kotz D, Michie S, West R. Real-world effectiveness of e-cigarettes when used to aid smoking cessation: a cross-sectional population study. Addiction. 2014;109(9):1531–40.

[124] Harrell PT, Marquinez NS, Correa JB, Meltzer LR, Unrod M, Sutton SK et al. Expectancies for cigarettes, e-cigarettes, and nicotine replacement therapies among e-cigarette users (aka vapers). Nicotine Tob Res. 2015;17(2):193–200.

[125] Liber AC, Drope JM, Stoklosa M. Combustible cigarettes cost less to use than e-cigarettes: global evidence and tax policy implications. Tob Control. 2017;26(2):158–63.

[126] Jorenby DE, Smith SS, Fiore MC, Baker TB. Nicotine levels, withdrawal symptoms, and smoking A global nicotine reduction strategy: state of the science reduction success in real world use: a comparison of cigarette smokers and dual users of both cigarettes and E-cigarettes. Drug Alcohol Depend. 2017;170:93–101.

[127] Majeed BA, Weaver SR, Gregory KR, Whitney CF, Slovic P, Pechacek TF et al. Changing perceptions of harm of e-cigarettes among U.S. adults, 2012–2015. Am J Prev Med.

2017;52(3):331–8.

[128] Chou SP, Saha TD, Zhang H, Ruan WJ, Huang B, Grant BF et al. Prevalence, correlates, comorbidity and treatment of electronic nicotine delivery system use in the United States. Drug Alcohol Depend. 2017;178:296–301.

[129] Lee JA, Kim SH, Cho HJ. Electronic cigarette use among Korean adults. Int J Public Health. 2016;61(2):151–7.

[130] Wang M, Wang JW, Cao SS, Wang HQ, Hu RY. Cigarette smoking and electronic cigarettes use: a meta-analysis. Int J Environ Res Public Health. 2016;13(1):120.

[131] Carroll Chapman SL, Wu LT. E-cigarette prevalence and correlates of use among adolescents versus adults: a review and comparison. J Psychiatr Res. 2014;54:43–54.

[132] Shiplo S, Czoli CD, Hammond D. E-cigarette use in Canada: prevalence and patterns of use in a regulated market. BMJ Open. 2015;5(8):e007971.

[133] Reid JL, Rynard VL, Czoli CD, Hammond D. Who is using e-cigarettes in Canada? Nationally representative data on the prevalence of e-cigarette use among Canadians. Prev Med. 2015;81:180–3.

[134] Kilibarda B, Mravcik V, Martens MS. E-cigarette use among Serbian adults: prevalence and user characteristics. Int J Public Health. 2016;61(2):167–75.

[135] Vardavas CI, Filippidis FT, Agaku IT. Determinants and prevalence of e-cigarette use throughout the European Union: a secondary analysis of 26 566 youth and adults from 27 Countries. Tob Control. 2015;24(5):442–8.

[136] Farsalinos KE, Poulas K, Voudris V, Le Houezec J. Electronic cigarette use in the European Union: analysis of a representative sample of 27 460 Europeans from 28 countries. Addiction. 2016;111(11):2032–40.

[137] E-cigarette use among youth and young adults: a report of the Surgeon General. Washington (DC): Department of Health and Human Services; 2016 (http://www.surgeongeneral.gov/library/2016ecigarettes/index.html#execsumm, accessed 15 May 2019).

[138] Rennie LJ, Bazillier-Bruneau C, Rouëssé J. Harm reduction or harm introduction? Prevalence and correlates of E-cigarette use among French adolescents. J Adolesc Health. 2016;58(4):440–5.

[139] Mays D, Smith C, Johnson AC, Tercyak KP, Niaura RS. An experimental study of the effects of electronic cigarette warnings on young adult nonsmokers' perceptions and behavioral intentions. Tob Induc Dis. 2016;14:17.

[140] Jiang N, Wang MP, Ho SY, Leung LT, Lam TH. Electronic cigarette use among adolescents: a cross-sectional study in Hong Kong. BMC Public Health. 2016;16:202.

[141] Lee JA, Lee S, Cho HJ. The relation between frequency of e-cigarette use and frequency and intensity of cigarette smoking among South Korean adolescents. Int J Environ Res Public Health. 2017;14(3):305.

[142] Chang HC, Tsai YW, Shiu MN, Wang YT, Chang PY. Elucidating challenges that electronic cigarettes pose to tobacco control in Asia: a population-based national survey in Taiwan. BMJ Open. 2017;7(3):e014263.

[143] White J, Li J, Newcombe R, Walton D. Tripling use of electronic cigarettes among New Zealand adolescents between 2012 and 2014. J Adolesc Health. 2015;56(5):522–8.

[144] Chaffee BW, Couch ET, Gansky SA. Trends in characteristics and multi-product use

among adolescents who use electronic cigarettes, United States 2011–2015. PLoS One. 2017;12(5):e0177073.

[145] Fotiou A, Kanavou E, Stavrou M, Richardson C, Kokkevi A. Prevalence and correlates of electronic cigarette use among adolescents in Greece: a preliminary cross-sectional analysis of nationwide survey data. Addict Behav. 2015;51:88–92.

[146] Cullen KA, Ambrose BK, Gentzke AS, Apelberg BJ, Jamal A, King BA. Notes from the field: use of electronic cigarettes and any tobacco product among middle and high school students – United States, 2011–2018. Morb Mortal Wkly Rep. 2018;67:1276–7.

[147] Chatterjee K, Alzghoul B, Innabi A, Meena N. Is vaping a gateway to smoking: a review of the longitudinal studies. Int J Adolesc Med Health. 2016;30(3). doi: 10.1515/ijamh-2016-0033.

[148] Lanza ST, Russell MA, Braymiller JL. Emergence of electronic cigarette use in US adolescents and the link to traditional cigarette use. Addict Behav. 2017;67:38–43.

[149] Schneider S, Diehl K. Vaping as a catalyst for smoking? An initial model on the initiation of electronic cigarette use and the transition to tobacco smoking among adolescents. Nicotine Tob Res. 2016;18(5):647–53.

[150] Zhong J, Cao S, Gong W, Fei F, Wang M. Electronic cigarettes use and intention to cigarette smoking among never-smoking adolescents and young adults: a meta-analysis. Int J Environ Res Public Health. 2016;13(5):465.

[151] Barrington-Trimis JL, Urman R, Berhane K, Unger JB, Cruz TB, Pentz MA et al. E-cigarettes and future cigarette use. Pediatrics. 2016;138(1). doi: 10.1542/peds.2016-0379.

[152] Wills TA, Sargent JD, Knight R, Pagano I, Gibbons FX. E-cigarette use and willingness to smoke: a sample of adolescent non-smokers. Tob Control. 2016;25(e1):e52–9.

[153] Conner M, Grogan S, Simms-Ellis R, Flett K, Sykes-Muskett B, Cowap L et al. Do electronic cigarettes increase cigarette smoking in UK adolescents? Evidence from a 12-month prospective study. Tob Control. 2017;27(4). doi: 10.1136/tobaccocontrol-2016-053539.

[154] Soneji S, Barrington-Trimis JL, Wills TA, Leventhal AM, Unger JB, Gibson LA et al. Association between initial use of e-cigarettes and subsequent cigarette smoking among adolescents and young adults: a systematic review and meta-analysis. JAMA Pediatr. 2017;171(8):788–97.

[155] Etter JF. Gateway effects and electronic cigarettes. Addiction. 2018;113(10):1776–83.

[156] Franck C, Filion KB, Kimmelman J, Grad R, Eisenberg MJ. Ethical considerations of e-cigarette use for tobacco harm reduction. Respir Res. 2016;17(1):53.

[157] Meier EM, Tackett AP, Miller MB, Grant DM, Wagener TL. Which nicotine products are gateways to regular use? First-tried tobacco and current use in college students. Am J Prev Med. 2015;48(1 Suppl 1):S86–93.

[158] Polosa R, Russell C, Nitzkin J, Farsalinos KE. A critique of the US Surgeon General's conclusions regarding e-cigarette use among youth and young adults in the United States of America. Harm Reduct J. 2017;14(1):61.

[159] Kozlowski LT, Warner KE. Adolescents and e-cigarettes: objects of concern may appear larger than they are. Drug Alcohol Depend. 2017;174:209–14.

[160] Surgeon General's advisory on e-cigarette use among youth. Atlanta (GA): Office of the Surgeon General, Office on Smoking and Health; 2018 (https://e-cigarettes.surgeongeneral.gov/documents/surgeon-generals-advisory-on-e-cigarette-use-among-youth-2018.pdf, accessed 1

January 2019).
[161] Azar AM, Gottlieb S. Opinion. We cannot let e-cigarettes become an on-ramp for teenage addiction. The Washington Post, 11 October 2018 (https://www.washingtonpost.com/opinions/we-cannot-let-e-cigarettes-become-an-on-ramp-for-teenage-addiction/2018/10/11/55ce424eccc6-11e8-a360-85875bac0b1f_story.html?utm_term=.3b243a1a45f5, accessed 1 January 2019).
[162] National Research Council. Understanding the US illicit tobacco market: characteristics, policy context, and lessons from international experiences. Washington (DC): National Academies Press; 2015 (https://doi.org/10.17226/19016).
[163] Lee JGL, Golden SD, Ribisl KM. Limited indications of tax stamp discordance and counterfeiting on cigarette packs purchased in tobacco retailers, 97 counties, USA, 2012. Prev Med Rep. 2017;8:148–52.
[164] Ribisl KM, Williams R, Gizlice Z, Herring AH. Effectiveness of state and federal government A global nicotine reduction strategy: state of the science agreements with major credit card and shipping companies to block illegal Internet cigarette sales. PLoS One. 2011;6(2):e16745.
[165] Hatsukami DK, Heishman SJ, Vogel RI, Denlinger RL, Roper-Batker AN, Mackowick KM et al. Dose–response effects of Spectrum research cigarettes. Nicotine Tob Res. 2013;15(6):1113–21.
[166] Wiley JL, Marusich JA, Thomas BF, Jackson KJ. Determination of behaviorally effective tobacco constituent doses in rats. Nicotine Tob Res. 2015;17(3):368–71.
[167] Grizzell JA, Echeverria V. New insights into the mechanisms of action of cotinine and its distinctive effects from nicotine. Neurochem Res. 2015;40(10):2032–46.
[168] Brennan KA, Laugesen M, Truman P. Whole tobacco smoke extracts to model tobacco dependence in animals. Neurosci Biobehav Rev. 2014;47:53–69.
[169] Costello MR, Reynaga DD, Mojica CY, Zaveri NT, Belluzzi JD, Leslie FM. Comparison of the reinforcing properties of nicotine and cigarette smoke extract in rats. Neuropsychopharmacology. 2014;39(8):1843–51.
[170] Gellner CA, Belluzzi JD, Leslie FM. Self-administration of nicotine and cigarette smoke extract in adolescent and adult rats. Neuropharmacology. 2016;109:247–53.
[171] Smith TT, Schaff MB, Rupprecht LE, Schassburger RL, Buffalari DM, Murphy SE et al. Effects of MAO inhibition and a combination of minor alkaloids, β-carbolines, and acetaldehyde on nicotine self-administration in adult male rats. Drug Alcohol Depend. 2015;155:243–52.
[172] Harris AC, Tally L, Schmidt CE, Muelken P, Stepanov I, Saha S et al. Animal models to assess the abuse liability of tobacco products: effects of smokeless tobacco extracts on intracranial selfstimulation. Drug Alcohol Depend. 2015;147:60–7.
[173] Truman P, Grounds P, Brennan KA. Monoamine oxidase inhibitory activity in tobacco particulate matter: Are harman and norharman the only physiologically relevant inhibitors? Neurotoxicology. 2017;59:22–6.
[174] Arnold MM, Loughlin SE, Belluzzi JD, Leslie FM. Reinforcing and neural activating effects of norharmane, a non-nicotine tobacco constituent, alone and in combination with nicotine. Neuropharmacology. 2014;85:293–304.
[175] Caine SB, Collins GT, Thomsen M, Wright C, Lanier RK, Mello NK. Nicotine-like behavioral effects of the minor tobacco alkaloids nornicotine, anabasine, and anatabine in male rodents.

Exp Clin Psychopharmacol. 2014;22(1):9–22.

[176] Plescia F, Brancato A, Venniro M, Maniaci G, Cannizzaro E, Sutera FM et al. Acetaldehyde selfadministration by a two-bottle choice paradigm: consequences on emotional reactivity, spatial learning, and memory. Alcohol. 2015;49(2):139–48.

[177] Smith TT, Rupprecht LE, Cwalina SN, Onimus MJ, Murphy SE, Donny EC et al. Effects of monoamine oxidase inhibition on the reinforcing properties of low-dose nicotine. Neuropsychopharmacology. 2016;41(9):2335–43.

[178] Biswas L, Harrison E, Gong Y, Avusula R, Lee J, Zhang M et al. Enhancing effect of menthol on nicotine self-administration in rats. Psychopharmacology. 2016;233(18):3417–27.

[179] Advisory note: banning menthol in tobacco products. WHO Study Group on Tobacco Product Regulation. Geneva: World Health Organization; 2016 (http://apps.who.int/iris/bitstream/10665/205928/1/9789241510332_eng.pdf, accessed 15 May 2019).

[180] Vagg R, Chapman S. Nicotine analogues: a review of tobacco industry research interests. Addiction. 2015;100:701–12.

[181] Desai RI, Doyle MR, Withey SL, Bergman J. Nicotinic effects of tobacco smoke constituents in nonhuman primates. Psychopharmacology. 2016;233(10):1779–89.

[182] Wong LP, Alias H, Agha Mohammadi N, Ghadimi A, Hoe VCW. E-cigarette users' attitudes on the banning of sales of nicotine e-liquid, its implication on e-cigarette use behaviours and alternative sources of nicotine e-liquid. J Community Health. 2017;42(6):1225–32.

[183] O'Brien EK, Nguyen AB, Persoskie A, Hoffman AC. US adults' addiction and harm beliefs about nicotine and low nicotine cigarettes. Prev Med. 2017;96:94–100.

[184] Denlinger-Apte RL, Joel DL, Strasser AA, Donny EC. Low nicotine content descriptors reduce perceived health risks and positive cigarette ratings in participants using very low nicotine content cigarettes. Nicotine Tob Res. 2017;19(10):1149–54.

[185] Mercincavage M, Saddleson ML, Gup E, Halstead A, Mays D, Strasser AA. Reduced nicotine content cigarette advertising: how false beliefs and subjective ratings affect smoking behavior. Drug Alcohol Depend. 2017;173:99–106.

[186] Bolcic-Jankovic D, Biener L. Public opinion about FDA regulation of menthol and nicotine. Tob Control. 2015;24(e4):e241–5.

[187] Kiviniemi MT, Kozlowski LT. Deficiencies in public understanding about tobacco harm reduction: results from a United States national survey. Harm Reduct J. 2015;12:21.

[188] Czoli CD, Fong GT, Mays D, Hammond D. How do consumers perceive differences in risk across nicotine products? A review of relative risk perceptions across smokeless tobacco, e-cigarettes, nicotine replacement therapy and combustible cigarettes. Tob Control. 2017;26(e1):e49–58.

[189] Darredeau C, Stewart SH, Barrett SP. The effects of nicotine content information on subjective and behavioural responses to nicotine-containing and denicotinized cigarettes. Behav Pharmacol. 2013;24(4):291–7.

[190] Mercincavage M, Smyth JM, Strasser AA, Branstetter SA. Reduced nicotine content expectancies affect initial responses to smoking. Tob Regul Sci. 2016;2(4):309–16.

[191] Harrell PT, Juliano LM. A direct test of the influence of nicotine response expectancies on the subjective and cognitive effects of smoking. Exp Clin Psychopharmacol. 2012;20(4):278–86.

[192] Gu X, Lohrenz T, Salas R, Baldwin PR, Soltani A, Kirk U et al. Belief about nicotine modulates subjective craving and insula activity in deprived smokers. Front Psychiatry. 2016;7:126.

[193] Harrell PT, Simmons VN, Piñeiro B, Correa JB, Menzie NS, Meltzer LR et al. E-cigarettes and expectancies: why do some users keep smoking? Addiction. 2015;110(11):1833–43.

[194] Pearson JL, Abrams DB, Niaura RS, Richardson A, Vallone DM. Public support for mandated nicotine reduction in cigarettes. Am J Public Health. 2013;103(3):562–7.

[195] Li J, Newcombe R, Walton D. Responses towards additional tobacco control measures: data from a population-based survey of New Zealand adults. N Z Med J 2016;129:87–92.

[196] Fraser T, Kira A. Perspectives of key stakeholders and smokers on a very low nicotine content cigarette-only policy: qualitative study. N Z Med J. 2017;130:36–45.

[197] Denlinger-Apte RL, Tidey JW, Koopmeiners JS, Hatsukami DK, Smith TT, Pacek LR et al. Correlates of support for a nicotine-reduction policy in smokers with 6-week exposure to very low nicotine cigarettes. Tob Control. 2018. pii: tobaccocontrol-2018-054622.

[198] Donny EC, Walker N, Hatsukami D, Bullen C. Reducing the nicotine content of combusted tobacco products sold in New Zealand. Tob Control. 2017;26(e1):e37–42.

[199] Laugesen M. Modelling a two-tier tobacco excise tax policy to reduce smoking by focusing on the addictive component (nicotine) more than the tobacco weight. N Z Med J. 2012;125:35–48.

[200] Yuke K, Ford P, Foley W, Mutch A, Fitzgerald L, Gartner C. Australian urban Indigenous smokers' perspectives on nicotine products and tobacco harm reduction. Drug Alcohol Rev. 2018;42(6):1225–32.

[201] Danielle LJ, Hatsukami DK, Hertsgaard L, Dermody SS, Donny EC. Gender differences in the perceptions of very low nicotine content cigarettes (abstract). Annual meeting of the College on Problems of Drug Dependence. Drug Alcohol Depend. 2014;140:e97.

[202] Mercincavage M, Wileyto EP, Saddleson ML, Lochbuehler K, Donny EC, Strasser AA. Attrition during a randomized controlled trial of reduced nicotine content cigarettes as a proxy for understanding acceptability of nicotine product standards. Addiction. 2017;112(6):1095–103.

5. 降低卷烟烟气中有害物质暴露的监管策略

Stephen S. Hecht and Dorothy K. Hatsukami, Masonic Cancer Center, University of Minnesota, Minneapolis (MN), USA

5.1 引　　言

2008年出版的《烟草制品管制科学基础报告》（WHO技术报告系列951）[1]详述了一种降低卷烟主流烟气中选定有害物质含量的方法。我们回顾了这些建议，并总结了烟草行业发表的几篇讨论监管策略方面的论文。我们还将讨论自2008年报告以来在评估卷烟烟气有害物质和致癌物的生物标志物方面取得的重大进展，以及这些生物标志物与卷烟烟气含量之间的关系，在某些情况下还将探讨其与疾病发病率之间的关系。我们考虑在产品监管策略中使用强制性有害物质含量，并根据卷烟中烟碱含量提出有关更新监管策略的建议。

5.2　WHO 技术报告系列 951 中所述的卷烟烟气成分监管

《烟草制品管制科学基础报告》（WHO技术报告系列951）认识到，在ISO标准抽吸模式下基于吸烟机测量的每支卷烟的焦油、烟碱和一氧化碳正通过误导吸烟者认为所谓的"低释放量卷烟"危害较小而造成危害。因此，该报告建议对某些烟气成分按照每毫克烟碱的方式设定限量值而不是按每支卷烟，因为人们吸烟是为了获得足够的烟碱来满足其生理和行为需要，并且卷烟烟气中的大多数有害物质与烟碱成比例地释放。这将重点转向标准条件下产生的烟气的潜在毒性。该报告还建议禁止将基于吸烟机测量的结果向消费者进行传播，因为这些信息很容易被误解。

该报告的结论是在监管策略中考虑生物标志物为时尚早，因为尽管当时存在暴露生物标志物，但没有经过验证的危害性生物标志物。如下所述，生物标志物用于评估卷烟烟气的潜在有害性和致癌性方面已经取得了重大进展。毫无疑问，某些暴露生物标志物的平均水平反映了卷烟烟气中前体化合物的释放量。

在WHO技术报告系列951中，TobReg基于多种因素选择了部分有害物质进行监管，包括确定的心血管和肺部毒性及致癌性，以及采用现有技术降低其浓度的可行性以及化学物质类别和品牌的变化，此外还考虑了在卷烟烟气的气相和粒相物中都包含的化合物。在这些特性中，最重要的是毒性证据。

加拿大卫生部的数据以及有关美国和其他地区菲利普·莫里斯国际品牌中的成分含量的论文可用于评估加拿大和美国市场上卷烟中选定的有害物质含量的差异[2]。这些数据可用作选择监管的有害成分的初始水平，作为从市场中淘汰含量较高的品牌的总体策略的第一步，从而降低了剩余品牌中有害物质的总体平均值。采用滚动方式最终将降低所有留在市场上的品牌中所选有害物质的水平，并阻止有害物质水平高于平均值的品牌进入市场。

该报告建议分阶段实施监管策略，首先是要求每年报告选定的有害物质含量，然后公布因为有害物质含量高于平均值而不能进行销售的卷烟品牌，最后实施设定的限量。在实现这个目标方面已经取得了一些进展，但是尚不清楚是否有足够的发展动力。

表5.1的第1~3列总结了建议强制降低的有害物质和最初的推荐水平。修改的加拿大深度抽吸模式被用来测量有害物质的含量。此抽吸模式是从ISO标准抽吸方法修改而来，将抽吸容量从35 mL增加到55 mL，并将抽吸间隔从60 s减少到30 s；此外，所有滤嘴通风孔均被封闭。该方法在报告中称为"改良的深度抽吸"方案。表5.1中"国际品牌"（即除美国品牌以外的所有其他品牌）的含量来自含有混合型烟草的美国风格卷烟品牌样品[2]。"加拿大品牌"中的含量是根据加拿大卫生部的报告得出的，适用于主要采用烤烟制造的卷烟。由于数据的可获得性和不同类型的烟草而选择了这些样本作为示例，并根据该市场中普遍存在的产品类型可以粗略地代表要监管的市场。除表5.1中列出的化合物外，建议报告的优先级有害物质为丙烯腈、4-氨基联苯、2-氨基萘、镉、邻苯二酚、巴豆醛、氰化氢、氢醌和氮氧化物[1]。2013年又增加了20种要考虑进行报告的化合物（请参阅第8章）。

表5.1　建议强制性降低的有害物质水平（加拿大深度抽吸模式，μg/mg烟碱）

有害成分	WHO技术报告系列951推荐含量[1]		出版的文献中的含量			选择建议含量的标准
	国际品牌[2]	加拿大品牌[a]	[3][b]	[4][c]	[5][d]	
NNK	0.072	0.047	0.080	0.050~0.072	0.0273	中位值
NNN	0.114	0.027	0.134	0.077~0.142	0.0916	中位值
乙醛	860	670	676	468~626	684.1	中位值的125%
丙烯醛	83	97	85.4	48.8~64.3	64.5	中位值的125%
苯	48	50	43.1	40.8~53.7	66.1	中位值的125%

续表

有害成分	WHO技术报告系列951推荐含量[1]		出版的文献中的含量			选择建议含量的标准
	国际品牌[2]	加拿大品牌[a]	[3][b]	[4][c]	[5][d]	
苯并[a]芘	0.011	0.011	0.0096	0.0074~0.0084	0.0085	中位值的125%
1,3-丁二烯	67	53	50.5	38.1~52.1	71.9	中位值的125%
一氧化碳	18 400	15 400	15 700	11 500~14 800	16 100	中位值的125%
甲醛	47	97	47.3	28.2~43.7	81.4	中位值的125%
烟碱（每支卷烟）	—	—	2.13mg	1.56~2.33 mg	1.6 mg	—

NNK：4-(甲基亚硝基氨基)-1-(3-吡啶基)-1-丁酮；NNN：N'-亚硝基降烟碱

a. 加拿大卫生部数据

b. 61个美国品牌平均值

c. 三个国际品牌的平均含量范围

d. 中国市场上20个品牌的平均值

表5.1中的化合物列表与监管高度相关，因为所有化合物均具有文献结果证明的毒性和/或致癌活性。FDA已发布了"烟草制品和卷烟烟气中有害和潜在有害成分（HPHC）"的清单，其中包括表5.1中的所有化合物以及建议用于报告的化合物[6]。FDA列出的HPHC重点关注与癌症、心血管疾病、生殖问题、成瘾和呼吸作用有关的化学物质，比此处列出的要长，有93种成分，因为它包括每一类别中每种已鉴定的有害或致癌化合物，而不是几种代表性物质（例如，表5.1中NNK和NNN作为烟草特有亚硝胺的代表，苯并[a]芘作为多环芳烃的代表）。

表5.1还列出了在WHO技术报告系列951[3-5]之后发表的三项企业研究中所报告的每毫克烟碱中选定有害物质的含量。一般而言，有害成分浓度的平均值和范围与世界卫生组织报告中的相似。一项企业研究报告了烟气中NNN和NNK的含量，并得出结论说它们正在降低[7]，但并未列在表5.1中。

WHO技术报告系列951还解决了有关成分水平和品牌之间的衡量标准的问题，并报告了各品牌成分水平之间的变异系数（一系列测量的标准偏差/该系列的平均值）。将其除以给定成分的多次测量的变化系数以得出比例（例如分析方法中的变化）。通过分析方法确定的比例高的成分（例如水平的较大差异）将是最合适的监管目标。报告的结论：菲利普·莫里斯国际品牌和加拿大品牌中个别有害物质的比例表明品牌间大多数有害物质的水平存在足够的差异，强制降低将对市场上剩余品牌有害成分的含量产生重大影响。

5.3 烟草行业对WHO技术报告系列951的响应

Bodnar等[3]对美国市售商品卷烟中选定的主流烟气成分进行了一项调查，根据"焦油"类别和卷烟设计参数将市场划分为13个层次。采用加拿大深度抽吸模式，并估算了在每毫克烟碱中主流烟气成分的释放量。根据ISO/FTC剑桥滤片方法的"焦油"类别，即"全味">"轻型">"超轻"，世界卫生组织《烟草控制框架公约》禁止使用这种说法，按每毫克烟碱校准后释放量与按每支卷烟计的成分释放量成反向排列。

Purkis等[8]审查了烟气成分的测量以及可能影响监管标准的变异性因素。他们讨论了方法开发、协调化和标准化，并审查了产品之间的差异、吸烟机抽吸模式和实验室测量值及其潜在后果等问题，例如数据误解。他们呼吁制定国际认可的测试标准和测量误差。

Haussmann[9]对使用危害性指数评估卷烟烟气成分进行了详细的讨论，包括WHO技术报告系列951中所述方法的优点和局限性。Haussmann得出结论，建议的方法的应用将导致50个品牌中的39个被禁止（在表5.1中称为"国际品牌"）。他证明排除一种品牌是因为一种成分的释放量超过了建议的上限，这可能意味着未考虑其他八种成分的释放量低于各自上限。结果将是增加而不是减少剩余品牌中这些成分的平均含量。因此，应仔细评估各种成分的推荐水平。

Piade等[10]测定了WHO技术报告系列951中262个商业品牌中18种优先级成分的含量（以每毫克烟碱表示），包括来自13个国家的多种混合型品牌。主成分分析用于识别负相关和其他模式。三个主要变量解释了大约75%的数据变异。第一个变量对气相和粒相物中化合物的相对含量比较敏感，而其他两个变量将美国混合型和维吉尼亚型卷烟归为一组，揭示了负相关性。例如，氮氧化物、氨或亚硝基芳香类化合物的含量与甲醛、丙烯醛、苯并[a]芘或二羟基苯负相关。他们得出的结论是监管方法可能会因这种负相关而混淆。

Belushkin等[11]研究了肯塔基参比卷烟和一个商业品牌在7年中多次分析的烟气成分释放量的变化。他们的研究结果表明某些烟气成分的释放量在统计学上有显著差异，但不一定与产品之间的差异相对应，因此应定义允差。他们得出结论，与t检验相比，使用可检测的最小差异和统计等效性比较更有意义。

Eldridge等[4]每月都检测参比卷烟和三种商品化卷烟烟气中的有害物质水平，共持续了10个月。大多数分析物的月变异率<15%，但当以烟碱水平的比例报告时，则有所增加。他们得出结论商品化卷烟的释放物的测量会有很大的变异，特别是对于低含量的有害物质。

Deng等[5]研究了测量不确定度对表5.1中所列出的9种成分监管建议的影响。在2012~2016年进行的共同实验中，对20个代表性卷烟样品的不确定度进行了评估。他们得出结论，测量不确定度将严重影响监管提案的实施。

总的来说，这些研究为每毫克烟碱中9种成分监管提案的实施提供了一些警示和实用说明。分析化学中测量的不确定性众所周知，但在大多数情况下相对较小，尤其是对于含量高的成分。已验证的方法可用于测量所有建议测量的化合物（请见参考文献[12-18]；也可参考http://www.who.int/tobacco/industry/product_regulation/TobLabNet/en/）。因此，可能不需要进行方法标准化的进一步研究，包括可能耗时数年的实验室间比对试验。此外，某些品牌的成分水平之间的负相关性可能会成为问题，因为降低一种成分的水平可能会增加其他成分的水平。几十年来，这一点已经为人们所知[19,20]，解决他们希望在市场上保持的产品中这些成分的含量水平是烟草行业的责任。总之，这些行业研究的主要目的是提出质疑和延迟卷烟烟气释放物产品标准的理由，没有有效的科学理由表明实施无法进行。

5.4 烟气成分含量与生物标志物之间的关系

烟草中的致癌物和有害物质的生物标志物，大多数是卷烟烟气成分的尿液代谢产物，是人体暴露于烟草制品成分的有力、有效和可靠的指标[21]。这些生物标记物中有几种与表5.1中列出的建议强制降低的有害物质直接相关。对于NNK，总的NNAL（游离NNAL及其葡萄糖苷酸总和）是可接受的暴露生物标志物[22]。对于丙烯醛，尿液中代谢物3-羟丙基巯基乙酸已被广泛用作丙烯醛的暴露生物标志物[23,24]。同样，对于苯和1,3-丁二烯，尿巯基酸、S-苯基巯基酸和羟丁基巯基尿酸已被定量分析[25-27]。芘的尿代谢物1-羟基芘已被广泛用作PAH的暴露生物标志物，PAH是包括苯并[a]芘的一类化合物[21,28]。呼出气中CO是吸烟者CO暴露的公认生物标志物[21]。总烟碱当量为尿液中烟碱、可替宁和3'-羟基可替宁及其葡萄糖苷的总和，在烟草使用者烟碱摄入剂量中占比很高，是烟碱剂量的极佳生物标志物[29]。

这些生物标记物中的大多数已在大型研究中进行了定量分析，例如加拿大的健康措施研究、美国的国家健康与营养检查调查、烟草行业赞助的总暴露量研究以及流行病学和临床研究[26,30-42]。当人们停止吸烟时，所有生物标志物的水平都会显著降低，并且所有尿液生物标志物都与总烟碱含量相关[24,26-28,43,44]。卷烟烟气中成分水平以及口腔中的丙烯醛、NNK、烟碱和芘的水平与各自的尿液生物标志物明显相关[33,34,38,41]。总体而言，这些研究表明，通过对生物标志物的评估可以建立烟气成分水平与实际人类暴露剂量之间的联系，可以支持法规中烟气成分的分析。此外，在前瞻性的流行病学研究中，烟碱生物标志物和总NNAL与肺癌呈正相关[45-47]，因此可被认为是潜在的危害生物标志物。

5.5 产品监管策略中有害物质水平的强制性降低

强制降低卷烟中有害物质含量对公共卫生的影响实际上是未知的。为减少因使用烟草制品而引起的疾病和死亡来改善公众健康，对产品标准的评估必须包括人群影响。根据美国《家庭吸烟预防和烟草控制法案》，该法案授予FDA在该国的烟草制品监管权，如果烟草制品标准"适合保护公众健康"，则可以采用该标准。评估产品标准对公众健康的影响时所考虑的科学证据包括：

- 推荐标准对包括烟草制品使用者和非使用者在内的整个人群的风险和利益；
- 现有烟草制品使用者停止使用此类产品的可能性增加或减少；
- 不使用烟草制品的人开始使用此类产品的可能性增加或减少。

迄今为止，一项研究表明，在卷烟中NNK和NNN含量较低的国家中，吸烟者尿中NNAL的含量较低[48]，但尚不清楚这些差异是否与特定国家/地区的癌症发病率有关。然而，已发现NNK和NNN暴露生物标志物与罹患肺癌[46,47,49]和食道癌[50]的风险之间存在剂量-效应关系。

为避免在实施强制性标准之前重复低焦油和低烟碱卷烟对公众健康负面影响的经验[51]，应确定其对吸烟行为、有害成分暴露和潜在危害的影响，以及消费者对受管制的烟草制品的认知和误解，以及对产品的接受和持续使用[52-54]，并根据需要采取适当的措施。WHO技术报告系列951中未作为监管目标，但在随后的咨询说明作为监管目标的一种成分[55]是烟碱。如上一节所述，对烟碱的管制可能会带来最大的公共卫生利益（见下文）。将烟碱降低到最低限度的成瘾水平将阻止对卷烟依赖性的发展并促进吸烟者戒烟，从而降低吸烟率。

值得注意的是，自发布WHO技术报告系列以来，某些市场已推出了诸如ENDS之类的产品，其有害成分含量和有害性均大大低于传统卷烟[56-58]。如果被证明有害性低得多的产品取代了卷烟，并且仅在吸烟者中使用，则公众健康影响可能是重大的，前提是不存在不良后果，例如儿童和/或年轻人的不当使用，以及与传统卷烟或其他烟草制品双重使用。

5.6 建议强制性降低的有害成分和建议的限量

Donny等[40]获得了可能会改变监管格局的高度相关的结果。他们对烟碱含量为15.8 mg/g烟草到0.4 mg/g不等的、美国典型的商品化卷烟品牌，进行了双盲、平行、随机临床试验。定期吸烟者被随机分配吸食低烟碱卷烟或他们通常的品牌卷烟6周。在第6周，被分配到每克烟草含2.4 mg、1.3 mg或0.4 mg烟碱的卷烟（分别为16.5支/天、16.3支/天和14.9支/天）的参与者每天吸烟数量明显低于随机分配到其通常品牌的卷烟或含15.8 mg/g烟草的卷烟（分别为22.2支/天和21.3支/天）的参与者。烟碱含量较低的卷烟减少了吸烟者对烟碱的接触和依赖性，并且在戒烟期间仍渴望吸烟，且没有增加呼出气体中的CO水平或总抽吸量，这表明补偿抽吸很小。这些结果与几个较小试验[30,59-63]和针对成年弱势人群的试验[64]的结果是一致的。

在Donny等[40]研究中，分配到每克烟草中≤2.4 mg烟碱的卷烟的参与者与分配到含15.8 mg/g烟草的参与者相比，第6周每天抽吸的卷烟量减少了23%~30%。结果显示，在含烟碱为5.2 mg/g卷烟与2.4 mg/g的卷烟之间每天吸烟量和尿液中烟碱含量之间有明显的断点。此外，最重要的是，在威斯康星州的吸烟依赖性评估实验中，在第6周时，抽吸含烟碱为0.4 mg/g卷烟的吸烟数量显著低于吸15.8 mg/g卷烟的（$P=0.001$）。在HCI条件下，含烟碱为0.4 mg/g的卷烟在烟气中烟碱的释放量约为0.04 mg[65]。

这些结果表明，产品标准对于拥有适当资源的国家可能具有重大的公共卫生益处（更多细节请参阅参考文献[55]）。卷烟烟丝中烟碱的含量应加以监管，以使在HCI条件下卷烟主流烟气中烟碱的含量不超过0.04 mg/支卷烟（比当前品牌低39~58倍，表5.1）。对于其他成分，根据最新的市场调查信息，新的推荐水平是表5.1中列出每毫克烟碱的中位值，该信息可从主要烟草公司和加拿大卫生部的数据中获得，例如在HCI条件下每支卷烟可提供0.04 mg烟碱。该建议比TobReg先前确定的"中位值的125%"强。我们还保留了报告丙烯腈、4-氨基联苯、2-氨基萘、镉、邻苯二酚、巴豆醛、氰化氢、氢醌和氮氧化物含量的需求。这些有害物质满足以下几项选择标准：充分的致癌性证据、已知的呼吸道或心脏毒性、不同国家不同品牌中有害成分含量水平的差异、可测量性和产品降低释放物的潜力[1]。

我们注意到基于上述临床研究中使用的Spectrum卷烟的数据当以每毫克烟碱表示时，有害成分含量将高于表5.2中列出的现行强制性有害成分含量[65]。显然，建议的烟碱含量0.04 mg每支卷烟是一个比较低的含量水平，而目前的含量水平为1.5~2 mg烟碱/支。然而，生物标志物的测量清楚地表明，这些低烟碱卷烟的吸烟者尿液中总烟碱含量和总NNAL含量显著降低，因此有害成分和致癌化合物的摄入量较少[40]。

表5.2 卷烟主流烟气中某些有害物质的含量（NRC 102，0.04 mg烟碱/支，加拿大深度抽吸模式）[65]

有害成分	μg/mg烟碱
乙醛	41 500
丙烯醛	2 070
苯并[a]芘	0.220
一氧化碳	720 000
甲醛	258
NNK	0.798
NNN	2.88

NNK：4-(甲基亚硝基氨基)-1-(3-吡啶基)-1-丁酮；NNN：N'-亚硝基降烟碱

5.7　强制性降低有害物质水平的实施

为了使监管机构有效地降低卷烟中有害物质和烟碱的含量，需要适当的程序以及充足的基础设施和资源。这些措施包括要求公开卷烟中有害成分的含量，并要求有独立的实验室来定期验证和监测这些有害成分。在上一份WHO报告中[1]，分析卷烟的组成被认为是减少有害物质的第一步。应建立标准化的测试程序，不仅包括实验室方法（例如，反映出人体暴露范围的机器测量方法和经过验证的分析化学方法），还应包括卷烟被测试的方式（例如，不同的地理位置、同一位置的不同零售商、在不同月份直接从制造商处购买等）。还应考虑跟踪合法和非法卷烟，以及减少非法卷烟进入市场。此外，应该设计教育运动和信息传播，以最大限度地减少任何与吸烟有关危害的误解。因此，无论是否要求降低有害物质的含量，都必须明确、清楚地传达燃烧产物对健康有害的信息。最后，应建立一个监测系统以监测任何意外后果，并应考虑纠正方法。

5.8　结论与建议

此处建议的监管策略保留了TobReg先前建议的原则，并对其进行了修订和增强，以使市场上的品牌不能超过表5.1中所列成分的中位值，还保留了报告其他几个成分的要求。此外，我们建议考虑一种管制策略：烟草中的烟碱含量不超过0.4 mg/g烟草（在HCI抽吸条件下，卷烟主流烟气中每支卷烟的烟碱含量为0.04 mg）。基于烟碱的新的监管是数十年研究的结果，这些研究表明烟碱是促使人们吸烟的主要化学物质[66,67]，以及临床试验的结果表明低烟碱的卷烟减少了对烟碱的接触和依赖性，并减少了吸烟量。因此，我们建议同时采用限制特定有害成分的先前策略和限制烟碱的新建议。减少有害物质的第一个关键步骤是通过WHO烟草实验室网络（TobLabNet）建立的方法收集当前产品的数据。

我们认识到在实施卷烟产品标准时可能会遇到挑战，并且可能会产生意想不到的后果（例如非法贸易）。因此，有必要加强监管，以将有害物质和烟碱含量超过既定含量的卷烟从市场上剔除。必须建立一个监测烟草和卷烟烟气中受管制成分的可靠系统。必须对国际市场上的代表性卷烟进行分析，以便报告其烟气中可能含有的有害物质的平均水平以及烟草中烟碱的平均含量，此外注意监测以确保新的监管措施不会造成损害。

5.9 参考文献

[1] The scientific basis of tobacco product regulation. Second report of a WHO study group (WHO Technical Report Series, No. 951). Geneva: World Health Organization; 2008: 45–277.

[2] Counts ME, Morton MJ, Laffoon SW, Cox RH, Lipowicz PJ. Smoke composition and predicting relationships for international commercial cigarettes smoked with three machine-smoking conditions. Regul Toxicol Pharmacol. 2005;41(3):185–227.

[3] Bodnar JA, Morgan WT, Murphy PA, Ogden MW. Mainstream smoke chemistry analysis of samples from the 2009 US cigarette market. Regul Toxicol Pharmacol. 2012;64(1):35–42.

[4] Eldridge A, Betson TR, Gama MV, McAdam K. Variation in tobacco and mainstream smoke toxicant yields from selected commercial cigarette products. Regul Toxicol Pharmacol. 2015;71(3):409–27.

[5] Deng H, Li Z, Bian Z, Yang F, Liu S, Fan Z et al. Influence of measurement uncertainty on the World Health Organization recommended regulation for mainstream cigarette smoke constituents. Regul Toxicol Pharmacol. 2017;86:231–40.

[6] Food and Drug Administration. Harmful and potentially harmful constituents in tobacco products and tobacco smoke; established list. Fed Reg. 2012;77:20034–7.

[7] Gunduz I, Kondylis A, Jaccard G, Renaud JM, Hofer R, Ruffieux L et al. Tobacco-specific N-nitrosamines NNN and NNK levels in cigarette brands between 2000 and 2014. Regul Toxicol Pharmacol. 2016;76:113–20.

[8] Purkis SW, Meger M, Wuttke R. A review of current smoke constituent measurement activities and aspects of yield variability. Regul Toxicol Pharmacol. 2012;62(1):202–13.

[9] Haussmann HJ. Use of hazard indices for a theoretical evaluation of cigarette smoke composition. Chem Res Toxicol. 2012;25(4):794–810.

[10] Piade JJ, Wajrock S, Jaccard G, Janeke G. Formation of mainstream cigarette smoke constituents prioritized by the World Health Organization – yield patterns observed in market surveys, clustering and inverse correlations. Food Chem Toxicol. 2013;55:329–347.

[11] Belushkin M, Jaccard G, Kondylis A. Considerations for comparative tobacco product assessments based on smoke constituent yields. Regul Toxicol Pharmacol. 2015;73(1):105–13.

[12] Eschner MS, Selmani I, Groger TM, Zimmermann R. Online comprehensive two-dimensional characterization of puff-by-puff resolved cigarette smoke by hyphenation of fast gas chromatography to single-photon ionization time-of-flight mass spectrometry: quantification of hazardous volatile organic compounds. Anal Chem. 2011;83(17):6619–27.

[13] Pang X, Lewis AC. Carbonyl compounds in gas and particle phases of mainstream cigarette smoke. Sci Total Environ. 2011;409(23):5000–9.

[14] Pappas RS, Fresquez MR, Martone N, Watson CH. Toxic metal concentrations in mainstream smoke from cigarettes available in the USA. J Anal Toxicol. 2014;38(4):204–11.

[15] Vu AT, Taylor KM, Holman MR, Ding YS, Hearn B, Watson CH. Polycyclic aromatic hydrocarbons in the mainstream smoke of popular US cigarettes. Chem Res Toxicol. 2015;28(8):1616–26.

[16] Ding YS, Yan X, Wong J, Chan M, Watson CH. In situ derivatization and quantification of seven

carbonyls in cigarette mainstream smoke. Chem Res Toxicol. 2016;29(1):125–31.

[17] Cecil TL, Brewer TM, Young M, Holman MR. Acrolein yields in mainstream smoke from commercial cigarette and little cigar tobacco products. Nicotine Tob Res. 2017;19(7):865–70.

[18] Edwards SH, Rossiter LM, Taylor KM, Holman MR, Zhang L, Ding YS et al. Tobacco-specific nitrosamines in the tobacco and mainstream smoke of U.S. commercial cigarettes. Chem Res Toxicol. 2017;30(2):540–51.

[19] Adams JD, Lee SJ, Hoffmann D. Carcinogenic agents in cigarette smoke and the influence of nitrate on their formation. Carcinogenesis. 1984;5:221–3.

[20] Ding YS, Zhang L, Jain RB, Jain N, Wang RY, Ashley DL et al. Levels of tobacco-specific nitro samines and polycyclic aromatic hydrocarbons in mainstream smoke from different tobacco varieties. Cancer Epidemiol Biomarkers Prev. 2008;17(12):3366–71.

[21] Hecht SS, Yuan JM, Hatsukami DK. Applying tobacco carcinogen and toxicant biomarkers in product regulation and cancer prevention. Chem Res Toxicol. 2010;23:1001–8.

[22] Hecht SS, Stepanov I, Carmella SG. Exposure and metabolic activation biomarkers of carcinogenic tobacco-specific nitrosamines. Acc Chem Res. 2016;49(1):106–14.

[23] Yan W, Byrd GD, Brown BG, Borgerding MF. Development and validation of a direct LCMS-MS method to determine the acrolein metabolite 3-HPMA in urine. J Chromatogr Sci. 2010;48(3):194–9.

[24] Park SL, Carmella SG, Chen M, Patel Y, Stram DO, Haiman CA et al. Mercapturic acids derived from the toxicants acrolein and crotonaldehyde in the urine of cigarettes smokers from five ethnic groups with differing risks for lung cancer. PLoS One. 2015;10:e0124841.

[25] Dougherty D, Garte S, Barchowsky A, Zmuda J, Taioli E. NQO1, MPO, CYP2E1, GSTT1 and GSTM1 polymorphisms and biological effects of benzene exposure – a literature review. Toxicol Lett. 2008;182(1–3):7–17.

[26] Roethig HJ, Munjal S, Feng S, Liang Q, Sarkar M, Walk RA et al. Population estimates for biomarkers of exposure to cigarette smoke in adult US cigarette smokers. Nicotine Tob Res. 2009;11(10):1216–25.

[27] Haiman CA, Patel YM, Stram DO, Carmella SG, Chen M, Wilkens L et al. Benzene uptake and glutathione S-transferase T1 status as determinants of S-phenylmercapturic acid in cigarette smokers in the Multiethnic Cohort. PLoS One. 2016;11(3):e0150641.

[28] Suwan-ampai P, Navas-Acien A, Strickland PT, Agnew J. Involuntary tobacco smoke exposure and urinary levels of polycyclic aromatic hydrocarbons in the United States, 1999 to 2002. Cancer Epidemiol Biomarkers Prev. 2009;18(3):884–93.

[29] Hukkanen J, Jacob P III, Benowitz NL. Metabolism and disposition kinetics of nicotine. Pharmacol Rev. 2005;57(1):79–115.

[30] Benowitz N, Hall SM, Stewart S, Wilson M, Dempsey D, Jacob P III. Nicotine and carcinogen exposure with smoking of progressively reduced nicotine content cigarette. Cancer Epidemiol Biomarkers Prev. 2007;16(11):2479–85.

[31] Le Marchand L, Derby KS, Murphy SE, Hecht SS, Hatsukami D, Carmella SG et al. Smokers with the CHRNA lung cancer-associated variants are exposed to higher levels of nicotine equivalents and a carcinogenic tobacco-specific nitrosamine. Cancer Res. 2008;68(22):9137–40.

[32] Lowe FJ, Gregg EO, McEwan M. Evaluation of biomarkers of exposure and potential harm in

smokers, former smokers and never-smokers. Clin Chem Lab Med. 2009;47(3):311–20.

[33] Shepperd CJ, Eldridge AC, Mariner DC, McEwan M, Errington G, Dixon M. A study to estimate and correlate cigarette smoke exposure in smokers in Germany as determined by filter analysis and biomarkers of exposure. Regul Toxicol Pharmacol. 2009;55(1):97–109.

[34] Morin A, Shepperd CJ, Eldridge AC, Poirier N, Voisine R. Estimation and correlation of cigarette smoke exposure in Canadian smokers as determined by filter analysis and biomarkers of exposure. Regul Toxicol Pharmacol. 2011;61(3 Suppl):S3–12.

[35] Xia Y, Bernert JT, Jain RB, Ashley DL, Pirkle JL. Tobacco-specific nitrosamine 4-(methylnitrosamino)-1-(3-pyridyl)-1-butanol (NNAL) in smokers in the United States: NHANES 2007–2008. Biomarkers. 2011;16(2):112–9.

[36] Appleton S, Olegario RM, Lipowicz PJ. TSNA exposure from cigarette smoking: 18 years of urinary NNAL excretion data. Regul Toxicol Pharmacol. 2014;68(2):269–74.

[37] Murphy SE, Park S-SL, Thompson EF, Wilkens LR, Patel Y, Stram DO et al. Nicotine N-glucuronidation relative to N-oxidation and C-oxidation and UGT2B10 genotype in five ethnic/racial groups. Carcinogenesis. 2014;35:2526–33.

[38] Sakaguchi C, Kakehi A, Minami N, Kikuchi A, Futamura Y. Exposure evaluation of adult male A regulatory strategy for reducing exposure to toxicants in cigarette smoke Japanese smokers switched to a heated cigarette in a controlled clinical setting. Regul Toxicol Pharmacol. 2014;69(3):338–47.

[39] Czoli C, Hammond D. TSNA exposure: levels of NNAL among Canadian tobacco users. Nicotine Tob Res. 2015;17(7):825–30.

[40] Donny EC, Denlinger RL, Tidey JW, Koopmeiners JS, Benowitz NL, Vandrey RG et al. Randomized trial of reduced-nicotine standards for cigarettes. N Engl J Med. 2015;373(14):1340–9.

[41] Theophilus EH, Coggins CR, Chen P, Schmidt E, Borgerding MF. Magnitudes of biomarker reductions in response to controlled reductions in cigarettes smoked per day: a one-week clinical confinement study. Regul Toxicol Pharmacol. 2015;71(2):225–34.

[42] Wei B, Blount BC, Xia B, Wang L. Assessing exposure to tobacco-specific carcinogen NNK using its urinary metabolite NNAL measured in US population: 2011–2012. J Expo Sci Environ Epidemiol. 2016;26(3):249–56.

[43] Carmella SG, Chen M, Han S, Briggs A, Jensen J, Hatsukami DK et al. Effects of smoking cessation on eight urinary tobacco carcinogen and toxicant biomarkers. Chem Res Toxicol. 2009;22(4):734–41.

[44] Park SL, Carmella SG, Ming X, Stram DO, Le Marchand L, Hecht SS. Variation in levels of the lung carcinogen NNAL and its glucuronides in the urine of cigarette smokers from five ethnic groups with differing risks for lung cancer. Cancer Epidemiol Biomarkers Prev. 2015;24:561–9.

[45] Church TR, Anderson KE, Caporaso NE, Geisser MS, Le C, Zhang Y et al. A prospectively measured serum biomarker for a tobacco-specific carcinogen and lung cancer in smokers. Cancer Epidemiol Biomarkers Prev. 2009;18:260–6.

[46] Yuan JM, Butler LM, Stepanov I, Hecht SS. Urinary tobacco smoke-constituent biomarkers for assessing risk of lung cancer. Cancer Res. 2014;74(2):401–11.

[47] Yuan JM, Nelson HH, Carmella SG, Wang R, Kuriger-Laber J, Jin A et al. CYP2A6 genetic

[48] polymorphisms and biomarkers of tobacco smoke constituents in relation to risk of lung cancer in the Singapore Chinese Health Study. Carcinogenesis. 2017;38(4):411–8.

[48] Ashley DL, O'Connor RJ, Bernert JT, Watson CH, Polzin GM, Jain RB et al. Effect of differing levels of tobacco-specific nitrosamines in cigarette smoke on the levels of biomarkers in smokers. Cancer Epidemiol Biomarkers Prev. 2010;19(6):1389–98.

[49] Church TR, Anderson KE, Caporaso NE, Geisser MS, Le CT, Zhang Y et al. A prospectively measured serum biomarker for a tobacco-specific carcinogen and lung cancer in smokers. Cancer Epidemiol Biomarkers Prev. 2009;18(1):260–6.

[50] Yuan JM, Knezevich AD, Wang R, Gao YT, Hecht SS, Stepanov I. Urinary levels of the tobaccospecific carcinogen N′-nitrosonornicotine and its glucuronide are strongly associated with esophageal cancer risk in smokers. Carcinogenesis. 2011;32(9):1366–71.

[51] Risks associated with smoking cigarettes with low machine-measured yields of tar and nicotine. Bethesda (MD): Department of Health and Human Services, National Institutes of Health, National Cancer Institute; 2001.

[52] Stratton K, Shetty P, Wallace R, Bondurant S, editors. Clearing the smoke: assessing the science base for tobacco harm reduction. Washington, DC: Institute of Medicine; 2001.

[53] Hatsukami DK, Biener L, Leischow SJ, Zeller MR. Tobacco and nicotine product testing. Nicotine Tob Res. 2012;14(1):7–17.

[54] Berman ML, Connolly G, Cummings KM, Djordjevic MV, Hatsukami DK, Henningfield JE et al. Providing a science base for the evaluation of tobacco products. Tob Regul Sci. 2015;1(1):76–93.

[55] WHO Study Group on Tobacco Product Regulation. Advisory note: global nicotine reduction strategy (WHO Technical Report Series No. 989). Geneva: World Health Organization; 2015.

[56] Nicotine without smoke: tobacco harm reduction. London: Royal College of Physicians; 2016.

[57] Benowitz NL, Fraiman JB. Cardiovascular effects of electronic cigarettes. Nat Rev Cardiol. 2017;14(8):447–56.

[58] Shahab L, Goniewicz ML, Blount BC, Brown J, McNeill A, Alwis KU et al. Nicotine, carcinogen, and toxin exposure in long-term e-cigarette and nicotine replacement therapy users: a crosssectional study. Ann Intern Med. 2017;166(6):390–400.

[59] Donny EC, Houtsmuller E, Stitzer ML. Smoking in the absence of nicotine: behavioral, subjective and physiological effects over 11 days. Addiction. 2007;102(2):324–34.

[60] Benowitz NL, Dains KM, Dempsey D, Herrera B, Yu L, Jacob P III. Urine nicotine metabolite concentrations in relation to plasma cotinine during low-level nicotine exposure. Nicotine Tob Res. 2009;11(8):954–60.

[61] Hatsukami DK, Kotlyar M, Hertsgaard LA, Zhang Y, Carmella SG, Jensen JA et al. Reduced nicotine content cigarettes: effects on toxicant exposure, dependence and cessation. Addiction. 2010;105:343–55.

[62] Benowitz NL, Dains KM, Hall SM, Stewart S, Wilson M, Dempsey D et al. Smoking behavior and exposure to tobacco toxicants during 6 months of smoking progressively reduced nicotine content cigarettes. Cancer Epidemiol Biomarkers Prev. 2012;21(5):761–9.

[63] Hatsukami DK, Hertsgaard LA, Vogel RI, Jensen JA, Murphy SE, Hecht SS et al. Reduced nicotine content cigarettes and nicotine patch. Cancer Epidemiol Biomarkers Prev.

2013;22(6):1015-24.

[64] Higgins ST, Heil SH, Sigmon SC, Tidey JW, Gaalema DE, Hughes JR et al. Addiction potential of cigarettes with reduced nicotine content in populations with psychiatric disorders and other vulnerabilities to tobacco addiction. JAMA Psychiatry. 2017;74(10):1056-64.

[65] Ding YS, Richter P, Hearn B, Zhang L, Bravo R, Yan X et al. Chemical characterization of mainstream smoke from SPECTRUM variable nicotine research cigarettes. Tob Regul Sci. 2017;3(1):81-94.

[66] The health consequences of smoking. Nicotine addiction. A report of the Surgeon General. Rockville (MD): Department of Health and Human Services; 1988.

[67] How tobacco smoke causes disease: the biology and behavioural basis for smoking-attributable disease. A report of the Surgeon General. Atlanta (GA): Centers for Disease Control and Prevention; 2010.

6. 烟草制品中调味剂的科学研究

Suchitra Krishnan-Sarin and Stephanie S. O'Malley, Yale School of Medicine, New Haven (CT), USA

Barry G. Green, John B. Pierce Laboratory and Yale School of Medicine, New Haven (CT), USA

Sven-Eric Jordt, Duke University School of Medicine, Durham (NC), USA

6.1 引 言

大多数常规、传统、新型烟草制品都包含调味剂，包括卷烟、雪茄、细支雪茄、无烟烟草制品（包括口含烟）、水烟和含烟碱（也称为电子烟碱传输系统，ENDS）和不含烟碱（也称为电子非烟碱传输系统，ENNDS）的电子烟。广义上讲，调味是指经口摄入或吸入某些东西时产生的感官体验[1]。烟草本身就有味道（例如"天然烟草"香味），这取决于烟草的类型和烘烤过程[2]，当然很多产品也会添加调味剂。一些调味剂是从天然产品中提取的，例如可可、甘草、蜂蜜和蔗糖，而其他调味剂则是人工合成的，例如有巧克力味的吡嗪类物质和有甜味的三氯蔗糖。在烟草制品监管的背景下，在一定程度上呈现出强烈的非烟草气味或味道被视为"特征性香味"。尽管"特征性"一词的使用存在很多争议，并且在全世界范围内并未使用，但欧盟烟草制品指令[3]将特征性香味定义为：

由一种或多种添加剂组成，包括但不限于水果、香辛料、药草、酒精、糖果、薄荷醇或香草，在烟草制品消费前或消费过程中具有明显不同于烟草的明显气味或味道。

FDA并未直接定义"特征性香味"，而是在2009年禁止风味卷烟的指南中使用了以下措辞[4]：

……卷烟或卷烟的任何组成部分（包括烟草、滤嘴或滤纸），作为成分（包括烟气成分）或添加剂，含有人工或天然香料（烟草或薄荷醇除外）或草药或香辛料，包括草莓、葡萄、柑橘、丁香、肉桂、菠萝、香草、椰子、甘草、可可、巧克力、樱桃或咖啡，这是烟草制品或卷烟烟气的特征风味。

烟草行业已经将调味剂引入烟草制品中以增加烟草制品的吸引力。人们认为向这些产品中添加了调味剂以减少刺激性,增加吸引力并改善适口性。此外,在包装和广告材料上显示调味剂的名称或图形可以增强产品的吸引力。

我们在此提供了在世界范围内使用调味烟草和烟碱产品的流行病学简要概述以及对这些产品的看法。我们还将讨论已应用于调味产品的技术创新,包括它们如何提高吸引力。我们提供了基于香味认知的感官过程,已知调味剂的药理学靶标以及调味剂潜在有害性证据的概述。自始至终突出了对研究和监管的影响。

6.2 调味烟草和烟碱产品使用的流行病学

调味烟草和烟碱产品在世界范围内使用。尽管有关这些产品的可用性和使用的系统证据有限,但偏好和使用通常是针对特定的国家和地区。例如,在印度尼西亚和其他南亚国家,含有丁香块、油脂和香料的传统丁香卷烟(kretek)十分受欢迎。在印度,含有混合香料的无烟烟草制品(pan masala、gutka)比较畅销,其中混合香料包含食品香料和油、调味料、槟榔和其他成分。水烟起源于印度和中东,现在在欧洲和北美的年轻人中越来越流行,它使用了很重的香料和加糖的烟草,即maassel。

烟草行业长期以来将烟草制品中的调味剂用作营销策略[5],包括面向年轻人[6]。例如,2013年在美国含有调味剂的非卷烟烟草制品包括52.3%的"卷烟大小的雪茄"、81.3%的"雪茄包衣"("小雪茄包衣"、"烟草包衣"和"包衣")、55.1%的湿鼻烟、86.1%的水烟和81.5%的可溶解烟草制品(例如Ariva、Camel Orb/Sticks)[7]。对市场趋势的纵向研究显示,2005~2011年间,无烟烟草销量增长的59.4%是由于调味剂的存在[8]。最近推出了多种口味的含烟碱电子烟[9,10]。

调查表明,调味烟草的使用率很高,尽管在世界范围内使用这些产品的系统证据很少。2013~2014年美国全国成人烟草调查显示,在过去30天里估计有1020万电子烟使用者(68.2%)、610万水烟使用者(82.3%)、410万雪茄使用者(36.2%)和400万无烟烟草制品使用者(50.6%)曾使用调味烟草制品[11]。还评估了多种口味的调味剂(如薄荷醇/薄荷、丁香/香辛料/香草、水果、酒精、糖果/巧克力/其他甜味)的添加。根据烟草制品类型,最普遍使用的调味剂是,无烟烟草制品:薄荷醇/薄荷(76.9%);水烟:水果(74.0%);雪茄、细支雪茄,有滤嘴的小雪茄:水果(52.4%)、糖果、巧克力和其他甜味香料(22.0%)、酒精(14.5%);电子烟:水果(44.9%)、薄荷醇/薄荷(43.9%)以及糖果、巧克力和其他甜味香料(25.7%);斗烟:水果(56.6%)、糖果、巧克力和其他甜味剂(26.5%)、薄荷/薄荷(24.8%)[11]。根据2009~2010年全球成人烟草调查的数据,波兰有26%的女性吸烟者和10.5%的男性吸烟者报告了目前使用了调味卷烟(薄荷醇、香草或其他调味剂)[12]。

调味烟草制品在年轻人中非常受欢迎[13-15]。在使用各种烟草制品(卷烟、斗烟、雪茄、细支雪茄、*bidis*、无烟烟草制品、水烟、小雪茄、自卷烟等）的加拿大年轻人中，有52%的人报告使用调味烟草制品[16]。同样，波兰进行的一项全国调查得出的证据表明，年轻的吸烟者更可能使用调味卷烟[12]。在美国进行的烟草健康人群评估研究的证据[15]表明使用调味烟草制品的年龄在12~17岁的儿童中最高（80%），在18~24岁的烟草使用者中使用率为73%，在≥65岁的使用者中最低（29%）。大多数儿童（81%）和年轻人（86%）以及年龄在25岁以上的成年人中只有54%的人表示他们的第一款产品含有调味剂。此外，在年轻人和成年人中，首次使用调味烟草制品与当前烟草使用的较高流行率有关。

总结：未来有关调味烟草制品的监管应按国家和人群的可用性（营销和获取）和使用方面的重大差异进行分类识别。不幸的是，当前的监测工具可能不足以监测这些产品的使用。例如，全球青少年烟草调查[17]是监测青少年烟草使用的重要全球工具，它并没有要求专门提供有关调味烟草制品使用的信息。在全球成人烟草调查中[18]，仅针对水烟研究了加香和不加香的使用情况，只有通过分析有关卷烟品牌的文字回复才能获得有关调味卷烟的信息。为了解决这一信息空缺，可以将有关含调味剂和不含调味剂的其他问题添加到这些调查的迭代中。监测工具还应评估各种调味剂的使用情况。鉴于薄荷醇通常与其他调味剂分开监管，因此建议将薄荷醇或薄荷与其他香料区分开来。

6.3 调味产品：认知、尝试、摄入和监管

人们普遍认为调味烟草制品比其他烟草制品危害小[15,19-21]。调味烟草和烟碱产品的使用者和非使用者对各种产品都有良好的看法（例如雪茄、电子烟、水烟、丁卷烟、*bidis*和无烟烟草制品）[22]。调味的非薄荷类烟草制品（水烟、细支雪茄、电子烟）被认为比传统卷烟危害性小[20]。

不足为奇的是，烟草和烟碱产品中存在调味剂与尝试这些产品的意愿更大相关[20,23,24]，因为这是年轻人重要的考虑因素。调味剂的可获得性与开始使用各种烟草制品有关，包括水烟、电子烟和雪茄[25-27]。调味剂的存在还可以促进从尝试到常规使用的转变。电子烟中存在的调味剂更多与青少年使用电子烟的频率较高相关，而在成年人中则不存在[28]。此外，显示出甜味和水果味的广告可以激活年轻人的大脑奖赏区域并干扰健康警示[29]。例如，在美国，尝试过调味烟草的青少年成为当前吸烟者的可能性为未尝试过这些产品的青少年的3倍[30]，并且当前吸烟或烟碱使用者中超过80%的人报告说首次使用了调味烟草制品[15]。

正如烟草法律控制协会[31]所述，为了减少吸引力、尝试和摄入，许多国家已经对调味烟草制品进行了管制。一些国家禁止使用除薄荷醇以外的具有特征性风味的卷烟（例如在欧盟国家），而其他一些国家则将禁令扩展至其他烟草制品，

例如美国、加拿大（小雪茄、小雪茄包衣）和埃塞俄比亚（所有产品）。一些国家已经立法，禁止在卷烟和其他烟草制品中使用薄荷醇作为调味剂。例如，土耳其禁止添加任何水平的薄荷醇和任何薄荷醇衍生物（例如薄荷）；加拿大最终禁止在加拿大市场上销售的卷烟、小雪茄包衣和大多数雪茄烟中添加任何剂量的薄荷醇。加拿大还禁止使用任何促销材料，包括描述包含糖果、甜点、大麻、汽水或能量饮料口味的包装。

总结：烟草和烟碱产品中特定调味剂的存在以及包装和广告中香味描述的存在都与烟草和烟碱产品的吸引力和摄入有关，特别是对青少年。鉴于这些联系，许多国家已经对调味烟草制品进行了管制。鉴于不同国家和地方法规在时间和特殊性上的差异，有必要进行监测研究（例如，在国际烟草控制项目或其他监测系统中）以研究监管对青少年摄入烟草、烟草销售和使用方式的影响，以及烟草行业对监管的响应。

6.4 烟草和电子烟行业的调味烟草发展简史

烟草行业在整合调味剂科学和技术以提高产品吸引力和销售能力方面一直处于最前沿。一个早期的例子是雷诺烟草公司在18世纪90年代将糖精添加到嚼烟中。这种高效能的人造甜味剂代替了更昂贵的糖，并延长了产品的保质期和一致性。烟草行业继续改进烟草育种、选种、烤制和制造方法，以控制烟草烟气的刺鼻性，通过烟丝加香、加料、保润剂（保湿剂），并设计包装纸和滤嘴系统，以提高吸引力和香味传递的均匀性。

几十年来，烟草公司一直是世界市场上可可和甘草的主要消费者。两者都被添加到卷烟中作为调味料，不是为了控制香味体验，而是要增加口味饱满度并降低烟气刺鼻感。相比之下，薄荷卷烟中的薄荷醇是一种特征性香料，是在产品标签上列出的非烟草香料，在香味体验中占主导地位[32]。薄荷卷烟于19世纪20年代首次在美国销售，并且仍然很受欢迎。许多未标有薄荷醇的卷烟也可能含有低含量的薄荷醇[33]，足以增强产品的吸引力[34]。

尽管大多数早期产品都包含天然香料，包括从薄荷提取的薄荷醇、水果提取物以及天然萜烯和醛类，但烟草行业对规范化的调味剂需求不断增加，而天然来源的供应有限，促使供应商开发和优化了化学合成工艺以实现大量关键调味剂的规模化生产。例如，当今大部分的薄荷醇供应是由三个主要的化学公司生产的，这些公司开发了立体选择性合成或纯化L-薄荷醇的工艺[35]。同样，非薄荷醇产品，包括香草醛（香草味）、吡嗪（咖啡和巧克力香味）、醛类（如苯甲醛，浆果和糖果香味）和许多其他烟草调味剂都是化学合成的，通常来自石油化工碳氢化合物前体物[35]。现代加香卷烟制品包含经过精炼和复配的调味剂混合物，以产生特征性香味，经过烟草工业调味剂研发小组和消费者的测试后，将其推向市场。

通常调味剂是通过加入烟草料液或喷洒在烟草上的保润剂、滤嘴以及烟草包装的内衬纸中的方法添加到烟草制品中。烟草和调味剂行业持续引入新的调味剂和调味剂传递系统，以使产品多样化并提高吸引力。例如，在过滤嘴中引入爆珠胶囊形成了一类新的卷烟，当捏碎滤嘴中的爆珠时它们会释放出香味。含有薄荷醇的爆珠胶囊卷烟2007年在日本首次推出，随后，2008年在欧洲和美国推出，2011~2012年在澳大利亚和墨西哥推出[36]。产品中的大多数爆珠都含有薄荷醇，但也可能含有醛或其他增香剂，以创造出薄荷、柠檬薄荷、苹果薄荷和草莓薄荷的香味[37]。一些品牌包含两个风味爆珠，一个包含薄荷醇，另一个包含浓郁的水果味或更多薄荷醇。爆珠可以单独压碎，使用者可以控制双重风味。含香料爆珠的存在证明其可以增强卷烟对年轻人的吸引力[37]。

在一些国家禁止具有特征性香味的卷烟后（薄荷味产品除外），烟草行业迅速开发了替代产品，例如小雪茄，其中一些雪茄具有基本的卷烟设计，并且包装纸中含有烟草。含特征香料的小雪茄在年轻人中非常受欢迎。烟草行业申请的专利描述了在雪茄包装纸和滤嘴上添加增香剂（特别是甜味剂）的创新技术，以确保产品被认为是甜的。在一项研究中，美国市场上许多流行的小雪茄或细支雪茄中发现了高含量的糖精和其他合成的高强度甜味剂。糖精的含量在小雪茄的嘴部和尖端非常高，其甜度远远超过了糖[38]。无论是否被标记为甜味，所有口味类别的小雪茄都含有糖精[39,40]。

烟草行业专利还描述了使用重组甜味模拟蛋白作为甜味剂，甜蛋白的甜度是食用糖的2000倍，合成薄荷衍生物作为创新的清凉剂。在加拿大禁止使用薄荷醇卷烟之后，烟草行业开始销售在包装上不标示含薄荷醇的卷烟，但其设计与以前销售的薄荷醇卷烟几乎相同，这意味着具有同样凉爽的感觉[41]。目前尚不清楚这些卷烟是否包含清凉剂来替代薄荷醇。

在亚洲发现的许多无烟烟草制品（例如zarda、quiwam、gutkha、khaini）都含有香味物质[42,43]。新型无烟烟草制品也已经在多个国家推出，包括口含烟（以前仅在瑞典销售）和湿鼻烟的新品种。口含烟和鼻烟有许多种口味，包括薄荷醇/薄荷、樱桃、香草和草莓，并且有人提出，湿鼻烟的销量显著增长与香料的引入有关[8,44]。还发现新上市的口含烟产品中含有极高含量的三氯蔗糖（Splenda®），这是一种合成的高强度甜味剂，其比鼻烟中仍在使用的糖精更像糖[45]。

加香水烟烟草（masssel）是通过将烟叶与糖蜜、甘油和水果香精发酵而制成的甜味的、加香的烟草。它是水烟中首选的烟草形式，尤其是在青少年和年轻人中[46]。

电子烟和电子烟碱传输系统（ENDS）2004年开始上市，2006年在欧洲上市，2007年在美国上市，电子烟液香料的种类在不断扩大，特殊的香味和烟碱组合对使用者具有非常强的吸引力。这些产品通过电加热来蒸发电子烟液，可用于各种设备，包括类似卷烟产品、雾化器或水烟笔，被称为"mods"或个性化雾化装置，以及独立荚式设备（如JUUL）。在2013~2014年，网上有466个电子烟品牌，有7000多种口味；每月增加242种新口味[9]。到2016~2017年，数量增加了1倍，超过15000种口味和口味组合[10]。口味选择促进吸烟者从可燃卷烟向电子烟或ENDS转换与增加青少年摄入风险之间的平衡是一个热门争论话题。

总结：烟草以及现在的电子烟行业都在不断引入新的添加剂，例如香料和甜味剂，以提高烟草制品的吸引力，包括水烟、无烟烟草制品、电子烟。监管监测和跟踪此类产品发展动态对于减少烟草制品的吸引力至关重要。

6.5　香味感知系统

要了解烟草制品中调味剂的吸引力，重要的是要了解有助于香味体验的感官系统的作用。味觉和嗅觉在香味感知中起着最大的作用。但是，香味还包括温度、触觉和化学感觉（即通过温度、触觉和疼痛等化学刺激产生的感觉）[47]。

味觉：味觉源自舌头上的味蕾，提供甜、咸、酸、苦、辣的感觉。由于烟碱有苦味[48]，因此苦味是烟草香味的主要特征。向烟草制品中添加调味剂可以掩盖苦味并改善其风味和吸引力。

嗅觉：嗅觉不仅提供有关环境中气味的信息（经鼻嗅觉），而且还提供有关从口腔和气道散发的气味（鼻后嗅觉）的信息，这些气味对烟草制品的香味有很大贡献。鼻腔后部与口腔后部[49,50]相连的开口允许产品的气味进入口腔或通过口腔吸入，从而在呼气时刺激嗅觉受体。

温度：温度至少可以通过两种方式对烟草的风味做出贡献，直接地通过可燃或烟草制品加热型烟草制品产生的温度和热感，以及间接地通过调节鼻后嗅觉感受到的挥发性香味分子的释放速率。

触感：触感可在吸烟或抽烟时给卷烟、雪茄、烟斗和烟嘴以口腔"感觉"，以及无烟烟草制品在口中的质感。触摸还可以感觉到吸入或口服烟草制品时引起的口腔黏膜表面变化，例如干燥和苦涩[51]。

化学感觉：除了味道和气味外，某些化学物质还可以通过刺激对机械或热刺激作出反应的黏膜受体而产生温感、触觉或疼痛感[47]。化学感觉在烟草制品的风味中起关键作用，包括卷烟烟气中烟碱、丙烯醛和其他化学刺激物的刺激性以及薄荷醇的凉爽和"燃烧"感。

香味的交互作用：味觉的一个决定性特征是它的多感官成分在口腔中整合成连贯的感知。一个特别重要的交互作用是将鼻后香味转至口腔[52,53]，这会导致将香味错误地标记为味道，最常见的是具有类似味道的，例如"甜"味，像香草、樱桃和草莓味[54]。这很重要，因为这意味着特征香味通常是由多个相互作用的感官系统的感觉的无缝结合而出现的。

烟草制品和电子烟产品的香味成分的掩盖或抑制作用也很重要，例如用甜味掩盖苦味[55]。一些研究发现甜味剂还可以抑制疼痛，特别是对于婴儿[56,57]。无烟烟草制品中的甜味剂含量超过糖果产品中的甜味剂含量[45]。烟草行业申请的专利描述了向雪茄包皮和过滤嘴中添加人造甜味剂以确保消费者感受产品的甜味。同样，含薄荷醇烟草制品的吸引力在很大程度上归功于清凉、薄荷味的吸引力，但薄荷醇还具有镇痛作用，可以降低烟碱[34,58]以及卷烟烟气[59]和无烟烟草制品中的其他成分对感官的刺激[58,60,61]。

总结：香料的多模式、综合特性使其成为烟草制品吸引力的重要贡献因素，包括从口腔和喉咙的基本触觉和热性质，到基于化学性质衍生而来的味觉和嗅觉成分。虽然管制薄荷醇和人造甜味剂等众所周知的单个香料可能对烟草制品的吸引力产生直接影响，但是香味成分之间通过甜味组合或掩蔽和镇痛等机制的相互作用来提高烟草制品的吸引力，这给设计监管策略带来了更大的困难。由于味觉综合了感官经验，是主观的，需要进行感官测试和化学分析，以确定"添加剂的浓度高于该浓度时会产生特征性的香味"[3]或通过掩盖厌恶的味道或感觉来增加产品的吸引力。对于味道可感知但不明确的产品，标签可能有助于增强对味道的感知和识别。

6.6　香味受体：香料感知和编码新科学

分子、遗传、药理和行为方法彻底改变了香味研究，包括发现香味受体。嗅觉受体，也称为气味受体，首先在鼻嗅上皮的神经末梢被发现。这些受体在鼻后感受烟草和烟碱产品中的挥发性香味物质中起主要作用。在舌和上颚的味觉乳突中鉴定出味觉受体。人类有一个单一的甜味受体，该受体由两个蛋白质亚基（TAS1R2和TAS1R3）组成，它们结合蔗糖（例如食用糖），并以更高的亲和力结合人工高效甜味剂（例如糖精和三氯蔗糖）。苦味受体，TAS2R受体，人类具有38种不同的表达亚型，信号传导存在潜在有毒化学物质，并且可能与感知烟碱和其他烟草生物碱的苦味有关。编码人类苦味受体TAS2R38的基因多态性与吸薄荷醇卷烟有关，这表明对苦味较敏感的吸烟者可能使用薄荷醇来掩盖烟碱或卷烟烟气的苦味[62,63]。

介导调味剂和其他烟草成分的化学感官特性的受体，瞬时受体电位（TRP）离子通道位于神经上，这些神经从口腔、鼻腔通道和气道传递有害的和无害的化学、机械和热刺激。TRPA1是卷烟烟气中有害醛类的受体，引起燃烧和刺激感。TRPA1还介导了许多水果味调味剂中存在的烟碱和调味醛类（如肉桂醛和苯甲醛）

的刺激作用。TRPM8是薄荷醇的受体,介导其清凉和舒缓作用。小鼠实验表明TRPM8对于抑制薄荷醇卷烟烟气和烟碱的刺激和厌恶作用至关重要[60,64]。

总结:有了有关香味受体及其药理学的更多知识,调味剂行业可以使用分子和药理学方法来开发高度优化的新型香味受体调节剂。这包括新型甜味强化剂、苦味阻滞剂(减少苦味)、咸味(非谷氨酸)强化剂和新型凉味剂。一些被批准用作食品添加剂,还有更多正在开发中。烟草行业已经试验了合成的凉味剂,该凉味剂有淡薄荷味,减少刺激性并提高稳定性,从而激活薄荷醇受体TRPM8。香味受体的鉴定提供了可能性,基于其在人体和动物模型系统中受体介导的药理和行为作用来调节香味。例如薄荷醇,它可以快速被凉味剂取代,监管者可以决定将烟草中的所有TRPM8受体激动剂作为受体特异性香料进行管制。

6.7 调味剂的毒理学效应

尽管法律法规因国家/地区而异,但在许多国家/地区,将调味剂作为食品添加剂需要进行上市前的科学审查。在欧盟,调味剂受欧洲食品安全局监管。在美国,FDA已将食品添加剂指定为一般认为安全(GRAS)类别,在该分类中,利益相关者提交有关安全性的数据,审查人员将根据添加剂的预期用途进行审查。但是,GRAS关于在糖果产品中使用调味剂的声明并不自动适用于烟草制品,尤其是吸入类的产品,包括卷烟、雪茄、水烟和电子烟。例如,肉桂醛是广泛应用于烘焙食品和糖果中的GRAS肉桂调味料,以非常高的浓度添加到一些电子烟液中[65]。在毒理学研究中,这些液体中的蒸汽损害了肺上皮细胞,并引起了小鼠的肺部炎症[66]。还发现其他甜味,包括具有香蕉和樱桃味的调味剂,会破坏细胞并使电子烟使用者接触苯甲醛,苯甲醛是许多浆果香料混合物的关键成分,其对呼吸系统具有已知的有害性[67,68]。丁香是黄油类香料中常见的香料,对呼吸系统的毒性作用也是众所周知的[69,70]。调味化学物质可与溶剂和电子烟液中的其他成分发生反应,形成具有未知毒理作用的刺激性化合物[71]。

尽管存在这些担忧,但一些电子烟和ENDS的制造商仍在其广告中使用了GRAS标签,表明其液体中的调味剂是安全的,因为它们先前已获准添加到食品中。此类健康声明遭到美国调味料和提取物制造协会的强烈反驳,该协会已向FDA提交了GRAS申请,该协会指出,在电子烟中使用调味剂不视为预期用途[72]。

缺乏用于评估调味剂安全性的监管程序,这也是对烟草制品(如口含烟、鼻烟、加香雪茄、卷烟和水烟)的限制,可能会导致消费者摄入或吸入毒性水平的调味剂。例如,分析研究表明冬青风味的鼻烟产品使普通使用者暴露于冬青风味的水杨酸甲酯水平比联合国粮食及农业组织和世界卫生组织确定的食品可接受的每日摄入量高12倍[73]。无烟烟草制品(包括zarda、quiwam、gutka和khaini)被发现含有高含量的调味化学物质，例如丁子香酚、香豆素、樟脑和二苯醚[43]。一些口含烟产品中含有高含量的甜味剂，如果使用者同时食用其他甜味产品，则经常使用的甜味剂可能超出建议的每日糖摄入量[45]。

关于燃烧型烟草制品（卷烟、雪茄、水烟）或加热的烟碱溶液（如电子烟）中调味剂的信息很少。烟草行业报告说，薄荷醇爆珠或kretek（丁香）卷烟和不含调味剂的同等卷烟烟中的有害物质含量没有重大差异[74,75]；但是，调味剂的有害作用可能被卷烟中的其他有害成分的压倒性贡献掩盖了。在其他研究中，调味卷烟的烟气中VOC含量增加，烟草行业文件表明该行业知道某些类型的调味卷烟的致癌性[76,77]。在这些研究中，仅测量了主要的烟气成分（例如TSNA和PAH），并没有检测调味剂燃烧产生的特定化学物质。对电子烟调味剂的递送规律进行分析可能会提供有关调味剂有害性的信息。当电子烟使用高电压参数时，吸烟者暴露于调味剂的机会增加，并且甲醛等氧化产物的水平会超过毒性水平。由于卷烟、雪茄和水烟比蒸汽型电子烟在更高的温度下燃烧，因此人们担心会形成大量的调味剂氧化产物和其他有害化学物质。有关电子烟液调味剂的证据表明调味化学物质可能具有毒性作用[65,78,79]。

总结：可接受的调味剂含量水平监管取决于烟草制品的类型、调味剂传输到消费者口腔和呼吸系统的水平，以及在储存、加热和燃烧过程中发生化学变化的可能性。调味剂的化学性质，尤其是在高温下加热时，应进一步研究。这可能会导致将调味剂及其化学产品添加到由国家监管机构（如FDA）和TobReg报告中关于烟草制品中有害物质的HPHC清单中。烟草制品的未来管制可能要求对调味化学物质的含量和有害性进行具体报告。

6.8　结　　论

关于调味剂在烟草和烟碱产品的吸引力和致瘾性中的作用仍然存在一些关键性问题。应当在全球范围内进行监测，以评估调味烟草制品的使用，以及对加香和非加香卷烟草制品的吸引力和致瘾性的看法。香料和甜味剂是一类具有独立吸引力的化学物质，也能通过提高烟碱和其他苦涩成分的适口性来增加烟草制品的使用。薄荷醇类香料可以通过药理作用减轻烟碱暴露的厌恶作用（例如咳嗽、刺激性、热度）来达到此目的，并且卷烟中薄荷醇的存在可能与常规吸烟的开始使用、成瘾发展和难以戒烟有关[80]。至于其他香味成分是否以及如何影响人们对烟草和烟碱产品的偏好和使用知之甚少。

由于香味复杂且主观，调味卷烟和烟碱产品的吸引力的测试需要对人类行为学和毒理学进行评估，并结合化学分析。确定香味物质存在和含量的方法应基于向消费者提供的产品（例如烟气或蒸汽产品），还应用于确定香料在储存、加热和燃烧过程中的潜在变化。调味剂的吸引力和有害性都应进行评估。确定产生这些效果的最佳调味剂剂量很重要，但了解调味剂浓度的影响也很重要。所有这些信息都应用于进一步完善"特征香味"的定义。

监测和测试方法必须跟上烟草行业不断创新的步伐，这些创新旨在提高加香烟草制品的吸引力，包括使用新的合成化合物以替换烟草制品中的香味物质。在烟草制品标签上标明调味剂及其浓度的问题应被关注。此外，监管工作还必须与烟草行业引入应对有关调味剂监管的替代营销策略相抗衡，例如操纵卷烟包装以继续传达与风味相关的品牌特征[41]。

未来的监管应考虑到某些香料和香味成分（例如甜味剂）本身也具有积极强化作用，并且当与烟碱结合使用时，可能会增强低剂量烟碱的奖赏作用，并促进向更高含量烟碱的过渡，从而导致成瘾。这些复杂问题的实验证据和监测对于确保旨在减少烟草制品吸引力和致瘾性的监管至关重要。为了监管这些物质，需要对香料科学以及它们如何增强烟草制品的吸引力和致瘾性潜力有更好的了解。

6.8.1 优先研究事项建议

- 系统地监控传统及新型烟草和烟碱产品的全球流行病学。
- 识别香味化学物质、其反应产物及其在烟草和烟碱产品以及气溶胶、蒸汽和卷烟烟气中的浓度。
- 确定产品包装和营销对香味的描述如何改变公众对烟草和烟碱产品的看法和吸引力，特别是对青少年。
- 评估烟草和烟碱产品中特定调味剂的存在及其浓度是否会改变其吸引力和滥用潜力。
- 在吸烟者中，调查调味产品的可获得性作为转向危害较小的烟草制品（相对减害程度）的动机的作用。
- 测定不同浓度的吸入性调味剂及其代谢物和加合物的有害性和健康影响。
- 监控取代禁用调味剂的替代性化学成分的使用，并确定其吸引力、有害性和健康影响。

6.8.2 政策建议

- 考虑获得关于加香烟草制品、电子烟和ENDS使用的系统性全球证据。
- 考虑禁止在有害的燃烧型卷烟产品中使用调味剂，包括薄荷醇。
- 考虑限制烟草和烟碱产品中允许的调味剂含量、数量和/或特定调味剂，因为有证据表明这样做会改变或降低风险，以减少青少年开始吸烟，并支持停止使用燃烧型烟草制品。
- 考虑要求在各种产品的成分表中列出调味剂的类型和浓度。

6.9 参考文献

[1] Small DM, Green BG. A model of flavor perception. In: Murray MM, Wallace MT, editors. The neural bases of multisensory processes. New York: CRC Press; 2012:717-38.

[2] Talhout R, van de Nobelen S, Kienhuis AS. An inventory of methods suitable to assess additiveinduced characterising flavours of tobacco products. Drug Alcohol Depend. 2016;161:9-14.

[3] Directive 2014/40/EU of the European Parliament and of the Council on the approximation of the laws, regulations and administrative provisions of the Member States concerning the manufacture, presentation and sale of tobacco and related products. Brussels: European Union; 2014 (http://ec.europa.eu/health/sites/health/files/tobacco/docs/dir_201440_en.pdf, accessed 15 May 2019).

[4] Federal Food, Drug, and Cosmetic Act. Section 907. Tobacco product standard. General questions and answers on the ban on cigarettes that contain certain characterizing flavors (edition 2). Silver Spring (MD): Center for Tobacco Products, Food and Drug Administration; 2009.

[5] Carpenter CM, Wayne GF, Pauly JL, Koh HK, Connolly GN. New cigarette brands with flavors that appeal to youth: tobacco marketing strategies. Health Affairs. 2005;24(6):1601-10.

[6] Connolly GN. Sweet and spicy flavours: new brands for minorities and youth. Tob Control. 2004;13(3):211-2.

[7] Morris DS, Fiala SC. Flavoured, non-cigarette tobacco for sale in the USA: an inventory analysis of Internet retailers. Tob Control. 2015;24(1):101-2.

[8] Delnevo CD, Wackowski OA, Giovenco DP, Manderski MT, Hrywna M, Ling PM. Examining market trends in the United States smokeless tobacco use: 2005-2011. Tob Control. 2014;23(2):107-12.

[9] Zhu SH, Sun JY, Bonnevie E, Cummins SE, Gamst A, Yin L et al. Four hundred and sixty brands of e-cigarettes and counting: implications for product regulation. Tob Control. 2014;23(Suppl 3):iii3-9.

[10] Hsu G, Sun JY, Zhu SH. Evolution of electronic cigarette brands from 2013-2014 to 2016-2017: analysis of brand websites. J Med Internet Res. 2018;20(3):e80.

[11] Bonhomme MG, Holder-Hayes E, Ambrose BK, Tworek C, Feirman SP, King BA et al. Flavoured non-cigarette tobacco product use among US adults: 2013-2014. Tob Control. 2016;25(Suppl 2):ii4-13.

[12] Kaleta D, Usidame B, Szosland-Faltyn A, Makowiec-Dabrowska T. Use of flavoured cigarettes in Poland: data from the global adult tobacco survey (2009-2010). BMC Public Health. 2014;14:127.

[13] Corey CG, Ambrose BK, Apelberg BJ, King BA. Flavored tobacco product use among middleThe science of flavour in tobacco products and high school students – United States, 2014. Morbid Mortal Wkly Rep. 2015;64(38):1066-70.

[14] King BA, Dube SR, Tynan MA. Flavored cigar smoking among US adults: findings from the 2009-2010 National Adult Tobacco Survey. Nicotine Tob Res. 2013;15(2):608-14.

[15] Villanti AC, Johnson AL, Ambrose BK, Cummings KM, Stanton CA, Rose SW et al. Flavored

tobacco product use in youth and adults: findings from the first wave of the PATH study (2013–2014). Am J Prev Med. 2017;53(2):139–51.

[16] Minaker LM, Ahmed R, Hammond D, Manske S. Flavored tobacco use among Canadian students in grades 9 through 12: prevalence and patterns from the 2010–2011 youth smoking survey. Prev Chronic Dis. 2014;11:E102.

[17] Global Youth Tobacco Survey (GYTS). Core questionnaire with optional questions, version 1.0. Geneva: World Health organization; Atlanta (GA): Centers for Disease Control and Prevention; 2012.

[18] Global Adult Tobacco Survey (GATS). Core questionnaire with optional questions, version 2.0. Atlanta (GA): Centers for Disease Control and Prevention; 2010.

[19] Ashare RL, Hawk LW Jr, Cummings KM, O'Connor RJ, Fix BV, Schmidt WC. Smoking expectancies for flavored and non-flavored cigarettes among college students. Addict Behav. 2007;32(6):1252–61.

[20] Kowitt SD, Meernik C, Baker HM, Osman A, Huang LL, Goldstein AO. Perceptions and experiences with flavored non-menthol tobacco products: a systematic review of qualitative studies. Int J Environ Res Public Health. 2017;14(4):338.

[21] Sokol NA, Kennedy RD, Connolly GN. The role of cocoa as a cigarette additive: opportunities for product regulation. Nicotine Tob Res. 2014;16(7):984–91.

[22] Feirman SP, Lock D, Cohen JE, Holtgrave DR, Li T. Flavored tobacco products in the United States: a systematic review assessing use and attitudes. Nicotine Tob Res. 2016;18(5):739–49.

[23] Choi K, Fabian L, Mottey N, Corbett A, Forster J. Young adults' favorable perceptions of snus, dissolvable tobacco products, and electronic cigarettes: findings from a focus group study. Am J Public Health. 2012;102(11):2088–93.

[24] Manning KC, Kelly KJ, Comello ML. Flavoured cigarettes, sensation seeking and adolescents' perceptions of cigarette brands. Tob Control. 2009;18(6):459–65.

[25] Hammal F, Wild TC, Nykiforuk C, Abdullahi K, Mussie D, Finegan BA. Waterpipe (hookah) smoking among youth and women in Canada is new, not traditional. Nicotine Tob Res. 2016;18(5):757–62.

[26] Kong G, Morean ME, Cavallo DA, Camenga DR, Krishnan-Sarin S. Reasons for electronic cigarette experimentation and discontinuation among adolescents and young adults. Nicotine Tob Res. 2015;17(7):847–54.

[27] Delnevo CD, Giovenco DP, Ambrose BK, Corey CG, Conway KP. Preference for flavoured cigar brands among youth, young adults and adults in the USA. Tob Control. 2015;24(4):389–94.

[28] Morean ME, Butler ER, Bold KW, Kong G, Camenga DR, Cavallo DR et al. Preferring more e-cigarette flavors is associated with e-cigarette use frequency among adolescents but not adults. PLoS One. 2018;4:13(1):e0189015.

[29] Garrison KA, O'Malley SS, Gueorguieva R, Krishnan-Sarin S. A fMRI study on the impact of advertising for flavored e-cigarettes on susceptible young adults. Drug Alcohol Depend. 2018;186:233–41.

[30] Farley SM, Seoh H, Sacks R, Johns M. Teen use of flavored tobacco products in New York City. Nicotine Tob Res. 2014;16(11):1518–21.

[31] How other countries regulate flavored tobacco products. Saint Paul (MN): Tobacco Control

Legal Consortium; 2015 (http://www.publichealthlawcenter.org/sites/default/files/resources/tclc-fs-global-flavored-regs-2015.pdf, accessed 15 May 2019).

[32] Henkler F, Luch A. European Tobacco Product Directive: How to address characterizing flavors as a matter of attractiveness? Arch Toxicol. 2015;89(8):1395–8.

[33] Giovino GA, Sidney S, Gfroerer JC, O'Malley PM, Allan JA, Richter PA et al. Epidemiology of menthol cigarette use. Nicotine Tob Res. 2004;6 Suppl 1:S67–81.

[34] Krishnan-Sarin S, Green BG, Kong G, Cavallo DA, Jatlow P, Gueorguieva R. Studying the interactive effects of menthol and nicotine among youth: an examination using e-cigarettes. Drug Alcohol Depend. 2017;180:193–9.

[35] Noyori R. Asymmetric catalysis: science and opportunities (Nobel lecture). Angew Chem Int Ed Engl. 2002;41(12):2008–22.

[36] Thrasher JF, Abad-Vivero EN, Moodie C, O'Connor RJ, Hammond D, Cummings KM et al. Cigarette brands with flavour capsules in the filter: trends in use and brand perceptions among smokers in the USA, Mexico and Australia, 2012–2014. Tob Control. 2016;25(3):275–83.

[37] Abad-Vivero EN, Thrasher JF, Arillo-Santillan E, Pérez-Hernández R, Barrientos-Guitierrez I, Kollath-Cattano C et al. Recall, appeal and willingness to try cigarettes with flavour capsules: assessing the impact of a tobacco product innovation among early adolescents. Tob Control. 2016;25(e2):e113–9.

[38] Erythropel HC, Kong G, deWinter TM, O'Malley SS, Jordt SE, Anastas PT et al. Presence of high intensity sweeteners in popular cigarillos of varying flavor profiles. JAMA. 2018;2:320(13):1380–3

[39] Sweeney WR, Marcq P, Pierotti J, Thiem D. Sweet cigar. Google Patents; 2013.

[40] Cundiff RH. Letter to Dr Paul Whiter, Wilkinson-Sword Ltd, Berkeley Heights (NJ)., 24 March 1975. Winston-Salem (NC): R.J. Reynolds Tobacco Co.; 1975 (Menthol Derivatives. RJ Reynolds Records. 1975 (https://www.industrydocumentslibrary.ucsf.edu/tobacco/docs/#id=ngdp0061, accessed January 2019).

[41] Brown J, DeAtley T, Welding K, Schwartz R, Chaiton M, Kittner DL et al. Tobacco industry response to menthol cigarette bans in Alberta and Nova Scotia, Canada. Tob Control. 2017;26(e1):e71–4.

[42] Stanfill SB, Croucher RE, Gupta PC, Lisko JG, Lawler TS, Kuklenyik P et al. Chemical characterization of smokeless tobacco products from South Asia: nicotine, unprotonated nicotine, tobacco-specific N´-nitrosamines, and flavor compounds. Food Chem Toxicol. 2018;118:626–34.

[43] Lisko JG, Stanfill SB, Watson CH. Quantitation of ten flavor compounds in unburned tobacco products. Anal Meth. 2014;6(13):4698–704.

[44] Kostygina G, Ling PM. Tobacco industry use of flavourings to promote smokeless tobacco products. Tob Control. 2016;25(Suppl 2):ii40–9.

[45] Miao S, Beach ES, Sommer TJ, Zimmerman JB, Jordt SE. High-intensity sweeteners in alternative tobacco products. Nicotine Tob Res. 2016;18(11):2169–73.

[46] Maziak W, Taleb ZB, Bahelah R, Islam F, Jaber R, Auf R et al. The global epidemiology of waterpipe smoking. Tob Control. 2015;24(Suppl 1):i3–12.

[47] Green BG. Chemesthesis: pungency as a component of flavor. Trends Food Sci Technol. 1996;7:415–20.

[48] Oliveira-Maia AJ, Phan THT, Melone PD, Mummalaneni S, Nicolelis MAL, Simon SA et al. Nicotinic acetylcholine receptors (NACHRS): novel bitter taste receptors for nicotine. Chem Senses. 2008;33(8):S113.

[49] Lim J, Johnson MB. Potential mechanisms of retronasal odor referral to the mouth. Chem Senses. 2011;36(3):283–9.

[50] Hummel T. Retronasal perception of odors. Chem Biodivers. 2008;5(6):853–61.

[51] Green BG. Oral astringency: a tactile component of flavor. Acta Psychol. 1993;84(1):119–25.

[52] Rozin P. "Taste-smell confusions" and the duality of the olfactory sense. Percept Psychophys.1982;31:397–401.

[53] Lim J, Green BG. Tactile interaction with taste localization: influence of gustatory quality and intensity. Chem Senses. 2008;33(2):137–43.

[54] Small DM, Prescott J. Odor/taste integration and the perception of flavor. Exp Brain Res. 2005;166(3–5):345–57. The science of flavour in tobacco products

[55] Green BG, Lim J, Osterhoff F, Blacher K, Nachtigal D. Taste mixture interactions: suppression, additivity, and the predominance of sweetness. Physiol Behav. 2010;101(5):731–7.

[56] Kassab M, Foster JP, Foureur M, Fowler C. Sweet-tasting solutions for needle-related procedural pain in infants one month to one year of age. Cochrane Database Syst Rev. 2012;12:CD008411.

[57] Schobel N, Kyereme J, Minovi A, Dazert S, Bartoshuk L, Hatt H. Sweet taste and chorda tympani transection alter capsaicin-induced lingual pain perception in adult human subjects. Physiol Behav. 2012;107(3):368–73.

[58] Rosbrook K, Green BG. Sensory effects of menthol and nicotine in an e-cigarette. Nicotine Tob Res. 2016;18(7):1588–95.

[59] Ferris Wayne G, Connolly GN. Application, function, and effects of menthol in cigarettes: a survey of tobacco industry documents. Nicotine Tob Res. 2004;6(Suppl 1):S43–54.

[60] Fan L, Balakrishna S, Jabba SV, Bonner PE, Taylor SR, Picciotto MR et al. Menthol decreases oral nicotine aversion in C57BL/6 mice through a TRPM8-dependent mechanism. Tob Control. 2016;25(Suppl 2):ii50–4.

[61] Liu B, Fan L, Balakrishna S, Sui A, Morris JB, Jordt SE. TRPM8 is the principal mediator of menthol-induced analgesia of acute and inflammatory pain. Pain. 2013;154(10):2169–77.

[62] Oncken C, Feinn R, Covault J, Duffy V, Dornela E, Kranzler HR et al. Genetic vulnerability to menthol cigarette preference in women. Nicotine Tob Res. 2015;17(12):1416–20.

[63] 6Risso D, Sainz E, Gutierrez J, Kirchner T, Niaura R, Drayna D. Association of TAS2R38 haplotypes and menthol cigarette preference in an African American cohort. Nicotine Tob Res. 2017;19(4):493–4.

[64] Willis DN, Liu B, Ha MA, Jordt SE, Morris JB. Menthol attenuates respiratory irritation responses to multiple cigarette smoke irritants. FASEB J. 2011;25(12):4434–44.

[65] Behar RZ, Davis B, Wang Y, Bahl V, Lin S, Talbot P. Identification of toxicants in cinnamonflavored electronic cigarette refill fluids. Toxicol In Vitro. 2014;28(2):198–208.

[66] Lerner CA, Sundar IK, Yao H, Gerloff J, Ossip D, McIntosh S et al. Vapors produced by electronic cigarettes and e-juices with flavorings induce toxicity, oxidative stress, and inflammatory response in lung epithelial cells and in mouse lung. PloS One. 2015;10(2):e0116732.

[67] Gerloff J, Sundar IK, Freter R, Sekera ER, Friedman AE, Robinson R et al. Inflammatory

response and barrier dysfunction by different e-cigarette flavoring chemicals identified by gas chromatography–mass spectrometry in e-liquids and e-vapors on human lung epithelial cells and fibroblasts. Appl In Vitro Toxicol. 2017;3(1):28–40.

[68] Kosmider L, Sobczak A, Prokopowicz A, Kurek J, Zaciera M, Knysak J et al. Cherry-flavoured electronic cigarettes expose users to the inhalation irritant, benzaldehyde. Thorax. 2016;71(4):376–7.

[69] Allen JG, Flanigan SS, LeBlanc M, Vallarino J, MacNaughton P, Stewart JH et al. Flavoring chemicals in e-cigarettes: diacetyl, 2,3-pentanedione, and acetoin in a sample of 51 products, including fruit-, candy-, and cocktail-flavored e-cigarettes. Environ Health Perspect. 2016;124(6):733–9.

[70] Klager S, Vallarino J, MacNaughton P, Christiani DC, Lu Q, Allen JG. Flavoring chemicals and aldehydes in e-cigarette emissions. Environ Sci Technol. 2017;51(18):10806–13.

[71] Erythropel HC, Jabba SV, deWinter TM, Mendizabal M, Anastas PT, Jordt SE et al. Formation of flavorant–propylene glycol adducts with novel toxicological properties in chemically unstable e-cigarette liquids. Nicotine Tob Res. 2018. doi: 10.1093/ntr/nty192.

[72] Safety assessment and regulatory authority to use flavors – focus on electronic nicotine delivery systems and flavored tobacco products. Washington (DC): Flavor and Extract Manufacturers Association; 2018 (https://www.femaflavor.org/safety-assessment-and-regulatory-authority-use-flavors-focus-electronic-nicotine-delivery-systems, accessed 15 May 2019).

[73] Chen C, Isabelle LM, Pickworth WB, Pankow JF. Levels of mint and wintergreen flavorants: smokeless tobacco products vs. confectionery products. Food Chemical Toxicol. 2010;48(2):755–63.

[74] Dolka C, Piade JJ, Belushkin M, Jaccard G. Menthol addition to cigarettes using breakable capsules in the filter. Impact on the mainstream smoke yields of the health Canada list constituents. Chem Res Toxicol. 2013;26(10):1430–43.

[75] Roemer E, Dempsey R, Hirter J, Deger Evans A, Weber S, Ode A et al. Toxicological assessment of kretek cigarettes. Part 6: the impact of ingredients added to kretek cigarettes on smoke chemistry and in vitro toxicity. Regul Toxicol Pharmacol. 2014;70(Suppl 1):S66–80.

[76] Gordon SM, Brinkman MC, Meng RQ, Anderson GM, Chuang JC, Kroeger RR et al. Effect of cigarette menthol content on mainstream smoke emissions. Chem Res Toxicol. 2011;24(10):1744–53.

[77] Hurt RD, Ebbert JO, Achadi A, Croghan IT. Roadmap to a tobacco epidemic: transnational tobacco companies invade Indonesia. Tob Control. 2012;21(3):306–12.

[78] Behar RZ, Luo W, McWhirter KJ, Pankow JF, Talbot P. Analytical and toxicological evaluation of flavor chemicals in electronic cigarette refill fluids. Sci Rep. 2018;8(1):8288.

[79] Bengalli R, Ferri E, Labra M, Mantecca P. Lung toxicity of condensed aerosol from e-cig liquids: influence of the flavor and the in vitro model used. Int J Environ Res Public Health. 2017;14(10):1254.

[80] Menthol cigarettes and public health: review of the scientific evidence and recommendations. Washington (DC): Tobacco Products Advisory Committee, Food and Drug Administration; 2011.

7. 烟草制品中的糖含量

Lotte van Nierop and Reinskje Talhout, Centre for Health Protection, National Institute for Public Health and the Environment (RIVM), Bilthoven, Netherlands

7.1 引　言

糖是烟草制品的一种成分，往往其含量还比较高。糖存在于卷烟烟叶中，也可以在制造过程中人为地添加[1-3]。未经加工的烟叶含有许多类型的糖，包括葡萄糖、果糖和蔗糖。叶片的干燥（烘烤）会影响糖的含量。风干烟叶几乎不含糖，但烤烟中糖含量可能占其总质量的25%[4,5]。另外，在制造过程中，可以将各种类型的糖和含糖成分（例如蜂蜜和水果糖浆）添加到烟丝中。添加量取决于烟草混合物和产品类型。添加的糖可以用作黏合剂、料液成分、香料、赋形助剂或保润剂[3,4,6]。在烟草制品燃烧过程中，糖会产生许多有害物质，据推测其中一些会影响烟草制品的致瘾性[7]。糖还对烟草制品的风味做出了重要贡献[3,8]。在无烟烟草制品中，糖本身会给烟草制品增香，而在燃烧型烟草制品中，糖会与烟草中的其他成分发生反应，产生焦糖味，并使烟气吸入时更温和[3,9]。某些含糖的料液被烟草行业称为"改良剂"，并声称"能消除刺鼻的苦味和/或消除烟草中的刺激性气味"[10]。

由于糖是烟草制品的主要成分之一，并且可能通过增加产品的有害性、致瘾性或吸引力来危害消费者的健康，因此糖可能是监管烟草制品含量的合适成分。我们回顾了关于糖在无烟烟草制品和水烟等不同烟草制品中的存在及其影响的证据。首先，我们描述了天然存在于烟叶中的各种糖，由烘烤产生的糖以及在制造过程中添加到烟草中的糖。我们还报告了产品的加糖百分比和加糖量。其次，我们总结了糖对烟草制品释放物含量影响的证据。第三，通过其有害性评估，我们总结了糖对烟草制品致瘾性和吸引力的健康影响。最后，我们强调研究和监管的意义。

7.2 不同类型烟草制品中的糖

烟草是一种天然植物,烟叶中含有高水平的碳水化合物,例如糖(单糖和双糖)、淀粉、纤维素和果胶(多糖)[4]。在将绿色烟叶干燥(烘烤)成可用的棕色干燥烟叶的过程中,酶将淀粉降解为糖。烘烤条件取决于烟草类型。弗吉尼亚烟草在潮湿的和高温条件下进行调制,这使淀粉降解为糖,并阻止糖的酶促降解,从而使烤烟中糖的含量较高(8%~30%)[11]。晾晒烟的调制过程与烘烤过程相似,但控制程度比较差,因此烟叶中的糖含量为10%~20%[2]。相比之下,白肋烟在室温下以缓慢的过程进行烘烤,导致淀粉降解为糖,并且糖被酶分解,导致最终烟叶中的糖含量较低(≤0.2%)[4,11]。由于烟草制品包含一种或多种烟草品种,因此烟草制品的糖含量差异很大,具体取决于混合烟叶的类型。通常,在烤烟叶片中发现最多的糖是葡萄糖、果糖和蔗糖[4,11]。

除单糖和二糖外,烟草还含有大量(添加或天然存在的)多糖,例如纤维素、果胶和淀粉[9,12]。传统的卷烟包含约10%的纤维素、10%的果胶和2%的淀粉[4,12]。因此,碳水化合物可占卷烟的40%以上,因此对卷烟烟气的化学组成具有实质性影响[3]。

7.2.1 糖和含糖添加剂的类型

除了天然烟叶中存在的糖以外,烟草制品还添加了各种类型的糖和含糖成分,主要在烟丝表面。糖的类型取决于来源和加工方法。最经常添加到燃烧型卷烟和无烟烟草制品中的糖是蔗糖和转化糖(果糖和葡萄糖的混合物)[2,13,14]。其他含大量糖的烟草添加剂包括果汁、玉米和枫糖浆、糖蜜提取物、蜂蜜和焦糖[3,4,6,15]。也可能是其他具有类似于糖的甜味和特性的物质,但并不属于糖,如乙酰磺胺酸钾、阿斯巴甜、乙基麦芽酚、甘油、丙二醇、麦芽糖醇、麦芽酚、糖精、山梨糖醇和索马甜。

从制造商的电子数据库EMTOC[16]中的成分列表中识别出的烟草制品中的糖是:红糖、焦糖、(玉米)糖浆(固体)、糊精、葡萄糖、(水果)浓缩液、果汁、葡萄糖、高果糖、蜂蜜、转化糖、乳糖、麦芽提取物、麦芽糊精、(枫)糖浆、糖蜜、(部分)转化糖、蔗糖(糖浆)、甘蔗(糖浆)、糖浆和其他未说明的糖。单糖是比较简单的糖,如果糖、半乳糖和葡萄糖,分子式为$C_6H_{12}O_6$。二糖是由两个单糖分子脱去一个水分子而形成的,包括蔗糖、乳糖和麦芽糖,分子式为$C_{12}H_{22}O_{11}$。

7.2.2 糖和含糖添加剂的量

糖通常以肠衣的形式添加到烟叶中,尽管在某些情况下少量的糖被添加到非烟材料，如纸张、过滤嘴和胶水中[16]。常规卷烟中添加的糖量占烟草质量的2%~18%[2,12,17,18]。水烟（*maassel*）中的含量更高，占产品质量的50%~70%[19]。关于烟草制品中糖的数量和类型的信息很少。在荷兰，制造商有义务每年向电子数据库（EMTOC）披露荷兰市场上所有烟草制品的成分[16]。对这些数据的分析表明，2015年，在所有类型烟草制品中，糖的添加到了相当大的比例[20]。

对数据的进一步分析表明，大多数自卷烟产品（56%）、卷烟（77%）、斗烟（86%）和水烟制品（100%）都含有添加的糖[图7.1(a)]，而少量的雪茄（7%）或口含烟草制品（22%）也含有添加的糖。所有烟草制品中最常用的糖是转化糖和葡萄糖，以及含糖成分，例如焦糖、蜂蜜、糖浆和果汁[图7.1(b)~(g)]。在雪茄中添加的糖的质量分数为2%，在自卷烟和卷烟中为3%，在斗烟中为12%。值得注意的是，水烟产品中所含糖分占烟草质量的164%，表明糖比最终产品中的烟草多。由于烟叶中天然糖的含量很高，添加到口含烟草制品中的糖量仅为烟草重量的10%[21]。这取决于不同国家/地区产品中使用的烟草类型。

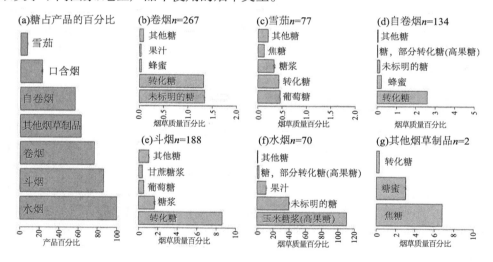

图7.1　烟草制品中糖含量

(a)根据每类烟草制品中含有添加剂的产品总数，添加到烟草中的含糖产品的百分比；(b)~(g)各类烟草制品最常使用的糖的平均量（每毫克烟草的百分比）；*n*，添加糖的产品数量；"其他糖合在一起"显示产品组中的所有剩余糖

资料来源：EMTOC数据库[16]，其中制造商每年在荷兰注册所有销售的烟草制品，并包含添加剂成分、添加量和功能的信息

7.2.3　添加和内源性糖的总水平

如前所述，天然烟草除了在生产过程中添加的糖外，还含有在生长和烘烤过程中形成的大量糖。因此，测量或报告已上市烟草制品中的最终糖含量比添加糖

清单上的含量信息更丰富,仅关注添加糖清单上的含量信息将低估最终糖含量[22]。很少有实验室对烟草制品中总糖水平进行分析。对最终烟草制品中最常见添加的碳水化合物(葡萄糖、果糖和蔗糖)的化学分析表明,总糖含量小于卷烟中烟丝质量的19%[23],雪茄为0.1%,鼻烟为0.03%,在嚼烟中为27.7%[24]。

7.2.4 烟草品种、产品和用途的区域和文化差异

市场上销售和使用的烟草制品的品种取决于区域和文化偏好。在一项针对32个国家/地区的青少年中烟草使用模式的研究中[25],目前吸烟的全国流行率从卢旺达的1.8%到拉脱维亚的32.9%,而目前无烟烟草制品使用的流行率从黑山的1.1%到莱索托的14.4%不等。与在东南亚等低收入和中等收入国家中观察到的模式相比,在大多数欧洲国家和美国,当前吸烟的流行率显著高于无烟烟草制品的使用。欧洲晴雨表(Eurobarometer)报道,无烟烟草制品在瑞典尤为流行,而水烟的使用在其他北欧国家和北非非常流行[26]。由于可获得性、可负担性和可接受性的差异,使用烟草的经验可能因文化而异[25]。

尽管对特定产品和品牌的含糖量知之甚少,但有关卷烟中使用的烟草品种和混合物的更多信息可用。由于国家的偏好,糖的使用因品牌而异,甚至在一个品牌中也有所不同[2]。烟草的品种也因地区而异。但是,总的来说,卷烟市场似乎由两个品种主导:烤烟和美国混合型卷烟。根据PMI的统计,2008年,弗吉尼亚风格(烤烟,高糖分和少量添加剂)卷烟在加拿大(99%)、新西兰(95%)、澳大利亚(92%)、英国(91%)、爱尔兰(87%)和南非(76%)比较受欢迎。美国混合型卷烟(弗吉尼亚烤烟、白肋烟和香料烟的混合物和许多添加剂,包括糖的混合物)在西欧大陆国家比较受欢迎,市场份额为85%~100%,在美国则为99%[1]。

7.3 糖对烟草制品释放物的影响

7.3.1 无烟烟草制品

无烟烟草制品的使用者暴露于咀嚼或吮吸产生的释放物。尽管无烟烟草制品中糖的含量很高(见上文),但在使用过程中释放的糖量以及由此导致的糖暴露量尚不清楚。

7.3.2 糖和简单混合物的热解产物

为了了解卷烟主流烟气中的前体物质与烟气的关系[1],卷烟的燃烧已经在热解实验中进行了模拟[27,28]。研究表明烟草中只有少量不挥发的糖(约0.5%)会完

整地转移到卷烟主流烟气中,而大多数糖会燃烧、热解或是其他热解过程的一部分[1,3]。焦糖化等燃烧过程会产生许多不同的化合物,包括醛(例如乙醛、丙烯醛、2-糠醛)、呋喃衍生物、VOC、有机酸、丙烯酰胺和PAH(通常在高温下)。这些研究中所使用的热解条件采用近似于卷烟燃烧时的温度和其他条件,但不能解释燃烧过程中其他烟草和/或烟气成分与糖的相互作用。例如,糖可以与烟草中的胺类反应(氨基化合物、氨基酸、蛋白质)形成美拉德产物(另请参见第7.2节),其可以分解成很多化学物质,例如醛、酮(例如双乙酰)、酸、丙烯酰胺、吡嗪和吡啶[1,3]。

7.3.3 卷烟和其他燃烧型烟草制品

2006年糖的添加对烟气成分影响的独立审查得出的结论是,向卷烟中添加糖增加了醛和酮的含量,尤其是甲醛、乙醛、丙酮、丙烯醛、2-糠醛和其他呋喃的含量[3]。另外,主流烟气中总酸的浓度增加会导致pH值偏低和游离态烟碱含量减少。Cheah等[29,30]研究还表明向几乎不含天然糖的白肋烟中添加糖会增加主流烟气中乙醛、丙烯醛、巴豆醛、丙醛和丁醛的浓度。这种增加是醛类所特有的,而焦油、烟碱和一氧化碳的增加要小得多。

尽管在烟草行业的许多论文中讨论了这些结论,并在随后发表的综述中进行了讨论,但关于糖是否会增加乙醛的含量以及是否会在较小程度上提高丙烯醛的含量仍存在很多争论。例如,Seeman等[12]指出纤维素而不是糖是烟气中乙醛的主要前体物,但这并不意味着糖也没有贡献。在对烟草添加剂的综述中,Klus等[1]得出结论(从许多研究结果中得出)向烟草中添加糖分会增加苯酚、呋喃、有机酸、2-糠醛和甲醛的浓度,并产生更多酸性物质。但他并没有报道糖和乙醛之间的联系。Baker[31,32]报道了甲醛的增加,发现糖增加了主流烟气中2-糠醛的含量,但并没有增加乙醛或丙烯醛的含量。Coggins等[33]发现向烟草中添加碳水化合物和含糖天然产物只会使烟气化学性质发生很小的变化,但与不含添加剂的卷烟相比,甲醛的含量会持续增加。Roemer等[2]报道虽然检测的烟气成分中的大多数未显示出任何变化(每毫克烟碱),但添加糖水平的增加却增加了甲醛、丙烯醛、2-丁酮、异戊二烯、苯、甲苯、苯并[k]荧蒽的含量或降低了4-氨基联苯、N-亚硝基二甲胺、NNN等成分含量。Hahn和Schaub[34]报道添加5%蔗糖后甲醛的浓度增加,但与由50%弗吉尼亚烟草(烤烟)、20%白肋烟、20%的烟梗和10%的香料烟组成的参比卷烟相比,乙醛的浓度并没有增加。由于弗吉尼亚烟草中含有大量的天然糖,添加糖后糖的相对增加较少、以及由此产生的醛含量相对较少。

O'Connor和Hurley[35]发现通过将乙醛释放量标准化为焦油或总粒相物,可以消除糖和乙醛含量之间的关系,这两者都直接关系到设计特征,例如滤嘴通风和烟草质量。他们重新分析了Zilkey等的一项研究[36],发现糖占烟气中乙醛含量变化的50%以上。对Phillpotts等的数据[37]进行重新分析，结果表明烟草混合物中糖占醛含量变化的11%，而占总粒相物变化的23%。

水烟是所有烟草制品糖含量最高的。尽管这可能会导致烟气中的挥发性醛含量更高[38,39]，但尚无水烟或其他任何燃烧烟草制品烟气中糖含量与烟气成分之间关系的数据。此外，已经观察到电子烟液中的糖含量与其释放物中醛含量之间存在相关性[40]。

7.4 糖对烟草制品毒性的影响

当用于食品中时糖是GRAS添加剂，但用于烟草则不是[3]。抽烟时，糖会燃烧产生许多有害和致癌反应产物。此外，化合物吸入时比摄入时毒性更大，因为在很大程度上呼吸系统缺乏消化系统的解毒代谢途径[3]。

7.4.1 无烟烟草

市售无烟烟草制品每片含糖约0.5~2 g[41]。该产品在口腔中放置约30 min，每天重复使用[42,43]。无烟烟草制品会导致多种类型的癌症，如心血管疾病，不利的生殖结果以及对口腔健康的不利影响，包括口腔黏膜病变、白斑和牙周疾病[44]，但关于无烟烟草制品中糖对这些疾病的贡献知之甚少。无烟烟草制品的使用已经与龋齿有关，可能的原因是糖含量高，尽管不能排除其他与烟草使用有关的原因。无烟烟草制品中的糖会形成一种酸，它会吞噬掉牙釉质、引起蛀牙并导致口腔溃疡[42]。此外，高含量的发酵糖可以刺激致龋细菌的生长[14,43]。

无烟烟草制品中的糖也可能影响血糖。一项案例研究表明，停止吞咽嚼烟所产生汁液的糖尿病患者的血糖下降了50%（从300~400 mg/dL降至160~200 mg/dL）[41]。

7.4.2 添加和不添加糖的卷烟

糖的许多热解产物都具有毒性作用[45,46]。这包括甲醛（刺激性、致癌性）、丙烯醛（刺激性）和PAH（致癌性）。它们对燃烧型烟草制品总毒性的贡献程度尚不清楚。

在对烟草添加剂的研究中，Klus等[1]得出结论："糖和用作添加剂的糖对卷烟主流烟气毒性（如果有的话）只有很小且不起眼的作用。"对含各种糖分的卷烟燃烧产生的主流烟气进行了研究，发现在体外的细胞毒性、致突变性和遗传毒性，体内吸入毒性的亚慢性毒性研究以及皮肤致瘤性的研究中，没有发现毒性标志物有显著差异[2,33]。然而，根据欧盟新兴及新鉴定健康风险科学委员会[47]：毒性比较试验策略……被认为不适合用当前可用的方法解决该范围中概述的特性。事实上，由于烟草制品的高本底毒性，目前这些研究缺乏辨别力，并且其结果不能推广到所有产品和品牌，因为烟叶类型、配方和添加剂不同。

7.5 糖对烟草制品致瘾性的影响

没有关于糖在无烟烟草制品致瘾性中作用的相关信息。但是，除了有据可查的烟碱致瘾作用外，糖还可能增加对燃烧型烟草制品的渴望和使用范围[7]。燃烧烟草制品中糖的致瘾性可能会通过几种直接和间接途径来提高，包括卷烟的pH值、游离态烟碱水平和药理活性化合物的形成。

7.5.1 pH值和游离态烟碱

到达大脑的烟碱的量取决于游离态烟碱（不带电荷的挥发性形式）的可用性，游离态烟碱是在特定pH值下形成的。高pH值会产生更多的游离态烟碱[48,49]，其容易穿过口腔和肺上皮的细胞膜，从而导致到达大脑的烟碱含量更高。碳水化合物或糖含量高的烟草制品产生的卷烟烟气酸性更强，从而导致游离态烟碱的浓度降低[50]。为了使游离态烟碱含量较低的卷烟保持令人满意的烟碱吸收水平，吸烟者应更深和/或更频繁地吸入烟气或增加吸烟频率[51,52]。这导致更多地暴露于致癌和有害化合物（关于烟气pH值变化的其他影响，请参见第7.7节）。然而，其他研究表明，当改变烟草或烟气的pH值时，肺上皮的pH缓冲能力减弱了烟碱生物利用度变化的影响[1]。

7.5.2 烟草和烟气中药理活性化合物的形成

尽管糖本身没有致瘾性，但是烟草制品中糖的燃烧会在卷烟烟气中产生几种可能具有致瘾性的化合物。其中，乙醛是最重要的[12,53,54]，因为它提高了腹侧被盖区多巴胺能神经元的放电频率[54]。在实验动物中，静脉内乙醛会上瘾，并协同增强烟碱的成瘾作用[53,56-60]。尽管乙醛明显增强了对啮齿动物的作用，但尚不清楚这些作用是否会在人体以及在卷烟烟气中发现的浓度下也会发生。

乙醛通过与烟草或人体中存在的氨基酸色胺和色氨酸反应，产生缩合产物哈尔满（当氨基酸与甲醛反应时产生去甲哈尔满），从而发挥其致瘾性[61,62]。哈尔满抑制单胺氧化酶，该酶降解参与药物成瘾的神经递质，如多巴胺、去甲肾上腺素和血清素[51,61,63]。因此，暴露于哈尔满会增加伏隔核中多巴胺和5-羟色胺的含量，并增强烟碱的作用[64-66]。有趣的是，唾液中的L-半胱氨酸锭剂与乙醛反应降低了可形成哈尔满的水平，并在最初的随机双盲、安慰剂对照试验中显示了用作戒烟治疗的潜力[67]。

多项研究表明，吸烟者的哈尔满和去甲哈尔满水平与吸烟量有关[68-70]，与非吸烟者相比，吸烟者的单胺氧化酶活性降低会增加烟碱的自我给药[57]和维持对烟碱的行为敏感性[56,71]。

7.6 依赖与戒烟

烟草制品中的糖可以通过增强依赖性使人难以戒烟，从而有助于维持烟草使用。烟草行业进行了几次国家间比较，以调查添加剂对烟草致瘾性的影响。对戒烟率临床研究的荟萃分析并未显示以弗吉尼亚烤烟（天然糖含量高，添加成分很少，不添加糖）和美国混合型卷烟为主的烟草市场的戒烟率有显著差异（包括糖在内有许多添加成分）[72,73]。然而，该研究中的一个重要假设是戒烟困难是衡量烟草致瘾性的有效方法，并且无需针对特定国家的因素（例如产品的可获得性和戒烟计划）进行纠正[74]。此外，在两个市场之间，在吸烟率、强度、某些依赖性标志物、烟碱摄取或因吸烟相关的肺癌或慢性阻塞性肺疾病引起的死亡率方面也没有发现相关差异[73]。据Klus等[1]的研究，鉴于美国混合型卷烟和弗吉尼亚烤烟的总糖含量（天然存在和添加的）是可比的，因此推测没有糖的作用。

对于不同糖含量特定类型烟草制品的吸烟者的依赖性和戒烟率知之甚少[72]。关于乙醛潜在滥用的许多研究都是基于乙醇的研究，因为乙醛也是乙醇的代谢产物[75]，并且尚未进行适当的实验来评估乙醛蒸汽截断作用效果[76]。然而，烟碱和乙醛的协同作用仅在幼年大鼠中观察到了，而在成年大鼠中则未观察到[58]，这可能支持以下观察结果：青少年似乎比成年人更容易上瘾[58,65,77]。

7.7 糖对烟草制品吸引力的影响

7.7.1 认知：感官特征

烟草制品的感官特性（例如味道和气味）会极大地影响其吸引力[7]。例如，最初的吸引可以通过诸如糖和其他甜味剂的存在来建立[3,9,78]。制造商根据某些化学标准（例如糖和烟碱含量）选择烟草，并且通过使用糖和其他添加剂来平衡烟叶成分的年度变化，以确保产品随时间的一致性[7]。无烟烟草制品中的糖在嗅觉刺激下被口腔中的味觉感受器感知。相比之下，燃烧型烟草制品的使用者只能感知到糖的热解产物。总体而言，烟草制品制造商会根据消费者的需求调整产品的口味和特性，并创造"愉悦的体验"[79]。

7.7.2 吸烟体验和行为：香味、适口性、易吸入性、使用频率

糖可以改善烟草制品的一些特性，例如掩盖烟的刺鼻感和改善味道。易挥发的成分（例如氨、烟碱和生物碱）使烟气具有难闻的味道，从而使吸烟者不想吸

入烟气[3,80]。吸烟期间糖的燃烧会产生酸，降低烟气的pH值[50]，从而降低其刺激性[4,9]，增加产品的适口性并促进吸入。通过添加糖可以更频繁、更深度地吸入，增加对烟碱和其他烟气化学物质的暴露。糖在烟草香料中也起着重要作用[4,6,81]。在无烟烟草制品中，糖本身被消耗。在烟草的调制、储存、加工和吸烟过程中，氨基酸或氨与糖发生反应（美拉德反应），生产具有高度多样结构和风味潜力的化学物质。其中许多是杂环化合物，包括芳香族吡嗪，这是导致某些卷烟品牌特征口味的重要物质。另外，糖的焦糖化改善了使用者和旁观者接触到的卷烟烟气的味道和气味[79]。与美拉德反应不同，这是糖的非酶棕色化反应（不涉及胺）。

7.7.3 开始吸烟

吸烟者对卷烟烟气的接受度与烟草中的糖含量成正比[3,8]。尚未研究烟草制品中糖的含量及其水平影响开始烟草消费的程度。由于糖降低了pH值，掩盖了卷烟烟气的苦味，新使用者受到的刺激性降低了。此外，糖的燃烧产生的焦糖味对青少年特别有吸引力[53,82]。因此，烟草制品中糖的存在可能会鼓励青少年提早开始吸烟，继续吸更长的时间，并增加他们的烟草使用量[3]。在卷烟中添加糖以刺激或增强卷烟烟气的感官特性，并鼓励开始和维持吸烟，这不仅在科学上是合理的，而且作为营销策略的一部分已被烟草业讨论过[74,83]。

7.8 不同管辖区域对糖的监管

有几个国际组织和国家机构对烟草制品中的添加剂进行监管。下面列出了其中最重要的法规和监管措施。

- WHO FCTC：第9条规定了对烟草制品的成分和释放物的测试和测量及其监管，第10条规定了允许WHO FCTC缔约方通过并实施有效的立法、执法、行政管理或其他措施，要求烟草制品制造商和进口商向政府机构披露有关烟草制品成分和释放物的信息。该框架公约促进了缔约国采取全面的烟草控制措施，其中包括报告和披露，例如要求提交有关烟草制品中糖含量（和与糖有关化合物的释放物）的信息[84]。
- 加拿大卫生部：自2009年以来，C-32法案[85]的修正案禁止在卷烟、小雪茄和小雪茄外包皮中使用多种成分，包括糖。此外，加拿大要求制造商和进口商提供《烟草报告条例》[86]中规定的有关烟草制品成分的信息。
- 欧盟烟草制品指令：根据2014/40/EU指令，禁止在卷烟和自卷烟产品（但未在其他烟草制品中）中标明风味的禁令于2014年生效[87]。此外，每个欧盟成员国的制造商都必须每年向电子数据库（EU-CEG）披露所有上市的烟草制品及其成分（例如添加糖的数量和类型、烟草品种或配方、调制方法）。该指令指出在叶片烘烤过程中损失的糖可以在制造过程中予以补充；但是，烟叶的初始含糖量和加工过程中的损失并未确定，这使得很难监管添加糖的量。

- 2003年，英国烟草行业自愿协议对多种糖（蔗糖、转化糖、糖浆和糖蜜）设置的最大限值为：卷烟产品、雪茄和自卷烟产品质量的10%，斗烟质量的15%[88]。确定这些限量标准的理由尚不清楚。自2014年以来，该协议已被经修订的《欧盟烟草制品指令》所取代[89]。

- 巴西国家卫生监督管理局（ANVISA）：具有法律效力的监管决议（RDC 14/2012）[90]限制了烟草制品中某些添加剂的使用。该决议禁止使用甜味剂、蜂蜜、糖蜜以及除糖以外可以赋予甜味的物质。糖例外，允许卷烟制造商补充烟叶烘烤过程中损失的糖分。2013年9月，该决议因国家工业联合会代表烟草制品制造商提起的诉讼而终止[90]。一个独立专家工作组建议修订RDC 14/2012，使糖被排除在添加剂禁令之外[91]。截至2015年6月，允许使用添加剂的最高法院禁令仍然有效，但最终裁决仍在进行中。

- FDA：2009年颁布的《家庭吸烟预防和烟草控制法》禁止含有某些特征性风味的卷烟[92]，对某些调味电子烟液的使用和销售也进行了限制[93]。对于卷烟或其他烟草制品中的糖含量没有特别的要求或限制。

7.9 结　　论

添加的糖是许多烟草制品中的主要成分之一。添加的量可能很大，占烟草质量的2%~164%，具体取决于产品的类型。由于烟叶本身可能含有高含量的天然糖，因此烟草制品中糖（天然糖和添加糖）的总量可能会很高。关于不同类型烟草制品中总糖的含量的可用数据非常有限。

除对牙齿健康和糖尿病的影响外，几乎没有证据表明糖对无烟烟草制品的毒性和致瘾性有贡献。然而，糖通过改善其味道而有助于产品的吸引力。

经燃烧的糖和未经过燃烧的糖对所有烟草制品的吸引力都有贡献，而烟草制品中糖的燃烧则可能导致有害、致癌和致瘾性物质的形成。

一些司法管辖区域，例如加拿大和巴西，对烟草制品中糖的添加进行监管，英国也已设定了限值。需要采取更严格的控制措施，以防止使用含有天然糖和添加糖的烟草制品对健康的不利影响。

7.10 建 议

7.10.1 下一步研究内容

正如 *Nicotine and Tobacco Research* 的编委所强调的那样，在理解烟草和烟草制品中糖的健康影响方面仍然存在很大差距[94]。尤其需要更多关于以下几个方面的信息：

- 水烟中糖与烟气中的有害物质之间的关系；
- 糖对健康影响的暴露生物标志物以及糖含量变化对这些生物标志物的影响；
- 糖在烟草制品对儿童和青少年的吸引力中的作用，例如通过根据糖含量监测市场份额和进行市场研究；
- 糖改变了释放物的总体甜度，并赋予一种明显的特征风味；
- 根据烟草制品的类型，糖对开始抽烟、致瘾、依赖性和戒烟总体速率的影响；
- 卷烟烟气中测量的乙醛量对烟草（烟碱）成瘾的作用，通过单胺抑制剂（例如哈尔满）和体内形成乙醛-生物胺缩合产物；
- 糖对烟草制品总体毒性的影响，包括：
  - 对于无烟烟草制品，更多有关糖本身的有害性作用的信息，例如，糖是否与糖尿病和龋齿有关；
  - 对于燃烧型烟草制品，更多有关糖在吸入毒性中作用的信息；
  - 糖与其他烟气成分的相互作用可能会增加烟气的毒性。

7.10.2 政策

可以考虑采取几种政策措施来减少糖（及其热解产物）对产品毒性、致瘾性和吸引力的负面影响。

糖含量的披露

烟草制品制造商应披露所有类型的烟草制品中糖（和其他成分）的含量。这将为未来的法规和决策提供有价值的信息。可采用简单快速的方法来验证报告的含量水平，例如萃取烟草，然后进行高效液相色谱-蒸发光散射检测[23]。

降低最终烟草制品中糖的含量

定义和监管烟草制品中的最高糖含量可最好地保护消费者，因为烟草中天然存在的或制造过程中添加的糖的实际来源并不能区分其对健康的影响。如果需要并有足够的科学证据支持，则最高糖含量可以降低至几乎为零，即白肋烟叶中的糖含量。

降低所有烟草制品中糖的含量

加拿大已禁止在卷烟、小雪茄和小雪茄外包皮中添加糖，但在其他产品（例如水烟和无烟烟草制品）中并未禁止添加糖。仅监管某些烟草制品中的糖可能会鼓励使用者转向其他产品。因此，加拿大和美国禁止在卷烟中添加调味添加剂，这增加了调味小雪茄的市场和消费量[86,95]。

要求披露和降低卷烟烟气中最有害的成分

为了减少吸烟的负面影响，可以为卷烟烟气中糖燃烧产生的最有害成分设定上限[3]。醛是糖燃烧所产生的危害性最大的释放物，WHO TobLabNet验证了一种醛的分析方法，并且TobReg已将醛列入其建议强制降低的烟气成分清单中[45]。缔约方会议确定这些为优先级有害物质，应对其检测方法进行验证。英美烟草公司的Cunningham等支持将醛类列为优先监管物质，在他们关于为风险评估和管理目的分离卷烟烟气中有害物质的论文中，乙醛、丙烯醛和甲醛的暴露限量小于10000，被认为是减少暴露研究的高度优先事项[96]。

支持糖的有害性、致瘾性和吸引力研究

监管机构可以支持对烟草制品中糖燃烧产生的化合物及其有害性、致癌性、致突变性、生殖毒性、致瘾性和吸引力的研究和报告。已经发布了用于评估烟草添加剂对有害性[97]、致瘾性[98]和吸引力[99]的影响的科学指南和建议。

改变"糖"的状态和定义

另一个需要更多关注和科学研究的重要手段是在下一轮修订中将其列入国家和国际目标成分和组成成分清单，例如欧洲优先级清单[100]。该清单是根据欧盟烟草制品指令[87]第13条建立的，其中包括需要对卷烟和自卷烟的添加剂进行更科学的实验研究，包括其未燃烧和燃烧形式的有害性、致瘾性和致癌性、致突变性或生殖毒性。尽管糖符合所有选择标准（指令6.1和6.2），但尚未包括在内。

要求披露并监测除糖以外的烟草制品中甜味剂的含量

除糖以外，烟草制品中的（高强度）甜味剂含量也可有效地用于控制产品的适口性并增加青少年开始抽烟的比例。高强度甜味剂的甜度是蔗糖的数百倍[101]，并且存在于许多替代烟草制品中。因此，调节甜味剂含量可能是控制多种产品的

适口性并减少未成年人开始使用烟草制品的一种手段。例如，加拿大已经禁止除糖以外的甜味剂和几种香料。

7.11　致　　谢

感谢Rob van Spronsen对文献的系统搜索，感谢Jeroen Pennings对EMTOC数据库中烟草制品中糖的文献进行分析。

7.12　参 考 文 献

[1] Klus H, Scherer G, Muller L. Influence of additives on cigarette related health risks. Contrib Tob Res. 2012;25(3):412–93.

[2] Roemer E, Schorp MK, Piadé JJ, Seeman JI, Leyden DE, Haussmann HJ. Scientific assessment of the use of sugars as cigarette tobacco ingredients: a review of published and other publicly available studies. Crit Rev Toxicol. 2012;42(3):244–78.

[3] Talhout R, Opperhuizen A, van Amsterdam JG. Sugars as tobacco ingredient: effects on mainstream smoke composition. Food Chem Toxicol. 2006;44(11):1789–98.

[4] Leffingwell JC. Leaf chemistry: basic chemical constituents of tobacco leaf and differences among tobacco types. In: Davis DL, Nielsen MT, editors. Tobacco: production, chemistry and technology. Oxford: Blackwell Science; 1999:265–84.

[5] Cahours X, Verron T, Purkis S. Effect of sugar content on acetaldehyde yield in cigarette smoke. Beitr Tabakforsch Int. 2014;25(2):381–95.

[6] Seeman JI, Laffoon SW, Kassman AJ. Evaluation of relationships between mainstream smoke acetaldehyde and "tar" and carbon monoxide yields in tobacco smoke and reducing sugars in tobacco blends of US commercial cigarettes. Inhal Toxicol. 2003;15(4):373–95.

[7] Addictiveness and attractiveness of tobacco additives. Brussels: European Commission Scientific Committee on Emerging and Newly Identified Health Risks; 2010 (https://ec.europa.eu/health/scientific_committees/emerging/docs/scenihr_o_029.pdf, accessed October 2018).

[8] Bernasek PF, Furin OP, Shelar GR. Sugar/nicotine study. Industry documents library, 1992 (ATP 92-210:22) (https://www.industrydocumentslibrary.ucsf.edu/tobacco/docs/#id=fxbl0037, accessed 1 January 2019).

[9] Rodgman A. Some studies of the effects of additives on cigarette mainstream smoke properties. II. Casing materials and humectants. Beitr Tabakforsch Int. 2002;20(4):83–103.

[10] Jenkins CR et al. British American Tobacco Company limited product seminar. Tobacco documents 760076123–408, 1997 (https://www.industrydocumentslibrary.ucsf.edu/tobacco/docs/#id=rjgf0205, accessed 1 January 2019).

[11] Fisher P. Tobacco blending. In: Davis DL, Nielsen MT, editors. Tobacco: production, chemistry and technology. Oxford: Blackwell Science; 1999:346–52.

[12] Seeman JI, Dixon M, Haussmann HJ. Acetaldehyde in mainstream tobacco smoke: formation and Sugar content of tobacco products occurrence in smoke and bioavailability in the smoker. Chem Res Toxicol. 2002;15(11):1331–50.

[13] Hsu SC, Pollack RL, Hsu AF, Going RE. Sugars present in tobacco extracts. J Am Dent Assoc. 1980;101(6):915–8.

[14] Going RE, Hsu SC, Pollack RL, Haugh LD. Sugar and fluoride content of various forms of tobacco. J Am Dent Assoc. 1980;100(1):27–33.

[15] Rustemeier K, Stabbert R, Haussmann HJ, Roemer E, Carmines EL. Evaluation of the potential effects of ingredients added to cigarettes. Part 2: chemical composition of mainstream smoke. Food Chem Toxicol. 2002;40(1):93–104.

[16] Electronic Model Tobacco Control (EMTOC). Amsterdam: Rijks Instituut voor Volksgezondheiden Milieu and Ministerium für Gesundheit und Frauen; 2015.

[17] Newson JR. Evaluation of competitive cigarettes. Liggett & Myers; 1975 (https://industrydocuments.library.ucsf.edu/tobacco/docs/#id=zslf0014, accessed 10 February 2016).

[18] Gordon DL. PM's global strategy: Marlboro product technology. A summary developed with input from Batco, Batcf, Souza Cruz & B&W. Winston Salem (NC): Brown & Williamson; 1992.

[19] Soussy S, El-Hellani A, Baalbaki R, Salman R, Shihadeh A, Saliba NA. Detection of 5-hydroxymethylfurfural and furfural in the aerosol of electronic cigarettes. Tob Control. 2016;25(Suppl 2):ii88–93.

[20] van Nierop LE, Pennings JLA, Schenk E, Kienhuis AS, Talhout R. Analysis of manufacturer's information on tobacco product additive use. Tob Regul Sci. (accepted).

[21] Lawler TS, Stanfill SB, Zhang L, Ashley DL, Watson CH. Chemical characterization of domestic oral tobacco products: total nicotine, pH, unprotonated nicotine and tobacco-specific N-nitrosamines. Food Chem Toxicol. 2013;57:380–6.

[22] van Nierop LE, Talhout R. Sugar as tobacco additive tastes "bitter". J Addict Res Ther. 2016;7(4):2.

[23] Jansen EHJM, Cremers J, Borst S, Talhout R. Simple determination of sugars in cigarettes. Anal Bioanal Tech. 2014;5(6):219.

[24] Clarke MB, Bezabeh DZ, Howard CT. Determination of carbohydrates in tobacco products by liquid chromatography–mass spectrometry/mass spectrometry: a comparison with ion chromatography and application to product discrimination. J Agric Food Chem. 2006;54(6):1975–81.

[25] Agaku IT, Ayo-Yusuf OA, Vardavas CI, Connolly G. Predictors and patterns of cigarette and smokeless tobacco use among adolescents in 32 countries, 2007–2011. J Adolesc Health. 2014;54(1):47–53.

[26] Attitudes of Europeans towards tobacco and electronic cigarettes (Special Eurobarometer 458). Brussels: European Commission; 2017.

[27] Busch C, Streibel T, Liu C, McAdam KG, Zimmermann R. Pyrolysis and combustion of tobacco in a cigarette smoking simulator under air and nitrogen atmosphere. Anal Bioanal Chem. 2012;403(2):419–30.

[28] Baker RR, Bishop LJ. The pyrolysis of tobacco ingredients. J Anal Appl Pyrolysis. 2004;71(1):223–311.

[29] Cheah NP. Volatile aldehydes in tobacco smoke: source, fate and risk. Maastricht: Maastricht University; 2016:2016846.

[30] Cheah NP, Borst S, Hendrickx L, Cremers H, Jansen E, Opperhuizen A et al., Effect of adding

sugar to Burley tobacco on the emission of aldehydes in mainstream tobacco smoke. Tob Regul Sci. 2018;4(2):61–72.

[31] Baker RR. Sugars, carbonyls and smoke. Food Chem Toxicol. 2007;45(9):1783–6.

[32] Baker RR. The generation of formaldehyde in cigarettes – overview and recent experiments. Food Chem Toxicol. 2006;44(11):1799–822.

[33] Coggins CR, Wagner KA, Weley MS, Oldham MJ. A comprehensive evaluation of the toxicology of cigarette ingredients: carbohydrates and natural products. Inhal Toxicol. 2011;23(Suppl 1):13–40.

[34] Hahn J, Schaub J. Influence of tobacco additives on the chemical composition of mainstream smoke. Beitr Tabakforsch Int. 2014;24(3):101–16.

[35] O'Connor RJ, Hurley PJ. Existing technologies to reduce specific toxicant emissions in cigarette smoke. Tob Control. 2008;17(Suppl 1):i39–48.

[36] Zilkey B, Court WA, Binns MR, Walker EK, Dirks VA, Basrur PK. Chemical studies on Canadian tobacco and tobacco smoke. 1. Tobacco, tobacco sheet and cigarette smoke. Chemical analyses on various treatments of Bright and Burley. Tob Int. 1982;184:83–9.

[37] Phillpotts DF, Spincer D, Westcott DT. The effect of the natural sugar content of tobacco upon the acetaldehyde concentration found in cigarette smoke. London: British American Tobacco; 1974 (Bates Number: 505242960).

[38] Shihadeh A, Salman R, Jaroudi E, Saliba N, Sepetdjian E, Blank MD et al. Does switching to a tobacco-free waterpipe product reduce toxicant intake? A crossover study comparing CO, NO, PAH, volatile aldehydes, tar and nicotine yields. Food Chem Toxicol. 2012;50(5):1494–8.

[39] Al Rashidia M, Shihadeh A, Saliba NA. Volatile aldehydes in the mainstream smoke of the narghile waterpipe. Food Chem Toxicol. 2008;46(11):3546–9.

[40] Fagan P, Pokhrel P, Herzog TA, Moolchan ET, Cassel KD, Franke AA et al. Sugar and aldehyde content in flavored electronic cigarette liquids. Nicotine Tob Res. 2018;20(8):985–92.

[41] Pyles ST, Van Voris LP, Lotspeich FJ, McCarty SA. Sugar in chewing tobacco. N Engl J Med. 1981;304(6):365.

[42] Vellappally, S, Fiala Z, Smejkalová J, Jacob V, Shriharsha P. Influence of tobacco use in dental caries development. Centr Eur J Public Health. 2007;15(3):116–21.

[43] Tomar SL, Winn DM. Chewing tobacco use and dental caries among US men. J Am Dent Assoc. 1999;130(11):1601–10.

[44] WHO Study Group on Tobacco Product Regulation. Report on the scientific basis of tobacco product regulation: fifth report of a WHO study group (WHO Technical Report Series, No. 989). Geneva: World Health Organization; 2015.

[45] Burns DM, Dybing E, Gray N, Hecht S, Anderson C, Sanner T et al. Mandated lowering of toxicants in cigarette smoke: a description of the World Health Organization TobReg proposal. Tob Control. 2008;17(2):132–41.

[46] Talhout, R, Schulz T, Florek E, van Benthem J, Wester P, Opperhuizen A. Hazardous compounds in tobacco smoke. Int J Environ Res Public Health. 2011;8(2):613–28.

[47] Opinion on additives used in tobacco products (opinion 2). Tobacco additives II. Brussels: European Commission Scientific Committee on Emerging and Newly Identified Health Risks; 2016.

[48] Willems EW, Rambali B, Vleeming W, Opperhuizen A, van Amsterdam JG. Significance of ammonium compounds on nicotine exposure to cigarette smokers. Food Chem Toxicol. 2006; 44(5):678–88.

[49] Elson LA, Betts TE. Sugar content of the tobacco and pH of the smoke in relation to lung cancer risks of cigarette smoking. J Natl Cancer Inst. 1972;48(6):1885–90.

[50] Wayne GF, Carpenter CM. Tobacco industry manipulation of nicotine dosing. Handb Exp Pharmacol. 2009;192:457–85.

[51] Hammond, D, Fong GT, Cummings KM, O'Connor RJ, Giovino GA, McNeill A. Cigarette yields and human exposure: a comparison of alternative testing regimens. Cancer Epidemiol Biomarkers Prev. 2006;15(8):1495–501.

[52] Jarvis MJ, Boreham R, Primatesta P, Feyerabend C, Bryant A. Nicotine yield from machinesmoked cigarettes and nicotine intakes in smokers: evidence from a representative population survey. J Natl Cancer Inst. 2001;93(2):134–8.

[53] Bates C, Jarvis M, Connolly G. Tobacco additives: cigarette engineering and nicotine addiction. A survey of the additive technology used by cigarette manufacturers to enhance the appeal and addictive nature of their product. London: Action on Smoking and Health, Imperial Cancer Research Fund; 1999. Sugar content of tobacco products

[54] Paschke T, Scherer G, Heller W. Effects of ingredients on cigarette smoke composition and biological activity: a literature overview. Beitr Tabakforsch Int. 2002;20(3):107–47.

[55] Foddai M, Dosia G, Spiga S, Diana M. Acetaldehyde increases dopaminergic neuronal activity in the VTA. Neuropsychopharmacology. 2004;29(3):530–6.

[56] Villegier AS, Blanc G, Glowinski J, Tassin JP. Transient behavioral sensitization to nicotine becomes long-lasting with monoamine oxidases inhibitors. Pharmacol Biochem Behav. 2003;76(2):267–74.

[57] Guillem K, Vouillac C, Azar MR, Parsons LH, Koob GF, Cador M et al. Monoamine oxidase inhibition dramatically increases the motivation to self-administer nicotine in rats. J Neurosci. 2005;25(38):8593–600.

[58] Belluzzi JD, Wang R, Leslie FM. Acetaldehyde enhances acquisition of nicotine self-administration in adolescent rats. Neuropsychopharmacology. 2005;30(4):705–12.

[59] Gray N. Reflections on the saga of tar content: why did we measure the wrong thing? Tob Control. 2000;9(1):90–4.

[60] Henningfield JE, Benowitz NL, Connolly GN, Davis RM, Gray N, Myers ML et al., Reducing tobacco addiction through tobacco product regulation. Tob Control. 2004;13(2):132–5.

[61] Herraiz T, Chaparro C. Human monoamine oxidase is inhibited by tobacco smoke: betacarboline alkaloids act as potent and reversible inhibitors. Biochem Biophys Res Commun. 2005;326(2): 378–86.

[62] Pfau W, Skog K. Exposure to beta-carbolines norharman and harman. J Chromatogr B. 2004;802(1):115–26.

[63] Koob GF. Drugs of abuse: anatomy, pharmacology and function of reward pathways. Trends Pharmacol Sci. 1992;13(5):177–84.

[64] Fowler JS, Logan J, Wang GJ, Volkow ND. Monoamine oxidase and cigarette smoking. Neurotoxicology. 2003;24(1):75–82.

[65] Talhout R, Opperhuizen A, van Amsterdam JG. Role of acetaldehyde in tobacco smoke addiction. Eur Neuropsychopharmacol. 2007;17(10):627–36.

[66] Berlin I, Anthenelli RM. Monoamine oxidases and tobacco smoking. Int J Neuropsychopharmacol. 2001;4(1):33–42.

[67] Syrjanen K, Salminen J, Aresvuo U, Hendolin P, Paloheimo L, Eklund C et al. Elimination of cigarette smoke-derived acetaldehyde in saliva by slow-release L-cysteine lozenge is a potential new method to assist smoking cessation. A randomised, double-blind, placebo-controlled intervention. Anticancer Res. 2016;36(5):2297–306.

[68] Spijkerman R, van den Eijnden R, van de Mheen D, Bongers I, Fekkes D. The impact of smoking and drinking on plasma levels of norharman. Eur Neuropsychopharmacol. 2002;12(1):61–71.

[69] Rommelspacher H, Meier-Henco M, Smolka M, Kloft C. The levels of norharman are high enough after smoking to affect monoamineoxidase B in platelets. Eur J Pharmacol. 2002;441(1-2):115–25.

[70] Breyer-Pfaff U, Wiatr G, Stevens I, Gaertner HJ, Mundle G, Mann K. Elevated norharman plasma levels in alcoholic patients and controls resulting from tobacco smoking. Life Sci. 1996;58(17):1425–32.

[71] Rose JE, Mukhin AG, Lokitz SJ, Turkington TG, Herskovic J, Behm FM et al. Kinetics of brain nicotine accumulation in dependent and nondependent smokers assessed with PET and cigarettes containing 11C-nicotine. Proc Natl Acad Sci U S A. 2010;107(11):5190–5.

[72] Sanders E, Weitkunat R, Utan A, Dempsey R. Does the use of ingredients added to tobacco increase cigarette addictiveness? A detailed analysis. Inhal Toxicol. 2012;24(4):227–45.

[73] Lee PN, Forey BA, Fry JS, Hamling JS, Hamling JF, Sanders EB et al. Does use of flue-cured rather than blended cigarettes affect international variation in mortality from lung cancer and COPD? Inhal Toxicol. 2009;21(5):404–30.

[74] Ferreira CG, Silveira D, Hatsukami DK, Paumgartten FJ, Fong GT, Glória MB et al. The effect of tobacco additives on smoking initiation and maintenance. Cad Saude Publica. 2015;31(2):223–5.

[75] Correa M, Salamone JD, Segovia KN, Pardo M, Longoni R, Spina L et al. Piecing together the puzzle of acetaldehyde as a neuroactive agent. Neurosci Biobehav Rev. 2012;36(1):404–30.

[76] Hoffman AC, Evans SE. Abuse potential of non-nicotine tobacco smoke components: acetaldehyde, nornicotine, cotinine, and anabasine. Nicotine Tob Res. 2013;15(3):622–32.

[77] Adriani W, Spijker S, Deroche-Gamonet V, Laviola G, Le Moal M, Smit AB et al. Evidence for enhanced neurobehavioral vulnerability to nicotine during periadolescence in rats. J Neurosci. 2003;23(11):4712–6.

[78] Leffingwell JC. Nitrogen components of leaf and their relationship to smoking quality and aroma. Recent Adv Tob Sci. 1976;2:1–31.

[79] Opinion on additives used in tobacco products (opinion 1). Brussels: European Commission Scientific Committee on Emerging and Newly Identified Health Risks; 2016.

[80] Hoffmann D, Hoffmann I. The changing cigarette, 1950–1995. J Toxicol Environ Health. 1997;50(4):307–64.

[81] Weeks WW. Relationship between leaf chemistry and organoleptic properties of tobacco smoke. In: Davis DL, Nielsen MT, editors. Tobacco: production, chemistry and technology. Oxford: Blackwell Science; 1999:304–12.

[82] Fowles J. Chemical factors influencing the addictiveness and attractiveness of cigarettes in New Zealand. Wellington: Institute of Environmental Science and Research; 2001 (http://www.moh.govt.nz/notebook/nbbooks.nsf/0/D0B68B2D9CB811ABCC257B81000D9959/$file/chemicalfactorsaddictivenesscigarettes.pdf, accessed 15 May 2019).

[83] Truth Tobacco Industry Documents. San Francisco (CA): University of California at San Francisco Library and Center for Knowledge Management (https://www.industrydocumentslibrary.ucsf.edu/tobacco/, accessed 15 May 2019).

[84] WHO Framework Convention on Tobacco Control. 2013 Guidelines for implementation. Article 5.3, Article 8, Articles 9 and 10, Article 11, Article 12, Article 13, Article 14. Geneva: World Health Organization; 2013 (http://apps.who.int/iris/bitstream/10665/80510/1/9789241505185_eng.pdf, accessed 1 October 2018).

[85] An act to amend the Tobacco Act, Bill C-32. Ottawa: Government of Canada; 2009.

[86] Tobacco Reporting Regulations (SOR/2000-273). Tobacco and Vaping Products Act, Tobacco Act. Ottawa: Government of Canada; 2006.

[87] Tobacco Product Directive (2014/40/EU) of the European Parliament and of the Council. Brussels: European Commission; 2014 (http://ec.europa.eu/health/tobacco/docs/dir_201440_en.pdf, accessed 15 May 2019).

[88] Permitted additives to tobacco products in the United Kingdom. London: Department of Health; 2003.

[89] Consultation on implementation of the revised Tobacco Products Directive (2014/40/EU), section 5.11. London: Department of Health; 2015 (https://assets.publishing.service.gov.uk/government/uploads/system/uploads/attachment_data/file/440991/TPD_Consultation_Doc.pdf, accessed 16 January 2019).

[90] Richter AP. Best practices in implementation of Article 9 of the WHO FCTC Case study: Brazil and Canada (National Best Practices Series, Paper No. 5). Geneva: World Health Organization; 2015.

[91] Report of the Working Group on Tobacco Additives. Brasilia: Agência Nacional de Vigilância Sanitária; 2014 (http://portal.anvisa.gov.br/documents/106510/106594/Report+Working+Group+Tobacco+Additives/b99ad2e7-23d9-4e88-81cd-d82c28512199, accessed 15 May 2019).

[92] Family Smoking Prevention and Tobacco Control Act. Washington (DC): Food and Drug Administration; 2009 (http://www.fda.gov/tobaccoproducts/labeling/rulesregulationsguidance/ucm246129.htm#ingredients, accessed 1 January 2019). Sugar content of tobacco products

[93] Gottlieb S. FDA statement from FDA Commissioner Scott Gottlieb, MD, on proposed new steps to protect youth by preventing access to flavored tobacco products and banning menthol in cigarettes. Washington (DC): Food and Agriculture Administration; 2018 (https://www.fda.gov/NewsEvents/Newsroom/PressAnnouncements/UCM625884.htm, accessed 10 January 2019).

[94] Munafo M. Understanding the role of additives in tobacco products. Nicotine Tob Res. 2016;18(7):1545.

[95] Corey CG, Ambrose BK, Apelberg BJ, King BA. Flavored tobacco product use among middle and high school students – United States, 2014. Morb Mortal Wkly Rep. 2015;64(38):1066–70.

[96] Cunningham FH Fiebelkorn S, Johnson M, Meredith C. A novel application of the Margin of Exposure approach: segregation of tobacco smoke toxicants. Food Chem Toxicol.

2011;49(11):2921–33.

[97] Kienhuis AS, Staal YC, Soeteman-Hernández LG, van de Nobelen S, Talhout R. A test strategy for the assessment of additive attributed toxicity of tobacco products. Food Chem Toxicol. 2016;94:93–102.

[98] van de Nobelen S, Kienhuis AS, Talhout R. An inventory of methods for the assessment of additive increased addictiveness of tobacco products. Nicotine Tob Res, 2016;18(7):1546–55.

[99] Talhout R, van de Nobelen S, Kienhuis AS. An inventory of methods suitable to assess additiveinduced characterising flavours of tobacco products. Drug Alcohol Depend. 2016;161:9–14.

[100] European Commission. Priority list of additives contained in cigarettes and roll-your-own. Off J Eur Union. 2016:C/2016/2923.

[101] Miao S, Beach ES, Sommer TJ, Zimmerman JB, Jordt SE. High-intensity sweeteners in alternative tobacco products. Nicotine Tob Res. 2016;18(11):2169–73.

8. 燃烧型烟草制品中有害成分的优先级清单更新

Irina Stepanov, Associate Professor, Division of Environmental Health Sciences and Masonic Cancer Center, University of Minnesota, Minneapolis (MN), USA

Marielle Brinkman, Senior Research Scientist, College of Public Health, Ohio State University, Columbus, OH, USA

8.1 引　　言

在世界卫生组织《烟草控制框架公约》缔约方大会第五次会议上，要求世界卫生组织汇编，向缔约方提供并更新烟草制品有害成分和释放物的非详尽清单，并就如何最好地利用这些信息提供建议。TobReg在2013年12月4~6日于巴西里约热内卢举行的会议上提供了用于报告和监管的优先级有害成分的最新清单。

在本章，我们对清单进行了重新评估，因为自发布最新清单以来已有四年了，并且有了有关该主题的新知识。特别是，该报告包括：

- 关于当前TobReg有害物质优先级清单的背景信息，包括用于选择特定成分和释放物的标准（8.2）；
- 与优先级清单有关的新数据的概述，包括有关特定成分有害性的新出版物，新型或改进的分析方法以及有关可燃烟草制品中有害物质含量变化的新数据（8.3~8.5）；
- 讨论清单上将来对有害物质进行重新评估的标准（8.6）；
- 讨论清单选择新有害物质的标准（8.7）；
- 用于测试燃烧型烟草制品中选定成分和释放物的研究需求和法规建议（8.8）。

主要在PubMed数据库和SciFinder中搜索文献，后者从Medline和CAplus数据库检索引文。还包括出版物中引用的相关文章。此外，美国疾病控制与预防中心、美国环境保护局、烟草科学研究合作中心（CORESTA）的网站以及其他相关网站提供了有关成分的毒理学信息和使用的方法。

8.2 优先级清单的编制背景

烟草制品的监管要求建立一个或一组评估烟草制品的指标。对于卷烟，最常见的测量方法是根据ISO标准抽吸方案和美国联邦贸易委员会（FTC）的方法对每支卷烟的焦油、烟碱和一氧化碳进行测量。但是，众所周知，此类措施不能提供对人体暴露或有害性的有效估计，从而通过误导吸烟者和大多数监管者而造成伤害[1,2]。因此，认为有必要采用新的产品评估方法来制定监管措施。

TobReg提出的新方法要求采用标准化的机器测试方法，量化与烟碱（吸烟者在烟气中寻找的主要致瘾性物质）有关的已知有害物质的含量。通过将特定的有害物质标准化为单位烟碱释放量，可以测量标准抽吸条件下产生的烟气的有害性，而不是产生的烟气量。标准化的有害物质释放量将使管理者能够减少烟气中已识别出的优先级有害物质的水平，这与要求减少产品中已知有害物质的其他监管方法相一致。选择用于产品评估的高优先级烟气有害物质是该策略中的关键步骤。

TobReg在评估了与癌症、心血管和肺部疾病相关的有害化学物质清单后，TobReg从卷烟烟气中发现的7000多种化学物质中制定了当前可燃型烟草制品有害成分和释放物的优先级清单，由多个监管机构发布。还考虑了烟气的粒相和气相以及不同化学类别中的成分。由于该清单仅占烟气中存在的化学物质的一小部分，因此，烟草制品释放物的总体有害性不一定以这些化学物质的有害性为特征。但是，对大量有害成分的管制将导致现有市场的严重扭曲，并增加监管的复杂性。将有害物质的数量限制在优先级列表中，这认识到了监管结构的实际情况。

8.2.1 有害物质优先级清单选择标准

以下标准用于选择进行测试、报告和未来监管的优先级烟气成分和释放物。

卷烟烟气中特定化学物质的存在对吸烟者是有害的，其水平由公认的科学毒性指数确定

毒性证据是最重要的标准。所考虑的毒理学重要成分是烟草行业2004年向加拿大卫生部报告的名单上的成分[3]，并由Counts等[4]通过测量Philip Morris的几个国际品牌进行比较。这份名单是由加拿大卫生部选出的，代表了烟气中对烟草毒性贡献最大的成分。

通过计算"有害物质动物致癌性指数"和非癌反应指数，并采用Fowles和Dybing[5]提出的简化系统进行了修改，获得了表征所审查有害物质危害性的定量数据。为了进行这些计算，将已公布的有害物质释放量（通过改良的深度抽吸方案获得）按每毫克烟碱进行标准化，并乘以癌症和非癌症效能因子。"癌性因子"定义为每mgT_{25}（$1/T_{25}$），其中T_{25}是在高于25%的动物本底率的特定组织部位会产生肿瘤的长期日剂量[6]。对于非癌症效力的因素，采用环境健康危害评估加州办公室（美国）在2005年2月公布的长期参考暴露水平（http：// www.oehha.ca.gov/air/chronic_rels/AllChrels.html）。采用改良的深度抽吸方案测得的有害成分释放量可从以下三个来源获得：Counts等[4]，一组加拿大品牌[3]和一组澳大利亚品牌[7]。

卷烟品牌中浓度的变化远大于单个品牌中有害物质重复测量的变化

对于中点附近水平变化范围最大的那些有害物质，强制性减少可能对降低每毫克烟碱的平均水平具有最大的影响。不同有害成分含量的变化可以表示为变异系数，变异系数是各品牌的测量标准偏差除以所有品牌的平均值。由于测试的可重复性对于不同的有害成分和使用不同的测试方法可能会发生很大的变化，因此重复测量可用于估计品牌的平均值，重复测量的变化将CI限定在该平均值附近。第二种方法是通过确定品牌数据集中每种有害成分的最大值和最小值（表示为该有害物质中值比），直接确定品牌中有害物质的范围。这不包括调整有害物质重复测量的变化。另一种评估有害物质水平变化的有用方法是比较不同数据集中有害物质的平均水平。该分析可以确定在不同市场销售的同一品牌或不同制造商出售的同一品牌有害物质的水平，这表明可以生产出特定有害物质含量较低的卷烟。

降低烟气中特定有害物质含量的技术，应该规定一个上限

烟草行业改变其产品以符合较低有害成分水平的能力是选择监管的有害物质时要考虑的另一个因素。例如，可以通过改变农业生产实践、烘烤和烟丝掺配来降低卷烟烟气中致癌的TSNA、NNK和NNN的水平[8]。可以通过减少添加到烟草中的糖的浓度或使用活性炭滤嘴或其他滤嘴方法来降低挥发性有害物质（例如乙醛、丙烯醛和甲醛）的水平[9]。可以通过处理、提取或改良烟草混合物[10]来选择性降低苯并[a]芘的释放量。

8.2.2 缔约方会议关于可燃型烟草优先成分和释放物的关键决定

表8.1总结了缔约方会议（COP）的主要决定以及TobReg在根据WHO FCTC第9条和第10条监管可燃型烟草制品的成分和释放物方面的进展。TobReg在第三次COP上确定了9种优先级释放物的初始清单，并验证了其测试方法，见表8.2。从43种与毒理学相关的化合物中选择了这些有害物质，针对这些化合物计算了危害性指数，并如上所述审查了其他标准。烟碱未包括在释放物清单中，但建议测试烟支中的含量。根据COP的授权，由WHO烟草实验室网络（TobLabNet）开发并验证了针对这些优先级释放物的分析方法。

表8.1 COP决议历史和TobReg关于释放物报告和测试的进程

缔约方会议	缔约方会议的决议	TobReg进展、成分和释放物报告
COP 1（日内瓦，2006年）	COP/1/INF.DOC./3 成立一个工作组，根据世界卫生组织《烟草控制框架公约》第9条和第10条准备指南。第一阶段包括测试和测量烟草制品的成分和释放物	
COP 2（曼谷，2007年）	COP/2/DIV/9 继续进行TobReg的工作，包括在产品特征（例如设计特征）影响世界卫生组织《烟草控制框架公约》目标的程度	COP/2/8 TobReg提交了一份进度报告，其中包括一项建议，即卷烟释放物的临时清单，包括44种"霍夫曼分析物"
COP 3（德班，2008年）	COP/3/DIV/3 与优先级清单有关的决定包括： • 在5年内，在两种吸烟方案（ISO和改良的深度抽吸方法，滤嘴通风孔被封闭）下，验证用于测试和测量卷烟成分和释放物的分析化学方法，这些方法被确定为工作组进度报告中优先事项。 • 适当时，设计和验证工作组进度报告中确定的用于测试和测量产品特性的方法	COP/3/6 TobReg提交的新进度报告确定： • 应优先考虑用于测试和测量方法内容（烟碱、氨、保润剂）； • 应优先验证9种释放物的测试和测量方法（NNK、NNN、乙醛、丙烯醛、苯、苯并[a]芘、1,3-丁二烯、CO、甲醛）； • 用于验证测试方法的吸烟方案：（i）ISO 3308：2000和（ii）滤嘴通风孔堵塞的改进的深度抽吸方法； • 用于测试和公开产品特性的临时清单
COP 4（埃斯特角城，2010）	COP/4/DIV/6 继续验证用于测试和测量卷烟成分和释放物的分析化学方法	COP/4/INF.DOC./2 验证了三种分析方法： • 一氧化碳释放量； • 释放物中的NNN和NNK； • 烟草中的烟碱含量
COP 5（首尔，2012年）	COP/5/DIV/5 继续验证用于测试和测量卷烟成分和释放物的分析化学方法； 编制烟草制品有害物质成分和释放物的非详尽清单	COP/5/INF.DOC./1 进行中的工作： • 验证卷烟烟丝中保湿剂和氨的方法； • 验证卷烟主流烟气中苯并[a]芘的方法
COP 6（莫斯科，2014年）	COP6（12） 完成测试和测量卷烟成分和释放物的化学分析方法的验证	COP/6/14 提议： • 烟草产品中39种有害物质的非详尽清单，用于监测和最终监管； • 强制降低的9种有害物质的较短列表； • TobLabNet开发了以下标准化方法：镉和铅含量；水烟（shisha）烟气中的烟碱；以及无烟烟草制品中的烟碱、NNN、NNK及苯并[a]芘
COP 7（德里，2016）	COP7（14） 完成卷烟释放物中醛类和挥发性有机化合物分析化学方法的验证； 评估FCTC/COP/6/14中报告的烟草制品成分和释放物中有害物质扩展清单经验证分析方法的可用性	COP/7/INF.DOC./1 除醛类和挥发性有机化合物外，所有规定成分和释放物的方法验证均已完成

续表

缔约方会议	缔约方会议的决议	TobReg进展、成分和释放物报告
COP 8（日内瓦，2018）	COP8（21）鼓励各方承认并实施TobLabNet方法	COP/8/8 已完成醛和挥发性有机化合物的方法验证。所有规定的内容和排现在都已验证 TobReg确定了以下扩展有害物质成分和释放物清单的机会： •扩大标准操作规程，以包括剩余的醛和挥发性有机化合物； •编制卷烟烟丝金属含量的SOP； •设计水烟烟草和木炭的分析方法

CO：一氧化碳；COP：缔约方会议；ISO：国际标准化组织；NNK：4-(甲基亚硝基氨基)-1-(3-吡啶基)-1-丁酮；NNN：N'-亚硝基降烟碱；SOP：标准操作规程；VOC：挥发性有机化合物

表8.2　考虑列入测试、报告和管制优先级的燃烧型烟草制品的释放物

| TobReg评估的有害物质 | 致癌性或有害性数据 | | 列入优先级清单 | | |
| | | | COP/3/6 | COP/6/14 | |
	TACI	TNCRI	选择用于测试和测量的有害物质	监控和监管的扩展优先级清单	建议强制降低
生物碱					
烟碱[a]	—	—	×	×	—
醛类					
乙醛	6.1	67.1	×	×	×
丙烯醛		1099	×	×	×
甲醛		19.8	×	×	×
巴豆醛[a]	—	—		×	—
丙醛	—	—		×	
丁醛	—	—		×	
芳香胺					
3-氨基联苯	—	—		×	—
4-氨基联苯[a]					
1-氨基萘	0.00036	—		×	
2-氨基萘	0.00068	—		×	
烃类					
苯	2.6	0.64	×	×	×
1,3-丁二烯	9.9	2.4	×	×	×
异戊二烯	3.7	—		×	
苯乙烯[b]	—	0.01			
甲苯	—	0.22		×	
多环芳烃					
苯并[a]芘[a]	0.0086	—	×	×	×
烟草特有亚硝胺[a]					
NNK	3.4	—	×	×	×

续表

TobReg评估的有害物质	致癌性或有害性数据		列入优先级清单		
				COP/3/6	COP/6/14
	TACI	TNCRI	选择用于测试和测量的有害物质	监控和监管的扩展优先级清单	建议强制降低
NNN	0.29	—	×	×	×
NAB	—	—	—	×	—
NAT	—	—	—	×	—
酚类					
儿茶酚	0.58	—	—	×	—
间甲酚和对甲酚	—	0.01	—	×	—
邻甲酚	—	0.01	—	×	—
苯酚	—	0.07	—	×	—
对苯二酚	1.2	—	—	×	—
间苯二酚	—	—	—	×	—
其他有机化合物					
丙酮	1.4	—	—	×	—
丙烯腈	—	2.1	—	×	—
喹啉	—	—	—	×	—
吡啶	—	—	—	×	—
金属和非金属					
砷[c]	—	0.16	—	—	—
镉	1.7	2.6	—	×	—
铬[b]	—	—	—	—	—
铅	0.00	—	—	×	—
汞	—	0.02	—	—	—
镍[b]	—	—	—	—	—
硒[b]	—	—	—	—	—
其他成分					
氨	—	0.07	—	×	—
CO[a]	—	1.3	×	×	×
氰化氢	—	17.2	—	×	—
氮氧化物	—	3.1	—	×	—

CO：一氧化碳；NAB：N-亚硝基假木贼碱；NAT：N-硝基新烟草碱；NNK：4-(甲基亚硝基氨基)-1-(3-吡啶基)-1-丁酮；NNN：N'-亚硝基降烟碱；PAH：多环芳烃；TACI：有害物质动物致癌指数；TNCRI：有害物质非癌反应指数；TSNA：烟草特有亚硝胺

a. 将某些有害物质列入优先清单（初始和/或扩展）的理由：烟碱不在初始优先级清单上，但建议用于烟草（含量）的测试，随后与其他有害物质一起列入扩展清单。苯并[a]芘虽然TACI含量低，但它是卷烟烟气中多环芳烃家族的代表，而且有大量证据表明这些多环芳烃具有致癌性。4-氨基联苯被列入是因为它是一种人体致癌物，尽管实验数据不允许正确计算T_{25}。NNK和NNN被包括在内，因为它们已经被列入第一份关于强制降低有害物质释放量的报告中[12]。尽管缺乏可耐受的水平值，巴豆醛因其具有反应性的α,β-不饱和醛结构而被包括在内。尽管CO具有相对较低的有害物质非癌反应指数，但也包括在内，因为它被认为与心血管疾病的机制相关

b. 最初考虑[11]但未包括在最终建议清单中，因为它发生在低水平，不被认为对危害指数有明显贡献

c. 在COP/6/14号文件报告扩展清单之后增加

为了响应随后的COP要求编制烟草制品有害物质成分和释放物的非详尽清单，TobReg在2013年12月4~6日于巴西里约热内卢举行的会议上，通过审查重新评估了有害物质优先级清单。包括加拿大卫生部、荷兰国家公共卫生与环境研究所（RIVM）和FDA，并审查了世界卫生组织早些时候关于烟草制品管制科学依据的技术报告中评估的有害物质清单[11]。随后，TobReg通过添加满足以下标准的有害物质来列出非详尽清单：满足足够的致癌性证据标准或已知的呼吸道或心脏毒性标准、不同国家不同品牌的含量差异、易于测量，以及降低产品释放量的可能性（表8.2）。这份清单由37种此类有害物质和烟碱组成，已提交缔约方第六次会议讨论。在最初评估的43种成分清单中，只有少数有害物质未包括在更新的清单中。例如，铬、镍和硒的含量很低，或者无法在三个批次中进行准确定量。因此，尽管这些元素有剧毒，它们不被认为对危害指数有明显的贡献，也不包括在内。

在同一份报告（COP/6/14）中，TobReg建议应考虑将最初选择用于方法验证的9种化合物强制降低。TobReg得出结论，这些化合物是卷烟烟气中可以降低的最有害的有害物质。它们代表不同族和烟气的不同相的有害物质，对肺和心血管系统有毒，并具有致癌性。9种有害物质是：乙醛、丙烯醛、苯、苯并[a]芘、1,3-丁二烯、CO、甲醛、NNK和NNN（表8.2）。39种优先级清单中剩余的也应该进行测量并报告。

8.3 关于有害性的新科学知识概述

8.3.1 醛（乙醛，丙烯醛，甲醛，巴豆醛，丙醛，丁醛）

醛对呼吸系统有毒[13]。乙醛和甲醛对心血管系统也有毒性，并且是呼吸道肿瘤的致癌物[14,15]。丙烯醛是一种强烈的刺激性物质，对肺纤毛有毒性，并已被提议作为一种肺致癌物[16,17]。巴豆醛是一种强刺激性物质，是一种弱的肝致癌物，会在人肺中形成DNA加合物[18]。

毒性和致癌性机理的新研究

γH2AX证明了乙醛、丙烯醛和甲醛在人肺细胞中的遗传毒性，该测定法可检测早期细胞对DNA双链断裂的反应，并在人类流行病学中用作DNA损伤的生物标志物[19]。所有这三种醛均具有剂量（或时间）依赖的遗传毒性，丙烯醛具有最强的诱导DNA损伤的潜力，其次是甲醛和乙醛。对模拟的人类口腔呼吸的计算流体动力学建模研究中，卷烟烟气中的代表性醛释放量和终生平均日剂量表明，人体暴露的顺序是丙烯醛>甲醛>乙醛[20]。对丙烯醛在下呼吸道系统中的毒性作用方式的分析表明吸烟者通过吸入烟气暴露丙烯醛，表明丙烯醛毒性的机制包括氧化应激、慢性炎症、坏死细胞死亡、坏死诱导炎症、组织重塑和破坏以及随后的肺弹性丧失和扩大的肺空腔[21]。这些过程与慢性阻塞性肺疾病中呼吸道下部的炎症和坏死相一致。另一项研究的结果表明在吸烟过程中丙烯醛会导致在肺部观察到功能异常的先天免疫反应[22]。此外，丙烯醛可能会导致吸烟者膀胱癌的发生。Lee等[23]分析了正常人尿路上皮黏膜和膀胱肿瘤组织中的丙烯醛-脱氧鸟苷加合物的含量，并测量了它们在人尿路上皮细胞中的致突变性。两种组织中的加合物水平均高于已知的膀胱致癌物4-氨基联苯所引起的加合物含量，并且它们诱导的突变特征和光谱似乎比由4-氨基联苯所引起的突变更显著。膀胱肿瘤组织中的丙烯醛-脱氧鸟苷水平比正常细胞高两倍。

对人肺癌细胞系中醛暴露的转录反应的研究显示出与上述研究不同的结果[24]。甲醛反应最强，66个基因（主要涉及细胞凋亡和DNA损伤）的差异表达超过1.5倍。乙醛使57个基因失调，而丙烯醛仅上调了一个与氧化应激有关的基因。

尽管DNA加合物的形成是醛的致突变性和致癌性的关键因素，但蛋白质修饰也可能起作用。例如，除DNA加合物N^2-亚乙基-2′-脱氧鸟苷外，乙醛还与丙二醛形成了杂合蛋白加合物，丙二醛是卷烟烟气中有毒成分和反应物引起的脂质过氧化作用的产物。肺中丙二醛-乙醛杂化蛋白加合物的形成已表明可引发多种病理状况，包括炎症和抑制伤口愈合[25]。涉及对这些加合物（即IgM、IgG和IgA）的免疫反应的抗体（免疫球蛋白）可以预测动脉粥样硬化疾病的进展以及心血管事件的发生，例如急性心肌梗死或冠状动脉搭桥术[26]。

考虑基因多态性对于人体对有害物质反应的影响也很重要。例如，迄今为止，乙醛的风险评估是基于动物毒理学研究中确定的阈值，而阈值并未考虑到遗传和生化证据，即缺乏ALDH2的人极易受到该化学物质的致癌作用的影响。在一项通过饮酒暴露于乙醛的研究中，ALDH2失活与头颈癌和食道癌的比值比高达7[27]。一项对日本心肌梗死和稳定型心绞痛患者的研究以及相匹配的对照研究表明，ALDH2处于非活动状态基因型可能增加吸烟者发生心肌梗死的风险[28]。

醛的毒性和致癌性的其他潜在机理也已经进行了研究。丙烯醛通过与负责其乙酰化的酶相互作用而影响芳香胺的代谢和命运[29]。丙烯醛与抗逆转录病毒药物齐多夫定之间潜在相互作用的研究已在资源有限的国家中广泛使用，但与肝毒性相关，强烈表明通过吸烟和/或饮酒接触丙烯醛可能是齐多夫定诱导肝毒性的主要机制[30]。另一项研究表明，暴露于乙醛和甲醛可能会导致BRCA2突变携带者发生癌变[31]。BRCA2是一种肿瘤抑制基因，其突变会增加患乳腺癌的风险，并且还与卵巢癌、胰腺癌和其他癌症的易感性有关。两种醛都通过蛋白酶体降解选择性地消耗BRCA2，这可能会触发DNA复制过程中的自发诱变[31]。

除呼吸毒性和致癌性以外的其他对健康影响的新研究

致瘾性：越来越多的证据表明乙醛也对烟草的致瘾性有贡献[32]。对实验动物自我给药的研究有助于了解乙醛给药的行为相关性，以及与神经递质相互作用的动机、奖励和压力相关反应，例如多巴胺和内源性大麻素[33]。另一种可能的机理是通过形成哈尔满，即唾液中乙醛和胺的缩合产物。哈尔满是一种单胺氧化酶抑制剂，可以帮助维持对烟碱的行为敏感性。为了支持这一假设，已有研究证明吸烟者服用含L-半胱氨酸（与乙醛反应的氨基酸）的锭剂比接受安慰剂治疗的人戒烟率更高[34]。但是，应当指出，烟草和卷烟烟气中也存在哈尔满。

心血管作用：丙烯醛的心血管作用已被综述[35]。在体外和体内，心血管组织似乎对丙烯醛的毒性特别敏感，丙烯醛可以在心脏中产生氧化应激，与心肌细胞和血管内皮细胞蛋白形成蛋白质加合物，并引起血管痉挛。因此，长期接触丙烯醛可能会导致人的心肌病和心力衰竭。长期暴露于丙烯醛的小鼠的研究结果支持了这一结论，该研究表明，即使相对较低的丙烯醛暴露量（例如二手烟或电子烟丙烯醛暴露），也会通过减少内皮细胞修复，抑制免疫细胞和血管生成而增加心血管风险[36]。在路易斯维尔健康心脏研究的211名参与者中，这些参与者处于心血管疾病的中高风险水平，同时对丙烯醛的暴露量进行了评估[37]。接触丙烯醛与血小板活化、循环血管生成细胞水平的抑制和心血管疾病风险增加有关。

葡萄糖和脂质代谢：大鼠接触丙烯醛会引起高血糖症和葡萄糖耐受异常，从而引起代谢障碍，同时皮质酮水平显著增加，而肾上腺素水平则微量增加[38]。在2027名参加了2005~2006年美国国家健康和营养检查的成年人中，对丙烯醛代谢产物和糖尿病与胰岛素抵抗的生物标志物之间的关系进行了研究[39]。在所分析的生物标志物之间发现正相关，表明丙烯醛在2型糖尿病的病因学和人胰岛素抵抗中的潜在作用。

其他影响：丙烯醛可能会影响肠上皮的完整性。小鼠口服丙烯醛后，发现肠上皮屏障受损，导致通透性增加和随后的细菌移位[40]。Song等[41]提供了更多的证据表明丙烯醛是中耳炎的危险因素。此外还研究了醛的发育和生殖毒性。Amiri和Turner-Henson[42]报道了一项140名健康孕妇的横断面研究的结果，其中研究了妊娠中期甲醛暴露与胎儿生长之间的关系。线性回归模型显示将甲醛暴露的二等分水平（百万分之0.03）双顶径百分位数的显著预测因子（$P<0.006$）。在一项动物模型研究中，子宫暴露于5 mg/kg的丙烯醛导致雄性后代睾丸激素合成显著降低[43]。

新的流行病学证据

在一项对2200多名吸烟者进行的研究表明，五个族群中丙烯醛和巴豆醛暴露的尿液生物标志物存在显著差异[44]。调整混杂因素后，夏威夷原住民的两种醛类尿液生物标志物的几何平均水平最高，而拉丁美洲人最低。这些研究结果与这些人的另一项流行病学研究结果一致。与白人相比，夏威夷原住民罹患肺癌的风险较高，而拉丁美洲人与白人的风险较低[45]。这些结果表明，丙烯醛和巴豆醛可能与吸烟者肺癌的病因有关。

8.3.2 芳香胺（3-氨基联苯，4-氨基联苯，1-氨基萘，2-氨基萘）

暴露于芳香胺类物质与膀胱癌有关[46]，而2-氨基萘和4-氨基联苯是已知的人膀胱致癌物[15]。在一项研究中，吸烟者（>20支烟/天）接触4-氨基联苯的平均含量是不吸烟者的4.8倍，表明卷烟烟气是接触该致癌物的主要来源[47]。

芳香胺可能会引发非泌尿系统癌症。1953~2011年，在同一家工厂中的224名男性工人中，与2-氨基萘相关的肺癌和膀胱癌风险很高[48]。对暴露于2-氨基萘的工人发生肺癌风险的系统评价和荟萃分析显示，罹患肺癌的风险显著增加，有或没有职业接触其他肺有害物质和致癌物的效果估计相似[49]。在这两项研究中均监测到了联苯胺和2-氨基萘的暴露。

8.3.3 烃类化合物（苯，1,3-丁二烯，异戊二烯，甲苯）

苯和丁二烯被列为已知的人体致癌物，会引起血液淋巴器官癌[15]。异戊二烯在实验动物的各个部位引起肿瘤[50]。1,3-丁二烯和甲苯是呼吸道有害物质，甲苯对中枢神经系统也有毒，同时有生殖毒性。这些化合物大量存在于卷烟烟气中，并可能在吸烟者的肺癌中起作用[14,15,51]。

从最敏感毒性终点与主流和二手烟气中1,3-丁二烯暴露的适当估计值之间的比来评估1,3-丁二烯对烟草烟气相关健康的影响以及风险降低策略的可能影响[52]。作者得出的结论是降低卷烟烟气中的1,3-丁二烯含量可以显著降低罹患癌症（白血病）和非癌症（卵巢萎缩）的风险。他们提出对暴露范围的分析是评估风险降低策略对人类健康影响的实用手段。在对苯的非癌症健康影响进行的综述中得出的结论是，接触苯可在生殖、免疫、神经、内分泌、心血管和呼吸系统中产生许多结果[53]。

心血管作用

在小鼠和人类实验中研究了苯暴露与心血管疾病风险增加之间的潜在关联[54]。通过直接吸入来评估苯对小鼠的影响，而通过测量尿液生物标志物来评估苯对210位患有轻度至高风险心血管疾病风险的人的影响。与呼吸过滤空气的对照小鼠相比，小鼠的循环血管生成细胞水平显著降低，血浆中低密度脂蛋白水平更高。在人实验中，吸烟者和血脂异常患者中苯的暴露量较高，这与循环血管生成细胞的数量呈负相关,并且与根据弗雷明汉风险评分评估的心血管疾病风险呈负相关。

新的流行病学证据

一项2012~2013年在美国海湾进行的长期随访研究结果表明,环境中苯和甲苯的暴露与血液学影响有关,包括血红蛋白降低,平均红细胞血红蛋白浓度减少、并增加了红细胞分布宽度[55]。有关苯的证据尤其有力。

在多种族队列研究中,通过分析尿中的S-苯基硫基尿酸(苯的暴露生物标志物)比较了五个不同种族的吸烟者的苯摄入[56]。与白人相比,非裔美国人的S-苯基硫基尿酸水平明显较高,而日裔美国人的水平明显较低。尽管一般不认为苯会引起肺癌,但这些差异与该人群中罹患肺癌风险的差异一致。

8.3.4 多环芳烃(苯并[a]芘)

多环芳烃(PAH)是在有机物不完全燃烧过程中形成的,总是以混合物形式出现。许多PAH在实验动物中是强致癌物或有害物质[57],许多PAH存在于卷烟烟气中,包括苯并[a]芘,苯并[a]芘被国际癌症研究机构归类为人体致癌物(IARC)[57,58]。多环芳烃被广泛认为是吸烟者肺癌的主要病因[57,59-61]。在A/J小鼠中对苯并[a]芘和TSNA NNK的研究显示,剂量依赖性肿瘤发生的剂量远低于先前报道的剂量[62]。

Zaccaria和McClure[63]分析了已发表的有关苯并[a]芘和其他PAH的研究,发现推导的对9种PAH的免疫抑制相对效能因子与它们对癌症的效能因子之间具有相关性,从而证实了先前关于PAH致癌性和免疫抑制之间关系的观察。

苯并[a]芘的环境暴露与成年人学习和记忆障碍以及儿童神经发育不良有关。对苯并[a]芘的神经毒性的潜在机制进行了全面的文献综述[64]。结果表明与癌症相关的影响较低时,可观察到神经毒性作用。苯并[a]芘与芳烃受体的结合会导致神经元活性的丧失和长期能力的下降,从而损害学习和记忆能力。

8.3.5 烟草特有亚硝胺

TSNA是烟草制品的重要成分,其中NNK和NNN可能与烟草使用者的肺癌、胰腺癌、口腔癌和食道癌有关[65,66]。两者均已被国际癌症研究机构归类为人体致癌物[66,67]。这些亚硝胺在烟草加工过程中由烟草生物碱形成。形成的数量取决于烟草类型、硝酸盐含量和烟草加工技术,导致其在各种品牌卷烟中的数量差异很大[65,68-70]。在吸烟者中,由于卷烟设计的变化(包括滤嘴通风)而导致的烟气TSNA释放量增加,伴随着肺腺癌(一种由NNK在实验动物中诱发的肺癌类型)的发生率增加[71,72]。

实验动物的新证据

在经过处理的70周雄性F-344大鼠中研究了NNK及其代谢产物NNAL的致癌性。两种化合物均引起肺肿瘤高发，并且观察到从原发性肺癌到胰腺的转移。结果清楚地证明了NNK和NNAL在大鼠中具有潜在的肺致癌性和DNA损伤活性[73]。在同一课题组的另一项研究中[74]，在20只雄性F-344大鼠饮用水中加入NNN，诱发了96例口腔肿瘤和153例食道肿瘤。这项研究首次显示了NNN在食道中的致癌性，并确定NNN是烟草中存在的强口腔致癌物。

人类的新证据

与动物致癌性数据一致，在上海的一项队列研究中，吸烟者NNN和NNK暴露与食道癌和肺癌风险之间呈正相关。对同一人群的其他分析表明，接触NNK与食道癌无关，接触NNN与吸烟者患肺癌的风险无关[75]。总之，这些结果重申了NNN和NNK分别对吸烟者食道和肺的器官特异性，这与F-344大鼠的发现一致。在多种族队列研究中，测量了2252名吸烟者NNK的摄取量[76]。在调整了尿液收集的年龄、性别、肌酐和总烟碱含量（总烟碱摄入量的标志）后，非裔美国人的NNK暴露量最高，而日裔美国人的最低。这些发现与这些人群中吸烟者患肺癌风险的发现一致。

机理研究和对实验动物的研究表明，DNA加合物的形成是NNN和NNK诱导致癌作用的关键步骤。在头颈鳞状细胞癌吸烟者的口腔细胞中发现这些加合物的水平高于未患癌症的吸烟者[77]。

8.3.6 生物碱（烟碱）

烟碱是已知的烟草和卷烟烟气中的主要致瘾剂[78]，是烟草使用的主要驱动力。烟碱含量可以极大地影响产品使用的程度和方式，还可以定义产品吸引使用者的类别。此外，它还会影响产品中其他有害物质和致癌物的暴露。

关于卷烟烟气中烟碱水平降低带来影响的新证据

TobReg已建议将卷烟烟碱含量降至最低或非成瘾水平，并由FDA提议作为减少或消除使用燃烧型烟草制品的一种方法[79,80]。许多研究结果证实了这种方法的可行性。2013年6月至2014年7月在美国10个地点进行的双盲、平行、随机临床试验中，将840名参与者随机分配吸食通常品牌卷烟或6种不同烟碱含量研究用卷烟中的一种，烟碱含量范围是从0.4 mg/g到15.8 mg/g的烟草（与商业品牌的烟碱含量相比），实验共持续6周[81]。在研究结束时，吸食含2.4 mg、1.3 mg或0.4 mg烟碱卷烟的参与者每天平均吸烟量少于吸食通常品牌或含15.8 mg/g的卷烟（$P<0.001$）。吸食比对照卷烟烟碱含量低的卷烟减少了烟碱的暴露和依赖性，并且在戒烟期间减少了对烟瘾的渴望，且呼出气体中CO水平或总抽吸容量并没有增加，这表明补偿抽吸最小。

在Hatsukami等的后续随机、平行的8周研究中[82]，将不愿戒烟的吸烟者随机分配到正常或非常低烟碱含量卷烟（VLNC），并评估了替代烟碱产品的使用、吸烟行为和卷烟暴露生物标志物。与继续使用正常烟碱的卷烟相比，低烟碱卷烟的使用降低了吸烟率，减少了烟气有害成分生物标志物，并更多地使用了替代烟草或烟碱产品。

除成瘾以外的健康影响

烟碱可能导致吸烟者发生急性心血管事件和加速动脉粥样硬化，这可能是由于交感神经系统的刺激、冠状动脉血流量减少、内皮功能受损和其他药理作用[83]。对人和动物进行的关于怀孕和青春期接触烟碱的健康影响的系统研究综述表明，烟碱对不良反应至关重要，包括降低肺功能、听觉功能缺陷和婴儿心肺功能受损，并且可能对认知和晚年的行为缺陷有贡献[84]。该研究还发现，青春期暴露于烟碱与记忆、注意力和听觉处理缺陷、冲动和焦虑增加有关，对动物的研究表明烟碱增加了对其他药物上瘾的可能性。

8.3.7 苯酚（邻苯二酚，间、对和邻甲酚，苯酚，对苯二酚，间苯二酚）

邻苯二酚是一种辅助致癌物[85]，存在于卷烟烟气中[66]。美国环境保护局在总结了间、对和邻甲酚遗传毒性的基础上将其归为可能致癌的物质，并在啮齿类动物中增加皮肤和鼻腔肿瘤发生率。甲酚也是呼吸道有害物质。对苯二酚在体外和体内均具有诱变作用，包括在γH2AX实验中证明对苯二酚具有明显遗传毒性[19]。经口给药的雄性F-344大鼠的肾脏重复地诱发良性肿瘤，但有关人类的数据不足。苯酚是呼吸道有害物质，会引起心血管疾病，并且是肿瘤的促进剂。据报道，在一系列研究中证明间苯二酚具有多种毒性作用，被认为是呼吸道有害物质。

8.3.8 其他有机化合物（丙酮，丙烯腈，吡啶，喹啉）

丙酮是呼吸道有害物质，可刺激呼吸道。丙烯腈是呼吸道有害物质，被国际癌症研究机构（IARC）分类为可能对人体致癌的物质（2B组）[14]。它容易与蛋白质形成加合物，吸烟者中这种加合物的水平高于非吸烟者[14,86]。丙烯腈在某些方法中也具有致突变性[14]。吡啶的毒性研究很少。一些研究表明其对呼吸道、中枢神经系统和肝脏有影响。喹啉在急性暴露后有刺激性，对动物具有肝毒性和致癌性。自2013年以来，没有发现关于这些有机化合物新的重要毒理学证据。

8.3.9 金属和类金属（砷，镉，铅，汞）

砷和镉是人体致癌物[87]。这些金属是肺致癌物，也可能在膀胱癌（砷）和肾癌（镉）中起作用。砷还具有心血管和生殖作用，镉是一种神经和呼吸道有害物质。铅是一种神经、生殖和心血管毒性物质，可能是人体致癌物[88]。汞被国际癌症研究机构（IARC）分类为2B组，并且也具有生殖毒性[89]。这些元素在卷烟烟气[66]和无烟烟草制品[90]中的含量不同。这可能受其在种植烟草的土壤中浓度的影响。Pinto等[91]发现吸烟者的肺组织中砷、铅和镉的含量显著高于非吸烟者。幼儿接触二手烟会影响血铅水平且与智商和认知能力下降有关[92]。

致癌性

在1989~1991年间参加强心研究的亚利桑那州、俄克拉荷马州和南北达科他州的3792名美洲印第安人中，研究了尿液中镉与癌症死亡率之间的关系，发现镉暴露与所有癌症以及肺癌和胰腺癌的死亡率相关[93]。在文献综述中，Feki-Tounsi和Hamza-Chaffai[94]从现有的体外和流行病学研究中得出结论镉暴露与膀胱癌的风险增加有关，并且可能与尿路上皮毒性和致癌作用有关。

神经毒性

镉和铅是卷烟烟气中的神经毒性成分，可能导致与吸烟有关的抑郁症。在2011~2012年美国国家健康与营养调查（$N = 3905$）中研究了血镉和铅水平与当前抑郁症状之间的关联。发现年龄在20~47岁之间的男性参与者中，血液中的镉与抑郁症状的概率较高，而在这一年龄段的女性参与者中，血铅、吸烟和肥胖与抑郁症状相关[95]。在暴露于镉和/或烟碱的雌性青春期小鼠中，对空间和非空间记忆、焦虑相关行为和运动活动的影响的发现支持了这些结论[96, 97]。与对照组相比，烟碱和镉增加了被治疗小鼠的代谢、食物摄入和体重。烟碱给药可增强运动功能，而镉可降低运动活性。两种化合物均导致记忆指数降低。烟碱和镉的联合处理可减轻体重和运动活动，增加焦虑并显著减少非空间工作记忆。

心血管疾病

据推测，镉通过损害血管内皮细胞而增加了与吸烟有关的心血管风险[98]。在关于镉暴露与心血管疾病之间关系的流行病学研究的系统综述中，心血管疾病、冠心病、中风和外周动脉疾病的合并相对风险（95%CI）为：1.36（1.11，1.66）、1.30（1.12，1.52）、1.18（0.86，1.59）和1.49（1.15，1.92）[99]。实验证据，支持镉暴露与心血管疾病，尤其是冠心病之间的联系。

其他健康影响

与对照组相比，暴露于亚砷酸钠90天的小鼠的肝脏、肾脏和肠组织中的微核多色红细胞增多，遗传毒性和生殖细胞毒性作用增加。与无烟烟草制品提取物的联合暴露导致精子头部异常地显著增加[100]。

通过测量150名健康孕妇的血铅水平，研究了卷烟烟气中铅的暴露对胎儿生长的影响[101]。母亲吸烟的婴儿出生体重显著低于母亲不吸烟的婴儿体重（$P<0.001$），并且与血浆（$r=-0.38$；$P<0.001$）和全血（$r=-0.27$；$P<0.001$）铅水平呈负相关。

据推测，接触汞会导致代谢综合征和糖尿病。文献综述表明，虽然流行病学数据表明生物基质中的总汞浓度与这些健康结果的发生率之间可能存在关联，但这种关系并不一致[102,103]。对汞接触相关的健康影响的更全面的研究表明即使长期暴露于低浓度的汞，也可能导致心血管、生殖和发育毒性、神经毒性、肾毒性、免疫毒性和致癌性[104]。

8.3.10 其他成分（氨，一氧化碳，氰化氢，氮氧化物）

氨是一种呼吸道刺激物和有害物质，在各种工业和农业环境中都会增加呼吸道症状、哮喘和肺功能受损的流行率[105-108]。有限的研究表明哮喘患者对氨气的呼吸作用更为敏感[108,109]。CO是一种公认的心血管有害物质，它与氧气竞争结合血红蛋白。在吸烟者中，它被认为可以减少氧气的传输，引起内皮功能障碍并促进动脉粥样硬化和其他心血管疾病的进程[110-112]。

氰化氢是一种众所周知的有害物质，其主要靶标是心血管、呼吸系统和中枢神经系统。它通过抑制呼吸链中的细胞色素氧化酶起作用。卷烟烟气可减少机体的氰化氢解毒功能，从而导致吸烟者长期暴露于这种毒素，从而导致弱视、球后神经炎、不育以及可能对伤口愈合的损害[112,113]。

氮氧化物是呼吸道和心血管有害物质。一氧化氮是新鲜卷烟烟气中的主要形式，会引起血管舒张并导致DNA链断裂和脂质过氧化，并可能导致癌变[114]。它也可能通过增加烟碱吸收、减轻压力症状和增加突触后多巴胺水平来促进烟碱成瘾[115]。二氧化氮是一种肺刺激性物质。

8.4 分析方法的可用性

8.4.1 WHO TobLabNet优先有害物质分析标准方法

为了确保实施WHO FCTC第9条和第10条，实验室能力必须满足卓越、透明、可靠和可信的最高标准[116]。实验室需要标准化、可靠、准确的分析方法来对全球烟草制品进行科学和严格的测试[117]。对一套方法达成共识可能部分取决于这种方法是否能成功地转移到其他实验室。表8.3总结了截至2018年10月已建立并经过验证的WHO TobLabNet标准操作规程（SOP）。

表8.3 用于分析优先级有害物质的标准化WHO TobLabNet标准操作规程

标准操作规程	标题	优先级有害物质	出版年份
SOP 01	卷烟深度抽吸方案	产生主流烟气所必需	2012
SOP 02	烟草制品成分和释放物分析方法验证	描述方法验证	2017
SOP 03	ISO和深度抽吸方案下卷烟主流烟气中烟草特有亚硝胺的测定	主流烟气NNK、NNN（释放物）	2014
SOP 04	卷烟烟丝中烟碱的测定	烟碱（成分）	2014
SOP 05	ISO和深度抽吸方案下卷烟主流烟气中苯并[a]芘的测定	主流烟气苯并[a]芘（释放物）	2015
SOP 06	卷烟烟丝中保润剂的测定	丙二醇、甘油（丙三醇）、三甘醇（成分）	2016
SOP 07	卷烟烟丝中氨的测定	氨（成分）	2016
SOP 08	ISO和深度抽吸方案下卷烟主流烟气中醛类化合物的测定	乙醛、丙烯醛、甲醛（释放物）	2018
SOP 09	ISO和深度抽吸方案下卷烟主流烟气中挥发性有机物的测定	1,3-丁二烯、苯（释放物）	2016
SOP 10	深度抽吸方案下卷烟主流烟气中烟碱和一氧化碳的测定	主流烟气烟碱和一氧化碳（释放物）	2016

CO：一氧化碳；ISO：国际标准化组织；NNK：4-(甲基亚硝基氨基)-1-(3-吡啶基)-1-丁酮；NNN：N'-亚硝基降烟碱；SOP：标准操作规程

8.4.2 其他优先有害物质分析方法概述

下文描述了尚无标准化的WHO TobLabNet标准操作规程的优先级有害物质，以及公开发表的烟草制品和主流烟气中定量分析的方法。

醛（丁醛，巴豆醛，丙醛）

Ding等[118]报告了一种烟气中乙醛、丙烯醛、丙酮、巴豆醛、甲醛、丙醛和2-丁酮的分析方法，与之前的分析方法相比，该法在易用性、效率和环境友好性方面都有着显著改善。该方法使用双重滤片（剑桥滤片，一个用二硝基苯肼预处理，另一个干燥）同时捕获和衍生主流烟气中的羰基化合物，然后通过超高压液相色谱（LC）和串联质谱（MS/MS）在4分钟内进行定量。该方法的准确度由低、中、高水平的加标样品确定，方法准确度为83%~106%，而其精密度由30个参比卷烟（3R4F）的重复测量确定，所有目标分析物的相对标准偏差均小于20%。乙醛的检出限为2.7 μg，丙烯醛的检出限为0.1 μg，巴豆醛的检出限为0.2 μg，甲醛的检出限为2.4 μg，丙醛的检出限为0.6 μg。该方法非常适合常规分析，因为它可以与直线吸烟机一起使用，从而大大提高了样品通量。

中国烟草总公司的研究人员建立了一种简单的方法来快速测定主流烟气中的丙烯醛、丙酮、丙醛、巴豆醛、丁酮和丁醛[119]。尽管分析时间很短（4 min），但必须进行全扫描和子离子扫描的数据挖掘，以克服使用大气压化学电离技术分离和定量丙酮-丙醛和丁酮-丁醛异构体的困难。用参比卷烟（3R4F，35 mL抽吸容量，8口）测得的结果与文献报道的结果一致。WHO优先级有害物质的检出限为：丙烯醛为0.007 μg/L，丙酮为0.021 μg/L，丙醛为0.008 μg/L，巴豆醛为0.004 μg/L，丁酮为0.012 μg/L，丁醛为0.006 μg/L。为了将气体样品直接引入电离区域，对化学电离源进行了改进。

在另一种报道的方法中，卷烟中的主流烟气被收集在硫酸（20%）和抗坏血酸（25 mmol/L）冷肼中[120]，并且采用分散液-液微萃取方法同时提取溶液并转化为苯甲醛、丁醛和糠醛衍生物，然后通过高效液相色谱法进行定量。基质加标回收率为88.0%~109%，日间和日内相对标准偏差均小于8.50%。WHO优先级有害成分的检出限为：苯甲醛为14.2 μg/L，丁醛为21.3 μg/L，糠醛为7.92 μg/L。与之前发表的液相微萃取方法相比，该方法更快、更简单且成本更低。

芳香胺（1-氨基萘，2-氨基萘，3-氨基联苯，4-氨基联苯）

在另一个方法中，收集卷烟主流烟气的气相物（0.6 mol/L盐酸于冷肼中）和粒相物（玻璃纤维滤片），然后用酸溶液超声提取玻璃纤维滤片[121]。萃取液通过两个极性不同的固相萃取柱进行净化，并在苯基己基柱上分离，采用LC-MS/MS定量了9种芳香胺。通过将目标分析物以三种水平（低、中、高）添加到参比卷烟（3R4F）的提取物中来验证该方法。WHO优先级有害成分的回收率为84.8%~97.8%，日内和日间精密度分别为<9%和<14%。WHO优先级有害成分的检出限：1-氨基萘为0.08 ng/支烟，2-氨基萘为0.09 ng/支烟，3-氨基联苯为0.05 ng/支烟，4-氨基联苯为0.03 ng/支烟。该方法与Xie等报道的方法相类似[122]，但目标芳族胺的分离度和基准分离度更好。

Deng等[123]使用一种新颖的分散液-液微萃取净化方法和更快的超高效合相色谱定量分析了主流烟气中的9种芳香胺。该方法的仪器运行时间为5分钟，与Zhang等的高性能液相方法所需的30分钟相比有了实质性的改进[121]。在这种方法中，超临界CO_2用作主要的流动相，可确保更高的流速和更低的色谱柱压降和更快的运行时间。该方法已用参比卷烟（3R4F）进行了验证。对于世界卫生组织重点有害物质，加标回收率为69.4%~120%。WHO优先级有害物质的检出限为：1-氨基萘为0.29 ng/支烟，2-氨基萘为0.21 ng/支烟，3-氨基联苯为0.02 ng/支烟，4-氨基联苯为0.03 ng/支烟。

碳氢化合物（异戊二烯，甲苯）

Sampson等[124]报道了一种自动化的高通量方法，用于准确定量主流烟气中的各种有害挥发性有机化合物，包括丙烯腈、苯、丁醛、1,3-丁二烯和甲苯。将气相收集到聚氟乙烯气体采样袋中，并添加同位素标记的类似物作为内标，以解决由于处理和陈化引起的目标分析物的损失。将粒相物收集到玻璃纤维过滤片上，并添加内标。加热后，通过自动固相微萃取对气袋和滤片顶部空间进行采样，并通过具有质量选择检测功能的气相色谱（GC）进行定量。该方法已采用参比卷烟（1R5F和3R4F）在ISO和HCI抽吸方案下进行了验证。除甲苯外，结果与其他报告的结果相当。批间精度和批内精度均小于20%。甲苯和对二甲苯、间二甲苯之间的相关性很高（0.97）。该方法后来被同一小组使用来检测50个美国商业卷烟品牌中的目标分析物[125]。上述方法均未报告世界卫生组织优先级有害成分的检出限；但是，对于世卫组织所有优先级有害物质所用的最低校准标准为每十亿分之0.14体积比，苯除外，苯是十亿分之0.1体积比。

苯酚（邻苯二酚，间、对、邻甲酚，苯酚，对苯二酚，间苯二酚）

Saha等报道了一种简单、精确的方法，用于定量主流烟气中WHO优先级的酚[126]。该方法采用单滴微萃取和LC-MS/MS进行分析，减少了有机溶剂的消耗，缩短了样品制备时间并消除之前发表的GC-MS方法所需的衍生步骤。WHO优先级有害物质的检出限：儿茶酚为0.30 ng/mL，邻甲酚和对甲酚为0.05 ng/mL，苯酚为0.15 ng/mL，对苯二酚为0.30 ng/mL，间苯二酚为0.20 ng/mL。Wu等在Labstat International[127]报道的LC-MS/MS方法，在加拿大卫生部的方法（T-114、T-211）上进行了改进，方法采用了较短和更小粒径的色谱柱，以完全解析三种甲酚异构体。

其他有机化合物（丙酮，丙烯腈，吡啶，喹啉）

在上面烃类中讨论了包含丙烯腈的多组分分析方法，在醛类中讨论了包含丙酮的多组分分析方法。烟草行业研究人员报道了全二维色谱飞行时间质谱法用于分析主流烟气的气相[128]和粒相物[129]中的有机化合物。这些方法不适合测试释放物，但可用于监管目的，以区分由不同产品设计引起的主流烟气释放或识别掺假。

金属和类金属（砷，镉，铅，汞）

对主流烟气中镉含量的传统样品收集方法进行了严格评估[130]。用铂阱测定了两种不同样品采集方法：静电沉淀法和剑桥滤片法。通过初级捕集阱到达铂捕集阱的镉的检出限为0.30 ng/支烟。剑桥滤片中的镉穿透量很大（占样品的4%~23%），但与静电沉淀相比（<1%）可忽略不计。Fresquez等使用此技术[131]建立了一种高通量分析ISO和HCI抽吸方案下卷烟烟气和雪茄烟气中的汞的方法。该方法比以前报道的方法更快，更简单并且对环境更友好，因为它消除了对冷肼、强氧化剂（例如高锰酸盐）和强酸（例如硫酸）的需求。烟草中汞的检出限为0.27 ng/g，烟气中汞的检出限为0.097 ng/支烟。同一研究小组使用高通量方法测定了美国市场上50个品牌卷烟的烟支（含量）中所有WHO优先级有害金属的含量[132]。除了汞直接进行分析外，微波消解用于样品制备。该分析仪器包括用于汞的直接燃烧分析仪和用于除砷和硒以外的所有其他元素的电感耦合等离子体质谱仪，这些元素分别通过不同的电感耦合等离子体质谱方法进行分析。该方法的准确性由标准烟草参比材料确定，结果在目标范围内，并且WHO优先级有害金属的结果（铅除外）比调味烟草制品标准范围的下限略低（4%）。烟草中世界卫生组织优先级有害物质的检出限：砷为0.082 μg/g（扇形磁场质谱）、0.25 μg/g（四极杆质谱），镉为0.23 μg/g，铅为0.16 μg/g，汞为0.00063 μg/g。

其他成分（氰化氢，一氧化氮）

通过在经过氢氧化钠处理的剑桥滤上收集烟气样品，对主流烟气中的氰化氢进行定量，并通过带脉冲安培检测器的离子色谱法进行定量[133]。与传统的连续流动分析仪相比，基于剑桥滤片的方法更便利，且有更高的准确性和更宽的线性范围。该方法的性能是通过将氰化氢加标到滤片上来评估的，该滤片包含由混合型和烤烟制成的卷烟粒相物。方法的平均回收率为97%，日内和日间相对标准偏差均小于6%。对于25 μL定量环，氰化氢的检出限为3 μg/L。Mahernia等[134]使用极谱法测量了在伊斯兰共和国购买的50支大雪茄和卷烟中氰化氢的浓度。用泵从每种烟草制品中吸出主流烟气，并使其通过装有氢氧化钠（0.1 mol/L）的玻璃管。用已知量的氰化物溶液并使用标准添加方法确定氰化物的浓度。烟草制品中的氰化氢含量为每支烟17.6~1550 μg。没有提供氰化氢的检出限。未发现用于分析氮氧化合物的最新方法。

8.5 品牌间有害成分变化的最新动态

表8.4总结了标准吸烟机吸烟方法、卷烟和小雪茄的HCI抽吸方法以及水烟的Beirut抽吸方法所产生的主流烟气中有害成分含量的变化。表8.5总结了卷烟、小雪茄、水烟和木炭中有害物质的变化。为了确定第10、25、75和90百分位数，需要对越来越流行的新型烟草制品（与卷烟相比）中的WHO优先级有害物质的释放进行更多研究，以便进行全球比较，为世界卫生组织的优先级有害成分设定限量迈出第一步。

表8.4 世界卫生组织优先级有害物质的变化，以可燃型卷烟制品主流烟气中每支或每袋（水烟）的质量表示

有害物质	卷烟	小雪茄	水烟
醛类			
乙醛（μg）	1198~1947[118]（美国）	NR	120~2520[135]
丙酮（μg）	385~724[118]（美国）	NR	20.2~118[135]
丙烯醛（μg）	107~169[118]（美国）	105~185[136]（美国）	10.1~892[135]
巴豆醛（μg）	25~72[118]（美国）	NR	NR
甲醛（μg）	55~108[118]（美国）	NR	36~630[135]
丙醛（μg）	116~232[118]（美国）	NR	5.71~403[135]
芳香胺			
1-氨基萘（ng）	4.31~33.37[121]（中国）	NR	6.20[135]
2-氨基萘（ng）	0.89~4.60[121]（中国）	NR	2.84[135]
3-氨基联苯（ng）	1.46~4.68[121]（中国）	NR	<3.30[137]（德国）
4-氨基联苯（ng）	0.38~5.95[121]（中国）	NR	
烃类			
苯	58.7~128.5[125]（美国）	NR	271[135]
1,3-丁二烯	67.8~118.3[125]（美国）	NR	存在a[138]
异戊二烯	NR	NR	4.00[135]
甲苯	81~178[125]（美国）	NR	9.92[135]
多环芳烃			
苯并[a]芘（ng）	NR	17~32[139]（美国）	LOD~307[135]
芘b（ng）	NR	NR	30~12 950[135]
烟草特有亚硝胺			
NAB（ng）	NR	NR	8.45[135]
NAT（ng）	NR	NR	103[135]
NNK（ng）	NR	438~995[139]（美国）	LOD~46.4[135]
NNN（ng）	NR	434~1550[139]（美国）	34.3[135]
生物碱			
烟碱，mg	NR	1.85~6.15[140]（美国）	>0.01~9.29[135]
酚类			
儿茶酚（μg）	49.6~118[127]（加拿大）	NR	166~316[135]
间甲酚（μg）	1.93~6.92[127]（加拿大）	NR	NR
对甲酚（μg）	5.28~17.7[127]（加拿大）	NR	NR
间+对甲酚（μg）	7.24~24.6[127]（加拿大）	NR	2.37~4.66[135]
邻甲酚（μg）	2.21~7.93[127]（加拿大）	NR	2.93~4.41[135]
苯酚（μg）	8.34~32.7[127]（加拿大）	NR	3.21~58.0[135]
对苯二酚（μg）	60.1~140[127]（加拿大）	NR	21.7~110.7[135]
间苯二酚（μg）	1.25~2.46[127]（加拿大）	NR	1.69~1.87[135]
其他有机化合物			
丙酮	NR	NR	20.2~118[135]
丙烯腈（μg）	19.7~37.7[127]（美国）	NR	存在a[138]
吡啶	NR	NR	4.76[135]

续表

有害物质	卷烟	小雪茄	水烟
喹啉	NR	NR	BLQ
金属和类金属			
砷（ng）	NR	NR	165 [135]
镉（ng）	NR	NR	<100 [141]（英国）
铬 c（ng）	0.60~1.03 [142]（美国）	NR	250~1340 [135]
铅（ng）	NR	NR	200~6870 [135]
汞（ng）	NR	5.2~9.6 [131]（美国）	<100 [141]（英国）
其他优先有害物质			
一氧化碳，mg	NR	NR	5.7~367 [135]
氰化氢	NR	NR	NR
一氧化氮，mg	140.9~266.8 [143]（俄罗斯联邦）	NR	0.325~0.440 [135]

BLQ：低于定量限；LOD：低于检测限；NAB：N'-亚硝基假木贼碱；NAT：N'-亚硝基新烟碱；NNK：4-(甲基亚硝基氨基)-1-(3-吡啶基)-1-丁酮；NNN：N'-亚硝基降烟碱；NR：未报告；UK：英国

a. 加拿大卫生部深度抽吸方案：55 mL抽吸容量，30秒/次，2 s持续抽吸时间，通风孔100%阻塞

b. 存在，根据尿液生物标志物推断出在主流烟气中存在

c. 当前不是TobReg优先级有害物质

表8.5 燃烧型烟草制品的烟丝或木炭（水烟）中WHO优先级有害物质含量的变化，表示为每克烟草中的化学物质质量（g）

有害成分	卷烟	小雪茄	水烟	木炭
烟碱（mg）	16.2~26.3 [144]（美国） 10.5~17.8 [145]（巴基斯坦）	10.3~19.1 [145]（美国）	0.48~2.28 [146]（美国） 1.8~41.3 [147]（约旦）	NA
苯并[a]芘（ng）	NR	NR	NR	1~26 [148,149]（美国黎巴嫩）
芘（ng）	NR	NR	NR	6~170 [148,149]（美国黎巴嫩）
砷（μg）	0.22~0.36 [132]（美国）	NA	0.062（0.023）b [150]（埃及）	0.018（0.013）[150]（埃及）
镉（μg）	1.0~1.7 [132]（美国）	NA	0.34（0.007）[150]（埃及）	0.005（0.017）b [150]（埃及）
铬 a（μg）	1.3~3.1 [132]（美国）	NA	0.15~0.37 [151]（美国，中东）	0.161~8.32 [152]（全世界）
铅（μg）	0.60~1.16 [132]（美国）	NR	0.15（0.008）b [150]（埃及）	0.97（0.01）b [150]（埃及）
汞（μg）	0.013~0.020 [132]（美国）	0.017~0.024 [132]（美国）	NR	NR

NR：没有报道；NA：不适用

a. 目前不是TobReg优先级有害物质

b. 平均值（相对标准偏差）

8.6　清单上有害物质的未来重新评估标准

选择优先级有害物质的标准和TobReg的总体建议指出了四个领域，这些领域应定期重新评估优先级清单。

具体成分和释放物的毒理学概况：确定了有害物质的优先级清单，以帮助WHO FCTC缔约方和WHO成员国满足WHO FCTC第9条和第10条的要求。监测优先级清单被视为迈向监管燃烧型烟草制品的成分和释放物的第一步。有关选定释放物有害性的最新科学知识对于确保优先级清单的持续相关性和为将来的监管措施至关重要。

验证方法：必须根据WHO TobLabNet SOP测量优先级有害物质的含量，以确保结果的准确、可重现，并作为监管措施的基础。TobLabNet的标准化方法的开发以功能强大、特征明确、已发布的方法为指导，并由TobLabNet对其进行审查、测试和进一步验证。TobLabNet会定期审查分析化学和毒理学数据的进展，以确定测试方法的优先级。

品牌差异：根据世界卫生组织《烟草控制框架公约》第9条和第10条的部分指南，烟草行业为每个品牌生成和报告有关优先级成分和释放物的数据负责。例如2018年10月世界卫生组织《烟草控制框架公约》第八次缔约方会议的决定（FCTC/COP8[21]）所述，此类数据可根据WHO TobLabNet SOP生成。许多优先级成分的释放量尚未在所有市售产品中系统地测量。因此，定期审查有关燃烧型烟草制品化学成分的新报告，为理解产品类型之间的释放差异和可实现的优先级有害物质最低水平建立了参考文献框架。

品牌中各成分之间的相关性：几种有害成分之间的正相关关系表明监管策略中一种有害成分的水平可以替代其他几种有害物质。负相关性表明强制降低一种有害成分可能导致负相关有害成分水平的增加。应根据这些假设谨慎行事，但密切监控各成分之间的关系可以为将来选择强制性减少有害成分提供依据。

8.7　新型有害物质的选择标准

将来应考虑将新有害物质添加到优先级列表中。下面列出了几种选择新有害物质的标准。

对人体健康构成危险的确实证据：例如Talhout等[153]根据吸入风险列出了98种有害烟气成分。此外，FDA确定了93种有害或者潜在有害成分清单（HPHC）[154]。HPHC清单包括烟草或烟气中摄入（吸入、摄入或吸收）并会对使用者和非使用者造成直接或间接伤害的化学物质和化合物。FDA选择了由IARC、美国环境保护局或美国健康和人类服务部国家毒理学计划,以及美国国家职业安全与健康研究所确定的已知或可能的人体致癌物。美国环境保护局或有害物质和疾病登记局认定为具有不良呼吸或心脏影响的成分,加利福尼亚环境保护局认定为生殖或发育有害物质的成分,以及文献中报告为造成滥用的成分也包括在内。40种有毒和致癌成分在FDA HPHC清单和由Talhout等均列出[153],但当前不在TobReg清单中(表8.6)。TobReg在将来可以优先考虑这些成分,例如,通过应用危害指数和生成品牌间含量水平变化的数据。

表8.6　将来要考虑列入TobReg优先级有害物质清单中的成分

组分	健康影响[154]	吸入相关风险（mg/m³）[153]
乙酰胺	CA	5.0×10^{-4}（癌症）
丙烯酰胺	CA	8.0×10^{-3}（癌症）
3-氨基-1,4-二甲基-5H-吡啶基[4,3-b]吲哚	CA	1.4×10^{-6}（癌症）
2-氨基-3-甲基-9H-吡啶并[2,3-b]吲哚	CA	2.9×10^{-5}（癌症）
2-氨基-6-甲基二吡啶并[1,2-a : 3′,2-d]咪唑	CA	7.1×10^{-6}（癌症）
2-氨基-3-甲基咪唑并[4,5-f]喹啉	CA	2.5×10^{-5}（癌症）
2-氨基-9H-吡啶并[2,3-b]吲哚	CA	8.8×10^{-5}（癌症）
2-氨基二吡啶并[1,2-a:3′,2′-d]咪唑	CA	2.5×10^{-5}（癌症）
苯并[a]蒽	CA, CT	9.1×10^{-5}（癌症）
铍	CA	4.2×10^{-6}（癌症）
铬	CA, RT, RTD	8.3×10^{-7}（癌症）
䓛	CA, CT	9.1×10^{-4}（癌症）
钴	CA, CT	5.0×10^{-4}（呼吸）
二苯并[a,h]蒽	CA	8.3×10^{-6}（癌症）
二苯并[a,e]芘	CA	9.1×10^{-6}（癌症）
二苯并[a,h]芘	CA	9.1×10^{-7}（癌症）
二苯并[a,i]芘	CA	9.1×10^{-7}（癌症）
二苯并[a,l]芘	CA	9.1×10^{-7}（癌症）
乙苯	CA	7.7×10^{-1}（肝和肾）
氨基甲酸乙酯	CA, RDT	3.5×10^{-5}（癌症）
环氧乙烷	CA, RT, RDT	1.1×10^{-4}（癌症）
肼	CA, RT	2.0×10^{-6}（癌症）
茚并[1,2,3-cd]芘	CA	9.1×10^{-5}（癌症）
2-丁酮	RT	5.0（发育）
1-甲基-3-氨基-5H-吡啶并[4,3-b]吲哚	CA	1.1×10^{-5}（癌症）
5-甲基䓛	CA	9.1×10^{-6}（癌症）
萘	CA, RT	3.0×10^{-3}（鼻）
镍	CA, RT	9.0×10^{-5}（肺纤维化）
N-亚硝基二乙醇胺	CA	1.3×10^{-5}（癌症）

续表

组分	健康影响[154]	吸入相关风险（mg/m³）[153]
N-亚硝基二乙胺	CA	2.3×10^{-7}（癌症）
N-亚硝基二甲胺	CA	7.1×10^{-7}（癌症）
N-亚硝基甲基乙胺	CA	1.6×10^{-6}（癌症）
N-亚硝基哌啶	CA	3.7×10^{-6}（癌症）
N-亚硝基吡咯烷	CA	1.6×10^{-5}（癌症）
钋-210	CA	925.9（癌症）
环氧丙烷	CA，RT	2.7×10^{-3}（癌症）
硒	RT	8.0×10^{-4}（呼吸）
苯乙烯	CA	9.2×10^{-2}（神经毒性）
醋酸乙烯酯	CA，RT	2.0×10^{-1}（鼻）
氯乙烯	CA	1.1×10^{-3}（癌症）

CA：致癌物；CT：心血管有毒物质；RDT：生殖或发育有毒物质；RT：呼吸道有毒物质

与当前优先级有害物质属于同一化学类别的成分：包含此类成分的理由是，可以使用相同的分析技术同时分析几种化合物。例如金属和其他痕量元素以及PAH。表8.6中列出的铍、铬、钴、镍、钋-210和硒元素可以通过多组分方法与TobReg优先级列表中的镉和其他金属进行分析。可以使用相同的分析技术分析至少23种不同的PAH[155]。表8.4中列出了8种PAH：苯并[a]蒽、二苯并[a,h]蒽、二苯并[a,e]芘、二苯并[a,h]芘、二苯并[a,i]芘、二苯并[a,l]芘、茚并[1,2,3-cd]芘和萘。已经证明，这些PAH的水平与硝酸盐和TSNA的含量负相关。但是，这种关系取决于烟草类型[156]。此外，在许多人体暴露实验中，分析了分别作为非致癌PAH芘和菲暴露生物标志物的1-羟基芘或菲四醇在尿液中的含量，以评估对PAH的暴露[157,158]。由于这两种化合物在烟气中的含量远高于苯并[a]芘和其他致癌性多环芳烃，同时分析芘、菲与苯并[a]芘可为今后监测烟气中这些成分减少而引起的人体暴露量的变化提供信息。

卷烟烟气中有害物质的前体物：例如，烟叶中的亚硝酸盐和硝酸盐是烟草和烟气中致癌的亚硝胺NNK和NNN以及烟气中氮氧化合物的前体。来自学术界和烟草行业的数据表明，烟草中的硝酸盐和亚硝酸盐会显著影响烟气中的成分[159,160]。此外，烟草中的硝酸盐含量会影响卷烟烟气中的氨含量，进而影响烟气的pH值和烟碱的生物利用度。因此，卷烟烟丝中亚硝酸盐和硝酸盐的含量是预测卷烟烟气毒理学性质的重要指标。当前的TobReg优先级列表包括卷烟烟气中的一氧化氮。但是，对烟气中此类成分的分析可能很复杂，实验室之间的结果可能不一致。因此，分析烟丝中的亚硝酸盐和硝酸盐将是一种可靠的、有用的替代方法。

有助于提高烟草制品适口性和/或致瘾性的成分：添加剂对烟草制品的吸引力和滥用倾向的影响已得到审查[161]。例如，燃烧型烟草制品中的糖是一种添加剂,可以考虑用于将来的监控和监管。烟草中的糖分导致烟气中乙醛的形成[162,163],可直接或间接地通过形成哈尔满影响烟气的成瘾能力[32,34]。

8.8 研究需求和监管建议

8.8.1 研究需求、数据差距和未来工作

对有害性、分析方法和可燃烟草制品中优先级有害物质含量的报告进行的审查表明,以下领域需要进一步的研究。

尽管有毒理学证据表明39种优先级有害成分的重要性,并且有可用的分析方法,但关于各种品牌和产品中有害成分释放量的信息仍缺乏。有关不同品牌和类型的燃烧型烟草制品(例如卷烟、细支雪茄、水烟)中有害成分含量变化的信息可以证明对特定释放物进行监管。鉴于全球广泛使用本地制造的可燃型烟草制品(例如bidis)或消费者自制的可燃型卷烟产品(例如自卷烟),因此必须更好地表征这些产品的特性。

在某些情况下,对用于分析烟丝(成分)中某些成分的标准化分析方法进行评估是比测量卷烟烟气更可靠的一种替代方法。例如,烟草中的硝酸盐和亚硝酸盐含量强烈影响烟气中的一氧化氮含量。分析烟丝中的这些成分将需要较少的时间和资源,可以提供来自实验室的更一致的数据。此外,已验证的方法可直接用于无烟烟草制品的分析。在烟草加工过程中形成了致癌性的TSNA,并且热解对其烟气水平的贡献很小。因此,烟草中的含量是烟气中含量的有力预测指标[68,86,164,165]。

尽管TobReg以前建议应根据既定的毒理学原理确定烟草制品中有害物质的释放上限[166],为了最大限度地减少烟草制品造成的危害,必须在多大程度上降低复杂混合物或单独减低某一种有害物质目前尚不清楚。需要进行体外和体内研究,以更好地了解烟草制品的复杂化学成分对致癌和毒性的影响,以及单个成分的毒性阈值。人体暴露和分子流行病学研究,包括前瞻性队列研究,将表明实现最大公共卫生利益所需的烟草有害物质水平最适合的降低。

需要进行研究以更好地了解通过增加烟碱的生物利用度,哪些成分对烟草制品的致瘾性和适口性有贡献。这些信息将对未来的监管方法产生影响。例如,减少烟草制品的吸引力和适口性的措施可能比减少有害物质释放对公众健康的影响更大[167]。

当前关于确定有害物质释放优先次序的工作和产品测试方法的开发应扩展到人体暴露和健康后果。吸烟者与卷烟的相互作用比任何一种基于吸烟机的方案都要复杂[1],尚不清楚减少每毫克烟碱释放是否会相应减少人体暴露。应建立实验室分析能力来测试人类生物样品中优先有害物质的生物标志物。对用过的卷烟过

滤嘴进行分析可能被认为是一种较便宜的人类有害物质摄入量测量方法[69, 168]。此外，应进行研究以识别潜在危害的合适生物标志物，以用于评估未来产品标准对健康的长期影响。

8.8.2 监管建议和对缔约方的支持

TobLabNet已完成对卷烟烟丝和主流烟气中选定释放物测量方法的验证，有关这些方法的SOP可在WHO烟草制品法规网站上获得。建议缔约方考虑要求卷烟制造商根据TobLabNet SOP进行释放物含量测试，并将结果报告给国家主管部门。该建议是第八次缔约方会议（FCTC/COP8[21]）决议的一部分，该决议鼓励缔约方酌情承认和实施WHO TobLabNet方法，并强调需要WHO进一步支持TobLabNet的能力建设行动。

对优先级有害物质扩展清单的毒性研究评估表明，这些有害物质在可燃型烟草制品的毒性和致瘾性潜力中持续发挥作用。如本报告所述，经过TobLabNet验证的方法可用于扩展优先级清单中的其他有害物质。因此，建议各缔约方考虑酌情要求制造商采用基于TobLabNet SOP的方法报告其他优先级有害物质的释放量。

根据TobReg[166]的建议，下一步应制定TobLabNet验证方法尚不可用的优先级成分和释放物分析方法SOP：烟草中的镉和铅；水烟烟气中的烟碱；和无烟烟草制品中的烟碱、TSNA和苯并[a]芘。

对当前不在优先级清单中的成分，建议缔约方考虑要求制造商报告烟丝中硝酸盐和亚硝酸盐的含量，以作为一氧化氮释放量的潜在替代指标。目前尚没有一种简便、经济高效的方法来分析烟气中的一氧化氮。可以通过采用TobLabNet方法对烟丝中的氨进行分析来检测水烟提取物中的硝酸盐和亚硝酸盐（SOP 07）。此外，鉴于它们（以及可能在表8.6中列出的其他致癌多环芳烃）对人类生物监测研究的重要性，可以通过TobLabNet方法同时监测苯并[a]芘、芘和菲。

建议向参与WHO方法验证的TobLabNet实验室提供参比烟草制品材料，以帮助确定每种方法的准确性、可重复性，及验证方法转移的成功与否。可以与测试材料同时分析合适的样品基质匹配的参比材料，作为质量控制的一种形式，用于评估实验室间测试的数据。表8.7列出了市售的认证标准参比材料。

表8.7　适用于样品质量控制的认证标准参比材料

名称	描述	发行者	认证值
SRM 3222	卷烟烟丝	NIST	烟碱，NNN，NNK，VOC
1R6F	参比卷烟	KTRDC	主流烟气释放物，ISO和HCI抽吸条件下的抽吸口数，烟丝成分，物理性质
RT5	高TSNA含量烟粉	KTRDC	烟碱，降烟碱，新烟草碱，假木贼碱，NNN，NAT，NAB，NNK，水分

续表

名称	描述	发行者	认证值
RT4	白肋烟		
KTRDC	深色晾晒烟粉		
RT3	土耳其香料烟粉		
RT2	烤烟烟粉		
RTDFC	深色烤烟粉		
1R6F（RT1）	烟丝粉末		
1R5F（RT7）	烟丝粉末		
STRP 1S1	嚼烟散叶	NCSU	含量值未经认证，但在科学文献中报告
STRP 2S1	嚼烟散叶		
STRP 1S2	干鼻烟		
STRP 1S3	湿鼻烟		
STRP 2S3	湿鼻烟		
CRP1	口含烟		
CRP2	湿鼻烟		
CRP3	干鼻烟		
CRP4	嚼烟散叶		

HCI：加拿大深度抽吸模式；ISO：国际标准化组织；KTRDC：肯塔基大学烟草参比产品中心；NAB：N'-亚硝基假木贼碱；NAT：N'-亚硝基新烟碱；NCSU：北卡罗来纳州立大学无烟气烟草参比产品；NIST：国家标准技术研究所；NNK：4-(甲基亚硝基氨基)-1-(3-吡啶基)-1-丁酮；NNN：N'-亚硝基降烟碱；VOC：挥发性有机化合物

建议制定一项研究议程，在此议程中明确哪种成分对烟草制品的致瘾性和适口性做出贡献，并确定或开发出其在烟草和烟气中的定量分析方法，并确定各种卷烟品牌中的释放范围。

8.9 参考文献

[1] Djordjevic MV, Stellman SD, Zang E. Doses of nicotine and lung carcinogens delivered to cigarette smokers. J Natl Cancer Inst. 2000;92:106–11.

[2] Jarvis MJ, Boreham R, Primatesta P, Feyerabend C, Bryant A. Nicotine yield from machinesmoked cigarettes and nicotine intakes in smokers: evidence from a representative population survey. J Natl Cancer Inst. 2001;93:134–8.

[3] Health Canada data for 2004 using the intense smoking protocol. Ottawa: Health Canada Tobacco – Reports and Publications; 2004.

[4] Counts ME, Morton MJ, Laffoon SW, Cox RH, Lipowicz PJ. Smoke composition and predicting relationships for international commercial cigarettes smoked with three machine-smoking conditions. Regul Toxicol Pharmacol. 2005;41:185–227.

[5] Fowles J, Dybing E. Application of toxicological risk assessment principles to the chemical constituents of cigarette smoke. Tob Control. 2003;12:424–30.

[6] Dybing E, Sanner T, Roelfzema H, Kroese D, Tennant RW. T25: a simplified carcinogenic potency index. Description of the system and study of correlations between carcinogenic potency and species/site specificity and mutagenicity. Pharmacol Toxicol. 1997;80:272–9.

[7] Cigarette emissions data, 2001. Canberra: Australian Department of Health and Aging; 2001 (http://www.health.gov.au/internet/main/publishing.nsf/content/tobacco-emis, accessed 15 May 2019).

[8] Stepanov I, Hatsukami D. Call to establish constituent standards for smokeless tobacco products. Tob Regul Sci. 2016;2:9–30.

[9] Talhout R, Richter PA, Stepanov I, Watson CV, Watson CH. Cigarette design features: effects on emission levels, user perception, and behavior. Tob Regul Sci. 2018;4:592–604.

[10] Rodgman A. Studies of polycyclic aromatic hydrocarbons in cigarette mainstream smoke: identification, tobacco precursors, control of levels: a review. Beitr Tabakforsch Int. 2001;19:361–79.

[11] WHO Study Group on Tobacco Product Regulation. The scientific basis of tobacco product regulation. Second report of a WHO study group (WHO Technical Report Series, No. 951). Geneva: World Health Organization; 2008:45–125.

[12] WHO Study Group on Tobacco Product Regulation. Setting maximal limits for toxic constituents in cigarette smoke. Geneva: World Health Organization; 2001 (http://www.who.int/tobacco/global_interaction/tobreg/tsr/en/index.html, accessed 15 May 2019).

[13] O'Brien PJ, Siraki AG, Shangari N. Aldehyde sources, metabolism, molecular toxicity mechanisms, and possible effects on human health. Crit Rev Toxicol. 2005;35:609–62.

[14] Re-evaluation of some organic chemicals, hydrazine and hydrogen peroxide (IARC Monographs on the Evaluation of the Carcinogenic Risk of Chemicals to Humans, Vol. 71). Lyon: International Agency for Research on Cancer; 1999.

[15] A review of human carcinogens: chemical agents and related occupations (IARC Monographs on the Evaluation of Carcinogenic Risks to Humans, Vol. 100F), Lyon: International Agency for Research on Cancer; 2012.

[16] Dry cleaning, some chlorinated solvents and other industrial chemicals (IARC Monographs on the Evaluation of Carcinogenic Risks to Humans, Vol. 63). Lyon: International Agency for Research on Cancer; 1995:337–72.

[17] Feng Z, Hu W, Hu Y, Tang MS. Acrolein is a major cigarette-related lung cancer agent. Preferential binding at p53 mutational hotspots and inhibition of DNA repair. Proc Natl Acad Sci USA. 2006;103:15404–9.

[18] Zhang S, Villalta PW, Wang M, Hecht SS. Analysis of crotonaldehyde- and acetaldehyde-derived 1,N2-propanodeoxyguanosine adducts in DNA from human tissues using liquid chromatography-electrospray ionization–tandem mass spectrometry. Chem Res Toxicol. 2006;19:1386–92.

[19] Zhang S, Chen H, Wang A, Liu Y, Hou H, Hu Q. Assessment of genotoxicity of four volatile pollutants from cigarette smoke based on the in vitro gH2AX assay using high content screening. Environ Toxicol Pharmacol. 2017;55:30–6.

[20] Corley RA, Kabilan S, Kuprat AP, Carson JP, Jacob RE, Minard KR et al. Comparative risks of aldehyde constituents in cigarette smoke using transient computational fluid dynamics/

[21] Yeager RP, Kushman M, Chemerynski S, Weil R, Fu X, White M et al. Proposed mode of action for acrolein respiratory toxicity associated with inhaled tobacco smoke. Toxicol Sci. 2016;151:347–64.

[22] Takamiya R, Uchida K, Shibata T, Maeno T, Kato M, Yamaguchi Y et al. Disruption of the structural and functional features of surfactant protein A by acrolein in cigarette smoke. Sci Rep. 2017;7:8304.

[23] Lee HW, Wang HT, Weng MW, Hu Y, Chen WS, Chou D et al. Acrolein- and 4-aminobiphenyl-DNA adducts in human bladder mucosa and tumor tissue and their mutagenicity in human urothelial cells. Oncotarget. 2014;5:3526–40.

[24] Cheah NP, Pennings JL, Vermeulen JP, van Schooten FJ, Opperhuizen A. In vitro effects of aldehydes present in tobacco smoke on gene expression in human lung alveolar epithelial cells. Toxicol In Vitro. 2013;27:1072–81.

[25] Sapkota M, Wyatt TA. Alcohol, aldehydes, adducts and airways. Biomolecules. 2015;5:2987–3008.

[26] Antoniak DT, Duryee MJ, Mikuls TR, Thiele GM, Anderson DR. Aldehyde-modified proteins as mediators of early inflammation in atherosclerotic disease. Free Radical Biol Med. 2015;89:409–18.

[27] Lachenmeier DW, Salaspuro M. ALDH2-deficiency as genetic epidemiologic and biochemical model for the carcinogenicity of acetaldehyde. Regul Toxicol Pharmacol. 2017;86:128–36.

[28] Morita K, Miyazaki H, Saruwatari J, Oniki K, Kumagae N, Tanaka T et al. Combined effects of current-smoking and the aldehyde dehydrogenase 2*2 allele on the risk of myocardial infarction in Japanese patients. Toxicol Lett. 2015;232:221–5.

[29] Bui LC, Manaa A, Xu X, Duval R, Busi F, Dupret JM et al. Acrolein, an a,ß-unsaturated aldehyde, irreversibly inhibits the acetylation of aromatic amine xenobiotics by human arylamine Nacetyltransferase 1. Drug Metab Dispos. 2013;41:1300–5.

[30] Chare SS, Donde H, Chen WY, Barker DF, Gobejishvilli L, McClain CJ et al. Acrolein enhances epigenetic modifications, FasL expression and hepatocyte toxicity induced by anti-HIV drug zidovudine. Toxicol In Vitro. 2016;35:76.

[31] Tan SLW, Chadha S, Liu Y, Gabasova E, Perera D, Ahmed K et al. A class of environmental and endogenous toxins induces BRCA2 haploinsufficiency and genome instability. Cell. 2017;169:1105–18. Updated priority list of toxicants in combusted tobacco products

[32] Hoffman AC, Evans SE. Abuse potential of non-nicotine tobacco smoke components: acetaldehyde, nornicotine, cotinine, and anabasine. Nicotine Tob Res. 2013;15:622–32.

[33] Brancato A, Lavanco G, Cavallaro A, Plescia F, Cannizzaro C. Acetaldehyde, motivation and stress: behavioral evidence of an addictive ménage à trois. Front Behav Neurosci. 2017;11:23.

[34] Syrjänen K, Salminen J, Aresvuo U, Hendolin P, Paloheimo L, Eklund C et al. Elimination of cigarette smoke-derived acetaldehyde in saliva by slow-release L-cysteine lozenge is a potential new method to assist smoking cessation. A randomised, double-blind, placebo-controlled intervention. Anticancer Res. 2016;36:2297–306.

[35] Henning RJ, Johnson GT, Coyle JP, Harbison RD. Acrolein can cause cardiovascular disease: a

[36] Conklin DJ, Malovichko MV, Zeller I, Das TP, Krivokhizhina TV, Lynch BH et al. Biomarkers of chronic acrolein inhalation exposure in mice: implications for tobacco product-induced toxicity. Toxicol Sci. 2017;158:263–74.

[37] DeJarnett N, Conklin DJ, Riggs DW, Myers JA, O'Toole TE, Hamzeh I et al. Acrolein exposure is associated with increased cardiovascular disease risk. J Am Heart Assoc. 2014;3:e000934.

[38] Snow SJ, McGee MA, Henriquez A, Richards JE, Schladweiler MC, Ledbetter AD et al. Respiratory effects and systemic stress response following acute acrolein inhalation in rats. Toxicol Sci. 2017;158:454–64.

[39] Feroe AG, Attanasio R, Scinicariello F. Acrolein metabolites, diabetes and insulin resistance. Environ Res. 2016;148:1–6.

[40] Chen WY, Wang M, Zhang J, Barve SS, McClain CJ, Joshi-Barve S. Acrolein disrupts tight junction proteins and causes endoplasmic reticulum stress-mediated epithelial cell death leading to intestinal barrier dysfunction and permeability. Am J Pathol. 2017;187(12):2686–97.

[41] Song JJ, Lee JD, Lee BD, Chae SW, Park MK. Effect of acrolein, a hazardous air pollutant in smoke, on human middle ear epithelial cells. Int J Pediatr Otorhinolaryngol. 2013;77:1659–64.

[42] Amiri A, Turner-Henson A. The roles of formaldehyde exposure and oxidative stress in fetal growth in the second trimester. J Obstet Gynecol Neonatal Nurs. 2017;46:51–62.

[43] Yang Y, Zhang Z, Zhang H, Hong K, Tang W, Zhao L et al. Effects of maternal acrolein exposure during pregnancy on testicular testosterone production in fetal rats. Mol Med Rep. 2017;16:491–8.

[44] Park SL, Carmella SG, Chen M, Patel Y, Stram DO, Haiman CA et al. Mercapturic acids derived from the toxicants acrolein and crotonaldehyde in the urine of cigarette smokers from five ethnic groups with differing risks for lung cancer. PLoS One. 2015;10:e0124841.

[45] Haiman CA, Stram DO, Wilkens LR, Pike MC, Kolonel LN, Henderson BE et al. Ethnic and racial differences in the smoking-related risk of lung cancer. N Engl J Med. 2006;354:333–42.

[46] Some aromatic amines, organic dyes, and related exposures (IARC Monographs on the Evaluation of Carcinogenic Risks to Humans, Vol. 99). Lyon: International Agency for Research on Cancer; 2010.

[47] Cai T, Bellamri M, Ming X, Koh WP, Yu MC, Turesky RJ. Quantification of hemoglobin and white blood cell DNA adducts of the tobacco carcinogens 2-amino-9H-pyrido[2,3-b]indole and 4-aminobiphenyl formed in humans by nanoflow liquid chromatography/ion trap multistage mass spectrometry. Chem Res Toxicol. 2017;30:1333–43.

[48] Tomioka K, Obayashi K, Saeki K, Okamoto N, Kurumatani N. Increased risk of lung cancer associated with occupational exposure to benzidine and/or beta-naphthylamine. Int Arch Occup Environ Health. 2015;88:455–65.

[49] Tomioka K, Saeki K, Obayashi K, Kurumatani N. Risk of lung cancer in workers exposed to benzidine and/or beta-naphthylamine: a systematic review and meta-analysis. J Epidemiol. 2016;26:447–58.

[50] Some industrial chemicals (IARC Monographs on the Evaluation of the Carcinogenic Risk of Chemicals to Humans, Vol. 60). Lyon: International Agency for Research on Cancer; 1994:215–32.

[51] Hayes RB, Yin SN, Dosemeci M, Li GL, Wacholder S, Chow WH et al. Mortality among benzeneexposed workers in China. Environ Health Perspect. 1996;104(Suppl 6):1349–52.

[52] Soeteman-Hernandez LG, Bos PM, Talhout R. Tobacco smoke-related health effects induced by 1,3-butadiene and strategies for risk reduction. Toxicol Sci. 2013;136:566–80.

[53] Bahadar H, Mostafalou S, Abdollahi M. Current understandings and perspectives on noncancer health effects of benzene: a global concern. Toxicol Appl Pharmacol. 2014;276:83–94.

[54] Abplanalp W, DeJarnett N, Riggs DW, Conklin DJ, McCracken JP, Srivastava S et al. Benzene exposure is associated with cardiovascular disease risk. PLoS One. 2017;12:e0183602.

[55] Doherty BT, Kwok RK, Curry MD, Ekenga C, Chambers D, Sandler DP et al. Associations between blood BTEXS concentrations and hematologic parameters among adult residents of the US Gulf States. Environ Res. 2017;156:579–87.

[56] Haiman CA, Patel YM, Stram DO, Carmella SG, Chen M, Wilkens LR et al. Benzene uptake and glutathione S-transferase T1 status as determinants of S-phenylmercapturic acid in cigarette smokers in the Multiethnic Cohort. PLoS One. 2016;11(3):0150641.

[57] Some non-heterocyclic polycyclic aromatic hydrocarbons and some related exposures (IARC Monographs on the Evaluation of Carcinogenic Risks to Humans, Vol. 92). Lyon: International Agency for Research on Cancer; 2010.

[58] Ding YS, Ashley DL, Watson CH. Determination of 10 carcinogenic polycyclic aromatic hydrocarbons in mainstream cigarette smoke. J Agric Food Chem. 2007;55:5966–73.

[59] Hecht SS. Tobacco carcinogens, their biomarkers, and tobacco-induced cancer. Nature Rev Cancer. 2003;3:733–44.

[60] Pfeifer GP, Denissenko MF, Olivier M, Tretyakova N, Hecht SS, Hainaut P. Tobacco smoke carcinogens, DNA damage and p53 mutations in smoking-associated cancers. Oncogene. 2002;21:7435–51.

[61] Hecht SS. Tobacco smoke carcinogens and lung cancer. J Natl Cancer Inst. 1999;91:1194–210.

[62] Onami S, Okubo C, Iwanaga A, Suzuki H, Iida H, Motohashi Y et al. Dosimetry for lung tumorigenesis induced by urethane, 4-(N-methyl-N-nitrosamino)-1-(3-pyridyl)-1-butanone (NNK), and benzo[a]pyrene (B[a]P) in A/JJmsSlc mice. J Toxicol Pathol. 2017;30:209–16.

[63] Zaccaria KJ, McClure PR. Using immunotoxicity information to improve cancer risk assessment for polycyclic aromatic hydrocarbon mixtures. Int J Toxicol. 2013;32:236–50.

[64] Chepelev NL, Moffat ID, Bowers WJ, Yauk CI. Neurotoxicity may be an overlooked consequence of benzo[a]pyrene exposure that is relevant to human health risk assessment. Mutat Res Rev Mutat Res. 2015;764:64–89.

[65] Hecht SS. Biochemistry, biology, and carcinogenicity of tobacco-specific N-nitrosamines. Chem Res Toxicol. 1998;11:559–603.

[66] Tobacco smoke and involuntary smoking (IARC Monographs on the Evaluation of Carcinogenic Risks to Humans, Vol. 83). Lyon: International Agency for Research on Cancer; 2004.

[67] Smokeless tobacco and tobacco-specific nitrosamines (IARC Monographs on the Evaluation of Carcinogenic Risks to Humans, Vol. 89). Lyon: International Agency for Research on Cancer; 2007.

[68] Wu W, Zhang L, Jain RB, Ashley DL, Watson CH. Determination of carcinogenic tobacco-specific nitrosamines in mainstream smoke from US-brand and non-US-brand cigarettes from

[69] Ashley DL, O'Connor RJ, Bernert JT, Watson CH, Polzin GM, Jain RB et al. Effect of differing levels of tobacco-specific nitrosamines in cigarette smoke on the levels of biomarkers in smokers. Cancer Epidemiol Biomarkers Prev. 2010;19:1389–98.

[70] Stepanov I, Knezevich A, Zhang L, Watson CH, Hatsukami DK, Hecht SS. Carcinogenic tobacco-specific N-nitrosamines in US cigarettes: three decades of remarkable neglect by the tobacco industry. Tob Control. 2012;21:44–8.

[71] Hoffmann D, Djordjevic MV, Hoffmann I. The changing cigarette. Prev Med. 1997;26:427–34.

[72] Burns DM, Anderson CM, Gray N. Do changes in cigarette design influence the rise in adeno201 Updated priority list of toxicants in combusted tobacco products carcinoma of the lung? Cancer Causes Control. 2011;22:13–22.

[73] Balbo S, Johnson CS, Kovi RC, James-Yi SA, O'Sullivan MG, Wang M et al. Carcinogenicity and DNA adduct formation of 4-(methylnitrosamino)-1-(3-pyridyl)-1-butanone and enantiomers of its metabolite 4-(methylnitrosamino)-1-(3-pyridyl)-1-butanol in F-344 rats. Carcinogenesis. 2014;35:2798–806.

[74] Balbo S, James-Yi S, Johnson CS, O'Sullivan MG, Stepanov I, Wang M et al. (S)-N′-Nitrosonornicotine, a constituent of smokeless tobacco, is a powerful oral cavity carcinogen in rats. Carcinogenesis. 2013;34:2178–83.

[75] Stepanov I, Sebero E, Wang R, Gao YT, Hecht SS, Yuan JM. Tobacco-specific N-nitrosamine exposures and cancer risk in the Shanghai Cohort Study: remarkable coherence with rat tumor sites. Int J Cancer. 2014;134:2278–83.

[76] Park SL, Carmella SG, Ming X, Stram DO, LeMarchand L, Hecht SS. Variation in levels of the lung carcinogen NNAL and its glucuronides in the urine of cigarette smokers from five ethnic groups with differing risks for lung cancer. Cancer Epidemiol Biomarkers. 2015;24:561–9.

[77] Khariwala SS, Ma B, Ruszczak C, Carmella SG, Lindgren B, Hatsukami DK et al. High level of tobacco carcinogen-derived DNA damage in oral cells is an independent predictor of oral/head and neck cancer risk in smokers. Cancer Prev Res. 2017;10:507–13.

[78] Hukkanen J, Jacob P III, Benowitz NL. Metabolism and disposition kinetics of nicotine. Pharmacol Rev. 2005;57:79–115.

[79] WHO Study Group on Tobacco Product Regulation. Global nicotine reduction strategy. Advisory note. Geneva: World Health Organization; 2015.

[80] Food and Drug Administration. Tobacco product standard for nicotine level of combusted cigarette. Fed Regist. 2018;83.

[81] Donny EC, Denlinger RL, Tidey JW, Koopmeiners JS, Benowitz NL, Vandrey RG et al. Randomized clinical trial of reduced-nicotine standards for cigarettes. N Engl J Med. 2015;373:1340–9.

[82] Hatsukami DK, Luo X, Dick L, Kangkum M, Allen SS, Murphy SE et al. Reduced nicotine content cigarettes and use of alternative nicotine products: exploratory trial. Addiction. 2017;112:156–67.

[83] Benowitz NL, Burbank AD. Cardiovascular toxicity of nicotine: implications for electronic cigarette use. Trends Cardiovasc Med. 2016;26:515–23.

[84] England LJ, Aagaard K, Bloch M, Conway K, Cosgrove K, Grana R et al. Developmental toxicity

of nicotine: a transdisciplinary synthesis and implications for emerging tobacco products. Neurosci Biobehav Rev. 2017;72:176–89.

[85] Van Duuren BL, Goldschmidt BM. Cocarcinogenic and tumor-promoting agents in tobacco carcinogenesis. J. Natl Cancer Inst. 1976;56:1237–42.

[86] Fennell TR, MacNeela JP, Morris RW, Watson M, Thompson CL, Bell DA. Hemoglobin adducts from acrylonitrile and ethylene oxide in cigarette smokers: effects of glutathione S-transferase T1-null and M1-null genotypes. Cancer Epidemiol Biomarkers Prev. 2000;9:705–12.

[87] A review of human carcinogens: arsenic, metals, fibres, and dusts (IARC Monographs on the Evaluation of Carcinogenic Risks to Humans, Vol. 100C). Lyon: International Agency for Research on Cancer; 2012.

[88] Inorganic and organic lead compounds (IARC Monographs on the Evaluation of Carcinogenic Risks to Humans, Vol. 87). Lyon: International Agency for Research on Cancer; 2006.

[89] Beryllium, cadmium, mercury, and exposures in the glass manufacturing industry (IARC Monographs on the Evaluation of Carcinogenic Risks to Humans, Vol. 58). Lyon: International Agency for Research on Cancer; 1993.

[90] Pappas RS, Stanfill SB, Watson CH, Ashley DL. Analysis of toxic metals in commercial moist snuff and Alaskan iqmik. J Anal Toxicol. 2008;32:281–91.

[91] Pinto E, Cruz M, Ramos P, Santos A, Almeida A. Metals transfer from tobacco to cigarette smoke: evidences in smokers' lung tissue. J Hazard Mater. 2017;5:31–5.

[92] Richter PA, Bishop EE, Wang J, Kaufmann R. Trends in tobacco smoke exposure and blood lead levels among youths and adults in the United States: the National Health and Nutrition Examination Survey, 1999–2008. Prev Chronic Dis. 2013;10:E213.

[93] Garcia-Esquinas E, Pollan M, Tellez-Plaza M, Francesconi KA, Goessler W, Guallar E et al. Cadmium exposure and cancer mortality in a prospective cohort: the Strong Heart Study. Environ Health Perspect. 2014;122:363–70.

[94] Feki-Tounsi M, Hamza-Chaffai A. Cadmium as a possible cause of bladder cancer: a review of accumulated evidence. Environ Sci Pollut Res Int. 2014;21:10561–73.

[95] Buser MC, Scinicariello F. Cadmium, lead, and depressive symptoms: analysis of National Health and Nutrition Examination Survey 2011–2012. J Clin Psychiatry. 2017;78:e515–21.

[96] Adeniyi PA, Olatunji BP, Ishola AO, Ajonijebu DC, Ogundele OM. Cadmium increases the sensitivity of adolescent female mice to nicotine-related behavioral deficits. Behav Neurol. 2014;2014:36098.

[97] Ajonijebu C, Adekeye AO, Olatunji BP, Ishola AO, Ogundele OM. Nicotine-cadmium interaction alters exploratory motor function and increased anxiety in adult male mice. Neurodegener Dis. 2014;359436. doi: 10.1155/2014/359436.

[98] Hecht EM, Landy DC, Ahn S, Hlaing WM, Hennekens CH. Hypothesis: cadmium explains, in part, why smoking increases the risk of cardiovascular disease. J Cardiovasc Pharmacol Ther. 2013;18:550–4.

[99] Tellez-Plaza M, Jones MR, Dominguez-Lucas A, Guallar E, Navas-Acien A. Cadmium exposure and clinical cardiovascular disease: a systematic review. Curr Atheroscler Rep. 2013;15:356.

[100] Das S, Upadhaya P, Giri S. Arsenic and smokeless tobacco induce genotoxicity, sperm abnormality as well as oxidative stress in mice in vivo. Genes Environ. 2016;38:4.

[101] Chelchowska M, Ambroszkiewicz J, Jablonka-Salach K, Gajewska J, Maciejewski TM, Bulska E et al. Tobacco smoke exposure during pregnancy increases maternal blood lead levels affecting neonate birth weight. Biol Trace Elem Res. 2013;155:169–75.

[102] Tinkov AA, Ajsuvakova OP, Skalnaya MG, Popova EV, Sinitskii AI, Nemereshina ON et al. Mercury and metabolic syndrome: a review of experimental and clinical observations. Biometals. 2015;28:231–54.

[103] Roy C, Tremblay PY, Ayotte P. Is mercury exposure causing diabetes, metabolic syndrome and insulin resistance? A systematic review of the literature. Environ Res. 2017;156:747–60.

[104] Genchi G, Sinicropi MS, Carocci A, Lauria G, Catalano A. Mercury exposure and heart diseases. Int J Environ Res Public Health, 2017;14:pii: E74.

[105] Rahman MH, Bratveit M, Moen BE. Exposure to ammonia and acute respiratory effects in a urea fertilizer factory. Int J Occup Environ Health. 2007;13:153–9.

[106] Casas L, Zock JP, Torrent M, Garcia-Esteban R, Garcia-Lavedan E, Hyvarinen A et al. Use of household cleaning products, exhaled nitric oxide and lung function in children. Eur Respir J. 2013;42:1415–8.

[107] Arif AA, Delclos GL. Association between cleaning-related chemicals and work-related asthma and asthma symptoms among healthcare professionals. Occup Environ Med. 2012;69:35–40.

[108] Loftus C, Yost M, Sampson P, Torres E, Arias G, Breckwich Vasquez V et al. Ambient ammonia exposures in an agricultural community and pediatric asthma morbidity. Epidemiology. 2015;26:794–801.

[109] Petrova M, Diamond J, Schuster B, Dalton P. Evaluation of trigeminal sensitivity to ammonia in asthmatics and healthy human volunteers. Inhal Toxicol. 2008;20:1085–92.

[110] The health consequences of smoking: a report of the Surgeon General. Rockville (MD): Department of Health and Human Services, Public Health Service, Centers for Disease Control, Center for Health Promotion and Education, Office on Smoking and Health; 2004.

[111] Ludvig J, Miner B, Eisenberg MJ. Smoking cessation in patients with coronary artery disease. Am Heart J. 2005;149:565–72.

[112] Scherer G. Carboxyhemoglobin and thiocyanate as biomarkers of exposure to carbon monox203 Updated priority list of toxicants in combusted tobacco products ide and hydrogen cyanide in tobacco smoke. Exp Toxicol Pathol. 2006;58:101–24.

[113] Silverstein P. Smoking and wound healing. Am J Med. 1992;93:22S–24S.

[114] Weinberger B, Laskin DL, Heck DE, Laskin JD. The toxicology of inhaled nitric oxide. Toxicol Sci. 2001;59:5–16.

[115] Vleeming W, Rambali B, Opperhuizen A. The role of nitric oxide in cigarette smoking and nicotine addiction. Nicotine Tob Res. 2002;4:341–8.

[116] WHO Study Group on Tobacco Product Regulation. Guiding principles for the development of tobacco product research and testing capacity and proposed protocols for the initiation of tobacco product testing: Recommendation 6. Geneva: World Health Organization; 2004.

[117] WHO Tobacco Laboratory Network standard operating procedure for validation of analytical methods of tobacco product contents and emissions. WHO TobLabNet Official Method SOP02. Geneva: World Health Organization; 2017.

[118] Ding YS, Yan X, Wong J, Chan M, Watson CH. In situ derivatization and quantification of seven

carbonyls in cigarette mainstream smoke. Chem Res Toxicol. 2016;29:125–31.

[119] Zhao W, Zhang Q, Lu B, Sun S, Zhang S, Zhang J. Rapid determination of six low molecular carbonyl compounds in tobacco smoke by the APCI-MS/MS coupled to data mining. J Anal Meth Chem. 2017:8260860. doi: 10.1155/2017/8260860.

[120] Ahad BT, Abdollahi A. Dispersive liquid–liquid microextraction for the high performance liquid chromatographic determination of aldehydes in cigarette smoke and injectable formulations. J Hazard Mater. 2013;254–255:390–6.

[121] Zhang J, Bai R, Zhou Z, Liu X, Zhou J. Simultaneous analysis of nine aromatic amines in mainstream cigarette smoke using online solid-phase extraction combined with liquid chromatography-tandem mass spectrometry. Anal Bioanal Chem. 2017;409:2993–3005.

[122] Xie F, Yu J, Wang S, Zhao G, Xia G, Zhang X et al. Rapid and simultaneous analysis of ten aromatic amines in mainstream cigarette smoke by liquid chromatography/electrospray ionization tandem mass spectrometry under ISO and "Health Canada intensive" machine smoking regimens. Talanta. 2013;115:435–41.

[123] Deng H, Yang F, Li Z, Bian Z, Fan Z, Wang Y et al. Rapid determination of 9 aromatic amines in mainstream cigarette smoke by modified dispersive liquid liquid microextraction and ultraperformance convergence chromatography tandem mass spectrometry. J Chromatogr A. 2017;1507:37–44.

[124] Sampson MM, Chambers DM, Pazo DY, Moliere F, Blount BC, Watson CH. Simultaneous analysis of 22 volatile organic compounds in cigarette smoke using gas sampling bags for highthroughput solid-phase microextraction. Anal Chem. 2014;86:7088–95.

[125] Pazo DY, Moliere F, Sampson MM, Reese CM, Agnew-Heard KA, Walters MJ et al. Mainstream smoke levels of volatile organic compounds in 50 US domestic cigarette brands smoked with the ISO and Canadian intense protocols. Nicotine Tob Res. 2016;18:1886–94.

[126] Saha S, Mistri R, Ray BC. A rapid and selective method for simultaneous determination of six toxic phenolic compounds in mainstream cigarette smoke using single-drop microextraction followed by liquid chromatography-tandem mass spectrometry. Anal Bioanal Chem. 2013;405:9265–72.

[127] Wu J, Rickert WS, Masters A. An improved high performance liquid chromatography-fluorescence detection method for the analysis of major phenolic compounds in cigarette smoke and smokeless tobacco products. J Chromatogr A. 2012;1264:40–7.

[128] Savareear B, Brokl M, Wright C, Focant JF. Thermal desorption comprehensive two-dimensional gas chromatography coupled to time of flight mass spectrometry for vapour phase mainstream tobacco smoke analysis. J Chromatogr A. 2017;1525:126–37.

[129] Brokl M, Bishop L, Wright CG, Liu C, McAdam K, Focant JF. Analysis of mainstream tobacco smoke particulate phase using comprehensive two-dimensional gas chromatography timeof-flight mass spectrometry. J Separation Sci. 2013;36:1037–44.

[130] Pappas RS, Fresquez MR, Watson CH. Cigarette smoke cadmium breakthrough from traditional filters: implications for exposure. J Anal Toxicol. 2014;39:45–51.

[131] Fresquez MR, Gonzalez-Jimenez N, Gray N, Watson CH, Pappas RS. High-throughput determination of mercury in tobacco and mainstream smoke from little cigars. J Anal Toxicol. 2015;39:545–50.

[132] Fresquez MR, Pappas RS, Watson CH. Establishment of toxic metal reference range in tobacco from US cigarettes. J Anal Toxicol. 2013;37:298–304.

[133] Zhang ZW, Xu YB, Wang CH, Chen KB, Tong HW, Liu SM. Direct determination of hydrogen cyanide in cigarette mainstream smoke by ion chromatography with pulsed amperometric detection. J Chromatogr A. 2011;1218:1016–9.

[134] Mahernia S, Amanlou A, Kiaee G, Amanlou M. Determination of hydrogen cyanide concentration in mainstream smoke of tobacco products by polarography. J Environ Health Sci Eng. 2015;13:57.

[135] Shihadeh A, Schubert J, Klaiany J, El Sabban M, Luch A, Saliba NA. Toxicant content, physical properties and biological activity of waterpipe tobacco smoke and its tobacco-free alternatives. Tob Control. 2015;24:i22–30.

[136] Cecil TL, Brewer TM, Young M, Holman MR. Acrolein yields in mainstream smoke from commercial cigarette and little cigar tobacco products. Nicotine Tob Res. 2017;19:865–70.

[137] Schubert J, Kappenstein O, Luch A, Schulz TG. Analysis of primary aromatic amines in the mainstream waterpipe smoke using liquid chromatography–electrospray ionization tandem mass spectrometry. J Chromatogr A. 2011;1218:5628–37.

[138] Jacob P III, Abu Raddaha AH, Dempsey D, Havel C, Peng M, Yu I et al. Comparison of nicotine and carcinogen exposure with water pipe and cigarette smoking. Cancer Epidemiol Biomarkers Prev. 2013;22:765–72.

[139] Hamad SH, Johnson NM, Tefft ME, Brinkman MC, Gordon SM, Clark PI et al. Little cigars vs 3R4F cigarette: physical properties and HPHC yields. Tob Regul Sci. 2017;3:459–78.

[140] Goel R, Trushin N, Reilly SM, Bitzer Z, Muscat J, Foulds J et al. A survey of nicotine yields in small cigar smoke: influence of cigar design and smoking regimens. Nicotine Tob Res. 2018;20(10):1250–7.

[141] Apsley A, Galea KS, Sanchez-Jimenez A, Semple S, Wareing H, van Tongeren M. Assessment of polycyclic aromatic hydrocarbons, carbon monoxide, nicotine, metal contents and particle size distribution of mainstream shisha smoke. J Environ Health Res. 2011;11:93–103.

[142] Fresquez MR, Gonzalez-Jimenez N, Gray N, Valentin-Blasini L, Watson CH, Pappas RS. Electrothermal vaporization-QQQ-ICP-MS for determination of chromium in mainstream cigarette smoke particulate. J Anal Toxicol. 2017;41:307–12.

[143] Ashley M, Dixon M, Prasad K. Relationship between cigarette format and mouth-level exposure to tar and nicotine in smokers of Russian king-size cigarettes. Regul Toxicol Pharmacol. 2014;70:430–7.

[144] Lawler TS, Stanfill SB, de Castro BR, Lisko JG, Duncan BW, Richter P et al. Surveillance of nicotine and pH in cigarette and cigar filler. Tob Regul Sci. 2017;3:101–16.

[145] Mahmood T, Zaman M. Comparative assessment of total nicotine content of the cigarette brands available in Peshawar, Pakistan. Pak J Chest Med. 2015;16.

[146] Kulak JA, Goniewicz ML, Giovino GA, Travers MJ. Nicotine and pH in waterpipe tobacco. Tob Regul Sci. 2017;3:102–7.

[147] Hadidi KA, Mohammed FI. Nicotine content in tobacco used in hubble-bubble smoking. Saudi Med J. 2004;25:912–7.

[148] Sepetdjian E, Saliba N, Shihadeh A. Carcinogenic PAH in waterpipe charcoal products. Food

Chem Toxicol. 2010;48:3242–5.

[149] Nguyen T, Hlangothi D, Matrinez RA, Jacob D, Anthony K, Nance H et al. Charcoal burning as a source of polyaromatic hydrocarbons in waterpipe smoking. J Environ Sci Health B. 2013;48:1097–102. Updated priority list of toxicants in combusted tobacco products

[150] Schubert J, Muller FD, Schmidt R, Luch A, Schultz TG. Waterpipe smoke: source of toxic and carcinogenic VOCs, phenols and heavy metals? Arch Toxicol. 2015;89:2129–39.

[151] Saadawi R, Figueroa JAL, Hanley T, Caruso J. The hookah series part 1: total metal analysis in hookah tobacco (narghile, shisha) – an initial study. Anal Meth. 2014;4:3604–11.

[152] Saadawi R, Hachmoeller O, Winfough M, Hanley T, Caruso JA, Figueroa JAL. The hookah series part 2: elemental analysis and arsenic speciation in hookah charcoals. J Anal Atom Spectrom. 2014;29:2146–58.

[153] Talhout R, Schulz T, Florek E, Van Benthem J, Wester P, Opperhuizen A. Hazardous compounds in tobacco smoke. Int J Environ Res Public Health. 2011;8:613–28.

[154] Harmful and potentially harmful constituents in tobacco products and tobacco smoke; established list. Fed Regist. 2012;77, No. 64:20034–7.

[155] Stepanov I, Villalta PW, Knezevich A, Jensen J, Hatsukami DK, Hecht SS. Analysis of 23 polycyclic aromatic hydrocarbons in smokeless tobacco by gas chromatography-mass spectrometry. Chem Res Toxicol. 2010;23:66–73.

[156] Ding YS, Zhang L, Jain RB, Jain N, Wang RY, Ashley DL et al. Levels of tobacco-specific nitrosamines and polycyclic aromatic hydrocarbons in mainstream smoke from different tobacco varieties. Cancer Epidemiol Biomarkers Prev. 2008;17:3366–71.

[157] Hecht SS, Chen M, Yoder A, Jensen J, Hatsukami D, Le C et al. Longitudinal study of urinary phenanthrene metabolite ratios: effect of smoking on the diol epoxide pathway. Cancer Epidemiol Biomarkers Prev. 2005;14:2969–74.

[158] Hochalter JB, Zhong Y, Han S, Carmella SG, Hecht SS. Quantitation of a minor enantiomer of phenanthrene tetraol in human urine: correlations with levels of overall phenanthrene tetraol, benzo[a]pyrene tetraol, and 1-hydroxypyrene. Chem Res Toxicol. 2011;24:262–8.

[159] Hoffmann D, Adams JD, Brunnemann KD, Hecht SS. Assessment of tobacco-specific N-nitrosamines in tobacco products. Cancer Res. 1979;39:2505–9.

[160] DeRoton C, Wiernik A, Wahlberg I, Vidal B. Factors influencing the formation of tobacco-specific nitrosamines in French air-cured tobaccos in trials and at the farm level. Beitr Tabakforsch Int. 2005;21:305–20.

[161] van de Nobelen S, Kienhuis AS, Talhout R. An inventory of methods for the assessment of additive increased addictiveness of tobacco products. Nicotine Tob Res. 2016;18:1546–55.

[162] Talhout R, Opperhuizen A, van Amsterdam JG. Sugars as tobacco ingredient: effects on mainstream smoke composition. Food Chem Toxicol. 2006;44:1789–98.

[163] O'Connor RJ, Hurley PJ. Existing technologies to reduce specific toxicant emissions in cigarette smoke. Tob Control. 2008;18:139–48.

[164] Fischer S, Spiegelhalder B, Eisenbarth J, Preussmann R. Investigations on the origin of tobaccospecific nitrosamines in mainstream smoke of cigarettes. Carcinogenesis. 1990;11:723–30.

[165] Djordjevic MV, Sigountos CW, Brunnemann KD, Hoffmann D. Formation of

4-(methylnitrosamino)-4-(3-pyridyl)butyric acid in vitro and in mainstream cigarette smoke. J Agric Food Chem. 1991;39:209–13.

[166] WHO Study Group on Tobacco Product Regulation. Report on the scientific basis of tobacco product regulation. Fifth report of a WHO study group (WHO Technical Report Series, No. 989). Geneva: World Health Organization; 2015.

[167] Gray N, Borland R. Research required for the effective implementation of the Framework Convention on Tobacco Control, Articles 9 and 10. Nicotine Tob Res. 2013;15:777–88.

[168] Pauly JL, O'Connor RJ, Paszkiewicz GM, Cummings KM, Djordjevic MV, Shields PG. Cigarette filter-based assays as proxies for toxicant exposure and smoking behavior – a literature review. Cancer Epidemiol Biomarkers Prev. 2009;18:3321–33.

9. 测量和降低无烟烟草制品中有害物质浓度的方法

Stephen B. Stanfill, Division of Laboratory Sciences, National Center for Environmental Health, Centers for Disease Control and Prevention, Atlanta (GA), USA

Patricia Richter, Office of Noncommunicable Diseases, Injury and Environmental Health, Centers for Disease Control and Prevention, Atlanta (GA), USA

Sumitra Arora, National Centre for Integrated Pest Management, Indian Council of Agricultural Research, New Delhi, India

9.1 引　　言

本章介绍2016年世界卫生组织《烟草控制框架公约》（WHO FCTC）缔约方会议第七次会议（COP7）的一致决议[1]，测量和降低无烟烟草制品中某些有害物质含量的合理方法。全球评价表明，世界卫生组织划分的六个区域的3.67亿无烟烟草使用者[2]，其使用量超过可燃烟草制品的消费量[3]，几乎占东南亚（主要是孟加拉国和印度）成年人烟草消费量的90%。本章以无烟烟草的几份研究报告为基础，这些报告提供了有关无烟烟草制品化学成分的更详细信息[4,5]。

无烟烟草制品成分复杂，包含了有机和无机化学物质，这些化学物质导致无烟烟草制品具有致瘾性、有害性或致癌作用[6]。在无烟烟草制品的有害化学物质中，烟草特有亚硝胺（TSNA）含量最高，包括：NNN和NNK，它们是公认的人体致癌物。众所周知NNN会导致口腔癌[6]。在种植和生产过程中，可通过土壤吸收、生物碱的生物合成和微生物活性（包括形成霉菌毒素、亚硝酸盐和TSNA）来增加这些物质或其前体的含量水平[4,7]。在某些类型的产品中，明火烤制工艺会增加有害成分（例如VOC和PAH）的含量。另外，某些添加剂（如槟榔、零陵香豆、khat、二苯醚、咖啡因、pH促进剂）以及可增加致瘾性和致癌性的化学物质[4,5]。最后，微生物在整个生产过程中也起着重要作用，可导致烟草制品生产过程中的化学成分变化。综上所述，在整个生产过程中都会产生有害和致癌物[8-10]。

无烟烟草制品的种类繁多，从简单到复杂，从手工制作到工业制品，还包括将烟草和各种非烟草植物原料以及各种化学添加剂进行混合。工业制成品和家庭手工业产品，种类包括烟叶、散装片状烟草、切碎的烟草、粉状烟草、压制的饼状烟草、焦油、凝胶状糊剂、含烟草的牙膏和压制的颗粒。根据关键成分，无烟烟草制品通常可分为四个主要类别（先前也提出过类似的分类组合）[4,6,11]：

- 第1类，含有很少或没有碱性改性剂的烟草；
- 第2类，包含烟草和大量的碱性试剂；
- 第3类，包含烟草，一种或多种碱性剂和槟榔；
- 第4类，包含烟草与其他具有额外的生物活性的化学或植物成分混合物，例如兴奋剂。

由于包含各种化学成分，每种类别的产品都会危害健康。这些产品中包含的烟草导致其具有有害性、致癌性和致瘾性。然而，已知约70种烟草中，只有极少数被用于制造无烟烟草制品。通常，不同种烟草具有不同水平的生物碱[12,13]；红花烟草（N. tabacum）是最常用的烟草种类（栽培烟草），其具有适中含量的烟碱，而另一种常用的黄花烟草（N. rustica）的烟碱含量则很高[14]。一些无烟烟草制品中含有光烟草（N. glauca）（树烟），其中含有高浓度的烟草生物碱新烟碱。在少数情况下，意外中毒和意外死亡的事件已被证明与使用该种类烟草有关[15,16]。产品中使用不同种类的烟草导致暴露于烟草生物碱的剂量不同，可导致致瘾性、有害性和致癌性，因为烟草生物碱是TSNA形成的必要前体物质[10,12,14,17,18]。由于烟碱的存在，通常情况下所有的无烟烟草制品都会促使消费者对其持续使用，这就导致了消费者重复（通常是每天）暴露于致癌物和有害物质[4]。

尽管烟草中包含数以千计的化学物质[19]，但无烟烟草制品中的许多致癌物在新移植的烟草中并不存在或浓度很低[17]，但在种植和成品制备的过程中形成并积累。烟草中的某些化学成分是由金属和硝酸盐共同作用形成的，这些金属和硝酸盐是在烟草生长过程中吸收的[18]。根据土壤特性和种植烟草环境的不同，某些微生物可能天然存在于植物内部（内生菌）或植物表面（附生菌）[20,21]。在耕种、收获和加工过程中，其他微生物可能会从空气、水、土壤或肥料（如果使用）中沉积在烟草上，或者通过人工处理或作为添加剂引入。生产过程中和最终产品中存在的微生物群落会影响产品成分。无烟烟草生产还包括新鲜烟叶的明火烤制等步骤，这些步骤可能会引入其他化学物质，例如VOC、酚类化合物和PAH[22-24]。在发酵、陈化和储存等过程中，微生物在烟草中依然具有活力并具有代谢活性[8-10,25]，同时，微生物的存在会导致生成反应剂，例如亚硝酸盐和其他有害的副产物（黄曲霉毒素、内毒素、TSNA、其他亚硝胺、氨基甲酸乙酯）[9,26,27]（表9.1）。烟草行业科学家发表的研究表明，通过改变种植方式、生产工艺和持续监控，可以降低有害物质的含量[19,28]。

表9.1 无烟烟草制品中潜在的致癌物、有害物质和生物活性化合物的来源，主要来自土壤、微生物作用、明火烤制和添加剂

物质类别	IARC分类的致癌物（1，2A，2B），有害物质或生物活性化合物	潜在来源或原因
金属和准金属	1：砷、铍、镉、镍化合物、Po 210 第2A组：无机铅化合物 2B：钴敏化作用：铝、铬、钴、镍 皮肤刺激物：钡、汞 槟榔中的铜可能有助于口腔黏膜下层纤维化	从土壤吸收或通过土壤颗粒在烟叶表面的沉积；也可能存在于烟草使用的其他成分（槟榔叶、熟石灰）中
亚硝化剂	2B：亚硝酸盐	通过微生物产生
霉菌毒素	1：黄曲霉毒素（混合物） 2B：黄曲霉毒素M1、曲霉毒素A、葡萄球菌毒素	通过真菌（曲霉菌）形成
亚硝胺烟草特有亚硝胺	1：N′-亚硝基降烟碱、4-(甲基亚硝基氨基)-1-(3-吡啶基)-1-丁酮和4-(甲基亚硝基氨基)-1-(3-吡啶基)-1-丁醇（NNAL）	通过微生物产生的亚硝酸盐，然后在调制、发酵和陈化（亚硝酸盐与生物碱反应）
挥发性N′-亚硝胺	2A：N-亚硝基二甲胺 2B：N-亚硝基吡咯烷、N-亚硝基哌啶、N-亚硝基吗啉、N-亚硝基二乙醇胺	通过微生物产生的亚硝酸盐形成，然后在调制、发酵和陈化过程中亚硝化形成（亚硝酸盐与某些仲胺和叔胺反应）
亚硝酸	2B：N-亚硝基肌氨酸	
氨基甲酸酯	2A：氨基甲酸乙酯	发酵过程中形成（尿素和乙醇的反应）
多环芳烃	1：苯并[a]芘 2A：二苯并[a,h]蒽 2B：苯并[a]蒽、苯并[b]荧蒽、苯并[j]荧蒽、苯并[k]荧蒽、二苯并[a,i]芘、二苯并[a,i]芘、茚并[1,2,3-cd]芘、5-甲基䓛、䓛	明火烤制过程中沉积在烟草上
挥发性醛	1：甲醛 2B：乙醛	明火烤制过程中沉积在烟草上
非烟草植物材料	1：槟榔 肝脏有害物质：零陵香豆 兴奋剂：khat、咖啡因	添加剂

资料来源：参考文献[29]

无烟烟草使用者会暴露于无机物、有机物和生物制剂及其相互作用物。由于不同的土壤成分（金属、硝酸盐、微生物群落）、种植的烟草种类、农业实践、产品加工步骤以及某些允许的添加剂，因此，关注的有害成分范围取决于产品类型和所属的区域。

9.2 产品组成

烟草制品是高度复杂的混合物，其中包含烟碱、第1类致癌物（经IARC工作组评估）[6]和有毒金属，例如砷、镉和铅[30]。这些产品还含有硝酸盐，可被微生

物代谢为亚硝酸盐；亚硝酸盐可引发TSNA和其他亚硝胺的形成[9,31]。某些气态的氮氧化合物（NO_x）在明火烤制时也可引起亚硝化作用[32]。尤其是，降烟碱的亚硝化反应会产生N'-亚硝基降烟碱，一种已知的人体致癌物（IARC 第1类），公认会导致口腔癌[6]。目前，对于各种无烟烟草制品中上述物质的浓度已进行了广泛的监测[4-6,33]。

不同类型的产品中总烟碱和游离态烟碱的浓度差异很大。无烟烟草制品中总烟碱的可检测浓度范围为<0.2~95 mg/g[11,14]。在诸如干鼻烟和苏丹toombak等产品中发现了高浓度的烟碱[14,34]。某些产品，例如toombak、gul、烟叶和zarda中烟碱的浓度比较高可能是因为使用了多种黄花烟草（$N.\ rustica$），即一种富含烟碱的烟草[11,14]。对于给定的产品，从产品的pH值和适当的烟碱pK_a（8.02），使用Henderson-Hasselbach方程计算出未质子化烟碱的百分比[35]。据报道，无烟烟草制品的pH值范围为4.6~11.8。在此范围内，<0.1%~99.9%的烟碱将以游离态呈现[3]。目前发现的pH值最高的是nass[36]和阿拉斯加州使用的iqmik[23]。由于烟碱转化为游离态烟碱，因此在使用过程中烟碱更容易从烟草中释放出来，并穿过生物膜[37-39]。限制总烟碱浓度和允许的pH值可以保持无烟烟草制品中较低的游离态烟碱浓度。

金属和准金属自然存在于土壤中[40]，其中许多金属会和其他无机化合物一起被烟草植物组织吸收[41]，因此，金属元素也存在于无烟烟草制品中[4,42-45]。Golia等的研究表明[41]，某些金属（例如镉、铅、锌和铜）在靠近地面的下部烟叶表面中的浓度高于上部烟叶。烤烟和白肋烟就是这种情况。含金属的土壤可能在下雨时溅到烟叶上，然后沉积在下部烟叶表面。

无烟烟草制品中的金属含量因国家而异。产品中的金属含量可能受土壤含量、pH值[46]和工业污染[43,47,48]的影响。金属的浓度因产品类型和原产国而异。研究报告表明，加纳、印度和巴基斯坦无烟烟草制品中的铅含量很高，巴基斯坦产品中铅、镍和铬的含量高于加纳和印度产品中的铅、镍和铬的含量。印度的无烟烟草制品中铁和铜含量较高，而加纳的无烟烟草制品中铜、铁和铝的含量高于其他国家[43]。由于仔细选择烟草原料以及生产工艺（如对烟叶进行清洗[28,50]），瑞典鼻烟的金属含量极低[49]。

总TSNA的浓度范围为<0.5~12630 μg/g，在苏丹toombak无烟产品中浓度最高。研究发现，总TSNA含量最高的产品一直是经过发酵的无烟烟草制品（例如toombak、khaini、干鼻烟、湿鼻烟）[11,14,27,51-53]。发酵工艺在烟草加工中用于增强口感，但其特点在于微生物繁殖和向亚硝酸盐的主动代谢，亚硝酸盐与天然烟草生物碱反应形成TSNA[8,9]。在鼻烟等产品中发现TSNA含量最低的是snus（瑞典口含烟）和不含发酵烟草的可溶解口含烟草。（"可溶烟草"是指本质上包含烟草和其他成分的产品，这些产品被压缩成可咀嚼的片剂、威化饼状的条状或细长的棒。）未经发酵但经过巴氏消毒的口含烟（snus）产品可消除微生物，使TSNA浓度较低，

例如湿鼻烟。但印度的某些标有"snus"的产品具有较高的TSNA水平[53]。

细菌（葡萄球菌、棒状杆菌、乳杆菌、肠杆菌、梭状芽孢杆菌、芽孢杆菌、沙雷氏菌和大肠埃希氏菌）和真菌（曲霉菌、镰刀菌、克拉多孢菌、念珠菌、链霉菌和顶头孢霉菌）等多种微生物均可产生亚硝酸盐。这些微生物中某些具有潜在的危害性或致病性[26,54,55]。微生物产生的亚硝酸盐是烟草中TSNA和其他亚硝胺的主要决定因素[9,36,56]。肠杆菌科的许多细菌，包括葡萄球菌、棒状杆菌、乳杆菌在无烟烟草制品中可产生含有硝酸盐还原酶和相关的转运蛋白，可导致无烟烟草制品中产生亚硝酸盐（在nar基因操纵子中编码）[57]。干鼻烟还包含一种周质硝酸还原酶（nap基因），主要存在于肠杆菌科的细菌中，这也可能导致产生亚硝酸盐[58]。在某些细菌中，当氧气水平降低时，nar基因就会被激活[59]。

随着烟草的加工处理，烟草细胞破裂并释放出硝酸盐[31]。硝酸还原微生物可以在低氧条件下将硝酸盐转化为亚硝酸盐。某些细菌可以在氧气存在下将硝基烷（如果存在）转化为亚硝酸盐[60]。由于亚硝酸盐具有毒性，通常亚硝酸盐不会被进一步代谢。发酵、陈化、存储和低氧的密闭包装提供了条件，在这些条件下，还原硝酸盐的细菌通过呼吸作用可生成和释放亚硝酸盐[8-10]。无论亚硝酸盐如何形成，一旦形成，它就可以与生物碱以及其他仲胺和叔胺结合形成亚硝胺[31]。某些仲胺或叔胺也可与亚硝酸盐反应形成挥发性亚硝胺，例如N-亚硝基二甲胺和亚硝基酸[31]。

9.3 农业实践和生产过程中导致的有害化合物形成和积累

烟草制品中的许多有害物质含量较低，或者刚移植的烟草中几乎没有有害物质[61]，但是有害物质可在烟草种植和烘烤的早期阶段就开始积累。种植烟草过程中涉及许多农艺学决策，如要种植的烟草类型或种类、收获时机和程序、化肥和农用化学品的类型以及施用率，以及土壤成分、气候和降雨等环境因素，这些因素共同决定了烟草制品的化学成分[4]。

无烟烟草制品的成分，包括烟碱（各种离子形态）、有害物质和致癌物。这是由无烟烟草制品中存在无机物、有机物和生物制剂之间相互作用引起的[4,19,31,40,42,43]。无烟烟草制品使用的烟草原料中含有的天然有机成分包括木质素、脂肪酸、糖、生物碱、萜类、多酚、萜类和类胡萝卜色素。其中某些成分的分解产物会产生挥发性香味物质以及使烤烟着色[18,19,61]。

根据地理位置的不同，烟草的栽种、收获和烘烤方法会有所不同，这可能会影响烟草制品的化学性质[19]。如上所述，在生长期，烟草可以吸收金属[41,43]。使用硝酸盐肥料不仅增加了植物的质量，而且还增加了叶片中硝酸盐、烟碱和其他生物碱的含量，这些都是TSNA的前体物质[18,31,61-63]。应用于烟草的农业植物保护化学品也可能在收获时持续存在于烟草上[64]。在耕种和收获期间暴露于土壤和大气中会导致微生物和真菌的产生，这些真菌会产生霉菌毒素。收割时，烟草放在地上[18]或长时间（≤45天）堆放在田间[65]。烟草与土壤的接触为将微生物和其他生物（昆虫）的引入提供了机会。

收获后，烟叶可以通过晒、晾、烟道或明火来烘烤调制，这是传统无烟烟草制品（可溶解烟草等产品除外）的四种主要调制手段。晒烟调制包括在阳光下晒干烟草，而空气晾烟则包括将烟草挂在通风良好的谷仓中进行晾干。烟熏烘烤是通过在密闭的、装有通风机和烟道的建筑中，将烟叶暴露于由木材、煤炭、石油或液态石油气推动的高温下进行烘烤[18]。在明火烤制过程中，当烟雾衍生的化学物质（例如PAH和VOC）积聚在烟叶上时，烟叶会因木材或锯末的燃烧而在烟气中干燥[22-24]。明火烤烟中的PAH含量高于空气晾晒烟中的PAH[23]（图9.1）。

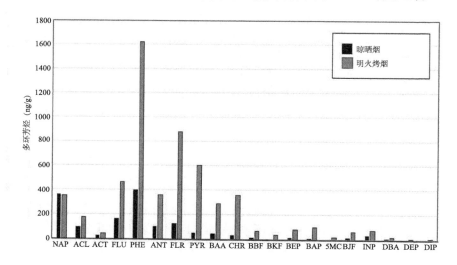

图9.1　烘烤类型对用于制造无烟产品的烟草中多环芳烃含量的影响

5MC：5-甲基蒄；ACL：苊烯；ACT：苊；ANT：蒽；BAA：苯并[a]蒽；BAP：苯并[a]芘；BBF：苯并[b]芴；BEP：苯并[e]芘；BJF：苯并[j]荧蒽；BKF：苯并[k]芴；CHR：䓛；DBA：二苯并[a,h]蒽；DEP：二苯并[a,e]芘；DIP：二苯并[a,i]芘；FLR：荧蒽；FLU：芴；INP：茚并[1,2,3-cd]芘；NAP：萘；PHE：菲；PYR：芘。BAA、BAP、BBF和BKF在FDA HPHC清单中[60]

资料来源：参考文献[23]

烘烤后，烟草可能会进一步陈化、发酵或长期保存，这些生产阶段可能会增加厌氧条件下的微生物繁殖。在这些工艺阶段中，烟草发生了剧烈的化学变化，包括糖的快速催化，pH值和温度的升高以及亚硝酸盐和TSNA浓度的增加[8]。在发酵过程中可能会故意添加微生物[9]。另外，在购买的无烟烟草制品中存在活的微生物[25,55]。黄曲霉毒素是由某些曲霉属霉菌产生的致癌物，可能在栽培或随后的生产步骤中积聚，并已在某些无烟烟草制品中发现[7,66]。微生物作用显然是无烟烟草制品中化学物质不断增高的驱动剂。

9.4 产品添加剂

在制造过程中，几乎所有产品都添加了一定程度的添加剂，包括不一定有毒但可能增加产品吸引力的香料化合物，从而促进开始或持续地使用。这些产品中使用的调味料来源于提取物、油性树脂、香料粉，单个化合物（例如薄荷醇、香兰素）以及60多种精油[67]。用作添加剂的已知对健康有不利影响的物质包括khat（致瘾性植物）、零陵香豆（肝毒素）和槟榔（致瘾性精神活性物质）；众所周知，槟榔会导致癌症和口腔畸形，例如口腔黏膜下层纤维化[68,69]。在某些情况下，槟榔和其他成分要大量添加到含烟草的无烟烟草制品（zarda、rapé）或手工制品（例如槟榔、tombol、dohra、moawa和mainpuri）中[3,7,70,71]。

9.5 无烟烟草制品中的有害物质及降低有害物质影响的方法

无烟烟草制品中的有害物质可以通过多种方式降低，包括通过控制无烟烟草产量[19]。其中一些措施可能有助于降低TSNA水平[9,28,72]。改变有害物质含量的方法包括以下几种。

- 肥料：减少使用含硝酸肥料或（在生长季节的晚期使用例如尿素或其他非硝酸盐肥料），使用其他肥料可以减小硝酸根的摄入，在收获期限制积累或形成亚硝胺[18,61]。
- 收获烟草的表面消毒：用稀释的漂白剂溶液（次氯酸盐-水溶液）洗涤收获的叶片原料可以对叶片表面进行消毒。该程序不仅可以去除微生物，而且可以去除土壤[50,56]。诸如绿叶蔬菜之类的食品进行消毒的方案可有效地用于消除烟草表面污染。对烟叶表面进行消毒已被证明可减少表面微生物的水平[28,50,56]。
- 在食品工业中有效使用的方法可以减少烟草被微生物和土壤污染。在瑞典，口含烟必须符合食品法规标准，从而使产品中有害物质含量降低具备可再现性[28]。
- 电子束技术：高能电子束辐照是一种非热、无化学物质的技术，其中袖珍的线性加速器产生高能（10 MeV）电子，已被用于对食品进行巴氏消毒和对医疗设备进行消毒。该技术通常称为"冷巴氏杀菌"，因为它辐照产品时不会产生过多的热量，因此就不会导致无烟烟草制品产生不良的变化[73]。尽管辐照剂量取决于产品类型，但电子束技术可以消除烟草中存在的产生真菌毒素或亚硝酸盐的活生物体，并去除无烟烟草中的活生物体和潜在有害生物。由于电子束技术不会产生热量，因此可以使用它对烟草、辅料、包装材料和最终产品进行灭菌。在生产初期（特别是在发酵和陈化之前）对烟草进行辐照可能会消除一些产生亚硝酸盐的微生物，与未辐照的烟草相比，TSNA较低。

- 与明火烤制和晾晒有关的变化：明火烤烟中的PAH和挥发性醛的含量更高[22,23,30,74]；如有可能，应省略明火烤制过程。在晾晒过程中，用淀粉芽孢杆菌DA9（一种有效积累亚硝酸盐的生物）进行微生物处理可能会降低亚硝酸盐和TSNA的含量[75]。在烟叶调制过程中，通风不良的调制设备中可能存在高湿度（70%）和10~32℃温度范围的条件因素，这些因素有利于霉菌的生长和形成潜在的霉菌毒素（如黄曲霉毒素和曲霉毒素）[11,76]。应努力防止上述情况，并进行监控以确保防止霉菌生长。

- 巴氏杀菌法：瑞典火柴公司在制备瑞典口含烟产品中使用巴氏杀菌，因此瑞典无烟烟草制品具有低浓度TSNA[38,77]。将磨碎的混合烟叶与水和氯化钠在密闭的混合器中混合，然后用热水和蒸汽注入其中进行热处理，温度达到80~100℃的条件下持续数小时。将混合物冷却，并在产品封装之前添加其他成分，例如香料和保湿剂。根据制造商的说法，此过程会改变口味，降低微生物活性，并在冷藏后具有14~30周的货架期[28,76]。瑞典口含烟产品的TSNA水平较低[28]，已经持续降低并维持了几十年。2017年，瑞典火柴公司产品中NNN加上NNK的平均含量为0.47 μg/g产品干重[28,77]。

- 改良发酵：发酵是烟草制品生产过程中的主要过程，这个过程会增加亚硝酸盐和TSNA的含量[8]，应加以改良[9]或完全避免。在发酵过程中，微生物种群可以迅速增殖或减少，并可以增加pH值、草酸盐、亚硝酸盐和TSNA的含量水平[8,9]。TSNA的形成步骤如图9.2所示。发酵前清洗发酵设备并在该阶段添加不产生亚硝酸盐的微生物可降低TSNA水平[9]。使用富氧吸热发酵[78]也可以减少烟草发酵过程中发生的厌氧呼吸硝酸盐还原[8,9]。辐照后可能需要添加不产生亚硝酸盐[9]的合适的发酵微生物。不含有微生物的产品（如瑞典口含烟）与含有未知的微生物产品相比通常被看作是有益的[8,9,28]，包括产生亚硝酸盐或对使用者有害的物质。

- 微波技术：在过去的50年中，该技术已安全地应用于食品中的烹饪、干燥、巴氏杀菌、消毒、细菌破坏和酶失活，但直到最近才在美国用于在线连续加工中。微波技术的原理很简单：当一种物质受到微波辐射时，材料中的水分子会吸收能量，并且分子的振动会产生热量。连续微波加热设备和技术于2008年在北卡罗来纳州的加工设备中首次用于食品的无菌加工。微波技术已用于生产药品和营养食品。在印度，商业化的带有包装的即食快餐使用专有的微波技术[79-83]。尽管目前尚未用于烟草,但已成功用于食品的微波技术在减少或消除烟草或烟草制品中的微生物方面的适用性是可以评估的。

- 亚硝酸盐清除剂:尽管最大限度地减少无烟烟草制品中亚硝酸盐的最有效方法是完全消除产生亚硝酸盐的生物,但已有的研究依然是使用亚硝酸盐清除剂[28]。某些多酚、维生素C、生育酚、绿茶提取物、绿茶成分(儿茶素、五倍子酸盐)和吗啉可作为潜在的亚硝酸盐清除剂来中和无烟烟草加工过程中产生的亚硝酸盐[28,84,85]。
- 产品冷藏:冷藏可以控制微生物种群的持续增长和产品中亚硝胺化合物的潜在形成。已有制造商鼓励对其产品进行冷藏,包括在销售点进行冷藏[28]。

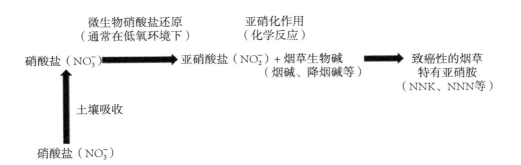

图9.2 增加无烟烟草产品中亚硝胺浓度的过程
NNK:4-(甲基亚硝基氨基)-1-(3-吡啶基)-1-丁酮;NNN:N'-亚硝基降烟碱
资料来源:参考文献[9,10,18,31,61]

无烟烟草生产会导致亚硝胺和微生物(包括细菌和真菌)的存在。烟草包含硝酸盐、烟草生物碱以及某些仲胺和叔胺,它们在亚硝胺的形成中起作用。烟草制品中最大量的致癌物是TSNA,但致癌物还包括其他亚硝胺,例如N-亚硝基氨基酸类和挥发性N-亚硝胺。亚硝胺可在无烟烟草生产的各个阶段形成。一种TSNA,NNN,是一种已知会导致口腔癌的人类致癌物[6,54]。

9.6 微生物检测

标准培养基中的细菌计数:微生物污染可以在培养基板上进行评估,但是这种方法很费时,因为它需要浇注培养基板、划线和结果判断;可能很难找到合适的培养基来测试烟草中的特定微生物[86]。烟草中生长着许多不同的微生物,可以在各种培养基上进行培养,包括胰蛋白酶大豆琼脂[55]、绵羊血琼脂、甘露醇盐琼脂和麦康凯琼脂平板[25]。根据"大板块异常"原理[87]，某些分类单元可能会被过度表达，从而使数量不多但重要的分类单元仍然无法确定，并且没有单一的介质能够捕获烟草制品的微生物多样性。在某些地方，专业知识、供应量以及对无菌设备的建设、接种、孵育和精确解读培养基板的要求可能在某些地方过于昂贵，因此较便宜、较简单的方法可能具有实际优势。

培养依赖一次性微生物膜：培养基的另一种替代性选择是小的两层膜。将上层膜提起以使接种物可在下层膜表面扩散，然后将其孵育并读取微生物生长程度。尽管可能存在其他品牌的双层膜产品，但用于测试多种细菌的双层膜由3M公司制造。

9.6.1 通过细胞活性快速检测活性微生物

酶标仪格式：这对了解烟草在种植和生产过程以及最终烟草制品中不同点上微生物污染程度将很有帮助。基于荧光素酶（例如BacTiter-Glo）的微生物生存力测定是通过定量测定ATP（存在于所有活细胞中的一种分子），其反映了存活微生物的数量[88]。单一试剂会引起细胞裂解并产生发光信号，该信号与ATP浓度成正比，与活细胞的数量成正比。该测定法检测多种细菌和真菌。在添加混合试剂后5分钟可以记录数据。该测定法在发光仪或CCD相机上最多可检测10个细菌细胞。该应用程序可与96孔板读数器[88]一起使用。

手持式光度计：至少有六家制造商制造的手持式阅读器已将ATP测量微型化。用于检测ATP的手持式光度计可在5~20 s内确定微生物污染的程度。多家制造商制造了ATP光度计，并对其功能进行了比较。一家制造商制造的设备可枚举特定细菌种类、总生存数、肠杆菌科、大肠菌群、大肠杆菌和李斯特菌。这些设备用于各种行业（例如医疗保健、制药、水处理、乳制品、肉类、农产品）。便携式装置可用于在几分钟之内就地测试烟草制品、制剂或相关配料[90,91]。

通过qPCR检测生物：还可以通过定量聚合酶链反应（qPCR）检测生物体，该技术是针对特定基因区域的探针结合、复制并可以定量检测的技术。特定生物或其他生物可以通过使用适当设计的分子探针来检测。可以设计用于qPCR的分子探针，这些探针对减少硝酸盐的生物和基因（例如narG、napA等）具有特异性，并可以通过qPCR进行定量[89]。

硝酸盐还原微生物的检测：除了列举微生物以评估污染程度外，还应确定烟草制品中实际硝酸盐还原生物。一种方法可能涉及在添加了亚硝酸盐指示剂的硝酸琼脂上培养微生物。另一种方法是使用Greise试剂盒（由磺胺与NO_2制备，然后加入萘乙二胺）[92]。

9.7 用于测量无烟烟草中有害物质的分析方法概述

测量烟草成分的分析方法（表9.2）主要取决于与各种类型的质谱仪（四极杆质谱和三重四极杆质谱；MS，MS/MS）和其他检测器（如电感耦合等离子体和火焰离子化检测器）相连的GC和LC。表9.2不是一个详尽的清单，但是显示了一些最近描述各个实验室用于烟草制品分析方法论文。

表9.2 烟草成分的测定方法

分析物	测量方式
烟碱[35,93]	GC-MS，GC/FID
次要生物碱[13]	GC-MS/MS
香豆素和樟脑等香料和非烟草植物成分[94]	GC-MS
多环芳烃[22-24]	GC-MS
挥发性有机化合物[22]	GC-MS
有毒金属[42]	ICP-MS
TSNA[11,52,53]	LC-MS/MS
黄曲霉毒素[7]	UHPLC-MS/MS
槟榔相关化合物[95]	LC/MS/MS
pH值[34,35]	离子探针
硝酸盐和亚硝酸盐[96]	离子色谱
真菌[76]	真菌特有的培养基
细菌[25]	细菌特有的培养基
金属和碱性试剂的鉴定[97]	X射线荧光（XRF）
烟草种类的鉴定（*Nicotiana tabacum*、*N. Rusta*、*N. glauca*），非烟草植物材料（例如槟榔、零陵香豆）和碱性剂（碳酸镁、熟石灰）[11,52,71]	傅里叶变换/红外光谱（FT/IR）

GC：气相色谱；LC：液相色谱；MS：质谱；PAH：多环芳烃；TSNA：烟草特有亚硝胺；VOC：挥发性有机化合物

如有必要，可以使用经过适当验证的其他用于测量这些分析物的技术来代替。对于资金或空间有限的监管机构，可以使用紧凑型或便携式仪器，包括GC-MS系统，该系统可以检测和定量烟碱、微量生物碱、槟榔碱（槟榔）、香料和非烟草植物成分（香豆素、二苯醚、樟脑）。一些GC-MS系统是小型化、可移动的，并且其成本低于传统的台式系统。

9.8 监管方法和对策

目前，多种监管措施和其他响应已被使用。尽管欧盟根据指令2001/37/EC[98]监管无烟烟草，但瑞典被豁免禁止销售某些类型的口用烟草。在美国，自2009年

《家庭吸烟预防和烟草控制法案》通过以来,无烟烟草已受到FDA的监管[99]。在其他地方,对无烟烟草使用的管制有限,例如禁止使用无烟烟草制品或管制其成分。尽管印度的安得拉邦、比哈尔邦、果阿、马哈拉施特拉邦和泰米尔纳德邦都曾试图禁止gutkha和pan masala作为食品销售[3],对于gutkha而言,制造商规避了禁令[100]的规定,将产品分成附属的包装,附属包装中包括薄片状烟草,而其他组成材料另外包装。然而,据报道,果阿通过《2005年果阿公共卫生(修订)法》[101]维持了对无烟烟草制品的禁令。

减少无烟烟草中有害物质针对性方法的一个例子是FDA在2017年1月针对成品无烟烟草制品中的NNN提出的规则[72],该规则基于NNN的致癌性是无烟烟草使用导致口腔癌风险升高的主要因素。该标准的要素包括:

- 成品无烟烟草制品要求有效期,以及(如适用)储存条件(例如,销售点冷藏)。
- 在有效期内的任何时间,成品无烟烟草制品中每批次的NNN ≤ 1 µg/g烟草(干重)平均水平。
- 测试评估成品无烟烟草制品中NNN含量的稳定性,并确定和验证产品的有效期和储存条件。
- 批量测试确定产品是否符合建议的NNN含量水平。

FDA还针对该法规提出了各种基于证据的、广泛的说明注意事项[72],可能对寻求减少无烟烟草制品中有害物质的其他法规机构有所帮助。

- 对无烟烟草有害物质的管制可能会影响消费者对无烟烟草使用危害的认识。
- 无烟烟草制品的使用者可能将一种或多种有害物质的减少解释为危害较小,可能会减少无烟烟草使用者戒烟的动机。然而,FDA并不认为产品标准提案会明显地妨碍无烟烟草产品的戒除,从而抵消癌症风险降低对公众健康的有益影响。
- 消费者必须持续地接受关于现有无烟烟草制品使用建议的健康信息教育。

另一种方法是由生产口含烟(snus)的瑞典火柴公司在90年代末建立的GothiaTek标准。它设定了亚硝酸盐、TSNA(NNN和NNK)、N-亚硝基二甲胺、苯并[a]芘、黄曲霉毒素、农用化学品和多种金属的最大允许浓度,这些在产品中已符合要求[28,102]。Snus使用的是经过巴氏消菌的晾烟烟草制成,制备过程中去掉了明火烤制和发酵过程,产品在销售点进行了冷藏,并且TSNA的浓度非常低[22]。

与FDA规定的NNN产品标准相反,GothiaTek限值是按"原样"(湿重)设定的[22]。其他还包括:

- 原材料标准,包括农药残留物的指导限量水平;
- 生产过程中的质量控制和质量保证措施;
- 烟草的热处理;

■ 降低含水量；

■ 与瑞典食品法规定的添加剂和香料相一致。

GothiaTek标准没有规定产品的pH值或烟碱含量[102]，但对于某些有害和致癌物，它可作为烟草制品制造商或管理者希望降低有害物质含量的参考。在2009年至2017年之间，大多数GothiaTek有害物质的水平降低或保持稳定。所含的物质与土壤含量、调制、微生物活性和农用化学品的使用有关，如表9.3所示。

表9.3　GothiaTek在2009年、2015年和2017年规定的有害物质水平的系统性降低

潜在来源和化学成分	2009年水平[a]	2015年水平[b]	2017年水平[b]	限量（2009年）	限量（2015年、2017年）
土壤金属含量					
砷（μg/g）	0.1	<0.06	0.06	0.5	0.25
镉（μg/g）	0.6	0.28	0.27	1个	0.5
铬（μg/g）	0.8	0.46	0.46	3	1.5
铅（μg/g）	0.3	0.15	0.15	2	1
汞（μg/g）	不适用	<0.02	<0.02	不适用	0.02
镍（μg/g）	1.3	0.87	0.82	4.5	2.25
调制[c]					
乙醛[d]（μg/g）	NA	6.5	6.3	NA	25
巴豆醛[d]（μg/g）	NA	<0.10	<0.10	NA	0.75
甲醛[d]（μg/g）	NA	2.3	2.3	NA	7.5
苯并[a]芘[e]（ng/g）	1.1	<0.6	<0.6	20	1.25
微生物（真菌）					
黄曲霉毒素（ng/g）	NA	<2.1	<2.1	NA	2.5
赭曲霉毒素A（ng/g）	NA	2.3	2	NA	10
微生物（细菌）					
亚硝酸盐（μg/g）	2.0	1.7	1.5	7	3.5
NNN+NNK（μg/g）	1.6	0.39	0.47	10	0.95
NDMA（ng/g）	0.7	<0.6	<0.6	10	2.5
农用化学品（μg/g）	NA	b	b	NA	a

NA：不适用；NDMA：N-亚硝基二甲胺；NNK：4-(甲基亚硝基氨基)-1-(3-吡啶基)-1-丁酮；NNN：N'-亚硝基降烟碱。这些分析物主要来自土壤吸收（有毒金属）、调制以及真菌和细菌的存在与活性。单位从每千克转换为每克

a. 根据瑞典火柴农药管理计划[28]

b. 低于瑞典火柴内部限量[102]

c. snus未明火烤制；这些分析物通常是在其他产品（干鼻烟和湿鼻烟）生产中的明火烤制过程中形成的

d. 这些醛是挥发性有机化合物

e. 该化合物是多环芳烃

9.9 政策建议和摘要

在本节中，我们描述了关于确定主要被关注特性的研究，以及检测和降低无烟烟草制品中某些有害物质含量的方法。我们同意TobReg在2015年关于为无烟烟草制品中的致癌物建立管制限量建议[29]，该限量仍然有效，并在下面进行重申。

- 减少毒性：减少或消除对黄花烟草（*N. rustica*）的使用；限制细菌污染，因为细菌可促进亚硝化和致癌物的形成；要求将烟草进行晾干或晒干，而不是火烤或风干，这样可以避免细菌滋生；通过巴氏杀菌法杀死细菌；改善存储条件，例如在销售前将产品冷藏；附上生产日期；配方中去除掉已知会致癌的成分，例如槟榔和零陵香豆。
- 强制实行产品标准：将NNN加NNK的上限设定为2 μg/g（干重），对苯并[a]芘的上限设定为5 ng/g（干重）；并监测烟草中的砷、镉和铅含量。
- 减少吸引力和致瘾性：采取措施减少烟草制品的吸引力和致瘾性，包括禁止添加甜味剂和香料（包括草药、香料和花卉），并限制游离态烟碱和pH值。
- 对跨境产品采用统一的标准：使出口的无烟烟草制品保持与制造国相同（或更高）的标准。

世界卫生组织TobReg还建议对无烟烟草制品的传播进行监管，以防止未经证实的有关暴露或减少疾病的说法[103]。

我们在下面总结了与无烟烟草制品中致癌的、有毒的和致瘾的物质有关的一些问题，并提出了降低这些成分浓度的方法。降低烟碱、游离态烟碱、砷、镉、铅、苯并[a]芘、NNN加NNK的含量并不会使无烟烟草制品安全；但是，谨慎地降低已知的致瘾的、有毒的或致癌的物质的水平。

与无烟烟草制品相关的担忧包括：

- 使用包括黄花烟草（*N. rustica*）在内的高烟碱烟草；
- 包含能显著提高pH值、增加游离态烟碱浓度并有助于提高血液中烟碱含量、增加心血管疾病风险和致瘾性的碱性试剂；
- 由于土壤吸收或受污染的土壤或环境沉积而在烟草中存在有毒金属；
- 使用明火烤制，会从烟雾中引入化学物质，例如多环芳烃（包括苯并[a]芘）、酚和挥发性醛；
- 烟叶上存在微生物污染（细菌和真菌），尤其是那些促进黄曲霉毒素、赭曲霉毒素A、亚硝酸盐或亚硝胺（尤其是TSNA）形成的生物，以及最有利于其形成的条件；
- 微生物（细菌和真菌）的存在会导致传染性疾病，对抗生素产生抗药性，改变口腔生物膜或取代胃肠道的健康菌群；
- 土壤施肥方法导致硝酸盐含量升高，导致收获时烟草中亚硝酸盐和亚硝胺的含量增加；收获时烟草上存在有害的农药残留；
- 购买产品中存在活的微生物（包括病原体），可以转移给使用者并可能在口腔或胃肠道中形成；
- 使用提供厌氧条件的发酵、陈化、产品存储或其他过程，这些条件可促进亚硝酸盐和TSNA的快速形成；
- 烟草成品中硝酸盐残留量高；
- 存在槟榔（IARC第1类人类致癌物）和其他具有公认毒性或致癌性的添加剂；
- 存在具有毒性作用的香料；
- 存在调味添加剂，通过增加吸引力、阻止戒烟或掩盖疾病症状的识别或严重性来增强烟草使用的开始。

还应注意无烟烟草制品中非烟草成分和植物材料的贡献。图9.3显示了可能降低无烟烟草制品中最丰富的致癌物质TSNA含量的步骤。

由于降烟碱是NNN的前体，因此从基因上筛选降烟碱含量较低烟草品种，以减少NNN的形成。
- 筛选土壤中的金属污染和硝酸盐含量。
- 研究农艺方法，以使收获时植物内或植物上的硝酸盐和农药的浓度降至最低。
- 在收获或处理烟草时，请戴上手套，以防止皮肤上的生物（例如葡萄球菌）转移到烟草上，并防止工人遭受"绿色烟草病"。
- 用稀释的漂白剂溶液（1∶250的次氯酸盐∶水）洗涤收获的烟草，以去除土壤、土壤金属、肥料、农药、昆虫和微生物。
- 确保调制设备清洁，以免引入其他生物（尤其是减少硝酸盐的生物）。
- 确定可以添加的无害化学或生物制剂，以防止微生物滋生。
- 在进一步处理之前，对烤烟中的亚硝酸盐含量升高或产生亚硝酸盐的生物（例如葡萄球菌、棒状杆菌）进行筛查。
- 对产品进行巴氏消毒或热处理。
- 使用前清洁发酵桶。
- 发酵前添加非硝酸盐还原发酵生物。
- 研究用于降低发酵过程中的硝酸盐含量方法（选择性过滤、反硝化作用、使发酵液循环通过亚硝酸盐与化学清除剂或反硝化生物反应的细胞）。
- 发酵前添加亚硝酸清除剂（维生素C、生育酚、绿茶提取物、昆仑茶提取物、半胱氨酸）。

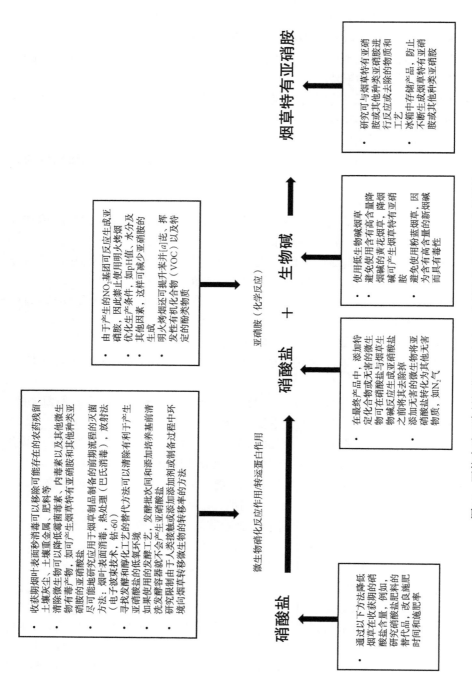

图9.3 可能有助于降低烟草制品中TSNA和其他有害物质浓度的措施

■ 在最终产品中测定微生物的生长。

■ 出售产品之前,请将其保存在冷藏库中。

这些观察结果为监测和控制无烟烟草制品中某些致癌物质和其他有害物质水平的可行性提供了额外支持。

9.9.1 免责声明

本章中的发现和结论是作者本人观点,并不一定代表疾病控制与预防中心的观点。商品名称的使用仅用于标识,并不意味着得到疾病控制与预防中心、公共卫生服务或美国卫生与公共服务部的认可。

9.10 参 考 文 献

[1] Decision FCTC/COP7(14). Further development of the partial guidelines for implementation of Articles 9 and 10 of the WHO FCTC (Regulation of the contents of tobacco products and Regulation of tobacco product disclosures). In: Report of the seventh session of the Conference of the Parties to the WHO Framework Convention on Tobacco Control. Geneva: World Health Organization; 2016) (http://www.who.int/fctc/cop/cop7/FINAL_COP7_REPORT_EN.pdf, accessed 15 May 2019).

[2] WHO global report on trends in prevalence of tobacco smoking 2000–2025, 2nd edition. Geneva: World Health Organization; 2018.

[3] Eriksen M, Mackay J, Schluger N, Gomeshtapeh FI, Drope J, editors. Tobacco atlas. Fifth edition. Atlanta (GA): American Cancer Society and World Lung Foundation; 2015 (https://tobaccoatlas.org/, accessed 16 July 2019).

[4] Smokeless tobacco and public health: a global perspective. Chapter 3. Global view of smokeless tobacco products: constituents and toxicity (NIH/CDC Monograph). Bethesda (MD): Department of Health and Human Services; 2014:75–114 (NIH Pub. 14-7983) (https://cancercontrol.cancer.gov/brp/tcrb/global-perspective/SmokelessTobaccoAndPublicHealth.pdf,accessed 16 July 2019).

[5] Stanfill SB. Chapter 7: Toxic contents and emissions in smokeless tobacco products. In: WHO Study Group on Tobacco Product Regulation. Report on the scientific basis of tobacco product regulation. Sixth report of a WHO study group (WHO Technical Report Series, No. 1001). Geneva: World Health Organization; 2017:23–39. (http://www.who.int/tobacco/publications /prod_regulation/trs1001/en/, accessed 16 July 2019).

[6] Smokeless tobacco and some tobacco-specific N-nitrosamines. Lyon: International Agency for Research on Cancer; 2007 (IARC Monographs on the Evaluation of Carcinogenic Risks to Humans, Vol. 89) (http://monographs.iarc.fr/ENG/Monographs/vol89/index.php, accessed 20 September 2018).

[7] Zitomer N, Rybak ME, Li Z, Walters MJ, Holman MR. Determination of aflatoxin B1 in smokeless tobacco products by use of UHPLC-MS/MS. J Agric Food Chem. 2015;63(41):9131–9138.

[8] Di Giacomo M, Paolino M, Silvestro D, Vigliotta G, Imperi F, Visca P et al. Microbial community structure and dynamics of dark fire-cured tobacco fermentation. Appl Environ Microbiol. 2007;73:825–37.

[9] Fisher MT, Bennett CB, Hayes A, Kargalioglu Y, Knox BL, Xu D et al. Sources of and technical approaches for the abatement of tobacco specific nitrosamine formation in moist smokeless tobacco products. Food Chem Toxicol. 2012;50:942–8.

[10] Andersen RA, Burton HR, Fleming PD, Hamilton-Kemp TR. Effect of storage conditions on nitrosated, acylated, and oxidized pyridine alkaloid derivatives in smokeless tobacco products. Cancer Res. 1989;49:5895–900.

[11] Stanfill SB, Connolly GN, Zhang L, Jia TL, Henningfield J, Richter P et al. Surveillance of international oral tobacco products: total nicotine, unionized nicotine and tobacco-specific nitrosamines. Tob Control. 2011;20:e2.

[12] Sisson VA, Severson RF. Alkaloid composition of Nicotiana species. Beitr Tabakforsch Int.1990;14:327–39.

[13] Lisko JG, Stanfill SB, Duncan BW, Watson CH. Application of GC-MS/MS for the analysis of toApproaches to measuring and reducing toxicant concentrations in smokeless tobacco products bacco alkaloids in cigarette filler and various tobacco species. Anal Chem. 2013;85:3380–4.

[14] Idris AM, Nair J, Ohshima H, Friesen M, Brouet I, Faustman EM et al. Unusually high levels of carcinogenic tobacco-specific nitrosamines in Sudan snuff (toombak). Carcinogenesis. 1991;12:1115–8.

[15] Steenkamp PA, van Heerden FR, van Wyk B. Accidental fatal poisoning by Nicotiana glauca: identification of anabasine by high performance liquid chromatography/photodiode array/mass spectrometry. Forensic Sci Int. 2002;128:208–17.

[16] Furer V, Hersch M, Silvetzki N, Breuer GS, Zevin S. Nicotiana glauca (tree tobacco) intoxication – two cases in one family. J Med Toxicol. 2011;7(1):47–51.

[17] Bhide SV, Nair J, Maru GB, Nair UJ, Kameshwar Rao BV, Chakraborty MK et al. Tobacco-specific N-nitrosamines [TSNA] in green mature and processed tobacco leaves from India. Beitr Tabakforsch Int. 1987;14:29–32.

[18] Burton HR, Dye NK, Bush LP. Distribution of tobacco constituents in tobacco leaf tissue. 1. Tobacco-specific nitrosamines, nitrate, nitrite, and alkaloids. J Agric Food Chem. 1992;40(6):1050–5.

[19] Rodgman A, Perfetti TA, editors. The chemical components of tobacco and tobacco smoke, 2nd edition. Boca Raton (FL): CRC Press; 2013.

[20] Chen Z, Xia Z, Lei L, Xu S, Wang M, Cun L et al. Characteristic analysis of endophytic bacteria population in tobacco. Acta Tabacaria Sinica. 2014;20:102–7.

[21] List of tobacco pests and diseases. In: Ephytia [website]. Paris: French National Institute for Agricultural Research; 2019 (http://ephytia.inra.fr/en/C/10749/Tobacco-List-of-tobacco-pestsand-diseases, accessed 16 July 2019).

[22] Stepanov I, Jensen J, Hatsukami D, Hecht SS. New and traditional smokeless tobacco: comparison of toxicant and carcinogen levels. Nicotine Tob. Res. 2008; 10:1773–82.

[23] Hearn BA, Ding YS, England L, Kim S, Vaughan C, Stanfill SB et al. Chemical analysis of Alaskan iq'mik smokeless tobacco. Nicotine Tob Res. 2013;15(7):1283–8.

[24] Stepanov I, Villalta PW, Knezevich A, Jensen J, Hatsukami D, Hecht SS. Analysis of 23 polycyclic aromatic hydrocarbons in smokeless tobacco by gas chromatography–mass spectrometry. Chem Res Toxicol. 2010;23:66–73.

[25] Han J, Sanad YM, Deck J, Sutherland JB, Li Z, Walters MJ et al. Bacterial populations associated with smokeless tobacco products. Appl Environ Microbiol. 2016;82:6273–83.

[26] Larsson L, Szponar B, Ridha B, Pehrson C, Dutkiewicz J, Krysińska-Traczyk E et al. Identification of bacterial and fungal components in tobacco and tobacco smoke. Tob Induced Dis. 2008;4:4.

[27] Pauly JL, Paszkiewicz G. Cigarette smoke, bacteria, mold, microbial toxins, and chronic lung inflammation. J Oncol. 2011:819129.

[28] Rutqvist LE, Curvall M, Hassler T, Ringberger T, Wahlberg I. Swedish snus and the GothiaTek standard. Harm Reduct J. 2011;8:11.

[29] WHO Study Group on Tobacco Product Regulation. Report on the scientific basis of tobacco product regulation: sixth report of a WHO study group (WHO Technical Report Series No. 1001). Geneva: World Health Organization; 2018 (https://www.who.int/tobacco/publications/prod_regulation/trs1001/en/, accessed 16 July 2019).

[30] Fact sheet, smokeless tobacco health effects. Atlanta (GA): Centers for Disease Control and Prevention; 2018 (https://www.cdc.gov/tobacco/data_statistics/fact_sheets/smokeless/, accessed 16 July 2019).

[31] Spiegelhalder B, Fischer S. Formation of tobacco-specific nitrosamines. Crit Rev Toxicol. 1991;21:241.

[32] Wang J, Yang H, Shi H, Zhou J, Bai R, Zhang M et al. Nitrate and nitrite promote formation of tobacco-specific nitrosamines via nitrogen oxides intermediates during postcured storage under warm temperature. J Chemistry. 2017;6135215 (https://doi.org/10.1155/2017/6135215, accessed 20 July 2019).

[33] Bhisey RA, Stanfill SB. Chapter 13. Chemistry and toxicology of smokeless tobacco. In: Smokeless tobacco and public health in India. New Delhi: Ministry of Health and Family Welfare; 2017 (https://www.researchgate.net/, accessed 23 September 2018).

[34] Lawler TS, Stanfill SB, Zhang L, Ashley DL, Watson CH. Chemical characterization of domestic oral tobacco products: total nicotine, pH, unprotonated nicotine and tobacco-specific N-nitrosamines. Food Chem Toxicol. 2013;57:380–6.

[35] Notice regarding requirement for annual submission of the quantity of nicotine contained in smokeless tobacco products manufactured, imported, or packaged in the United States. Washington (DC): Centers for Disease Control and Prevention; Department of Health and Human Services; 1999 (https://www.gpo.gov/fdsys/pkg/FR-1999-03-23/pdf/99-7022.pdf, accessed 16 July 2019).

[36] Brunnemann KD, Genoble L, Hoffmann D. N-Nitrosamines in chewing tobacco: an international comparison. J Agric Food Chem. 1985;33:1178–81.

[37] Tomar SL, Henningfield JE. Review of the evidence that pH is a determinant of nicotine dosage from oral use of smokeless tobacco. Tob Control. 1997;6(3):219–25.

[38] Fant RV, Henningfield JE, Nelson RA, Pickworth WB. Pharmacokinetics and pharmacodynamics of moist snuff in humans. Tob Control. 1999;8:387–92.

[39] Alpert HR, Koh H, Connolly GN. Free nicotine content and strategic marketing of moist snuff tobacco products in the United States: 2000–2006. Tob Control. 2008;17:332–8.

[40] Alloway BJ, editor. Heavy metals in soils: trace metals and metalloids in soils and their bioavailability. Dordrecht: Springer Netherlands; 2013.

[41] Golia EE, Dimirkou A, Mitsios IK. Accumulation of metals on tobacco leaves (primings) grown in an agricultural area in relation to soil. Bull Environ Contam Toxicol. 2007;79:158–62.

[42] Pappas RS, Stanfill SB, Watson C, Ashley DL. Analysis of toxic metals in commercial moist snuff and Alaskan iqmik. J Anal Toxicol. 2008;32(4):281–91.

[43] Pappas RS. Toxic elements in tobacco and in cigarette smoke: inflammation and sensitization. Metallomics. 2011; 3(11):1181–98.

[44] Zakiullah SM, Muhammad N, Khan SA, Gul F, Khuda F, Khan H. Assessment of potential toxicity of a smokeless tobacco product (naswar) available on the Pakistani market. Tob Control. 2012;(4):396–401.

[45] Gupta P, Sreevidya S. Laboratory testing of smokeless tobacco products. Final report to the India Office of the WHO (allotment No: SE IND TOB 001.RB.02). New Delhi; 2004.

[46] Zaprjanova P, Ivanov K, Angelova V, Dospatliev L. Relation between soil characteristics and heavy metal content in Virginia tobacco. In: World Congress of Soil Science. Brisbane, Australia. 2010:205–8 (https://www.iuss.org/19th%20WCSS/Symposium/pdf/2027.pdf, accessed 16 July 2019).

[47] Mulchi CL, Adamu CA, Bell PF, Chaney RL. Residual heavy metal concentrations in sludge amended coastal plain soils – II. Predicting metal concentrations in tobacco from soil test information. Commun Soil Sci Plant Anal.1992;23(9–10):1053–69.

[48] Adamu CA, Bell PF, Mulchi CL, Chaney RL. Residual metal levels in soils and leaf accumulations in tobacco a decade following farmland application of municipal sludge. Environ Pollut. 1989;56:113–26.

[49] New limit values for GothiaTek. In: Swedish Match [website]. Stockholm: Swedish Match; 2016 (https://www.swedishmatch.com/Media/Pressreleases-and-news/News/New-limit-values-for-GOTHIATEK/, accessed 16 July 2019).

[50] Wahlberg I, Wiernik A, Christakopoulos A, Johansson L. Tobacco-specific nitrosamines: a multidisciplinary research area. Agro Food Industry Hi Tech. 1999;10(4);23–8.

[51] Richter P, Hodge K, Stanfill S, Zhang L, Watson C. Surveillance of moist snuff: total nicotine, moisture, pH, un-ionized nicotine, and tobacco-specific nitrosamines. Nicotine Tob Res. 2008; Approaches to measuring and reducing toxicant concentrations in smokeless tobacco products 10(11):1645–52.

[52] Stanfill SB, Croucher R, Gupta P, Lisko J, Lawler TS, Kuklenyik P. Chemical characterization of smokeless tobacco products from South Asia: nicotine, unprotonated nicotine, tobacco-specific N-nitrosamines and flavor compounds. Food Chem Toxicol. 2018;118:626–34.

[53] Stepanov I, Gupta PC, Dhumal G, Yershova K, Toscano W, Hatsukami D et al. High levels of tobacco-specific N-nitrosamines and nicotine in Chaini Khaini, a product marketed as snus. Tob Control. 2015;24,e271–4.

[54] Balbo S, James-Yi S, Johnson CS, O'Sullivan MG, Stepanov I, Wang M et al. (S)-N'-nitrosonornicotine, a constituent of smokeless tobacco, is a powerful oral cavity carcinogen in

rats. Carcinogenesis. 2013;34:2178.

[55] Smyth EM, Kulkarni P, Claye E, Stanfill SB, Tyx RE, Maddox C et al. Smokeless tobacco products harbor diverse bacterial microbiota that differ across products and brands. Appl Microbiol Biotechnol. 2017;101:5391.

[56] Wiernik A, Christakopoulos A, Johansson L, Wahlberg I. Effect of air-curing on the chemical composition of tobacco. Recent Adv Tob Sci. 1995;21:39–80.

[57] González PJ, Correia C, Moura I, Brondino CD, Moura JJG. Bacterial nitrate reductases: molecular and biological aspects of nitrate reduction. J Inorg Biochem. 2006;100:1015–23.

[58] Tyx RE, Stanfill SB, Keong LM, Rivera AJ, Satten GA, Watson CH. Characterization of bacterial communities in selected smokeless tobacco products using 16 S rDNA analysis. PLoS One 2016;11:e0146939.

[59] Nishimura T, Vertes AA, Shinoda Y, Inui M, Yukawa H. Anaerobic growth of Corynebacterium glutamicum using nitrate as a terminal electron acceptor. Appl Microbiol Biotechnol. 2007;75:889–97.

[60] Gadda G, Francis K. Nitronate monooxygenase, a model for anionic flavin semiquinone intermediates in oxidative catalysis. Arch Biochem Biophysics. 2010;493:53–61.

[61] Burton HR, Dye NK, Bush LP. Distribution of tobacco constituents in tobacco leaf tissue. 1. Tobacco-specific nitrosamines, nitrate, nitrite, and alkaloids. J Agric Food Chem. 1992;40(6):1050–5.

[62] Burton HR, Dye NK, Bush LP. Relationship between tobacco-specific nitrosamines and nitrite from different air-cured tobacco varieties. J Agric Food Chem. 1994;42:2007–11.

[63] Bush LP. Alklaloid biosynthesis. In: Davis DL, Nielsen MT, editors. Tobacco: production, chemistry and technology. Oxford: Blackwell; 1999;285–91.

[64] Ghosh RK, Khan Z, Rao CVN, Banerjee K, Reddy DD, Murthy TGK et al. Assessment of organochlorine pesticide residues in Indian flue-cured tobacco with gas chromatography-single quadrupole mass spectrometer. Environ Monit Assess. 2014;186:5069–75.

[65] Idris AM, Ibrahim SO, Vasstrand EN, Johannessen AC, Lillehaug JR, Magnusson B et al. The Swedish snus and the Sudanese toombak: are they different? Oral Oncol. 1998;34(6):558–66.

[66] Harmful and potentially harmful constituents (HPHCs) [website].Washington (DC): Food and Drug Administration; 2018 (https://www.fda.gov/TobaccoProducts/Labeling/ProductsIngredientsComponents/ucm20035927.htm, accessed 17 January 2019).

[67] Smokeless tobacco ingredient list as of April 4, 1994. Report to the Subcommittee on Health and the Environment, Committee on Energy and Commerce. Washington (DC): House of Representatives; 1994 (http://legacy.library.ucsf.edu/tid/pac33f00/pdf, accessed 16 July 2019).

[68] Garg A, Chaturvedi P, Gupta PC. A review of the systemic adverse effects of areca nut or betel nut. Indian J Med Paediatr Oncol. 2014;35(1):3–9.

[69] Prabhu RV, Prabhu V, Chatra L, Shenai P, Suvarna N, Dandekeri S. Areca nut and its role in oral submucous fibrosis. J Clin Exp Dent. 2014;6(5):e569–75.

[70] Bao P, Huang H, Hu ZY, Haggblom MM, Zhu YG. Impact of temperature, CO2 fixation and nitrate reduction on selenium reduction, by a paddy soil clostridium strain. J Appl Microbiol. 2013;114:703–12.

[71] Stanfill SB, Oliveira-Silva AL, Lisko J, Lawler TS, Kuklenyik P, Tyx R et al. Comprehensive

chemical characterization of rapé tobacco products: flavor constituents, nicotine, tobacco-specific nitrosamines and polycyclic aromatic hydrocarbons. Food Chem Toxicol. 2015;82:50–8.

[72] Tobacco product standard for N-nitrosonornicotine level in finished smokeless tobacco products. Washington (DC): Department of Health and Human Services, Food and Drug Administration; 2017 (https://www.fda.gov/downloads/aboutfda/reportsmanualsforms/reports/economicanalyses/ucm537872.pdf, accessed 16 July 2019).

[73] Pillai SD, Shayanfar S. Electron beam technology and other irradiation technology applications in the food industry. Top Curr Chem. 2017;375:6. doi: 10.1007/s41061-016-0093-4.

[74] Idris AM, Ibrahim SO, Vasstrand EN, Johannessen AC, Lillehaug JR, Magnusson B et al. The Swedish snus and the Sudanese toombak: are they different? Oral Oncol. 1998;34(6):558–66.

[75] Wei W, Deng X, Cai D, Ji Z, Wang C, Li J et al. Decreased tobacco-specific nitrosamines by microbial treatment with Bacillus amyloliquefaciens DA9 during the air-curing process of Burley tobacco. Agric Food Chem. 2014;62:12701–6.

[76] Saleem S, Naz A, Safique M, Jabeen N, Ahsan SW. Fungal contamination in smokeless tobacco products traditionally consumed in Pakistan. J Pak Med Assoc. 2018;68(10):1471–77.

[77] How long is the shelf life of snus. Stockholm: Swedish Match; 2018 (https://www.swedishmatch.ch/en/what-is-snus/qa/shelf-life-of-snus/, accessed 16 July 2019).

[78] Brenik W, Rudhard H. Method of endothermic fermentation of tobacco. US Patent 04343318. Washington (DC): Patent Office; 1982 (https://pubchem.ncbi.nlm.nih.gov/patent/US4343318#section=Patent-Submission-Date, accessed 16 July 2019).

[79] Chandrasekaran S, Ramanathan S, Basak T. Microwave food processing—a review. Food Res Intl. 2013;52:243–61 (https://doi.org/10.1016/j.foodres.2013.02.033, accessed 16 July 2019).

[80] Industrial Microwave Systems LLC [website]. New Orleans (LA): Industrial Microwave Systems LLC; 2019 (http://www.industrialmicrowave.com, accessed 16 July 2019).

[81] Brinley TA, Dock CN, Truong VD, Coronel P, Kumar P, Simunovic J et al. Feasibility of utilizing bioindicators for testing microbial inactivation in sweet potato purees processed with a continuous-flow microwave system. J Food Sci. 2007;72(5):235–42.

[82] David JRD, Graves RH, Szemplenski T. Handbook of aseptic processing and packaging, 2nd edition. Boca Raton (FL): CRC Press; 2013.

[83] Parrott DL. Microwave technology sterilizes sweet potato puree. Food Technol. 2010:66–70 (https://ucanr.edu/datastoreFiles/608-672.pdf, accessed 20 July 2019).

[84] Lee S, Kim S, Jeong S, Park J. Effect of far-infrared irradiation on catechins and nitrite scavenging activity of green tea. J Agric Food Chem. 2006;54(2):399–403.

[85] Rundlöf T, Olsson E, Wiernik A, Back S, Aune M, Johansson L et al. Potential nitrite scavengers as inhibitors of the formation of N-nitrosamines in solution and tobacco matrix systems. J Agric Food Chem. 2000;48(9):4381–8.

[86] Cockrell WT, Roberts JS, Kane BE, Fulghum S. Microbiology of oral smokeless tobacco products. Tob Int. 1989;191:55–7.

[87] Harwani D. The great plate count anomaly and the unculturable bacteria. Int J Sci Res. 2012;2(9):350–1.

[88] Technical bulletin: BacTiter-Glo™ microbial cell viability assay (document TB337). Madison (WI): Promega Corporation; 2019 (https://www.promega.com/, accessed 20 July 2019).

[89] Law AD, Fisher C, Jack A, Moe LA. Tobacco, microbes, and carcinogens: correlation between tobacco cure conditions, tobacco-specific nitrosamine content, and cured leaf microbial community. Microb Ecol. 2016;72:1–10.

[90] Omidbakhsh N, Ahmadpour F, Kenny N. How reliable are ATP bioluminescence meters in assessing decontamination of environmental surfaces in healthcare settings? PLoS One. 2014;9(6):e99951.

[91] Approaches to measuring and reducing toxicant concentrations in smokeless tobacco products performance evaluation of various ATP detecting units (Food Science Center report RPN 13922). Chicago (IL): Silliker Inc; 2010.

[92] Sigma-Aldrich [website] (https://www.sigmaaldrich.com, accessed 16 July 2019).

[93] Stanfill SB, Jia LT, Watson CH, Ashley DL. Rapid and chemically-selective quantification of nicotine in smokeless tobacco products using gas chromatography/mass spectrometry, J Chrom Sci. 2009;47(10):902–9.

[94] Lisko JG, Stanfill SB, Watson CH. Quantitation of ten flavor compounds in unburned tobacco products. Anal Methods. 2014;6(13):4698–704. doi: 10.1039/C4AY00271G.

[95] Jain V, Garg A, Parascandola M, Chaturvedi P, Khariwala SS, Stepanov I. Analysis of alkaloids in areca nut-containing products by liquid chromatography-tandem mass spectrometry. J Agric Food Chem. 2017;65, 1977–1983 (https://doi.org/10.1021/acs.jafc.6b05140, accessed 20 July 2019).

[96] Stepanov I, Hecht SS, Ramakrishnan S, Gupta PC. Tobacco-specific nitrosamines in smokeless tobacco products marketed in India. Carcinogenesis. 2015;116:16–9 (https://doi.org/10.1002/ijc.20966, accessed 16 July 2019).

[97] Dalipi R, Margui E, Borgese L, Depero E. Multi-element analysis of vegetal foodstuff by means of low power total reflection X-ray fluorescence (TXRF) spectrometry. Food Chem. 2017; 218:348–55.

[98] Directive 2001/37/EC of the European Parliament and of the Council (L194/26–L194/34). Brussels: European Commission; 2001 (https://ec.europa.eu/health/sites/health/files/tobacco/docs/dir200137ec_tobaccoproducts_en.pdf, accessed 22 July 2019).

[99] Family Smoking Prevention and Tobacco Control Act 2009. Washington (DC): United States Congress (Public Law 111-31 111th Congress) (https://www.gpo.gov/fdsys/pkg/PLAW-111publ31/pdf/PLAW-111publ31.pdf, accessed 16 July 2019).

[100] Vidhubala E, Pisinger C, Basumallik B, Prabhakar DS. The ban on smokeless tobacco products is systematically violated in Chennai, India. Indian J Cancer. 2016;53:325–30.

[101] Arora M, Madhu R. Banning smokeless tobacco in India: policy analysis. Indian J. Cancer. 2012;49(4):336–41.

[102] GothiaTek limits for undesired components. Stockholm: Swedish Match; 2017 (https://www.swedishmatch.com/Snus-and-health/GOTHIATEK/GOTHIATEK-standard/, accessed 16 July 2019).

[103] Gray N, Hecht S. Smokeless tobacco – proposals for regulation. Lancet. 2010;375:1589–91

10. 水烟抽吸：流行性、健康影响和减少使用的干预措施

Mohammed Jawad and Christopher Millett, Public Health Policy Evaluation Unit, Imperial College London, London, England

10.1 引　　言

在《烟草控制框架公约》第七次缔约方会议上，文件FCTC/COP/7/10介绍了与《烟草控制框架公约》有关的控制水烟制品使用的政策选择和最佳做法[1]。本章以文件FCTC/COP/7/10中提出的证据为基础，为第八次缔约方会议做准备。目的是提供对水烟烟草控制的全面分析，并概述改善水烟烟草使用的预防和控制的挑战和建议。目的是分析有关水烟的研究，包括：

- 总结水烟使用的区域和全球模式，包括流行率的变化；
- 评估使用水烟的急性和慢性健康影响；
- 描述与启动和维持使用有关的文化习惯；
- 解释调味剂对启动、维持使用和增加使用的影响；
- 探索降低烟碱产品依赖性的可能性；
- 评估与文化相关的水烟特殊干预措施的证据基础，以防止摄入和促进戒烟；
- 根据概念框架概述有效的政策；
- 为研究和监管提供建议。

本部分仅限于水烟及其配件。未描述有"水烟"同义词的电子烟碱设备（例如shisha或hookah笔、电子shisha或hookah）。我们专注于最新文献，以最大限度地减少与文件FCTC/COP/7/10和WHO关于水烟抽吸咨询意见的重复[2]。

10.2　流行性、健康影响和减少使用的有效干预措施

关于水烟烟草的文献相对有限，但呈指数增长。2016年，在标题或摘要中包含"waterpipe"同义词的五个临床数据库（Medline、Embase、Web of Science、

10. 水烟抽吸：流行性、健康影响和减少使用的干预措施

PsychInfo、Global Health）中的引用次数是2006年的十倍以上（285 *vs.* 25）（图10.1）。学术界不断为预防和控制水烟烟草使用提供强有力的证据。

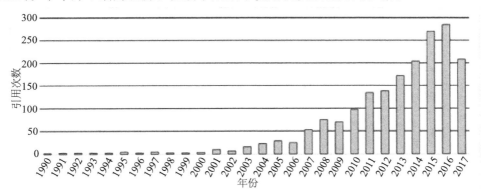

图10.1　1990~2017年，五个临床数据库中标题或摘要中带有"waterpipe"同义词的引用次数

资料来源：数据库Embase、Medline、Global Health、PsychInfo、Web of Science；重复条目已删除；2017年1月1日至8月27日

10.2.1　水烟使用的区域和全球模式

文件FCTC/COP/7/10指出，在世界卫生组织所有区域中的部分国家可以获得流行率估计值[1]。一般而言，尽管只有少数几个国家对这些人群进行了调查，但研究报告称13~15岁的儿童和大学生中的流行率很高，而且许多研究在全国范围内都不具有代表性。

据我们所知，131项研究报告了84个国家水烟使用率的典型性估算[3]。大约三分之一的估计值来自三个国际调查：全球成年人烟草调查、全球青年烟草调查和欧洲特别民意调查。来自最新一轮全球成年人烟草调查和特别欧洲民意调查（第87.1次，2017年）[4,5]的合并数据既包括成年人，也包括使用类似方法的其他研究结果[6-9]，揭示了两个重要发现：世界卫生组织欧洲区域的流行率不断增加，而世界卫生组织东地中海地区缺乏使用可比调查工具的研究，该地区的水烟烟草使用是全球最高的。

世界卫生组织东地中海地区缺乏关于成年人流行率的数据，这与世界卫生组织参与全球青年烟草调查形成了鲜明对比。最新一轮全球青年烟草调查[5]和其他方法学上类似的全国学校调查[10-13]数据显示，世界卫生组织所有区域中许多国家的流行率估计为5.0%~14.9%，这是一个令人关切的问题，需要进一步调查和干预。估计流行率最高的地区是东地中海地区（黎巴嫩，37.0%；西岸，35.2%）和塞浦路斯（33.2%）（图10.2）。

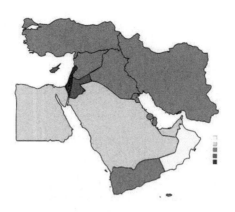

图10.2　东地中海地区年轻人过去30天水烟的使用[5-10]

很少有研究报道水烟的使用趋势。2009年和2017年分别调查了28个欧盟成员国的平均流行率，经常使用水烟成年人平均流行率增加了11.2%[4]。

水管烟草的使用似乎在13~15岁的儿童中越来越流行，尤其是在世界卫生组织地中海东部地区。在美国，一项全国性的调查报告说，在过去的30天内，18~24岁的年轻人使用率情况几乎没有变化或变化很小，2011~2016年期间一直保持在2%~3%[14]。相比之下，2008~2015年之间在美国年轻人中进行的许多其他调查（包括"全国青年烟草调查"）显示，过去30天内水烟使用的绝对增长为每年0.3%~1.0%[15-22]。在加拿大和黎巴嫩进行的学校调查中，也看到了类似的增长[23-26]，而在约旦，则看到了更大比例的绝对增长，2008~2011年之间每年约2.9%[27]。在约旦进行的一项研究中，随着时间的推移使用率的增加与受教育程度高、经常进行体育锻炼、曾经吸烟和同龄人使用水烟有关[28]。在土耳其，全球成人烟草调查显示水烟使用率从2008年的2.3%下降到2010年的0.8%[29]。

虽然水烟使用的最大使用特征是间歇性使用，但与其他烟草制品的相互作用可能需要引起人们的关注。在约旦的一个关于小学生的纵向研究中，吸烟者抽吸水烟的风险明显高于从未吸烟者[30]。一项针对美国青少年的研究得出了类似的结论，该研究采用相同的方法设计，但对吸烟基线倾向进行了额外的控制[31]。在一项针对美国北卡罗来纳州大学生的研究中，研究发现那些首选烟草制品是水烟的人在调查时更可能是双重或多重烟草使用者[32]。

与抽吸卷烟相比，没有明显模式表明社会地位较低的吸烟者会增加水烟的使用。然而，随着公众对水烟烟草的认知和监管环境的转变，社会经济地位的不平等情况可能发生变化。

10.2.2　急性和慢性健康影响

水烟使用者会暴露于来自烟草和加热水烟的木炭所产生的有害物质。有害释放物的研究发现，在水烟烟草烟气中有大量的焦油、烟碱、一氧化碳和致癌化学物质[33]。世界卫生组织地中海东部地区13~15岁人群接触对卷烟和水烟有害物质人群水平建模研究[34]包括估算烟草有害物质人群摄入量，同时考虑了使用频率和水烟共享。结果表明，水烟使用者暴露量相当于所有烟草产生的一氧化碳和苯的约70%，而卷烟使用者暴露量仅相当于30%[34]。高含量的CO和苯可能来自加热烟草所用的木炭。

在急性阶段，吸入了水烟烟草和木炭燃烧的成分会导致不利于心血管和呼吸系统的变化。考虑到水烟烟草中含有大量和变化范围较大的烟碱，可以预期使用水烟可导致血压和心率升高[35-37]，肺功能下降也常有报道[38]。特别是对于水烟烟草，随着木炭燃烧，CO含量急剧上升，并且有报道说水烟烟草使用者中发生了CO中毒事件。应该注意的是，非烟草"草药"水烟产品也要用木炭加热，除了没有烟碱外，其毒理学特征与烟草水烟产品相似[39,40]。

越来越多的证据表明，二手水烟烟气的风险，尤其是在水烟咖啡厅中水烟二手烟的暴露风险。一项对15名被动暴露于水烟的人暴露前后研究表明，他们坐在4~5名吸烟者旁边30分钟后，碳氧血红蛋白平均从0.8%（±0.2）增加到1.2%（±0.8）[41]。另一项研究表明，在水烟咖啡厅的非吸烟者或家庭中的水烟吸烟事例中，呼吸道有害物质丙烯醛水平升高[42]。二手水烟烟气对其他心肺和毒性参数的影响尚不确定[41,43,44]。在空气质量分析中，水烟咖啡厅中无烟室的细颗粒物浓度大于其他场所的无烟室的10倍[45]。在英国的研究也重现了水烟咖啡馆内空气质量差的评估[46]。

新出现的证据表明，水烟烟草对健康具有长期有害影响，类似于其他形式的抽吸型烟草。表10.1总结了几篇已发表的综述[36,47-50]以及随后发表的另外九项研究[51-59]，从而提供了长期危害的最新证据。该表显示，水烟的使用与呼吸系统疾病和心血管疾病、五种癌症、妊娠不良并发症（低出生体重和子宫内生长受限）以及许多其他疾病有关。事实证明，使用水烟烟草产生的二手烟会引起儿童哮喘[60,61]。更进一步的两项群体研究表明，使用水烟与总死亡率增加相关[56,62]。其中之一也显示出癌症死亡率的增加[56]。然而，由于水烟和卷烟的双重使用的习惯做法，很难从这些研究中得出明确的结论。因此，许多对此主题的系统评价都偏向将风险定为很高。因此，关于健康和水烟烟草使用的关联性的流行病学研究的质量仍然很低，应进行进一步的高质量研究，尤其是具有足够的统计能力，以确定从未吸烟者中抽吸水烟与健康结果之间的关系以帮助确认两者之间关联。

表10.1 与使用水烟烟草有关的死亡率和疾病

结果	研究数量	优势比（95%CI）
死亡率	—	—
总体	2	1.25（1.03, 1.51）[a]
心血管疾病	1	0.94（0.43, 2.07）
癌症	1	2.82（1.30, 6.11）[a]
心血管疾病	—	—
冠状动脉疾病	3	1.47（1.06, 2.04）[a]
高血压	1	1.95（1.54, 2.48）[a]
呼吸系统疾病	—	—
慢性阻塞性肺疾病	5	2.33（1.96, 2.77）[a]
支气管炎	2	3.85（1.92, 7.72）[a]
儿童哮喘（主动暴露）	2	1.32（1.20, 1.46）[a]
儿童喘息（被动暴露）	2	1.61（1.25, 2.07）[a]
癌症		
膀胱	6	1.28（1.10, 1.48）[a]
结直肠	3	1.20（0.79, 1.82）
胃	4	2.35（1.47, 3.76）[a]
头颈	7	3.00（2.39, 3.76）[a]
肝	1	1.13（0.62, 2.78）
肺	6	3.96（2.96, 5.30）[a]
食道	5	2.61（2.12, 3.26）[a]
胰腺	1	1.60（0.91, 2.82）
前列腺	1	7.00（0.88, 55.66）
怀孕相关疾病		
低出生体重	4	1.54（1.16, 2.04）[a]
宫内生长受限	1	3.50（1.10, 12.60）[a]
其他疾病		
糖尿病	1	0.52（0.27, 1.00）
胃食管反流病	2	1.31（1.13, 1.53）[a]
丙型肝炎	3	0.98（0.76, 1.25）
不孕症	1	2.50（1.00, 6.30）
精神健康	2	1.33（1.25, 1.42）[a]
多发性硬化症	1	1.77（1.36, 2.31）[a]
牙周疾病	3	3.65（2.22, 6.01）[a]

a. 在随机效应荟萃分析中，水烟烟草使用与该疾病之间统计显著正相关

10.2.3　文化习俗以及开始和持续使用

水烟的文化习俗广泛地融入了家庭、同龄人和社区各级的社会认可网络中。使用者和非使用者都对水烟的使用持积极态度，这源于水烟的消费所处的社会环境，对各种口味和精心设计的器具的吸引力以及营销媒体对水烟的描述[63-65]。

世界卫生组织关于水烟烟草使用的第二份咨询说明[2]中的研究建议之一是关于文化习俗及其如何影响吸烟者开始和持续使用。对于政策，MPOWER政策包[66]旨在协助各国实施有效的干预措施，以减少对烟草制品的需求。MPOWER的六个组成部分是：监控烟草使用和预防政策，保护人们免受烟草烟雾的侵害，提供帮助戒烟的警告，警告烟草危害，强制执行烟草广告、促销和赞助禁令，以及提高烟草税。世界卫生组织《烟草控制框架公约》的几个缔约方，例如印度、伊朗伊斯兰共和国、肯尼亚、马来西亚、巴基斯坦、阿拉伯联合酋长国和坦桑尼亚联合共和国，已经超过了MPOWER的水平并得到了执行（尽管在某些情况下，后来又被推翻），即在州或国家/地区的层面上全面禁止公共场所消费水烟。该政策可能基于社会规范，因为水烟的使用根植于这些缔约方的文化之中。禁止水烟使用禁令的执行程度及其对吸烟率的影响仍然不确定，这应该是这些缔约方评估的关键领域。建议在国际烟草控制政策评估项目中进行评估[67]。

在欧洲和北美，文化的互相渗透可能会决定水烟的使用。在美国进行的一项研究表明，来自阿拉伯国家的移民对北美身份的适应程度较低，因此更有可能使用水烟[68]。此外，加拿大的定性研究表明，来自阿拉伯国家的移民认为自己是通过抽水烟来表达其文化传统，而在英国，非阿拉伯使用者则认为自己正在体验另类文化[69]。

水烟特有的使用方式差异很大。但是，尚不清楚文化内和文化间变异的程度。例如，在黎巴嫩的一项水烟吸烟行为研究中，使用者平均每17 s吸一次烟[70]，而在邻国约旦，使用完全相同方法的研究中，该频率为8 s，导致更多地暴露于有害物质。在这些研究中还注意到水烟吸烟方式的性别差异，男性倾向于比女性抽吸每口的抽吸量更大以及持续时间更长[71]。

文化差异在商业上也很明显。在商业水烟场所，消费的水烟大约占全部的一半[72,73]，咖啡馆文化包括从安静的咖啡馆式氛围到繁忙、喧闹的酒吧式环境[74]。咖啡馆文化很可能与使用方式有关，例如，对美国商业水烟场所的秘密观察表明，一些咖啡馆老板坚称进入该场所的所有顾客都抽水烟[74]。在吸引许多顾客的喧闹的咖啡馆中，他们很可能将水烟与社交联系起来，而在那些安静的咖啡馆中可能更倾向于单人的、烟碱驱动型使用。

10.2.4 调味剂的影响

绝大部分水烟烟草消耗过程中都是经过调味的[75,76]。水烟烟草业积极地推广香精香料,对水烟贸易展览会上52种营销材料的回顾发现香精香料是最常见的主题之一[77]。令人震惊的是,最普遍的主题是水烟比卷烟更安全[77]。水烟烟草业的市场营销实践表明,为了减少消费税,可以单独出售烟草和香料成分以逃避香精香料禁令或减轻烟草重量[77]。

水烟烟草香精香料在对其安全性和吸引力的认知中起着主导作用,尤其是对年轻人而言。2016年4月对10项关于水烟烟草中香料作用的定性研究发现,香料是有吸引力的[78]。在四项研究中(在加拿大、英国和美国),年轻人报告说使用水烟是因为其特殊的口味,因为他们不想使用其他烟草制品[78]。也有人认为加香水烟烟草的危害不及卷烟。在同一篇综述中,加拿大和黎巴嫩的成年人解释说年轻人使用水烟烟草是因为有其风味,而另一位解释者则说风味是年轻人开始使用水烟烟草的原因[78]。在美国对青少年进行的一项大规模调查中,过去30天中有将近80%的水烟使用者报告说,使用特有的香料是使用水烟的原因[79]。

这些定性研究得到其他研究的支持。在美国的一项实验中,要求367名成年水烟吸烟者从菜单中进行选择,这些菜单假设了不同类型产品组合[80]。结果表明:加香的水烟烟草制品比未加香的水烟产品更受青睐,而且与价格或烟碱含量相比,加香对抽吸水烟的决策影响更大[80]。对于女性和不吸烟的人,这种关联更为紧密。在美国进行的另一项实验中,有36名成年水烟烟民以随机交叉设计完成了两次水烟抽吸体验,其中一个水烟是喜欢的口味,另一种水烟的口味不喜欢[81]。那些抽吸了喜欢的味道的人报告了主观上更好的抽吸体验,例如对持续使用的兴趣增加、吸烟带来更多乐趣、喜欢和享受的增加以及愿意继续使用的意愿[81]。这项研究是在水烟使用者偏爱风味的前提下进行的,这本身就是一个重要发现。总之,这些研究表明,香精香料限制或禁止加香在阻止水烟使用方面可能是有效的。

10.2.5 低烟碱产品的依赖性

世界卫生组织关于水烟吸烟的第二份咨询意见明确界定了烟碱依赖性在水烟使用中的作用[2]。定期的水烟使用者吸收足够的烟碱以达到依赖性阈值,并表现出典型的依赖性症状,例如渴望、戒断症状和戒烟困难。但是,在东地中海地区以外,定期使用水烟的情况相对较少。在一项研究中,估计使用者必须每周至少抽水烟三次才判定为具有依赖性[82],而在美国,过去30天内抽水烟的年轻人中只有11.7%吸烟频率达到或超过这个水平[83,84]。

关于低烟碱水烟烟草制品的依赖性有两个问题。首先是几乎没有水烟制造的法规,这导致了不同产品之间烟碱含量的差异很大。在一项实验中,有110名水烟使用者参与了在45分钟吸烟环节实验，标有"0.05%烟碱"的品牌比标有"0.5%烟碱"的品牌的平均血浆烟碱峰值水平更高[37]。由于缺乏对水烟烟草政策的关注，有关制造的讨论应优先考虑香料禁令，因为已证明，香料在水烟购买中比低烟碱发挥更大的作用[80]。

第二个问题是，尽管水烟烟草中的烟碱含量已经比卷烟少，但是吸烟人群对水烟的依赖已被充分证明了，虽然依赖人群的数量仍然未知。按照每分钟吸烟的烟碱含量标准化，水烟每分钟吸烟所含的烟碱含量比卷烟少2~13 ng（表10.2）。由于烟碱水平已经很低，而且对人群水平的依赖性了解相对较少，因此为低烟碱水烟辩护的理由尚不清楚。

表10.2　水烟和卷烟中每分钟的烟碱释放量

参考文献	烟碱释放量（ng/min）	比卷烟中浓度低的倍数[a]
[34]	0.11	1.9~5.6
[70]	0.08	2.7~8.0
[84]	0.05	4.4~13.2

a. 根据Hoffmann & Hoffmann[85]的数据，他们估计烟碱的释放量为1~3 ng/支卷烟，平均抽烟时间为5分钟

10.2.6　干预

对减少水烟使用量干预措施的回顾发现，只有四项个人级别的和五项小组级别的干预研究[86]。对本部分进一步搜索未发现更多的干预研究，表明该领域缺乏研究。在五项随机对照试验中，干预组中只有两项显示出统计学上更高的戒除率[87,88]；在一项试验中，对水烟使用者使用了卷烟特定干预[87]。表10.3列出了这些干预措施的详细信息。非随机研究在戒烟、行为和认知效果方面的结果参差不齐，而且方法学质量通常较低。

这些随机研究中使用的行为干预措施差异很大，但广泛地基于与卷烟行为干预措施相同的原理。在Asfar等的研究中[89]。该干预研究包括由一名训练有素的医生进行的3个45分钟的单独和面对面咨询环节，以及在拟议的戒烟日期之前和之后的五个10分钟的电话访问。Dogar等[87]基于WHO "5As"方法提供了两次结构化的行为训练，前一次持续30分钟，第二次10分钟，并使用了被认为对戒烟有效的行为改变技术。Lipkus等[90]向参与者展示了20张幻灯片，这些幻灯片提供了有关水烟使用的事实信息，包括其对健康的影响。Mohlman等[88]在12个月内为村庄提供了健康促进服务，包括在小学进行消除烟草使用吸引力教育，并描述了其对小学、清真寺和家庭的健康危害；在预科学校和中学中，还应对同伴吸烟的压力进行了教育。Nakkash等[91]为学生提供了10节课，其中四节是关于知识的，六节是关于技能培养的（例如媒体读写能力、决策和拒绝技能）。

表10.3 水烟戒除的随机干预

参考文献	实验设计	实验时间	样本量性别平均年龄	国籍	干预	对照	结果	生化检验	效应量 AOR (95% CI)
[89]	双臂平行实验	2007-08	N=50 94%男性 30岁	叙利亚	行为	常规护理	长期戒烟	是	1.46（0.69,3.09）
[89]	双臂平行实验	2007-08	N=50 94%男性 30岁	叙利亚	行为	常规护理	戒烟的七日点流行率	是	1.34（0.57,3.35）
[89]	双臂平行实验	2007-08	N=50 94%男性 30岁	叙利亚	行为	常规护理	连续戒烟	是	1.07（0.27,4.42）
[87]	三臂整群实验	2011-12	N=1955 79%男性 52岁	巴基斯坦	行为	常规护理	连续戒烟	是	2.20（1.30,3.80）
[87]	三臂整群实验	2011-12	N=1955 79%男性 52岁	巴基斯坦	行为和安非他酮	常规护理	连续戒烟	是	2.50（1.30,4.70）
[90]	双臂实验	2009-10	N=91，76%男性 20岁	美国	行为（健康培训）	行为（非健康培训）	过去30天无水烟使用	否	1.46（0.81,2.62）
[88]	双臂整群实验	2004-06	N=7657 45%男性 36岁	埃及	行为	无干预	当前无使用水烟	否	3.25（1.39,8.89）a
[91]	双臂整群实验	2011-12	N=1857 6~8年级学生	黎巴嫩和卡塔尔	行为	无干预	过去30天无水烟使用	否	1.50（0.89,2.53）

AOR:调整后优势比；CI：置信区间

a. 分析仅限于男性

10.3 未来需要的研究

根据本节的研究结果，应解决有关水烟烟草的四个研究领域。

（1）应继续对水烟烟草使用进行监测：尽管如此，研究人员应考虑使用改进的标准化工具来衡量流行率，以便可以在国家之间比较估算值。一个简单的估计是地中海东部地区的成年人、中欧和东南亚区域的年轻人的流行率。对于研究人员而言，提供了一个调查项目工具箱，其中涵盖了使用模式、依赖性、暴露程度和政策[92]。一个经常被忽略但重要的细节是所消耗的水烟烟草类型。本报告主要针对mo'assel型水烟进行了总结研究，这是一种通常在商业场所销售的调味型水烟烟草。世界卫生组织地中海东部和东南亚地区经常消费的其他类型水烟烟草与使用方式和社会人口统计学相关[93]。

（2）应对水烟烟草的长期健康影响进行高质量的流行病学研究：尤其是，对于卷烟和水烟的间歇性或不经常使用或重复使用的长期健康影响知之甚少[34]。此外，由于水烟烟草危害还包括有可能成为引起其他烟草使用的途径，这将为政策辩论和执行提供强有力的信息。建模研究可以弥补传统的流行病学方法的不足，从而填补这一研究的差距。在收集有关使用水烟烟草对长期健康影响证据的同时，按木炭和烟草类型更好地表征产品，将是更好地了解其潜在危害的重要步骤。

（3）随着世界卫生组织《烟草控制框架公约》缔约方继续制定和执行有关水烟烟草的政策，这些应对政策应该进行正式评估：重要的是评估政策对水烟烟草的使用和态度的影响以及任何意想不到的后果，并进行行业合规性监测。分享经验将对减少水烟使用对健康的影响效果具有指导作用。关键考虑因素是水烟咖啡馆的执法是否会因吸烟者在家中增加使用量而被抵消，以及减少水烟使用量的政策是否会增加其他烟草制品的使用量。

（4）需要有个人和团体干预措施有效性的证据来支持水烟戒烟：尽管缺乏有效的个人和团体干预措施的研究，但这不应延迟已实施的人群水平干预措施的实施和评估。个体和团体干预的主要差距包括戒烟辅助药的有效性[94]以及针对卷烟使用者的干预措施可直接用于水烟使用者的程度。考虑进行个人或团体干预的研究人员，应查阅水烟使用行为改变技术清单[95]。MPOWER框架中描述的人群层面方法应立即实施。

10.4　政　策　建　议

FCTC/COP/7/10[1]和FCTC/COP7[4,96]中提供了与WHO FCTC一致的广泛政策建议清单。我们支持这些建议，将其作为预防和控制水烟使用的重要措施。下面，我们提供了基于MPOWER框架成功的政策选择清单，可以被认为是最相关的。鉴于尚未完全实施MPOWER的国家对水烟的使用率较高，这是谨慎的做法[97]。尽管MPOWER适用于所有烟草制品，但由于水烟的特殊性，我们要求重新评估，明确实施细节，例如主要使用香料、长时间、固定地使用烟草以及在商业场所定期使用。我们将此修订称为MPOWER-W。

在执行下面列出的政策时，应记住，水烟烟草替代品（例如"草药"物质和蒸汽石）仍需要木炭作为热源[76]，并且由于已知的化学成分而应归类为烟草制品。它们产生的烟气，与水烟烟草制品一起销售的事实以及声称无烟的产品可能仍包含烟草的事实[39,98]。对于用作无烟烟草替代品或模仿无烟烟草的产品也提出了类似的要求[99]。

（1）监测烟草使用和预防政策（WHO FCTC第20条）。在过去的十年中，对水烟烟草使用的监督有了很大的改善，但是许多国家仍然没有机制来估算水烟烟草使用的流行程度、二手烟的暴露程度以及水烟烟草行业的活力。本章表明，地中海东部地区无法获得成人可比流行率估计值，中欧和东南亚地区也无法获得年轻人的流行率估计值。高质量的监测对于预防和控制水烟烟草的使用至关重要。流行率应使用标准化工具[92]进行衡量，也许还应补充例行管理数据或水烟咖啡馆目录[100]。

（2）保护人们免受二手烟暴露（WHO FCTC第8条）。允许客户在其场所内抽吸水烟的商业机构应被无烟法律全面纳入。有证据表明水烟抽吸产生的二手烟有害[41,43,44]。在商业场所吸烟的研究表明，全面的无烟法律减少了二手烟的危害，帮助吸烟者戒烟并减少了年轻人的吸烟[101]。水烟的吸烟时间可能会很长（例如长达几个小时），并且吸烟者在户外的移动会增加他们的公众可见度，并且可能很吵闹，并给附近的居民带来麻烦[98]。土耳其和东地中海地区的一些国家/地区制定了分区法律，以禁止在居住区和教育机构一定距离内的水烟咖啡馆[102]。分区法律应与无烟法律共同执行，以最大限度地保护公众。

（3）为戒烟提供帮助（WHO FCTC第14条）。定期抽水烟的吸烟者依赖烟碱。因此，世界卫生组织《烟草控制框架公约》的缔约方应将水烟戒烟纳入传统的戒烟服务中。尽管有效的个体干预的文献很少，但它们显示出可喜的结果，应扩大规模并进行评估，以供各国之间交流。解决使用水烟的强大社会因素也很重要。

（4）警示烟草的危险（WHO FCTC第11条和第12条）。与卷烟一样，水烟烟盒上的健康警示应至少覆盖主要展示区域的50%，并应包括图形图像，这对于接触识字率较低的人群更为有效。健康警示也应用于水烟烟具配件，例如设备和木炭。有充分的证据表明，使用水烟对健康有害[36,47-50]，但烟草行业对世界卫生组织《烟草控制框架公约》关于健康警示的建议的遵守程度很差[103,104]。由于水烟壶设备精巧、装饰精美，并且是水烟吸烟的积极影响的组成部分，因此应将其调节至标准尺寸和样式，以降低其吸引力。

（5）强制禁止烟草广告、促销和赞助（WHO FCTC第13条）。应该执行全面禁止水烟烟草广告、促销和赞助的禁令，这在水烟咖啡馆及其周围以及社交媒体上很常见。与跨国烟草公司的产品相比，大多数水烟烟草是由零售商或在水烟咖啡馆进行广告和促销[77,105,106]。学生优惠、折扣和在商店橱窗中展示水烟斗设备是鼓励使用水烟的常用方法。

（6）提高烟草税（WHO FCTC第6条）。应提高对水烟烟草的税收，至少与卷烟的税率保持一致。烟草税是烟草控制中最有效的政策之一，但是水烟烟草的税率远低于卷烟。黎巴嫩的一项全国调查表明，水烟税增加10%将使家庭消费减少14%[107]。尽管这种税收对水烟咖啡馆使用的影响尚不清楚，但政策制定者应注意，当一个团体使用水烟烟草时，由于每位使用者的成本较低，因此税收的作用减弱了。因此，将需要增加对水烟烟草的税收，以达到与卷烟税收类似的效果。

10.4.1 水烟使用的相关政策

上面列出的6项政策是全球烟草控制政策的基础，是世界卫生组织《烟草控制框架公约》全面政策建议的一部分；但是，有其他两项建议未纳入MPOWER框架中。MPOWER-W的引入将指导政策特定优先地解决水烟问题。

首先，由于大多数水烟烟草都带有调味剂，因此禁止使用调味剂可能会对水烟烟草行业产生深远的影响[108]。它将减少水烟的吸引力，促进戒烟并防止摄入，尤其是年轻人。但是，政策制定者应该保持警惕，可以通过单独出售瓶装调味剂并由使用者将其添加到无味的烟草混合物中来规避调味剂禁令。这样的瓶子很可能是针对水烟烟草使用的，通过严格监管、强制执行的政策，可以防止烟草行业向这种方法的任何转变。也可以将调味剂添加到木炭和水中，而不是添加到烟草中，并且该设备的某些部件可以用水果基质来替代（例如从水果上部挖去内心的水果壳并放置烟草来使用）。决策者应禁止在零售环境中使用烟草调味剂，以确保在整个供应链中维持任何调味剂禁令。

其次，由于许多此类政策都是针对水烟咖啡馆的客户，因此许可授权可能是减轻地方政府执法负担并确保水烟零售商了解其立法责任的最有效方法[98]。除上述政策外，许可授权框架还可纳入与烟草不直接相关的更广泛的保护，例如健康和安全要求，促进良好卫生的质量控制措施（例如一次性软管和烟嘴以及设备清洁程序）以及进入年龄的限制。通过在该许可框架中包括烟草控制政策，执法可以具有成本效益。此外，根据国家/地区的不同，水烟烟草许可的条款可能会受到地方政府的管辖，这将有助于实施当地政策，例如禁止使用调味剂以及将非烟草（"草药"）产品归类为烟草制品，无须更改国家许可授权框架。

10.5 结 论

关于水烟烟草使用的研究呈指数增长，提供的证据表明，这种有害烟草制品的使用在所有大洲的许多国家中都很普遍，尤其是在13~15岁的儿童中。尽管水烟使用者可能会依赖烟碱，但大多数人并未依赖烟碱，因此与个人或团体干预相比，他们更容易受到人口政策的影响。在许多国家，水烟吸烟的流行率继续上升，这主要是由于缺乏对该行业的监管。公共卫生对水烟的关注主要集中在以下事实：水烟主要是调味产品，而产生的香气又会鼓励水烟使用，在决定吸水烟的因素中，香精香料可能比价格更重要。禁止加香的水烟烟草可能会降低这些产品的吸引力，从而减少需求并最终改善公共卫生。但是，该禁令应与烟草控制政策相辅相成，例如更高的税收，全面的无烟法律以及消除对危害误解的公众教育。结合水烟吸烟的特殊性，重新实施MPOWER可以促进预防和控制水烟的使用。同时，研究重点包括持续的标准化监测、对危害的流行病学研究、政策评估以及预防和戒断干预措施的设计。

10.6 参 考 文 献

[1] Control and prevention of waterpipe tobacco products (FCTC/COP/7/10). Geneva: World Health Organization; 2016 (http://bit.ly/2xGIxSx, accessed 28 August 2017).

[2] Advisory note. Waterpipe tobacco smoking: health effects, research needs and recommended actions for regulators. 2nd edition. Geneva: World Health Organization; 2015 (http://bit.ly/2i-7nAwr, accessed 10 September 2017).

[3] Jawad M, Charide R, Waziry R, Darzi A, Ballout RA, Akl EA. The prevalence and trends of waterpipe tobacco smoking: A systematic review. PLoS One. 2018;13(2):e0192191.

[4] Filippidis FT, Jawad M, Vardavas CI. Trends and correlates of waterpipe use in the European Union: analysis of selected Eurobarometer surveys (2009–2017). Nicotine Tob Res. 2017. doi:10.1093/ntr/ntx255.

[5] Yang Y, Jawad M, Filippidis FT, Millett C. The prevalence, correlates and trends of waterpipe tobacco smoking: a secondary analysis of the Global Youth Tobacco Survey (under review).

[6] Agaku IT, King BA, Husten CG, Bunnell R, Ambrose BK, Hu SS et al. Tobacco product use among adults – United States, 2012–2013. Morb Mortal Wkly Rep. 2014;63(25):542–7.

[7] Ward KD, Ahn S, Mzayek F, Al Ali R, Rastam S, Asfar T et al. The relationship between waterpipe smoking and body weight: population-based findings from Syria. Nicotine Tob Res. 2015;17(1):34–40.

[8] Aden, B, Karrar S, Shafey O, Al Hosni F. Cigarette, water-pipe, and medwakh smoking prevalence among applicants to Abu Dhabi's pre-marital screening program, 2011. Int J Prev Med. 2013;4(11):1190–5.

[9] Baron-Epel O, Shalata W, Hovell MF. Waterpipe tobacco smoking in three Israeli adult populations. Israel Med Assoc J. 2015;17(5):282–7.

[10] Szklo AS, Autran Sampaio MM, Masson Fernandes E, de Almeida LM. Smoking of non-cigarette tobacco products by students in three Brazilian cities: Should we be worried? Cad Saúde Pública. 2011;27(11):2271–5.

[11] Minaker LM, Shuh A, Burkhalter RJ, Manske SR. Hookah use prevalence, predictors, and perceptions among Canadian youth: findings from the 2012/2013 Youth Smoking Survey. Cancer Causes Control. 2015;26(6):831–8.

[12] Kuntz B, Lampert T. Waterpipe (shisha) smoking among adolescents in Germany: results of the KiGGS study: first follow-up (KiGGS wave 1) [in German]. Bundesgesundheitsblatt Gesundheitsforsch Gesundheitsschutz. 2015;58(4–5):467–73.

[13] Singh T, Arrazola RA, Corey CG, Husten CG, Neff LJ, Homa DM et al. Tobacco use among middle and high school students – United States, 2011–2015. Morbid Mortal Wkly Rep. 2016;65(14):361–7.

[14] Cohn AM, Johnson AL, Rath JM, Villanti AC. Patterns of the co-use of alcohol, marijuana, and emerging tobacco products in a national sample of young adults. Am J Addict. 2016;25(8):634–40.

[15] Arrazola RA, Neff LJ, Kennedy SM, Holder-Hayes E, Jones CD. Tobacco use among middle and high school students – United States, 2013. Morb Mortal Wkly Rep. 2014;63(45):1021–6.

[16] Arrazola RA, Singh T, Corey CG, Husten CG, Neff LJ, Apelberg BJ et al. Tobacco use among middle and high school students – United States, 2011–2014. Morb Mortal Wkly Rep. 2015;64(14):381–5.

[17] Singh T, Arrazola RA, Corey CG, Husten CG, Neff LJ, Homa DM et al. Tobacco use among middle and high school students – United States, 2011–2015. Morb Mortal Wkly Rep. 2016;65(14):361–7.

[18] Centers for Disease Control and Prevention. Tobacco product use among middle and high school students –United States, 2011 and 2012. Morb Mortal Wkly Rep. 2013;62(45):893–7.

[19] Bover Manderski MT, Hrywna M, Delnevo CD. Hookah use among New Jersey youth: associations and changes over time. Am J Health Behav. 2012;36(5):693–9. Waterpipe tobacco smoking: prevalence, health effects and interventions to reduce use

[20] Haider MR, Salloum RG, Islam F, Ortiz KS, Kates FR, Maziak W. Factors associated with smoking frequency among current waterpipe smokers in the United States: findings from the National College Health Assessment II. Drug Alcohol Depend. 2015;153:359–63.

[21] Primack BA, Fertman CI, Rice KR, Adachi-Mejia AM, Fine MJ. Waterpipe and cigarette smoking among college athletes in the United States. J Adolesc Health. 2010;46(1):45–51.

[22] Sidani JE, Shensa A, Primack BA. Substance and hookah use and living arrangement among fraternity and sorority members at US colleges and universities. J Community Health. 2013;38(2):238–45.

[23] Chan WC, Leatherdale ST, Burkhalter R, Ahmed R. Bidi and hookah use among Canadian youth: an examination of data from the 2006 Canadian Youth Smoking Survey. J Adolesc Health. 2011;49(1):102–4.

[24] Minaker LM, Shuh A, Burkhalter RJ, Manske SR. Hookah use prevalence, predictors, and perceptions among Canadian youth: findings from the 2012/2013 Youth Smoking Survey. Cancer Causes Control. 2015;26(6):831–8.

[25] Jawad M, Lee JT, Millett C. Waterpipe tobacco smoking prevalence and correlates in 25 eastern Mediterranean and eastern European countries: cross-sectional analysis of the Global Youth Tobacco Survey. Nicotine Tob Res. 2016;18(4):395–402.

[26] Saade G, Warren CW, Jones NR, Asma S, Mokdad A. Linking Global Youth Tobacco Survey (GYTS) data to the WHO Framework Convention on Tobacco Control (FCTC): the case for Lebanon. Prev Med. 2008;47(Suppl 1):S15–9.

[27] McKelvey KL, Attonito J, Madhivanan P, Jaber R, Yi Q, Mzayek F et al. Time trends of cigarette and waterpipe smoking among a cohort of school children in Irbid, Jordan, 2008–11. Eur J Public Health. 2013;23(5):862–7.

[28] Jaber R, Madhivanan P, Khader Y, Mzayek F, Ward KD, Maziak W. Predictors of waterpipe smoking progression among youth in Irbid, Jordan: a longitudinal study (2008–2011). Drug Alcohol Depend. 2015.153:265–70.

[29] Erdol C, Ergüder T, Morton J, Palipudi K, Gupta P, Asma S. Waterpipe tobacco smoking in Turkey: policy implications and trends from the Global Adult Tobacco Survey (GATS). Int J Environ Res Public Health. 2015;12(12):15559–66.

[30] Jaber R, Madhivanan P, Veledar E, Khader Y, Mzayek F, Maziak W. Waterpipe a gateway to cigarette smoking initiation among adolescents in Irbid, Jordan: a longitudinal study. Int J Tuberc Lung Dis. 2015;19(4):481–7.

[31] Soneji S, Sargent JD, Tanski SE, Primack BA. Associations between initial water pipe tobacco smoking and snus use and subsequent cigarette smoking: results from a longitudinal study of us adolescents and young adults. JAMA Pediatrics. 2015;169(2):129–36.

[32] Sutfin EL, Sparks A, Pockey JR, Suerken CK, Reboussin BA, Wagoner KG et al. First tobacco product tried: associations with smoking status and demographics among college students. Addict Behav 2015;51:152–7.

[33] Shihadeh A, Schubert J, Klaiany J, El Sabban M, Luch A, Saliba NA. Toxicant content, physical properties and biological activity of waterpipe tobacco smoke and its tobacco-free alternatives. Tob Control. 2015;24:i22–30.

[34] Jawad M, Eissenberg T, Salman R, Alzoubi KH, Khabour OF, Karaoghlianian N et al. Real-time in situ puff topography and toxicant exposure among singleton waterpipe tobacco users in Jordan. Tob Control. 2018 [Epub ahead of print]

[35] Haddad L, Kelly DL, Weglicki LS, Barnett TE, Ferrell AV, Ghadban R. A systematic review of effects of waterpipe smoking on cardiovascular and respiratory health outcomes. Tob Use Insights. 2016;9:13–28.

[36] El-Zaatari ZM, Chami HA, Zaatari GS. Health effects associated with waterpipe smoking. Tob Control. 2015;24:i31–43.

[37] Vansickel AR, Shihadeh A, Eissenberg T. Waterpipe tobacco products: nicotine labelling versus nicotine delivery. Tob Control. 2012;21(3):377–9.

[38] Raad D, Gaddam S, Schunemann HJ, Irani J, Abou Jaoude P, Honeine R et al. Effects of water-pipe smoking on lung function: a systematic review and meta-analysis. Chest. 2011;139(4):764–74.

[39] Hammal F, Chappell A, Wild TC, Kindzierski W, Shihadeh A, Vanderhoek A et al. "Herbal" but potentially hazardous: an analysis of the constituents and smoke emissions of tobaccofree waterpipe products and the air quality in the cafes where they are served. Tob Control. 2015;24(3):290–7.

[40] Shihadeh A, Salman R, Jaroudi E, Saliba N, Sepetdjian E, Blank MD, Cobb CO et al. Does switching to a tobacco-free waterpipe product reduce toxicant intake? A crossover study comparing CO, NO, PAH, volatile aldehydes, "tar" and nicotine yields. Food Chem Toxicol. 2012;50(5):1494–8.

[41] Bentur L, Hellou E, Goldbart A, Pillar G, Monovich E, Salameh M et al. Laboratory and clinical acute effects of active and passive indoor group water-pipe (narghile) smoking. Chest. 2014;145(4):803–9.

[42] Kassem NOF, Kassem NO, Liles S, Zarth AT, Jackson SR, Daffa RM et al. Acrolein exposure in hookah smokers and non-smokers exposed to hookah tobacco secondhand smoke: implications for regulating hookah tobacco products. Nicotine Tob Res. 2018;20(4):492–501.

[43] Azar RR, Frangieh AH, Mroué J, Bassila L, Kasty M, Hage G et al. Acute effects of waterpipe smoking on blood pressure and heart rate: a real-life trial. Inhal Toxicol. 2016;28(8):339–42.

[44] Kassem NOF, Kassem NO, Liles S, Jackson SR, Chatfield DA, Jacob P 3rd et al. Urinary NNAL

in hookah smokers and non-smokers after attending a hookah social event in a hookah lounge or a private home. Regul Toxicol Pharmacol. 2017;89:74–82.

[45] Cobb CO, Vansickel AR, Blank MD, Jentink K, Travers MJ, Eissenberg T. Indoor air quality in Virginia waterpipe cafes. Tob Control. 2013;22(5):338–43.

[46] Gurung G, Bradley J, Delgado-Saborit JM. Effects of shisha smoking on carbon monoxide and PM2.5 concentrations in the indoor and outdoor microenvironment of shisha premises. Sci Total Environ. 2016;548–9:340–6.

[47] Akl EA, Gaddam S, Gunukula SK, Honeine R, Jaoude PA, Irani J. The effects of waterpipe tobacco smoking on health outcomes: a systematic review. Int J Epidemiol. 2010;39(3):834–57.

[48] Waziry R, Jawad M, Ballout RA, Al Akel M, Akl EA. The effects of waterpipe tobacco smoking on health outcomes: an updated systematic review and meta-analysis. Int J Epidemiol. 2017;46(1):32–43.

[49] Awan KH, Siddiqi K, Patil S, Hussain QA. Assessing the effect of waterpipe smoking on cancer outcome – a systematic review of current evidence. Asian Pac J Cancer Prev. 2017;18(2):495–502.

[50] Mamtani, R, Cheema S, Sheikh J, Al Mulla A, Lowenfels A, Maisonneuve P. Cancer risk in waterpipe smokers: a meta-analysis. Int J Public Health. 2017;62(1):73–83.

[51] Nasrallah MP, Nakhoul NF, Nasreddine L, Mouneimne Y, Abiad MG, Ismaeel H et al. Prevalence of diabetes in greater Beirut area; worsening over time. Endocr Pract. 2017;23(9):1091–1100.

[52] Lotfi MH, Keyghobadi N, Javahernia N, Bazm S, Tafti AD, Tezerjani HD. The role of adverse life style factors in the cause of colorectal carcinoma in the residents of Yazd, Iran. Zahedan J Res Med Sci. 2016;18(9):e8171.

[53] Larsen K, Faulkner GEJ, Boak A, Hamilton HA, Mann RE, Irving HM, To T. Looking beyond cigarettes: Are Ontario adolescents with asthma less likely to smoke e-cigarettes, marijuana, waterpipes or tobacco cigarettes? Respir Med. 2016;120:10–15.

[54] Lai HT, Koriyama C, Tokudome S, Tran HH, Tran LT, Nandakumar A et al. Waterpipe tobacco smoking and gastric cancer risk among Vietnamese men. PLoS One. 2016;11(11):e0165587.

[55] Fedele DA, Barnett TE, Dekevich D, Gibson-Young LM, Martinasek M, Jagger MA. Prevalence of and beliefs about electronic cigarettes and hookah among high school students with asthma. Ann Epidemiol. 2016;26(12):865–9. Waterpipe tobacco smoking: prevalence, health effects and interventions to reduce use

[56] Etemadi A, Khademi H, Kamangar F, Freedman ND, Abnet CC, Brennan P et al. Hazards of cigarettes, smokeless tobacco and waterpipe in a Middle Eastern population: a cohort study of 50000 individuals from Iran. Tob Control. 2017;26(6):674–82.

[57] Khodamoradi Z, Gandomkar A, Poustchi H, Salehi S, Imanieh MH, Etemadi A et al. Prevalence and correlates of gastroesophageal reflux disease in southern Iran: Pars Cohort Study. Middle East J Dig Dis. 2017;9(3):129–38.

[58] Bandiera FC, Loukas A, Wilkinson AV, Perry CL. Associations between tobacco and nicotine product use and depressive symptoms among college students in Texas. Addict Behav. 2016;63:19–22.

[59] Abdollahpour I, Nedjat S, Sahraian MA, Mansournia MA, Otahal P, van der Mei I. Waterpipe smoking associated with multiple sclerosis: a population-based incident case–control study.

Mult Scler. 2017;23(10):1328–35.

[60] Tamim H, Akkary G, El-Zein A, El-Roueiheb Z, El-Chemaly S. Exposure of pre-school children to passive cigarette and narghile smoke in Beirut. Eur J Public Health. 2006;16(5):509–12.

[61] Mohammad Y, Shaaban R, Hassan M, Yassine F, Mohammad S, Tessier JF et al. Respiratory effects in children from passive smoking of cigarettes and narghile: ISAAC phase three in Syria. Int J Tuberc Lung Dis. 2014;18(11):1279–84.

[62] Wu F, Chen Y, Parvez F, Segers S, Argos M, Islam T et al. A prospective study of tobacco smoking and mortality in Bangladesh. PLoS One. 2013;8(3):e58516.

[63] Akl EA, Ward KD, Bteddini D, Khaliel R, Alexander AC, Lotfi T et al. The allure of the waterpipe: a narrative review of factors affecting the epidemic rise in waterpipe smoking among young persons globally. Tob Control. 2015;24(Suppl 1):i13–21.

[64] Akl EA, Jawad M, Lam WY, Co CN, Obeid R, Irani J. Motives, beliefs and attitudes towards waterpipe tobacco smoking: a systematic review. Harm Reduction J. 2013;10(1):12.

[65] Griffiths MA, Ford EW. Hookah smoking: behaviors and beliefs among young consumers in the United States. Soc Work Public Health. 2014;29(1):17–26.

[66] MPOWER. Advancing the WHO Framework Convention on Tobacco Control (WHO FCTC). Geneva: World Health Organization; 2018 (http://www.who.int/cancer/prevention/tobacco_implementation/mpower/en/, accessed 1 November 2018).

[67] International Tobacco Control Policy Evaluation Project. Waterloo, Ontario; 2017 (https://bit.ly/1pe5ELk, accessed 3 August 2018).

[68] El Hajj DG, Cook PF, Magilvy K, Galbraith ME, Gilbert L, Corwin M. Tobacco use among Arab immigrants living in Colorado: prevalence and cultural predictors. J Transcult Nurs. 2017;28(2):179–86.

[69] Roskin J, Aveyard P. Canadian and English students' beliefs about waterpipe smoking: a qualitative study. BMC Public Health. 2009;9:10.

[70] Katurji M, Daher N, Sheheitli H, Saleh R, Shihadeh A. Direct measurement of toxicants inhaled by water pipe users in the natural environment using a real-time in situ sampling technique. Inhal Toxicol. 2010;22(13):1101–9.

[71] Alzoubi KH, Khabour OF, Azab M, Shqair DM, Shihadeh A, Primack B et al. CO exposure and puff topography are associated with Lebanese waterpipe dependence scale score. Nicotine Tob Res. 2013;15(10):1782–6.

[72] Bejjani N, l Bcheraoui C, Adib SM. The social context of tobacco products use among adolescents in Lebanon (MedSPAD-Lebanon). J Epidemiol Global Health. 2012;2(1):15–22.

[73] Alzyoud S, Weglicki LS, Kheirallah KA, Haddad L, Alhawamdeh KA. Waterpipe smoking among middle and high school Jordanian students: patterns and predictors. Int J Environ Res Public Health. 2013;10(12):7068–82.

[74] Carroll MV, Chang J, Sidani JE, Barnett TE, Soule E, Balbach E et al. Reigniting tobacco ritual: waterpipe tobacco smoking establishment culture in the United States. Nicotine Tob Res. 2014;16(12):1549–58.

[75] Joudrey PJ, Jasie KA, Pykalo L, Singer ST, Woodin MB, Sherman S. The operation, products and promotion of waterpipe businesses in New York City, Abu Dhabi and Dubai. East Mediterr Health J. 2016;22(4):237–43.

[76] Amin TT, Amr MAM, Zaza BO, Suleman W. Harm perception, attitudes and predictors of waterpipe (shisha) smoking among secondary school adolescents in Al-Hassa, Saudi Arabia. Asian Pac J Cancer Prev. 2010;11(2):293–301.

[77] Jawad M, Nakkash RT, Hawkins B, Akl EA. Waterpipe industry products and marketing strategies: analysis of an industry trade exhibition. Tob Control. 2015;24(e4):e275–9.

[78] Kowitt SD, Meernik C, Baker HM, Osman A, Huang LL, Goldstein AO. Perceptions and experiences with flavored non-menthol tobacco products: a systematic review of qualitative studies. Int J Environ Res Public Health. 2017;14(4):338.

[79] Ambrose BK, Day HA, Rostron B, Conway KP, Borek N, Hyland A et al. Flavored tobacco product use among US youth aged 12–17 years, 2013–2014. JAMA. 2015;314(17):1871–3.

[80] Salloum RG, Maziak W, Hammond D, Nakkash R, Islam F, Cheng X et al. Eliciting preferences for waterpipe tobacco smoking using a discrete choice experiment: implications for product regulation. BMJ Open. 2015;5(9):e009497.

[81] Leavens EL, Driskill LM, Molina N, Eissenberg T, Shihadeh A, Brett EI, Floyd E et al. Comparison of a preferred versus non-preferred waterpipe tobacco flavour: subjective experience, smoking behaviour and toxicant exposure. Tob Control. 2018;27(3):319–24.

[82] Salameh P, Waked M, Aoun Z. Waterpipe smoking: construction and validation of the Lebanon Waterpipe Dependence Scale (LWDS-11). Nicotine Tob Res. 2008.10(1):149–58.

[83] Haider MR, Salloum RG, Islam F, Ortiz KS, Kates FR, Maziak W. Factors associated with smoking frequency among current waterpipe smokers in the United States: findings from the National College Health Assessment II. Drug Alcohol Depend. 2015;153:359–63.

[84] Shihadeh A. Investigation of mainstream smoke aerosol of the nargileh water pipe. Food Chem Toxicol. 2003;41(1):143–52.

[85] Hoffmann D, Hoffmann I. Tobacco smoke components. Beitr Tabakforsch Int. 1998;18(1):49–52.

[86] Jawad M, Jawad S, Waziry RK, Ballout RA, Akl EA. Interventions for waterpipe tobacco smoking prevention and cessation: a systematic review. Sci Rep. 2016;6:25872.

[87] Dogar O, Jawad M, Shah SK, Newell JN, Kanaan M, Khan MA et al. Effect of cessation interventions on hookah smoking: post-hoc analysis of a cluster-randomized controlled trial. Nicotine Tob Res. 2014;16(6):682–8.

[88] Mohlman MK, Boulos DN, El Setouhy M, Radwan G, Makambi K, Jillson I et al. A randomized, controlled community-wide intervention to reduce environmental tobacco smoke exposure. Nicotine Tob Res. 2013;15(8):1372–81.

[89] Asfar T, Al Ali R, Rastam S, Maziak W, Ward KD. Behavioral cessation treatment of waterpipe smoking: the first pilot randomized controlled trial. Addict Behav. 2014;39(6):1066–74.

[90] Lipkus IM, Eissenberg T, Schwartz-Bloom RD, Prokhorov AV, Levy J. Affecting perceptions of harm and addiction among college waterpipe tobacco smokers. Nicotine Tob Res. 2011;13(7):599–610.

[91] Nakkash RT, Al Mulla A, Torossian L, Karhily R, Shuayb L, Mahfoud ZR et al. Challenges to obtaining parental permission for child participation in a school-based waterpipe tobacco smoking prevention intervention in Qatar. BMC Med Ethics. 2014;15:70.

[92] Maziak W, Ben Taleb Z, Jawad M, Afifi R, Nakkash R, Akl EA et al. Consensus statement on

assessment of waterpipe smoking in epidemiological studies. Tob Control. 2017;26(3):338–43.

[93] Jawad M, Lee JT, Millett C. The relationship between waterpipe and cigarette smoking in low and middle income countries: cross-sectional analysis of the global adult tobacco survey. PLoS One. 2014;9(3):e93097. Waterpipe tobacco smoking: prevalence, health effects and interventions to reduce use

[94] Maziak W, Jawad M, Jawad S, Ward KD, Eissenberg T, Asfar T. Interventions for waterpipe smoking cessation. Cochrane Database Syst Rev. 2015;7:CD005549.

[95] O'Neill N, Dogar O, Jawad M, Kellar I, Kanaan M, Siddiqi K. Which behaviour change techniques may help waterpipe smokers to quit? An expert consensus using a modified delphi technique. Nicotine Tob Res. 2018;20(2):154–60.

[96] Decision FCTC/COP7(4). Control and prevention of waterpipe tobacco products. Geneva: World Health Organization; 2016 (https://bit.ly/2vbWd9s, accessed 3 August 2018).

[97] Heydari, G, EbnAhmady A, Lando HA, Chamyani F, Masjedi M, Shadmehr MB. Third study on WHO MPOWER tobacco control scores in Eastern Mediterranean countries 2011–2015. East Mediterr Health J. 2017;23(9):598–603.

[98] Jawad M. Legislation enforcement of the waterpipe tobacco industry: a qualitative analysis of the London experience. Nicotine Tob Res. 2014;16(7):1000–8.

[99] Mukherjea A, Modayil MV, Tong EK. Paan (pan) and paan (pan) masala should be considered tobacco products. Tob Control. 2015;24(e4):e280–4.

[100] Cawkwell PB, Lee L, Weitzman M, Sherman SE. Tracking hookah bars in New York: utilizing Yelp as a powerful public health tool. JMIR Public Health Surveill. 2015;1(2):e19.

[101] Protect people from exposure to second-hand tobacco smoke. Geneva: World Health Organization, Tobacco Free Initiative; 2018 (https://bit.ly/2vbkUCU, accessed 3 August 2018).

[102] Jawad M, El Kadi L, Mugharbil S, Nakkash R. Waterpipe tobacco smoking legislation and policy enactment: a global analysis. Tob Control. 2015;24:i60–5.

[103] Jawad M, Darzi A, Lotfi T, Nakkash R, Hawkins B, Akl EA. Waterpipe product packaging and labelling at the 3rd international Hookah Fair; does it comply with Article 11 of the Framework Convention on Tobacco Control? J Public Health Policy. 2017;19:19.

[104] Jawad M, Choaie E, Brose L, Dogar O, Grant A, Jenkinson E et al. Waterpipe tobacco use in the United Kingdom: a cross-sectional study among university students and stop smoking practitioners. PLoS One. 2016;11 (1):e0146799.

[105] Primack BA, Walsh M, Bryce C, Eissenberg T. US hookah tobacco smoking establishments advertised on the internet. Am J Prev Med. 2012;42(2):150–6.

[106] Griffiths MA, Harmon TR, Gilly MC. Hubble bubble trouble: the need for education about and regulation of hookah smoking. J Public Policy Market. 2011;30(1):119–32.

[107] Jawad M, Lee JT, Glantz S, Millett C. Price elasticity of demand of non-cigarette tobacco products: a systematic review and meta-analysis. Tob Control. 2018; doi: 10.1136/tobaccocontrol-2017-054056.

[108] Jawad M, Millett C. Impact of EU flavoured tobacco ban on waterpipe smoking. BMJ. f2014;348:2698.

11. 总体建议

世界卫生组织烟草制品管制研究小组发布报告，为烟草制品管制提供科学依据。根据世界卫生组织《烟草控制框架公约》第9条和第10条，这些报告确定了基于证据的烟草制品监管方法。

在第九次会议上，研究组讨论了：水烟吸烟的流行和健康影响以及减少使用的干预措施；降低包括卷烟和无烟烟草在内的烟草制品中有害物质浓度的方法；全球烟碱降低策略的科学现状；ENDS中烟碱的临床药理学；烟草制品中的糖含量；减少卷烟烟气中有害物质暴露的监管策略；用于监管目的的烟草制品中优先毒性物质的最新清单；加热型烟草制品；烟草制品中的调味剂科学。讨论的目的是更新这些领域的知识，以便为全球政策提供信息并促进烟草制品监管。

本报告通过执行委员会向成员国提供指导。正如FCTC/COP7(4)、FCTC/COP7(9)和FCTC/COP7(14)号决定所阐明的那样，本报告主要侧重于缔约方会议通过公约秘书处在2016年第七届会议上向世界卫生组织提出的对《烟草控制框架公约》的要求。这些决定提供了上述领域背景文件的内容，成员国已要求为这些领域提供技术援助，以此作为国家政策的基础。由10人组成的研究小组邀请了主题专家，他们起草了背景文件，为讨论做出了贡献，并提供了有关所审议主题的最新经验数据。报告第2~10节提供了科学信息和政策建议，以指导成员国解决烟草制品监管中的难题。本报告进一步就弥合烟草控制方面的监管差距和制定协调一致的烟草制品监管框架以指导国际政策的最有效的循证手段，向成员国提供了指导。此外，确定了进一步工作和未来研究的领域，重点放在成员国的监管需求上；考虑到了地区差异，从而为成员国提供了持续、有针对性的技术支持的战略。

11.1 主要建议

本报告对决策者和所有其他有关方面的主要建议如下：

- 监测和收集关于加热型烟草制品和替代产品可靠的、独立的数据，以便了解行为和对使用者和二手烟暴露者的潜在风险，并验证减少暴露和降低风险的声明。
- 考虑并检查确定电子烟碱传输系统中烟碱通量的设计特征以及这些产品

可促进或阻止吸烟的程度，并投资研究电子烟碱传输系统的适当政策和法规。
- 考虑减少烟草制品烟气中的有害物质暴露的管制策略，该策略应使烟草中的烟碱含量不超过0.4 mg/g烟草（在HCI抽吸条件下，主流烟气中每个燃烧型产品中的烟碱为0.04 mg）。应当伴随一个可靠的系统，以监测烟草和烟气中的受管制成分、全面的烟草控制以及国家和国际为防止黑市所作的共同努力。
- 如果相关当局批准并经相关部门批准，并考虑采取适当的保障措施，则应考虑与烟碱降低的政策相协调，该政策应允许充分获得烟碱替代疗法和其他产品。这应通过人群监督、产品监控和测试、产品标准的执行以及对保护儿童和年轻人的高度重视来支持。
- 考虑禁止或限制在烟碱传输系统和烟草制品中使用调味剂，以减少年轻人的摄入，并考虑禁止或限制燃烧的烟草制品中的调味剂以促进戒烟。
- 要求制造商披露有关糖含量的相关信息，并考虑降低烟草制品中糖的含量，以减少其对产品毒性、致瘾性和吸引力的影响。
- 要求制造商在可应用的和恰当的情况下报告使用基于WHO TobLabNet SOP方法分析的优先毒性物质。
- 降低烟草制品（包括卷烟和无烟烟草）中的致瘾、有毒和致癌物质的含量，并且承认降低这些物质的含量不会使这些产品变得安全。
- 考虑将WHO FCTC规定应用于水烟，以预防和控制水烟的使用。

需要进行持续研究以监测烟草及相关行业的产品开发和使用、促销策略以及其他活动，以建立保护公众健康的情报。有关本报告中考虑的每个主题的具体建议，请参见第2.9、3.8、4.5~4.6、5.8、6.8、7.10、8.8、9.9和10.3~10.5节。

11.2　对公共卫生政策的意义

研究组的报告为了解选定产品（例如无烟烟草、水烟、加热型烟草制品和电子烟）的成分、释放物和设计特征提供了有益的指导，并描述了这些产品和特征对公众健康的影响。近年来，非传统烟碱和烟草制品已经渗透到多个市场，这是没有先例的，这些市场给成员国带来了独一无二的监管挑战。此外，由于知识的进步，人们对传统烟草制品的科学认知、不利影响、特征、成分和释放物有了更深入的了解；因此，本报告为成员国提供了关于新型烟草制品和烟碱传递系统的最新信息，以支持制定管理烟草和烟碱产品的有效战略。

由于研究小组的独特组成，拥有监管、技术和科学专家，因此可以浏览和分发复杂的数据和研究结果，并将其综合为国家、地区和全球各级的政策制定建议。

这些建议努力促进了国际协作监管工作和采用烟草制品监管最佳方法，增强了世界卫生组织所有地区的烟草制品监管能力，并为成员国提供了现成的、基于科学的资源。

11.3　对 WHO 计划的影响

本报告完成了世界卫生组织烟草制品管制研究小组的任务，向总干事就烟草制品管制问题提供科学合理的、基于证据的建议。烟草制品监管是烟草控制的一个高科技领域，成员国在其中面临着复杂的监管挑战。研究组的审议结果及其主要建议将提高成员国对烟草和烟碱产品的了解。本报告对烟草制品管制知识体系的贡献将在向世界卫生组织非传染性疾病预防司的烟草规划工作提供信息方面，特别是向成员国提供技术支持方面发挥关键作用。

1. Introduction

Effective tobacco product regulation is an essential component of a comprehensive tobacco control programme. It includes regulation of contents and emissions by mandated testing, disclosure of test results, setting limits, as appropriate, and imposing standards for product packaging and labelling. Tobacco product regulation is covered under Articles 9, 10 and 11 of the WHO Framework Convention on Tobacco Control (WHO FCTC) *(1)* and in the partial guidelines on implementation of Articles 9 and 10 *(2)*. Other WHO resources, including the basic handbook on tobacco product regulation *(3)* and the handbook on building laboratory testing capacity *(4)*, support Member States in this area.

The WHO Study Group on Tobacco Product Regulation (TobReg) was formally constituted by the Director-General of WHO in 2003 to address gaps in the regulation of tobacco products. Its mandate is to provide evidence-based policy recommendations on tobacco product regulation to the Director-General. TobReg is composed of national and international scientific experts on product regulation, treatment of tobacco dependence, toxicology and laboratory analyses of tobacco product ingredients and emissions. The experts come from countries in all six WHO regions. As a formalized entity of WHO, TobReg submits technical reports to the WHO Executive Board through the Director-General to draw the attention of Member States to the Organization's work in tobacco product regulation. The technical reports are based on unpublished background papers that have been discussed, evaluated and reviewed by TobReg.

The ninth meeting of TobReg took place in Minneapolis, United States of America, on 5–7 December 2017, and was generously hosted by the Masonic Cancer Center, University of Minnesota, USA. The participants discussed priorities in the regulation of nicotine and novel and tobacco products and addressed requests from the WHO FCTC Conference of the Parties (COP) made at its seventh session, as outlined in documents FCTC/COP7(4), FCTC/COP7(9) and FCTC/COP7(14). The requests included the following.

- Continue to monitor and examine market developments and usage of novel and emerging tobacco products, such as heated tobacco products;
- Collect scientific information on the chemicals in the contents and emissions of smokeless tobacco products that contribute to their toxicity, addictiveness and attractiveness, on analytical methods for measuring them and on the levels in products on the market; and identify technical approaches for reducing toxicants in smokeless tobacco;
- Promote research on culturally relevant interventions to prevent the uptake of waterpipe tobacco smoking and to promote quitting (cessation); the epidemiology of use; acute and chronic health risks; cultural

practices; initiation and maintenance of use; the influence of flavourings on initiation, maintenance of use and increasing use; risk of dependence on low-nicotine tobacco products; and effective policies based on concepts such as information technology and communications.

In response to these requests, WHO commissioned the following background papers:

- Heated tobacco products (section 2);
- Clinical pharmacology of nicotine in electronic nicotine delivery systems (section 3);
- A global nicotine reduction strategy: state of the science (section 4);
- A regulatory strategy for reducing exposure to toxicants in cigarette smoke (section 5);
- The science of flavour in tobacco products (section 6);
- Sugar content of tobacco products (section 7);
- Updated priority list of toxicants in combusted tobacco products (section 8);
- Approaches to measuring and reducing toxicant concentrations in smokeless tobacco products (section 9);
- Waterpipe tobacco smoking: prevalence, health effects and interventions to reduce use (section 10).

1.1 References

1. WHO Framework Convention on Tobacco Control. Geneva: World Health Organization; 2003 (http://www.who.int/fctc/en/, accessed 14 May 2019).
2. Partial guidelines on implementation of Articles 9 and 10. Geneva: World Health Organization; 2012 (https://www.who.int/fctc/guidelines/Guideliness_Articles_9_10_rev_240613.pdf, accessed 14 January 2019).
3. Tobacco product regulation: basic handbook. Geneva: World Health Organization; 2018 (https://www.who.int/tobacco/publications/prod_regulation/basic-handbook/en/, accessed 14 January 2019).
4. Tobacco product regulation: building laboratory testing capacity. Geneva: World Health Organization; 2018 (https://www.who.int/tobacco/publications/prod_regulation/building-laboratory-testing-capacity/en/, accessed 14 January 2019).

2. Heated tobacco products

Dr Richard J. O'Connor, Professor of Oncology, Department of Health Behavior, Roswell Park Comprehensive Cancer Center, Buffalo, New York, USA

"Nowadays, tobacco companies continue reassuring health concerned smokers by offering with their new products the illusion of safety." (World No Tobacco Day 2006)

Contents
- 2.1 Introduction
- 2.2 The science of heated tobacco products
- 2.3 A brief history of heated tobacco products
- 2.4 Recent products
    - 2.4.1 Emissions
    - 2.4.2 Biomarkers of exposure
- 2.5 Consumer perceptions of heated tobacco products
- 2.6 Uptake in selected markets where products are available
- 2.7 Application by Philip Morris International for status as a "modified risk tobacco product" in the USA
- 2.8 Implications for regulation and tobacco control policies
- 2.9 Recommendations for research and policy
- 2.10 References

2.1 Introduction

Heated tobacco products (HTPs), also sometimes referred to as "heat-not-burn" products, a term coined by the tobacco industry, are an emerging class of "potentially reduced exposure products" (PREPs) or "modified risk tobacco products" (MRTPs). The concept emerged in the 1980s from the tobacco companies Philip Morris and RJ Reynolds in the USA, which marketed Accord and Premier, the first generation of these products, respectively. Since then, these and conceptually similar products have continued to evolve and may now be poised to capture a significant market share. The introduction, aggressive marketing and growing popularity of electronic cigarettes may have facilitated the success of such products, partly by changing social norms and perceptions about conventional cigarette smoking and about the use of devices to deliver nicotine. There is substantial published literature on certain HTPs marketed in the 1990s and 2000s and emerging literature on newer products, although much of it comes from the tobacco industry. Further, there are few studies on prevalence and substitution in the marketplace, as many of these products were test-marketed rather than made broadly available. Still, laboratory and field research studies can provide information on the likelihood of such substitution.

This review is based on the literature on HTPs available through October 2017, including their history, design, delivery of nicotine to the user and toxicants and marketing (including online advertising and sales); HTP technology; manufacturer's claims of reduced toxicity, harm, risk and exposure; comparison with conventional cigarettes; consumer perceptions of these alternative products; and the implications of these products for regulatory, product and market policy. The review focuses on HTPs marketed by tobacco companies but also covers handheld or portable products for use as "dry herb" vaporizers (often for marijuana) that could be used with tobacco. Desktop "vaporizers" usually used to administer marijuana were excluded, as were hookah-, waterpipe- or narghile-type products. The review is based primarily on published literature and also on news reports and press releases, stockholder reports, scientific presentations and internet blogs, as necessary.

The PubMed search terms were: heat not burn; heat-not-burn; heated tobacco; tobacco heating; Accord; Eclipse; Heatbar; Premier; THS [tobacco heating system]; vaporizer; PREP; MRTP; THP [tobacco heating product]; iQOS; and glo.

2.2 The science of heated tobacco products

HTPs are based on the principle that most of the harm associated with tobacco smoking is due to the combustion process. In a conventional cigarette, the temperature of the burning cone can reach up to 900 °C, and the median temperature along the rod is 600 °C. This can result in combustion, pyrolysis, pyrosynthesis and myriad other reactions that result in the > 7000 compounds identified as components of tobacco smoke (1). Polycyclic aromatic hydrocarbons (PAHs), heterocyclic aromatic amines and some volatile organic compounds (VOCs) (e.g. benzene, 1,3-butadiene, acrolein, toluene) are formed primarily as a result of combustion. Tobacco-specific N'-nitrosamines (TSNAs) such as N'-nitrosonornicotine (NNN) and 4-(methylnitrosamino)-1-(3-pyridyl)-1-butanone (NNK) are present in cured tobacco and are partially transferred to smoke in a near-linear fashion at typical cigarette temperatures. Some TSNAs are formed during tobacco combustion. Toxic metals such as cadmium present in tobacco may also be transferred to smoke at typical tobacco combustion temperatures. Burning of tobacco is, however, ultimately unnecessary to "volatilize" nicotine (although it is efficient), and alternative means of liberating nicotine from tobacco in an inhalable form without combustion are preferable from the point of view of both toxicological risk and consumer acceptability. One way of extracting nicotine while maintaining something visually similar to smoking behaviour is to heat tobacco to a temperature that volatilizes nicotine but does not combust the plant material. Volatilized nicotine without combustion would, in principle, produce a less complex aerosol with fewer toxic constituents.

Nicotine is not efficiently delivered as a gas; to deliver nicotine to the user's lungs, an aerosol-forming agent must be included to suspend nicotine on aerosol particles. This has been achieved by four main approaches through the decades. The first is in a cigarette-like device with an embedded heat source that can be used to aerosolize nicotine – this is the general principle underlying Premier and Eclipse and Philip Morris International (PMI)'s "Platform 2" product. The second approach is to use an external heat source to aerosolize nicotine from specially designed cigarettes (rolls of tobacco in paper). This is the basic design of Accord, Heatbar, iQOS and glo. The tobacco used in PMI's HTP is apparently not typical tobacco cut-filler but rather a reinforced web of cast-leaf tobacco (a type of reconstituted tobacco), which includes 5–30% by weight of compounds that form aerosols, such as polyols, glycol esters and fatty acids. The examples given include glycerine, erythritol, 1,3-butylene glycol, tetraethylene glycol, triethylene glycol, triethyl citrate, propylene carbonate, ethyl laurate, triactin, meso-erythritol, a diacetin mixture, a diethyl suberate, triethyl citrate, benzyl benzoate, benzyl phenyl acetate, ethyl vanillate, tributyrin, lauryl acetate, lauric acid, myristic acid and propylene glycol *(2)*. This composition is advantageous as an aerosol-forming substrate for use with a heating system. The sticks (45 mm long, 7 mm diameter) contain approximately 320 mg of tobacco material – much less than a conventional cigarette (~700 mg). In iQOS, the tobacco is heated by a blade in the heater device inserted into the end of the heat stick, so that the heat dissipates through the tobacco plug on a puff *(3)*. The aerosol then passes through a hollow acetate tube and a polymer film filter on the way to the mouth. The product is designed not to exceed 350 °C, at which point the energy supplied to the blade is cut off at a maximum of 14 puffs or 6 min *(3)*. British American Tobacco describes its glo product as a heating tube consisting of two separately controlled chambers, which are activated by a button on the device to reach the operating temperature (240 °C) within 30–40 s *(4)*. The 82-mm long, 5-mm diameter stick inserted into the heating chamber contains approximately 260 mg of reconstituted sheet tobacco with 14.5% glycerol as an aerosolizing agent. The vent holes on the stick are described as necessary to "…provide the right amount of drawing effort and to encourage the…vapour to coagulate and condense…." *(4)*. The stick consists of a tobacco rod, a tubular cooling section and a filter and mouthpiece.

A third approach is to use a heated sealed chamber to aerosolize nicotine from tobacco leaf directly – this is the principle underlying personal dry-herb vaporizers such as Pax; however, the prevalence of use of such devices for tobacco is unknown. A fourth approach is use of electronic nicotine delivery systems (ENDS) to derive flavour elements from small amounts of tobacco (see section 3 for further information). British American Tobacco's iFuse product appears to be a hybrid ENDS–tobacco product, in which the aerosol is passed over tobacco to pick up the flavour and is then inhaled by the user *(5)*. The vapour appears to lose a small amount of heat when it passes over the tobacco chamber (from

35 °C to 32 °C), indicating that some tobacco is heated. As the e-liquid contains nicotine at 1.86 mg/mL, however, with a machine delivery of 20–40 μg per puff, it is difficult to estimate the contribution, if any, of the tobacco in the device to delivery of nicotine. The delivery of toxicants under machine smoking is reported to be nearly identical to that of an ENDS without a tobacco chamber, implying a minimal contribution of tobacco *(5)*. The Japan Tobacco International Ploom TECH operates in a similar manner, except that the ENDS-like component appears not to contain nicotine.

2.3 A brief history of heated tobacco products

A historical perspective is important for understanding the current situation. Table 2.1 lists the HTPs known to have been introduced into at least a test market. The timeline in Fig. 2.1 places the introduction (and withdrawal) of various HTPs into context. Activity in this market space has been clustered, 2006–2008 and 2015–2017 being particularly active periods.

Table 2.1. Heated tobacco products by manufacturer

Company	Trade name*	Brief description	Current status
Philip Morris USA / Altria	Accord	THS consisting of small cigarettes placed in an external heating device. Available in tobacco and menthol flavours. Introduced in 1998 in test markets in Richmond, Virginia, USA, and Japan. Advertising focused on reduced second-hand smoke rather than reduced potential health risks.	Withdrawn in 2006. No longer sold.
Philip Morris International	Heatbar	THS, consists of small cigarettes (Heatsticks) placed in an external heating device with puff detection technology. Four blends available. Devices came in a range of colours. Introduced in Switzerland in 2006.	Withdrawn in 2008. No longer sold.
	iQOS	THS, consists of small cigarettes (HEETS) placed in an external heating device. Available in Marlboro brand. Launched in Japan in 2015.	Currently marketed in nearly 40 countries including Canada, Italy, Japan and the United Kingdom. Expanding to other markets. MRTP application to FDA filed in May 2017 and reviewed by the Tobacco Products Scientific Advisory Committee in January 2018.
	THS 2.2	THS described in a series of published papers in scientific journals.	Marketed as iQOS.
	Platform 2	Pressed carbon heat source (conceptually similar to Eclipse).	Unclear whether available on the market. Branded as TEEPS in several PMI presentations.

2. Heated tobacco products

Japan Tobacco International	Ploom	Ploom began as independent company in 2007. Entered into marketing and commercialization agreement with Japan Tobacco International in 2011, which purchased patents and designs for device and pods (Ploom name and ModelTwo device) outright in 2015. Pods currently have the Mevius cigarette brand name. Former Ploom company became Pax, which produces dry herb vaporizers.	Currently marketed in Japan and Switzerland.
RJ Reynolds / Reynolds American	Premier	Cigarette-like device designed for heating and aerosolizing tobacco flavour. Cigarette column made of aluminium capsules containing tobacco pellets, with a carbon element at the tip to warm the tobacco column. Test-marketed in St Louis, Phoenix and Tucson, USA, in 1988. Subject of complaints to the FDA as a drug delivery device.	Withdrawn in 1989 because of poor sales. Not currently on the market.
	Eclipse	Cigarette-like device. Rod consists of reconstituted tobacco, with carbon element to heat and volatilize nicotine. Test-marketed in several versions between 1996 and 2000, when wider sales began, with accompanying advertisements and health claims. Subject of lawsuit over claims by the State of Vermont, USA.	Withdrawn from United States market in 2007; remained in limited distribution thereafter. Announced "substantial equivalence" filing with FDA in October 2017 to bring improved version to United States market.
	Revo	Eclipse-like product test-marketed in Wisconsin, USA, in 2014–2015.	Withdrawn in 2015.
British American Tobacco	glo	"glo™ heats proprietary Kent Neostiks™ to approximately 240 °C to provide a highly satisfying taste, similar to that of a cigarette, with around 90% less toxicants. glo™ also offers a number of added features, including: no burning or ash, less odour on your hands, hair, clothes and surroundings." Neostiks come in three flavours.	Available in Canada, Japan, Republic of Korea, Russian Federation and Switzerland.
	iFuse	Hybrid of e-cigarette and HTP. Cartridges contain e-liquid with flavourings and a chamber containing tobacco. Heating element aerosolizes the liquid, which passes through the tobacco chamber before reaching the user. Launched in 2015.	Test-marketed in Romania.
Dry herb vaporizers			
Pax	Pax 2	Rectangular device. "Magnetic oven" used to heat plant material, "thin film Kapton heater flex". Cannot be used with liquids, waxes or concentrates. Capacity, 1.6 mL. Comes in four colours.	Available online and from authorized retailers.
	Pax 3	Rectangular device. "Magnetic oven" used to heat plant material. Can also be used with concentrates with special insert. Oven starts when device senses lips on mouthpiece. 3500-mAh battery. Capacity, 0.17–0.35 g. Comes in four colours.	Available online and from authorized retailers.
V2	Pro Series 3	Cylindrical device. Three-in-one vaporizer, with different cartridges and mouthpieces for e-liquid, loose leaf or wax. Conduction oven. Pre-set temperature, 160–180 °C. 650-mAh battery. Comes in three colours.	Available online.

	Pro Series 7	Rectangular device. Three-in-one vaporizer, with different cartridges and mouthpieces for e-liquid, loose leaf or wax. Larger capacity, more customization options. Higher-temperature conduction oven with three levels (200, 215, 225 ºC). 1800-mAh battery; 3.7–4.7 V. Comes in three colours.	Available online.
Vapor Fi	Orbit	Cylindrical device. 1.7-mL capacity cartridge for dry plant material. Temperature range, 182–216 ºC. 2200-mAh battery. Two colours (black, red).	Available online.
	Atom	Rectangular device. Temperature range, 182, 210, 240 ºC. Plant material packed into an external oven cartridge. 3000-mAh battery; 3.3–4.2 V. Motion sensing. Superficially similar to Pax device.	Available online.
Atmos	Jump	Cylindrical device. Maximum temperature, 200 ºC; not variable. 1300-mAh battery; 3.4 V. Available in four colours.	Available online.

FDA: United States Food and Drug Administration; HTP: heated tobacco product; MRTP: modified risk tobacco product; PMI: Philip Morris International; THS: tobacco heating system.

Fig. 2.1. Timeline of introduction of heated tobacco products, 1988–2016

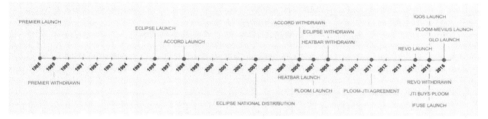

Eclipse

Eclipse, the successor of Premier, launched at the end of the 1980s, was available in various test markets in the USA between 1996 and 2007 and was extensively studied *(6–13)*. The product had a carbon-heating element embedded in the tip of a product resembling a cigarette to heat reconstituted tobacco and glycerol in the rod and generate a nicotine aerosol. To use it, the smoker would light the heating element and puff, much like a conventional cigarette. In two separate five-day trials, Breland and colleagues *(6, 7)* found that Eclipse reduced exposure to nicotine and NNK from that in the smoker's own brand but increased exposure to carbon monoxide (CO) and required a more intensive puffing pattern. Lee et al. *(10)* found a similar pattern of results for Eclipse in the laboratory. Fagerstrom and colleagues *(9)* showed that smokers' CO levels over two weeks of using Eclipse increased from 21.0 to 33.0 parts per million. In a subsequent eight-week study *(8)*, in which 10 participants were selected to continue using Eclipse, no difference in plasma nicotine levels was seen from baseline, but the exhaled CO levels were higher with Eclipse than with the smokers' own brand (32.5 versus 22.5 parts per million). Stewart and colleagues *(13)* reported that a small group of smokers who used Eclipse for 4 weeks showed less alveolar epithelial injury but increased levels of carboxyhaemoglobin and other markers of oxidative stress. Pauly and colleagues *(14)* identified a potentially unique health risk of Eclipse due to contamination with glass fibre. They noted that the insulation surrounding the carbon-heating element could fray and become loose during handling of the packaging or the product itself and expose smokers to inhalation of these fibres. Previous research had shown evidence of inhalation of cellulosic and plastic fibres in 87% of 114 human lung specimens *(15)*, indicating that this was not simply a theoretical concern. A subsequent survey of consumers *(16)* showed that they viewed this as a health risk. RJ Reynolds vigorously objected to these observations and published several studies to refute them, arguing that the fibres were too large to be inhaled and that the number of loose fibres was not as large as reported *(17–20)*.

The first health claims were made for Eclipse when it was launched nationally in 2003, citing potentially reduced risks for cancer, bronchitis and emphysema *(21)*. The manufacturer's advertising claimed that Eclipse reduced emissions of harmful smoke constituents and suggested that, next to quitting, Eclipse was the smoker's "next best option". These claims resulted in a lawsuit by the Attorney-General of the State of Vermont, USA, in 2005, which resulted in the finding in 2010 that the advertising had violated consumer protection laws and the Master Settlement Agreement. This was followed by a judgement against RJ Reynolds for US$ 8 million in 2013. A repackaged version of Eclipse, called Revo, was briefly test-marketed in Wisconsin, USA, in 2015. In November 2017, RJ Reynolds announced that they would market an improved version of Eclipse in 2018; this had not been launched yet as of January 2019.

Accord

Accord was first marketed in 1998 in the USA and later in Japan. This product, a predecessor of iQOS, had an external battery-powered heating device, into which specifically designed "cigarettes" were inserted for smoking. Puffing on the cigarette activated the heating bars in the device and generated an aerosol. Accord was marketed as a "cleaner" cigarette, without ash or second-hand smoke; no health claims were made. Review of an early study indicated that Accord use resulted in less CO intake and tachycardia than conventional cigarettes *(22, 23)*. Accord was, however, associated with poor suppression of withdrawal *(22, 23)*. In a study of concurrent use of Accord and the smokers' own cigarettes, Accord suppressed cigarette smoking and exposure to CO dose-dependently after six weeks; i.e. the more often Accord was used, the less participants smoked their own cigarettes. Furthermore, the participants did not increase their puff intensity when they reduced the number of cigarettes per day, and smoke from Accord contained lower levels of CO *(24)*. Analysis of participant exit interviews showed that study participants believed Accord to be a "safer cigarette". A study by Roethig et al. *(25)* of a second-generation electrically heated cigarette showed reductions in selected biomarkers of exposure (e.g. nicotine, 1-hydroxypyrene) of 43–85% relative to conventional cigarette smoking.

2.4 Recent products

2.4.1 Emissions

Table 2.2 summarizes the published literature on emissions of certain harmful and potentially harmful constituents (HPHCs) *(26)* under machine smoking conditions from the three HTPs for which there are the most published data – Eclipse (International Organization for Standardization (ISO) conditions), electrically heated cigarette smoking system (EHCSS; ISO conditions) and THS 2.2/iQOS (Health Canada Intense (HCI) conditions) – cognizant that the data were published by the manufacturers. PMI experiments showed that THS 2.2 aerosol did not contain solid carbon-derived particles, consistent with its claim of no combustion *(27)*. In general, the levels of constituents were lower than in a comparison reference cigarette (1R6F, which has certified levels of many of the emissions of concern *(28)*). Studies conducted within *(29)* and outside the industry *(30, 31)* largely replicated these findings, in particular the lower levels of nicotine in Eclipse and EHCSS than in a reference cigarette. British American Tobacco presented two posters on its HTP at the 2017 meeting of the Society for Research on Nicotine and Tobacco *(5, 32)* and in late 2017 had eight studies published in a supplement to a journal *(4, 33–40)*. The product, called THP1.0, is commercially marketed as glo. It heats tobacco (reconstituted sheet with 14.5% glycerol contained in a superslim cigarette) to a maximum of 250 °C *(32)*. Thermographic profiling suggested that the product releases moisture until 100 °C and then glycerol through 240 °C, followed by decomposition at 350 °C *(32)*. Data from machine smoking under modified HCI conditions generally showed lower emissions of TobReg priority toxicants *(41)* than in a reference cigarette (Table 2.2). THP1.0 was machine-smoked under HCI conditions except for the blocking of vent holes on the basis of a study that showed no vent blocking under actual smoking conditions (according to lip imprint measurements). A second poster *(42)* gave details of effects on indoor air and claimed that THP1.0 resulted in lower emissions than conventional cigarettes. In a laboratory comparison of British American Tobacco's iFuse with other HTPs (called cTHP in the paper; description fits iQOS) *(43)*, cTHP emitted higher levels of acetaldehyde (125 versus 35.9 μg per 10 puffs), while iFuse emitted more formaldehyde (< 4.18 versus 38.7 μg per 10 puffs). Both products emitted substantially less acetaldehyde than the 3R4F reference cigarette (> −88%), but iFuse emitted levels of formaldehyde comparable to those of a conventional cigarette (−8%). Bekki et al. *(31)* showed that the transfer rate of nicotine and nitrosamines from tobacco filler to aerosol in iQOS was comparable to, if not slightly higher than, that in conventional cigarettes.

Table 2.2. Harmful and potentially harmful constituents emitted from reference cigarette 1R6F, Eclipse, an electrically heated cigarette smoking system (EHCSS), a tobacco heating system (THS) and a tobacco heating product (THP)

Constituent	1R6F (µg/cigarette) ISO (28)	1R6F (µg/cigarette) HCI (28)	Eclipse (µg/cigarette) ISO (21)	ECHSS (µg/cigarette) ISO (44)	THS 2.2 (µg/cigarette) HCI (45)	THP1.0 (µg/cigarette) HCI (32)
Acetaldehyde	522	1552	84.2	179	219	111
Acrolein	43	154	11.5	27.3	11.3	2.22
Acrylonitrile	7.0	24		0.44	0.258	< 0.032
4-Aminobiphenyl	1.2	2.3		0.058	< 0.051	< 0.005
2-Naphthol	8	14		0.123	0.046	< 0.012
Ammonia	9.0	30		4.37	14.2	4.01
Benzene	33	88		0.363	0.649	< 0.056
Benzo[a]pyrene	6.8	15	1.2	< 0.19	< 1.0	< 0.35
Carbon monoxide (mg)	10.1	28.0	7.5	0.465	0.531	< 0.223
Formaldehyde	27	104		12.9	5.53	3.29
Isoprene	320	881		34.3	2.35	< 0.135
Nicotine	0.721	1.90	0.18	0.313	1.32	0.462
NNK (ng)	71	187	31.8	6.18	6.7	6.61
NNN (ng)	85	212	26	19.8	17.2	24.7
Toluene	53	150		1.48	2.59	< 0.204
Tar (mg)	8.58	29.1	3.2	3.1	10.3	13.6

EHCSS: electrically heated cigarette smoking system; HCI: Health Canada intense smoking regimen; ISO: International Organization for Standardization smoking regimen; NNK: 4-(methylnitrosamino)-1-(3-pyridyl)-1-butanone; NNN: N′-nitrosonornicotine; THP: tobacco heating product; THS: tobacco heating system.

2.4.2 Biomarkers of exposure

In 2008, Philip Morris USA and PMI published a series of papers on its EHCSS, including studies of toxicology, emissions (also second-hand smoke) and short- and long-term clinical exposure, and randomized trials (44, 46–48). A series of papers followed in 2012 (49–56). In 2016, PMI published another series of studies, this time on THS 2.2 (3, 45, 57–59), which reported the results of a broad range of product evaluations. The data on biomarkers after short-term exposure (~ 1 week, generally in residence) shown in Tables 2.3 and 2.4 were extracted from these papers (47, 49, 51–54, 58, 59). In studies in several populations, the levels of key toxicants, except nicotine, were lower than with continued cigarette smoking. As the participants in these studies were generally confined for the duration of the study, however, the findings may not be generalizable to real conditions of use (e.g. pattern of product use, use with cigarettes or other combusted or non-combusted products).

Table 2.3A. Biomarker levels in short-term studies of electrically heated cigarette smoking systems, with end-of-study values for numbers of conventional cigarettes and heated tobacco products and for abstinence when available, Japan and USA

Biomarker	HPHC	Japan (6 days) (53)			USA (8 days) (47)			Japan (8 days) (52)			
		Cigarettes	EHCSS	Abstinence	Cigarettes	EHCSS	Abstinence	Cigarettes	EHCSS-K6	EHCSS-K3	Abstinence
Nicotine equivalent	Nicotine (mg/24 h)	7.2	3.4	0.6	16.1	8.2	0.2	9.6	5.1	3.9	0.2
Nicotine	Nicotine	NR	NR	NR	NR	NR	NR	19.0	10.1	8.1	0.5
Cotinine	Nicotine (ng/mL)	150.5	79.4	11.1	NR	NR	NR	177.9	111.2	79.5	7.4
NNAL	NNK (ng/24 h)	188.0	95.0	80.0	601.3	169.1	117.8	216.0	100.0	102.0	83.0
COHb	Carbon monoxide (%)	5.1	2.1	1.9	6.4	2.1	1.1	5.4	2.1	2.2	1.7
MHBMA	1,3-Butadiene (µg/24 h)	1.5	0.6	0.4	4.8	1.5	0.2	1.5	0.7	0.7	0.4
3-HPMA	Acrolein (mg/24 h)	1.2	0.8	0.5	1859.0	1056.7	302.3	1.4	1.1	1.0	0.5
S-PMA	Benzene (µg/24 h)	2.3	0.4	0.5	5.3	1.1	0.4	2.2	0.6	0.5	0.3
1-OHP	Benzo[a]pyrene (ng/24 h)	106.5	38.4	37.4	149.6	58.1	41.1	135.3	56.0	59.0	42.0
3-OH-Benzo[a]pyrene	Benzo[a]pyrene	NR	NR	NR	NR	NR	NR	NR	NR	NR	NR
4-Aminobiphenyl	4-Aminobiphenyl (ng/24 h)	9.4	4.4	4.7	12.6	3.1	2.0	15.0	4.2	3.9	4.6
1-Naphthylamine	1-Naphthylamine	NR	NR	NR	NR	NR	NR	NR	NR	NR	NR
2-Naphthylamine	2-Naphthylamine (ng/24 h)	4.6	4.4	4.7	20.0	4.2	2.2	27.2	6.9	6.1	5.9
o-Toluidine	Toluene (µg/24 h)	66.2	33.6	33.8	87.6	58.5	40.9	103.9	31.1	27.7	33.8
HMPMA	Croton-aldehyde (mg/24 h)	0.5	0.2	0.1	2320.7	750.4	299.0	1.3	0.6	0.6	0.5

COHb: carboxyhaemoglobin; EHCSS: electrically heated cigarette smoking system; HMPMA: hydroxy methyl propylmercapturic acid; HPHC: harmful and potentially harmful constituent; 3-HPMA: 3-hydroxypropyl mercapturic acid; MHBMA: monohydroxy-butenylmercapturic acid; NNAL: 4-(methylnitrosamino)-1-(3-pyridyl)-1-butanol; NNK: 4-(methylnitrosamino)-1-(3-pyridyl)-1-butanone; NR: not recorded; 1-OHP: 1-hydroxypyrene; S-PMA: S-phenylmercapturic acid. ª K6 and K3 are two versions of the system tested.

Table 2.3B. Biomarker levels in short-term studies of electrically heated cigarette smoking systems, with end-of-study values for numbers of conventional cigarettes and heated tobacco products and for abstinence when available, United Kingdom and Republic of Korea

Biomarker	HPHC	Republic of Korea (8 days) (51)			United Kingdom (8 days) (54)		
		EHCSS-K3[a]	Abstinence	Cigarette	EHCSS-K6[a]	EHCSS-K3[a]	Abstinence
Nicotine equivalent	Nicotine (mg/24 h)	4.2	0.1	13.1	8.6	6.5	0.0
Nicotine	Nicotine	8.6	0.1	NR	NR	NR	NR
Cotinine	Nicotine (ng/mL)	85.8	0.1	234.3	158.3	125.0	0.0
NNAL	NNK (ng/24 h)	80.4	45.1	293.6	100.6	104.3	59.2
COHb	Carbon monoxide (%)	0.9	0.5	5.7	1.4	1.9	0.5
MHBMA	1,3-Butadiene (μg/24 h)	0.6	0.3	5.1	1.5	2.6	0.3
3-HPMA	Acrolein (mg/24 h)	2.4	1.7	1.8	1.3	1.2	0.5
S-PMA	Benzene (μg/24 h)	1.6	1.4	6.2	0.9	1.3	0.2
1-OHP	Benzo[a]pyrene (ng/24 h)	143.6	127.0	181.6	71.9	73.1	75.1
3-OH-Benzo[a]pyrene	Benzo[a]pyrene	NR	NR	NR	NR	NR	NR
4-Aminobiphenyl	4-Aminobiphenyl (ng/24 h)	5.5	5.5	NR	NR	NR	NR
1-Naphthylamine	1-Naphthylamine	NR	NR	NR	NR	NR	NR
2-Naphthylamine	2-Naphthylamine (ng/24 h)	5.4	5.5	NR	NR	NR	NR
o-Toluidine	Toluene (ng/24 h)	29.1	30.2	135.0	58.0	49.3	47.6
HMPMA	Crotonaldehyde (mg/24 h)	2.0	1.7	5.2	2.6	2.6	1.7

COHb: carboxyhaemoglobin; EHCSS: electrically heated cigarette smoking system; HMPMA: hydroxy methyl propylmercapturic acid; HPHC: harmful and potentially harmful constituent; 3-HPMA: 3-hydroxypropyl mercapturic acid; MHBMA: monohydroxy-butenylmercapturic acid; NNAL: 4-(methylnitrosamino)-1-(3-pyridyl)-1-butanol; NNK: 4-(methylnitrosamino)-1-(3-pyridyl)-1-butanone; NR: not reported; 1-OHP: 1-hydroxypyrene; S-PMA: S-phenylmercapturic acid. [a] K6 and K3 are two versions of the system tested.

Table 2.4. Levels of biomarkers reported in published short-term studies with tobacco heating system 2.2, with end-of-study values for conventional cigarettes, heated tobacco products and abstinence, when available

Biomarker	HPHC	Japan (58, 60)			Poland (59)		
		THS	Cigarettes	Abstinence	THS	Cigarettes	Abstinence
Nicotine equivalent	Nicotine (mg/g creatinine)	5.44	5.52	0.15	10.60	9.76	0.14
Nicotine	Nicotine	19.13	21.34	0.10	20.74	19.01	0.10
Cotinine	Nicotine	161.00	164.30	2.96	239.99	219.73	2.05
NNAL	NNK (pg/mg creatinine)	37.77	76.55	28.63	49.65	107.04	41.51
NNN	N'-nitrosonornicotine (pg/mg creatinine)	1.31	4.64	0.18	1.55	5.99	0.16
COHb	Carbon monoxide (%)	2.39	5.14	2.37	1.06	4.51	0.99
MHBMA	1,3-Butadiene (pg/mg creatinine)	107.39	450.19	92.18	192.93	2399.40	163.17
3-HPMA	Acrolein (ng/mg creatinine)	311.08	599.67	199.04	402.26	931.01	245.69
S-PMA	Benzene (pg/mg creatinine)	143.77	850.02	126.34	164.45	2922.81	143.70
1-OHP	Benzo[a]pyrene (pg/mg creatinine)	73.02	149.62	62.99	81.22	182.85	85.13
3-OH-benzo[a]pyrene	Benzo[a]pyrene (fg/mg creatinine)	29.52	96.42	24.47	37.07	130.29	33.64
4-Aminobiphenyl	4-Aminobiphenyl (pg/mg creatinine)	1.53	8.57	1.49	1.90	12.58	1.60
1-Naphthylamine	1-Naphthylamine (pg/mg creatinine)	2.47	57.08	2.45	3.30	89.37	2.56

2-Naphthylamine	2-Naphthylamine (pg/mg creatinine)	2.33	13.38	2.27	2.96	25.32	2.52
o-Toluidine	Toluene (pg/mg creatinine)	50.40	98.18	48.91	51.15	121.16	41.64
CEMA	Acrylonitrile (ng/mg creatinine)	10.61	54.19	9.04	13.18	99.48	12.60
HEMA	Ethylene oxide (pg/mg creatinine)	997.76	2099.41	806.29	1342.40	4504.00	1248.27
HMPMA	Crotonaldehyde (ng/mg creatinine)	59.51	157.83	47.84	86.65	376.78	63.25
S-BMA	Toluene (ng/mg creatinine)	2098.09	2354.17	2192.86	NR	NR	NR

CEMA: 2-chloroethylmethacrylate; COHb: carboxyhaemoglobin; HEMA: hydroxyethylmethacrylate; HMPMA: hydroxy methyl propylmercapturic acid; 3-HPMA: 3-hydroxypropyl mercapturic acid; MHBMA: monohydroxy-butenylmercapturic acid; NNAL: 4-(methylnitrosamino)-1-(3-pyridyl)-1-butanol; NNK: 4-(methylnitrosamino)-1-(3-pyridyl)-1-butanone; NNN: N´-nitrosonornicotine; NR: not reported; 1-OHP: 1-hydroxypyrene; S-BMA: S-benzylmercapturic acid; S-PMA: S-phenylmercapturic acid; THS: tobacco heating system.

PMI also published the results of longer-term clinical studies of EHCSS (12 weeks) and THS 2.2 (90 days). After use of EHCSS for 12 weeks by 60 participants in the USA, the level of 4-(methylnitrosamino)-1-(3-pyridyl)-1-butanol (NNAL, a biomarker of exposure to NNK) dropped by 63% ($P < .0001$), that of carboxyhaemoglobin (a biomarker of exposure to CO) by 23% ($P < .0001$) and that of S-phenylmercapturic acid (a biomarker of exposure to benzene) by 49% ($P < .0001$) (*48*). (The paper did not provide mean estimates at each time.) After 90 days of use of THS 2.2 by 76 participants in Japan, the level of NNAL dropped by 73%, that of carboxyhaemoglobin by 42% and that of S-phenylmercapturic acid by 86% (*61*). A number of other biomarkers of exposure were measured (Table 2.5), and participants were followed up for 90 days for continued smoking and abstinence. Exposure to certain constituents was reduced with use of the THS rather than conventional cigarettes, although exposure to a number of constituents, including nitrosamines and acrolein, remained substantially higher than during smoking abstinence.

Table 2.5. End-of-study biomarker levels in participants in a 90-day study of random assignment to THS 2.2, continued cigarette smoking or abstinence, Japan

Biomarker	HPHC	THS	Cigarettes	Abstinence	Difference from cigarettes (%)	Difference from abstinence (%)
Nicotine equivalent	Nicotine (mg/g creatinine)	7	6	0	8	1751
NNAL	NNK (pg/mg creatinine)	23	95	14	–76	67
NNN	NNN (pg/mg creatinine)	1	4	0	–67	438
COHb	CO (%)	3	6	3	–48	–2
MHBMA	1,3–Butadiene (pg/mg creatinine)	142	785	137	–82	4
3-HPMA	Acrolein (ng/mg creatinine)	386	696	276	–44	40
S-PMA	Benzene (pg/mg creatinine)	146	1157	144	–87	1
1-OHP	Benzo[a]pyrene (pg/mg creatinine)	85	167	88	–49	–3
3-OH-Benzo[a]pyrene	Benzo[a]pyrene (fg/mg creatinine)	30	87	29	–65	4
4-Aminobiphenyl	4-Aminobiphenyl (pg/mg creatinine)	2	10	2	–78	–12
1-Naphthylamine	1-Naphthylamine (pg/mg creatinine)	4	55	4	–94	–16

2-Naphthylamine	2-Naphthylamine (pg/mg creatinine)	2	149	3	−98	−11
o-Toluidine	Toluene (pg/mg creatinine)	68	126	78	−46	−12
CEMA	Acrylonitrile (ng/mg creatinine)	8	84	8	−91	−6
HEMA	Ethylene oxide (pg/mg creatinine)	1742	3739	1633	−53	7
HMPMA	Crotonaldehyde (ng/mg creatinine)	154	299	159	−48	−3

CEMA: 2-chloroethylmethacrylate; CO: carbon monoxide; COHb: carboxyhaemoglobin; HEMA: hydroxyethylmethacrylate; HMPMA: hydroxy methyl propylmercapturic acid; HPHC: harmful and potentially harmful constituent; 3-HPMA: 3-hydroxypropyl mercapturic acid; MHBMA: monohydroxy-butenylmercapturic acid; NNAL: 4-(methylnitrosamino)-1-(3-pyridyl)-1-butanol; NNK: 4-(methylnitrosamino)-1-(3-pyridyl)-1-butanone; NNN: *N*′-nitrosonornicotine; 1-OHP: 1-hydroxypyrene; S-PMA: S-phenylmercapturic acid.

When a compensation formula
$1 - ((\ln(marker1) - \ln(marker0)) / (\ln(yield1) - \ln(yield0)))$
is used to compare changes in machine-measured emissions (ISO yields for EHCSS, HCI yields for THSs, each compared with 1R6F which has certified values for specific HPHCs in smoke) with levels of biomarkers of exposure, there is clearly potentially substantial compensation, including significant, near-total compensation for nicotine with the THS (Table 2.6). The ability to obtain increased levels of nicotine may partly explain differences in adoption of previous-generation HTPs and of iQOS.

Table 2.6. Potential compensation for selected harmful and potentially harmful constituents (HPHCs) in Philip Morris International heated tobacco products relative to a reference cigarette

HPHC	ECHSS (Japan), %	EHCSS (USA), %	THS (Japan), %	THS (Poland), %
Acrolein	79	67	75	68
Acrylonitrile	NR	NR	64	55
4-Aminobiphenyl	79	62	55	50
2-Naphthol	99	67	69	62
Benzene	66	71	64	41
Benzo[*a*]pyrene	77	78	74	70
Carbon monoxide (mg)	78	73	81	63
Nicotine	58	63	96	123
NNK (ng)	80	63	79	77
Toluene	85	91	84	79

ECHSS: electrically heated cigarette smoking system; NNK: 4-(methylnitrosamino)-1-(3-pyridyl)-1-butanone; NR: not reported; THS: tobacco heating system.

No published reports of biomarkers of exposure to the THP1.0 product were identified, although a trial protocol for such a study has been published (62). British American Tobacco has published one study of human use of the glo product to determine usage patterns (38). Three groups took three products (glo-menthol, glo-tobacco, glo + iQOS) home for up to 14 days, with up to four laboratory visits, while a fourth group used glo in the laboratory only. The measures included puffing topography, mouth-level exposure and mouth insertion depth. Participants who took the products home completed daily diaries of product use.

Overall, the puff volume was about 60 mL with the glo product, with an average of 10–12 puffs per session, a duration of 1.8–2.0 s and a mean interval of 8 s. The volume per puff and the total volume were significantly higher than with comparison cigarettes but comparable to those for the iQOS product. At baseline, participants reported using 12–15 cigarettes per day and about 8–12 units of the glo and iQOS per day during the 4 days at home. The mouth-level exposure per product used was lower with glo than with conventional cigarettes, especially for nicotine. The average mouth insertion depth was 7.7 mm, and the company used this as an argument to omit the vent-blocking procedure from the HCI machine smoking regimen *(38)*.

Thus far, consumer use of personal vaporizers to heat tobacco has been examined in only one published study *(63)*, in which Pax (loaded with 1 g of roll-your-own cigarette tobacco) was compared with the smokers' own brand of cigarette and an e-GO-type e-cigarette among 15 participants in a laboratory paradigm with fixed puffing patterns. The vaporizer increased plasma nicotine to a lesser extent than the smokers' own brand (14.3 ng/mL vs 24.4 ng/mL) and was equivalent to an e-cigarette, with no substantial exposure to CO. No other biomarkers were reported.

2.5 Consumer perceptions of heated tobacco products

Population surveys and internal industry marketing studies reveal strong consumer demand for products that are claimed or implied to carry reduced health risks *(64–70)*. University students rated Eclipse less positively but also less negatively than Marlboro Lights after seeing a single advertisement for each product, although none had ever tried the product *(71)*. Shiffman and colleagues showed in two studies *(67, 69)* that smokers perceived Eclipse as reducing harm and that interest in using Eclipse was associated with reduced intention to quit smoking. Hamilton and colleagues *(64)* showed advertisements for several PREPs and conventional tobacco products to smokers and found that, even though the advertisements contained no health claims, PREPs were still perceived as less of a health risk and conveyed positive messages about health and safety. Data from Japan *(72)* suggested that a significant minority of consumers were aware of and interested in HTPs *(72)*. Although 48% of the 8240 respondents to the survey were aware of e-cigarettes and/or HTPs, actual use of any products was fairly low; 6.6% had used any of the products, and the vast majority had used e-cigarettes. These are not legally sold in Japan but may be imported by individuals for personal use. Only 0.51% of the respondents had ever used Ploom, and 0.55% had used iQOS (7.8% and 8.4%, respectively, among users of any HTP). A later study *(73)* showed a dramatic growth in use between 2016 and 2017, the prevalence of iQOS use rising from 0.6% in 2016 to 3.6% in 2017.

A common reason given for the failure of novel products is rejection by consumers because of their poor taste *(74,75)*. Sensory characteristics are an important part of cigarette design *(76, 77)*. A consistent finding of clinical studies of subjective effects of Eclipse and Accord is that the products are perceived negatively relative to their own brand, with generally low ratings of satisfaction and higher ratings of dislike *(6, 7, 10, 22, 23)*. Established smokers who took part in focus groups reported significant dislike of the PREP cigarettes that they tried, including Eclipse and Accord, and almost all reported that they would not recommend the products to other smokers *(78, 79)*. Caraballo and colleagues *(78)* found that established smokers who had tried Eclipse generally disliked the product and would not recommend it to other smokers. Most smokers who reported that they disliked Eclipse cigarettes considered them too mild and said that they did not deliver enough nicotine to satisfy their craving; many reported that they disliked the taste. Hughes and colleagues *(80)* reported that smokers who tried Eclipse generally did not like it, although they believed it to be safer than conventional cigarettes. A study of the Pax vaporizer *(63)* indicated that the vaporizer did partially suppress symptoms of abstinence but was considered significantly less satisfying and less tasty than the participants' own brand of cigarettes.

Studies published by PMI on its THS device include data on subjective responses to the product, which were not provided in reports of studies of EHCSS and previous-generation products. These data are often important for understanding why smokers use a product and how effective it might be as a substitute *(81, 82)*. The studies were conducted with the widely used modified cigarette evaluation questionnaire *(83)* and the brief questionnaire on smoking urges to measure craving *(84)*. The cigarette evaluation questionnaire consists of 12 items (seven-point scales) and resolves to five scales of: smoking satisfaction, aversion, craving reduction, enjoyment of respiratory tract sensation and psychological reward *(83)*. The questionnaire on smoking urges consists of 10 items and provides a single score on a seven-point scale *(84)*.

In a laboratory study of a THS in Japan, the mean satisfaction scores decreased more for the THS than for conventional cigarettes over the course of the study (mean = −0.69; 95% confidence interval (CI), −0.34, −1.04) *(58)*. Other scores in the cigarette evaluation questionnaire did not change or differ as substantially from those for cigarettes. No difference in the scores on the questionnaire on smoking urges was observed for the THS and conventional cigarettes, both of which were lower than that for smoking abstinence (as expected). In a similar study in Poland, the observed differences in subjective effects between a THS and cigarettes were substantially larger and statistically significant for satisfaction (mean = −1.26; 95% CI, −0.85, −1.68), craving reduction (mean = −1.12; 95% CI, −0.66, −1.58), sensation (mean = −1.00; 95% CI, −0.64, −1.36) and reward (mean = − 0.72; 95% CI, −0.39, −1.06). No significant difference in craving scores was seen between the THS and cigarettes, and both were significantly lower than those for abstinence. These studies suggest that Polish smokers viewed the THS less positively than Japanese smokers; this may have implications for the generalization of observations from one market to another. An earlier study of THS 2.1 *(85)* showed a similar pattern of results, with satisfaction scores on day 5 an average of 1.4 points lower for the THS than for conventional cigarettes ($P < .001$). Significant differences were also seen on the reward, sensation and craving subscales, the THS scoring lower than cigarettes in all cases. A study on longer-term use in Japan *(61)* suggested that the difference in satisfaction fades with continued use over 90 days. These data suggest that scores on the questionnaire on smoking urges increase for the THS over time, as do scores for withdrawal (as measured on the Minnesota nicotine withdrawal scale *(86)*). This may suggest some dissatisfaction with the product as a longer-term substitute for smoking over time.

2.6 Uptake in selected markets where products are available

Three studies have been conducted of population uptake of HTPs in Italy and Japan. In Italy, 1% of "never smokers", 0.8% of former smokers and 3% of current cigarette smokers had tried iQOS *(87)*. Tabuchi and colleagues *(73)* showed a rising prevalence of current use of iQOS (from 0.3% to 3.6%) and Ploom (from 0.3% to 1.2%) after 2015, with comparable prevalence for use of glo in 2017 (0.8%). Predictors of use of HTPs were current cigarette smoking (stronger effect with intention to quit), living in a more deprived area and having seen television promotion of iQOS. Dual use was common (72%). Unlike for conventional cigarette smoking, there does not appear to be an inverse relation between education and HTP use in Japan *(88)*.

One reason that HTPs are gaining a market share in Japan is that nicotine-containing ENDS are not permitted for sale. Thus, HTPs may fill a market niche in a country with a relatively high smoking rate and relatively weak tobacco control laws. iQOS has become available in several markets in the European Union (which permit sale of ENDS) and in Canada (which did not permit such sales until 2018). In a presentation to the Consumer Analyst Group of New York in 2017 *(89)*, PMI reported that the attained market share of iQOS had reached 4.9% in Japan in the fourth quarter of 2016 but was substantially lower in marketing focus areas in Switzerland (1.7%), Portugal (0.7%), Romania (0.6%), Italy (0.4%) and the Russian Federation (0.3%), where the product had been available for at least one year. High rates of conversion to iQOS were claimed on the basis of their user panels, ranging from 54% in Switzerland to 72% in Japan.

2.7 Application by Philip Morris International for status as a "modified risk tobacco product" in the USA

In May 2017, PMI submitted an MRTP application for its THS/iQOS to the United States Food and Drug Administration (FDA) for review. The three health claims under consideration were:

- "Switching completely from cigarettes to the iQOS system can reduce the risks of tobacco-related diseases."
- "Switching completely to iQOS presents less risk of harm than continuing to smoke cigarettes."
- "Switching completely from cigarettes to the iQOS system significantly reduces your body's exposure to harmful and potentially harmful chemicals."

In compliance with the Tobacco Control Act, redacted[2] versions of the full application were made available for public review and comment on the website of the FDA (https://www.fda.gov/tobaccoproducts/labeling/marketingandadvertising/ucm546281.htm). Much of the technical information about the device has been redacted. The findings cited in publicly available parts of the application that are not discussed above (i.e. unpublished studies) are discussed below.

The online material includes an additional human trial with THS 2.2, conducted in the USA in 2013–2014 (NCT01989156). Of the 160 participants who were randomized, 88 completed the full 90-day study. As in the studies in Japan and Poland, key biomarkers of exposure were significantly reduced as compared with those for cigarette smoking after 5 and 90 days. Compliance with abstinence was, however, substantially lower in this study than in that in Japan during the ambulatory phase, suggesting that experience in Japan cannot necessarily be generalized to the USA.

The MRTP documents also report on a series of studies to test the specific claims requested for iQOS on the United States market. A series of qualitative and quantitative studies is described, concluding with three "assessment phase" studies on comprehension of claims and risk perception in a variety of contexts (package, brochure, direct mail). PMI developed a multi-factor risk perception scale for use in this study. The studies were conducted with three samples of approximately 2500 adult current smokers (with and without intention to quit), 2500 former smokers and 2500 nonsmokers. iQOS was rated as being of intermediate risk between cigarettes and quitting, comparable to the risk rating for e-cigarettes. Comprehension of claims was good, although there was evidence that about 25% of the participants extrapolated reductions in exposure to reduced harm

---

2   Applications are redacted to remove information considered to be commercially confidential and/or trade secrets under United States law.

(which is potentially erroneous). It was not evident that the claims shifted risk perceptions of iQOS or led to increased intention to try iQOS among smokers.

A series of observational studies were reported on product-switching in Germany, Italy, Japan, the Republic of Korea, Switzerland and the USA. "Whole offer tests" were conducted with 2089 daily smokers in the first five countries. iQOS was made available for free for 4 weeks, and product use was recorded in electronic diaries. At the end of the 4-week trial, the rate of switching to a THS ranged from 10% in Germany to 37% in the Republic of Korea, and the rate of dual use ranged from 32% in Japan to 39% in the Republic of Korea. The study in the USA comprised 1106 current daily smokers who, after a 1-week baseline, were given free access to iQOS for 4 weeks. Product use (cigarettes and Heatsticks) was recorded in an electronic diary. At the end of the study, approximately 15% of subjects had switched to the THS, as defined in the study (> 70% of total consumption as Heatsticks), while 22% were dual users (30–70% of consumption as Heatsticks).

Limited post-marketing data were presented, primarily from Japan, drawing on PMI's registry of iQOS purchasers. They reported that the proportion of exclusive iQOS use (> 95% of total consumption) increased from 52% to 65% between January and July 2016. Markov modelling of transition in two cohorts of iQOS purchasers in Japan (September 2015 and May 2016) suggested that smokers who converted to exclusive iQOS use were unlikely to return to exclusive cigarette use (although scant details of the modelling are given in the executive summary).

At the meeting of the FDA Tobacco Products Scientific Advisory Committee in January 2018,[3] PMI presented evidence in support of iQOS, including much of the data contained in this report. The FDA presented its preliminary evaluation of the submission, and the Committee expressed concern about portions of the evidence base, in particular the definitions of "complete switching" and limitations to the studies that underlie consumer understanding of the claims. The Committee did not support the claims of risk modification, expressed support for the claim of exposure modification and expressed concern that the claims as worded would not be effective in communicating risk.[1] The Committee's recommendations to the FDA are nonbinding.

2.8 Implications for regulation and tobacco control policies

Predicting the uptake of unconventional tobacco products can be difficult. Analyses of trial testimony by the tobacco industry on older-generation PREPs showed that the industry sought to shift the focus from their responsibility to produce "safer" products to failure of smokers to adopt the "safer" products they

---

3   See https://www.fda.gov/downloads/AdvisoryCommittees/CommitteesMeetingMaterials/TobaccoProductsScientificAdvisoryCommittee/UCM599235.pdf.

have been offered *(75)*. Nonetheless, this market segment appears to be growing rather than shrinking and is capturing a significant market share in at least one country, Japan, where PMI claims to have sold more than 1 million iQOS devices *(90)*. The human studies available (all performed by the manufacturer) indicated that iQOS deliver fewer toxicants than cigarettes and may serve as an effective short-term substitute for conventional cigarettes, as assessed by nicotine delivery and subjective effects measured on the cigarette evaluation questionnaire; one 90-day study suggested similar outcomes. It must be remembered, however, that the participants in these research studies were compensated, and the results may therefore not reflect real usage patterns, including concurrent use of multiple products. In December 2017, committees on the Toxicity, Carcinogenicity and Mutagenicity of Chemicals in Food, Consumer Products and the Environment evaluated two HTPs on the market in the United Kingdom (iQOS and iFuse) and concluded *(91)* that:

> "While there is a likely reduction in risk for smokers switching to 'heat-not-burn' tobacco products, there will be a residual risk and it would be more beneficial for smokers to quit smoking entirely. This should form part of any long-term strategy to minimize risk from tobacco use."

"Product drift" may be another important consideration. In their presentation to the Consumer Analyst Group of New York *(89)*, PMI described certain modifications made to the device from the original product pilot-tested in Japan in 2014–2017, including better aesthetics, blade self-cleaning, better user interface, faster charging, Bluetooth connectivity and an accompanying mobile application. Colours are being used to increase the appeal of the device. Thus, a product may be a "moving target" after its introduction. These practices are not unlike those of the tobacco industry of making minor adjustments to their cigarette products over time and by market *(92–94)*, such that a Marlboro cigarette in 2017 is not necessarily identical to one in 2010, and a Marlboro sold in France is not necessarily the same as one sold in the USA. In addition, data on a product from research studies may not fully reflect the product currently available to consumers. Thus, research to support a product is not necessarily conducted with the final marketed product but with a prototype or even a series of different prototypes. While this practice is not in itself nefarious, any differences in design, function or presentation between the studied and marketed product should be established, including their impact on consumer use. The FDA requires reporting of changes to an existing product for marketing authorization and may monitor them, with greater scrutiny and requirements for changes that impinge on public health (e.g. changes to delivery of HPHCs; substantial design changes). European Union Member States have similar provisions, in line with the Tobacco Products

Directive. Regulators in other countries should consider a similar requirement for notice and justification of changes to products.

Side-stream emissions and second-hand exposure are further concerns with the introduction of unconventional tobacco products. Some tobacco manufacturers claim that certain HTPs result in minimal exposure to side-stream smoke *(42, 46, 95, 96)*, whereas some studies indicate substantial levels *(97, 98)*. The emissions may depend in part on the design of the product. Side-stream emissions have health and regulatory implications, and further research is needed to characterize environmental exposure from HTPs. Depending on the wording of claims, HTPs may or may not be covered by smoke-free legislation, for example. Nevertheless, the precautionary principle would support covering such emissions in regulations.

2.9 Recommendations for research and policy

Independent scientific evidence is required to verify the claims of industry scientists for reduced exposure and risk. The studies published in the peer-reviewed literature up to October 2017, primarily by PMI on their THS 2.2 product, focused on a subset of HPHCs. Biomarkers of exposure to metals, such as cadmium, were not reported in these studies. Metals are a concern both as carcinogens and potentiators (cadmium, nickel, cobalt, arsenic) and as toxicants in their own right (lead, copper). While most of the studies indicated reduced exposure to HPHCs as compared with cigarette smoking, the studies did not address the possibility of novel exposure from the THS, either from the heating system itself or from the additives used in the tobacco. In its MRTP presentation to the FDA Tobacco Products Scientific Advisory Committee, when pressed on these issues, PMI acknowledged that 50 constituents in iQOS aerosol were present at higher levels than in conventional cigarette smoke, three of which are unique to iQOS; about 750 constituents occur at equal or lower levels in iQOS than in conventional cigarette smoke, and > 4000 are unique to cigarette smoke. Of the 50 constituents, four were identified as of toxicological concern (glycidol, 2-furanemethanol, 3-monochloro-1,2-propanediol and furfural).

Few published data are available on British American Tobacco's HTPs, although a study protocol for a randomized trial has been published *(62)*, suggesting ongoing work in this area by the company. No published studies on Japan Tobacco International's Ploom product were identified, and data on exposure to tobacco HPHCs from personal vaporizers are similarly lacking. Data on usage and sales of personal vaporizers for tobacco use are extremely difficult to locate, and comparisons with this market segment cannot be made. Use of such devices to administer cannabis appears to be increasing *(99–101)*.

Some suggestions for near- and long-term research priorities are:

- monitoring of product availability, sales and marketing with validated tools (e.g. *(102)*);
- monitoring of product use, including complete switching, dual (or poly) use, use as compared with ENDS and initiation by non-tobacco users, with a particular focus on low-risk young people who would not otherwise have smoked;
- verification of reported product contents and emissions of the 39 priority toxicants and/or the FDA HPHC list;
- evaluation of potential novel toxicants produced by heated tobacco products that are not covered by commonly accepted lists (e.g. Hoffman analytes, HPHCs);
- evaluation of aerosol particle size distribution;
- assessment of device function and safety (e.g. batteries);
- cross-market comparisons of products, e.g. whether iQOS is the same in all markets and differences in the characteristics, contents and emissions of products and how they have changed over time;
- independent clinical and biomarker analyses of users according to typical patterns of use, including dual or concurrent use of heat-not-burn products and cigarettes;
- public perceptions of products (awareness, intention to use, risk) among users and non-users of tobacco;
- information on who is purchasing a product, reasons for purchasing and use patterns;
- research on rates of conversion of smokers to HTPs and vaping (ENDS) products, to determine whether these products discourage smoking in a way that is acceptable to smokers;
- modelling of potential population-level effects of the introduction and use of products (e.g. SimSmoke; microsimulation);
- investigation of the influence of marketing strategies on user behaviour, including whether these products are marketed as complementary or alternative products;
- effect of heat treatment on components in products other than tobacco (e.g. flavours); and
- studies of exposure to second-hand emissions (including effects on children and pregnant women) and their contribution to background air quality.

2.10 References

1. Rodgman A, Perfetti TA. The chemical components of tobacco and tobacco smoke. 2nd ed. Boca Raton (FL): CRC Press; 2013.
2. Batista RN. Reinforced web of reconstituted tobacco. Google Patents WO2015193031A1; 2015 (https://patents.google.com/patent/WO2015193031A1/en, accessed 1 October 2018).
3. Smith MR, Clark B, Ludicke F, Schaller P, Vanscheeuwijck P, Hoeing J et al. Evaluation of the tobacco heating system 2.2. Part 1: Description of the system and the scientific assessment program. Regul Toxicol Pharmacol. 2016;81(Suppl 2):S17–26.
4. Eaton D, Jakaj B, Forster M, Nicol J, Mavropoulou E, Scott K et al. Assessment of tobacco heating product THP1.0. Part 2: Product design, operation and thermophysical characterisation. Regul Toxicol Pharmacol. 2018;93:4–13.
5. Scott JK, Poynton S, Margham J, Forster M, Eaton D, Davis P et al. Controlled aerosol release to heat tobacco: product operation and aerosol chemistry assessment. Poster. Society for Research on Nicotine and Tobacco; 2–5 March 2016; Chicago (IL); 2016 (https://www.researchgate.net/publication/298793405_Controlled_aerosol_release_to_heat_tobacco_product_operation_and_aerosol_chemistry_assessment, accessed October 2018).
6. Breland AB, Buchhalter AR, Evans SE, Eissenberg T. Evaluating acute effects of potential reduced-exposure products for smokers: clinical laboratory methodology. Nicotine Tob Res. 2002;4(Suppl 2):S131–40.
7. Breland AB, Kleykamp BA, Eissenberg T. Clinical laboratory evaluation of potential reduced exposure products for smokers. Nicotine Tob Res. 2006;8(6):727–38.
8. Fagerstrom KO, Hughes JR, Callas PW. Long-term effects of the Eclipse cigarette substitute and the nicotine inhaler in smokers not interested in quitting. Nicotine Tob Res. 2002;4(Suppl 2):S141–5.
9. Fagerstrom KO, Hughes JR, Rasmussen T, Callas PW. Randomised trial investigating effect of a novel nicotine delivery device (Eclipse) and a nicotine oral inhaler on smoking behaviour, nicotine and carbon monoxide exposure, and motivation to quit. Tob Control. 2000;9(3):327–33.
10. Lee EM, Malson JL, Moolchan ET, Pickworth WB. Quantitative comparisons between a nicotine delivery device (Eclipse) and conventional cigarette smoking. Nicotine Tob Res. 2004;6(1):95–102.
11. Rennard SI, Umino T, Millatmal T, Daughton DM, Manouilova LS, Ullrich FA et al. Evaluation of subclinical respiratory tract inflammation in heavy smokers who switch to a cigarette-like nicotine delivery device that primarily heats tobacco. Nicotine Tob Res. 2002;4(4):467–76.
12. Stapleton JA, Russell MA, Sutherland G, Feyerabend C. Nicotine availability from Eclipse tobacco-heating cigarette. Psychopharmacology. 1998;139(3):288–90.
13. Stewart JC, Hyde RW, Boscia J, Chow MY, O'Mara RE, Perillo I et al. Changes in markers of epithelial permeability and inflammation in chronic smokers switching to a nonburning tobacco device (Eclipse). Nicotine Tob Res. 2006;8(6):773–83.
14. Pauly JL, Lee HJ, Hurley EL, Cummings KM, Lesses JD, Streck RJ. Glass fiber contamination of cigarette filters: an additional health risk to the smoker? Cancer Epidemiol Biomarkers Prev. 1998;7(11):967–79.
15. Pauly JL, Stegmeier SJ, Allaart HA, Cheney RJ, Zhang PJ, Mayer AG et al. Inhaled cellulosic and plastic fibers found in human lung tissue. Cancer Epidemiol Biomarkers Prev. 1998;7(5):419–28.
16. Cummings KM, Hastrup JL, Swedrock T, Hyland A, Perla J, Pauly JL. Consumer perception of risk associated with filters contaminated with glass fibers. Cancer Epidemiol Biomarkers Prev. 2000;9(9):977–9.
17. Higuchi MA, Ayres PH, Swauger JE, Morgan WT, Mosberg AT. Quantitative analysis of potential transfer of continuous glass filament from eclipse prototype 9-014 cigarettes. Inhal Toxicol.

2000;12(11):1055–70.
18. Higuchi MA, Ayres PH, Swauger JE, Morgan WT, Mosberg AT. Analysis of potential transfer of continuous glass filament from Eclipse cigarettes. Inhal Toxicol. 2000;12(7):617–40.
19. Swauger JE, Foy JW. Safety assessment of continuous glass filaments used in Eclipse. Inhal Toxicol. 2000;12(11):1071–84.
20. Higuchi MA, Ayres PH, Swauger JE, Deal PA, Guy T, Morton M et al. Quantification of continuous glass filaments on eclipse cigarettes retrieved from the test market. Inhal Toxicol. 2003;15(7):715–25.
21. Slade J, Connolly GN, Lymperis D. Eclipse: does it live up to its health claims? Tob Control. 2002;11(Suppl 2):ii64–70.
22. Buchhalter AR, Eissenberg T. Preliminary evaluation of a novel smoking system: effects on subjective and physiological measures and on smoking behavior. Nicotine Tob Res. 2000;2(1):39–43.
23. Buchhalter AR, Schrinel L, Eissenberg T. Withdrawal-suppressing effects of a novel smoking system: comparison with own brand, not own brand, and de-nicotinized cigarettes. Nicotine Tob Res. 2001;3(2):111–8.
24. Hughes JR, Keely JP. The effect of a novel smoking system – Accord – on ongoing smoking and toxin exposure. Nicotine Tob Res. 2004;6(6):1021–7.
25. Roethig HJ, Zedler BK, Kinser RD, Feng S, Nelson BL, Liang Q. Short-term clinical exposure evaluation of a second-generation electrically heated cigarette smoking system. J Clin Pharmacol. 2007;47(4):518–30.
26. United States Food and Drug Administration. Harmful and potentially harmful constituents in tobacco products and tobacco smoke; established list. Fed Reg. 2012:20034–7.
27. Pratte P, Cosandey S, Goujon-Ginglinger C. Investigation of solid particles in the mainstream aerosol of the Tobacco Heating System THS2.2 and mainstream smoke of a 3R4F reference cigarette. Hum Exp Toxicol. 2017;36(11):1115–20.
28. Center for Tobacco Reference Products. Certificate of Analysis 1R6F Certified Reference Cigarette. Vol. 2016-002CTRP. Lexington (KY): University of Kentucky; 2016.
29. Li X, Luo Y, Jiang X, Zhang H, Zhu F, Hu S et al. Chemical analysis and simulated pyrolysis of tobacco heating system 2.2 compared to conventional cigarettes. Nicotine Tob Res. 2018. doi:10.1093/ntr/nty005.
30. Farsalinos KE, Yannovits N, Sarri T, Voudris V, Poulas K. Nicotine delivery to the aerosol of a heat-not-burn tobacco product: comparison with a tobacco cigarette and e-cigarettes. Nicotine Tob Res. 2018;27:e76–8.
31. Bekki K, Inaba Y, Uchuyama S, Kunugita N. Comparison of chemicals in mainstream smoke in heat-not-burn tobacco and combustion cigarettes. J Univ Occup Environ Health Japan. 2017;39(3):201–7.
32. Jakaj B, Eaton D, Forster M, Nicol T, Liu C, McAdam K et al. Characterizing key thermophysical processes in a novel tobacco heating product THP1.0(T). Poster. Society for Research on Nicotine and Tobacco; 7–11 March 2017, Florence, Italy; 2017 (http://www.bat-science.com/groupms/sites/BAT_9GVJXS.nsf/vwPagesWebLive/DOAKAK4V/$FILE/SRNT%202017%20poster%2036-Liu%20et%20al_JMcA%20KM.pdf?openelement, accessed 1 October 2018).
33. Forster M, Fiebelkorn S, Yurteri C, Mariner D, Liu C, Wright C et al. Assessment of novel tobacco heating product THP1.0. Part 3: Comprehensive chemical characterisation of harmful and potentially harmful aerosol emissions. Regul Toxicol Pharmacol. 2018;93:14–33.
34. Murphy J, Liu C, McAdam K, Gaça M, Prasad K, Camacko O et al. Assessment of tobacco heating product THP1.0. Part 9: The placement of a range of next-generation products on an emissions continuum relative to cigarettes via pre-clinical assessment studies. Regul Toxicol Pharmacol. 2018;93:92–104.

35. Taylor M, Thorne D, Carr T, Breheny D, Walker P, Proctor C et al. Assessment of novel tobacco heating product THP1.0. Part 6: A comparative in vitro study using contemporary screening approaches. Regul Toxicol Pharmacol. 2018;93:62–70.
36. Thorne D, Breheny D, Proctor C, Gaca M. Assessment of novel tobacco heating product THP1.0. Part 7: Comparative in vitro toxicological evaluation. Regul Toxicol Pharmacol. 2018;93:71–83.
37. Forster M, McAughey J, Prasad K, Mavropoulou E, Proctor C. Assessment of tobacco heating product THP1.0. Part 4: Characterisation of indoor air quality and odour. Regul Toxicol Pharmacol. 2018;93:34–51.
38. Gee J, Prasad K, Slayford S, Gray A, Nother K, Cunningham A et al. Assessment of tobacco heating product THP1.0. Part 8: Study to determine puffing topography, mouth level exposure and consumption among Japanese users. Regul Toxicol Pharmacol. 2018:93:84–91.
39. Jaunky T, Adamson J, Santopietro S, Terry A, Thorne D, Breheny D et al. Assessment of tobacco heating product THP1.0. Part 5: In vitro dosimetric and cytotoxic assessment. Regul Toxicol Pharmacol. 2018;93:52–61.
40. Proctor C. Assessment of tobacco heating product THP1.0. Part 1: Series introduction. Regul Toxicol Pharmacol. 2018;93:1–3.
41. WHO Study Group on Tobacco Product Regulation. Report on the scientific basis of tobacco product regulation. Seventh report of a WHO study group. Geneva: World Health Organization; 2015.
42. Forster M, McAughey J, Liu C, McAdam K, Murphy J, Proctor C. Characterization of a novel tobacco heating product THP1.0(T): Indoor air quality. In: Society for Research on Nicotine and Tobacco, Florence, Italy, 7–11 March 2017.
43. Breheny D, Adamson J, Azzopardi D, Baxter A, Bishop E, Carr T et al. A novel hybrid tobacco product that delivers a tobacco flavour note with vapour aerosol (Part 2): In vitro biological assessment and comparison with different tobacco-heating products. Food Chem Toxicol. 2017;106(Pt A):533–46.
44. Werley MS, Freelin SA, Wrenn SE, Gerstenberg B, Roemer E, Schramke H et al. Smoke chemistry, in vitro and in vivo toxicology evaluations of the electrically heated cigarette smoking system series K. Regul Toxicol Pharmacol. 2008;52(2):122–39.
45. Schaller JP, Keller D, Poget L, Pratte P, Kaelin E, McHugh D et al. Evaluation of the tobacco heating system 2.2. Part 2: Chemical composition, genotoxicity, cytotoxicity, and physical properties of the aerosol. Regul Toxicol Pharmacol. 2016;81(Suppl 2):S27–47.
46. Frost-Pineda K, Zedler BK, Liang Q, Roethig HJ. Environmental tobacco smoke (ETS) evaluation of a third-generation electrically heated cigarette smoking system (EHCSS). Regul Toxicol Pharmacol. 2008;52(2):118–21.
47. Frost-Pineda K, Zedler BK, Oliveri D, Feng S, Liang Q, Roethig HJ. Short-term clinical exposure evaluation of a third-generation electrically heated cigarette smoking system (EHCSS) in adult smokers. Regul Toxicol Pharmacol. 2008;52(2):104–10.
48. Frost-Pineda K, Zedler BK, Oliveri D, Liang Q, Feng S, Roethig HJ. 12-week clinical exposure evaluation of a third-generation electrically heated cigarette smoking system (EHCSS) in adult smokers. Regul Toxicol Pharmacol. 2008;52(2):111–7.
49. Martin Leroy C, Jarus-Dziedzic K, Ancerewicz J, Lindner D, Kulesza A, Magnette J. Reduced exposure evaluation of an electrically heated cigarette smoking system. Part 7: A one-month, randomized, ambulatory, controlled clinical study in Poland. Regul Toxicol Pharmacol. 2012;64(2 Suppl):S74–84.
50. Schorp MK, Tricker AR, Dempsey R. Reduced exposure evaluation of an electrically heated cigarette smoking system. Part 1: Non-clinical and clinical insights. Regul Toxicol Pharmacol. 2012;64(2 Suppl):S1–10.
51. Tricker AR, Jang IJ, Martin Leroy C, Lindner D, Dempsey R. Reduced exposure evaluation of

52. an electrically heated cigarette smoking system. Part 4: Eight-day randomized clinical trial in Korea. Regul Toxicol Pharmacol. 2012;64(2 Suppl):S45–53.
52. Tricker AR, Kanada S, Takada K, Leroy CM, Lindner D, Schorp MK et al. Reduced exposure evaluation of an electrically heated cigarette smoking system. Part 5: 8-day randomized clinical trial in Japan. Regul Toxicol Pharmacol. 2012;64(2 Suppl):S54–63.
53. Tricker AR, Kanada S, Takada K, Martin Leroy C, Lindner D, Schorp MK et al. Reduced exposure evaluation of an electrically heated cigarette smoking system. Part 6: 6-day randomized clinical trial of a menthol cigarette in Japan. Regul Toxicol Pharmacol. 2012;64(2 Suppl):S64–73.
54. Tricker AR, Stewart AJ, Leroy CM, Lindner D, Schorp MK, Dempsey R. Reduced exposure evaluation of an electrically heated cigarette smoking system. Part 3: Eight-day randomized clinical trial in the UK. Regul Toxicol Pharmacol. 2012;64(2 Suppl):S35–44.
55. Urban HJ, Tricker AR, Leyden DE, Forte N, Zenzen V, Feuersenger A et al. Reduced exposure evaluation of an electrically heated cigarette smoking system. Part 8: Nicotine bridging – estimating smoke constituent exposure by their relationships to both nicotine levels in mainstream cigarette smoke and in smokers. Regul Toxicol Pharmacol. 2012;64(2 Suppl):S85–97.
56. Zenzen V, Diekmann J, Gerstenberg B, Weber S, Wittke S, Schorp MK. Reduced exposure evaluation of an electrically heated cigarette smoking system. Part 2: Smoke chemistry and in vitro toxicological evaluation using smoking regimens reflecting human puffing behavior. Regul Toxicol Pharmacol. 2012;64(2 Suppl):S11–34.
57. Dayan AD. Investigating a toxic risk (self-inflicted) the example of conventional and advanced studies of a novel tobacco heating system. Regul Toxicol Pharmacol. 2016;81 Suppl 2:S15–6.
58. Haziza C, de La Bourdonnaye G, Merlet S, Benzimra M, Ancerewicz J, Donelli A et al. Assessment of the reduction in levels of exposure to harmful and potentially harmful constituents in Japanese subjects using a novel tobacco heating system compared with conventional cigarettes and smoking abstinence: a randomized controlled study in confinement. Regul Toxicol Pharmacol. 2016;81:489–99.
59. Haziza C, de La Bourdonnaye G, Skiada D, Ancerewicz J, Baker G, Picavet P et al. Evaluation of the tobacco heating system 2.2. Part 8: 5-day randomized reduced exposure clinical study in Poland. Regul Toxicol Pharmacol. 2016;81(Suppl 2):S139–50.
60. Haziza C, de La Bourdonnaye G, Skiada D, Ancerewicz J, Baker G, Picavet P et al. Biomarker of exposure level data set in smokers switching from conventional cigarettes to tobacco heating system 2.2, continuing smoking or abstaining from smoking for 5 days. Data Brief. 2017;10:283–93.
61. Lüdicke F, Picavet P, Baker G, Haziza C, Poux V, Lama N et al. Effects of switching to the menthol tobacco heating system 2.2, smoking abstinence, or continued cigarette smoking on clinically relevant risk markers: a randomized, controlled, open-Label, multicenter study in sequential confinement and ambulatory settings (Part 2). Nicotine Tob Res. 2018;20(2):173–82.
62. Gale N, McEwan M, Eldridge AC, Sherwood N, Bowen E, McDermott S et al. A randomised, controlled, two-centre open-label study in healthy Japanese subjects to evaluate the effect on biomarkers of exposure of switching from a conventional cigarette to a tobacco heating product. BMC Public Health. 2017;17(1):673.
63. Lopez AA, Hiler M, Maloney S, Eissenberg T, Breland AB. Expanding clinical laboratory tobacco product evaluation methods to loose-leaf tobacco vaporizers. Drug Alcohol Depend. 2016;169:33–40.
64. Hamilton WL, Norton G, Ouellette TK, Rhodes WM, Kling R, Connolly GN. Smokers' responses to advertisements for regular and light cigarettes and potential reduced-exposure tobacco products. Nicotine Tob Res. 2004;6(Suppl 3):S353–62.
65. O'Connor RJ, Hyland A, Giovino GA, Fong GT, Cummings KM. Smoker awareness of and beliefs about supposedly less-harmful tobacco products. Am J Prev Med. 2005;29(2):85–90.

66. Parascandola M, Augustson E, O'Connell ME, Marcus S. Consumer awareness and attitudes related to new potential reduced-exposure tobacco product brands. Nicotine Tob Res. 2009;11(7):886–95.
67. Shiffman S, Pillitteri JL, Burton SL, Di Marino ME. Smoker and ex-smoker reactions to cigarettes claiming reduced risk. Tob Control. 2004;13(1):78–84.
68. Hund LM, Farrelly MC, Allen JA, Chou RH, St Claire AW, Vallone DM et al. Findings and implications from a national study on potential reduced exposure products (PREPs). Nicotine Tob Res. 2006;8(6):791–7.
69. Shiffman S, Jarvis MJ, Pillitteri JL, Di Marino ME, Gitchell JG, Kemper KE. UK smokers' and ex-smokers' reactions to cigarettes promising reduced risk. Addiction. 2007;102(1):156–60.
70. Levy DT, Cummings KM, Villanti AC, Niaura R, Abrams DB, Fong GT et al. A framework for evaluating the public health impact of e-cigarettes and other vaporized nicotine products. Addiction. 2017;112(1):8–17.
71. O'Connor RJ, Ashare RL, Fix BV, Hawk LW, Cummings KM, Schmidt WC. College students' expectancies for light cigarettes and potential reduced exposure products. Am J Health Behav. 2007;31(4):402–10.
72. Tabuchi T, Kiyohara K, Hoshino T, Bekki K, Inaba Y, Kunugita N. Awareness and use of electronic cigarettes and heat-not-burn tobacco products in Japan. Addiction. 2016;111(4):706–13.
73. Tabuchi T, Gallus S, Shinozaki T, Nakaya T, Kunugita N, Colwell B. Heat-not-burn tobacco product use in Japan: its prevalence, predictors and perceived symptoms from exposure to secondhand heat-not-burn tobacco aerosol. Tob Control. 2018;27(e):e25–33.
74. Fairchild A, Colgrove J. Out of the ashes: the life, death, and rebirth of the "safer" cigarette in the United States. Am J Public Health. 2004;94(2):192–204.
75. Wayne GF. Potential reduced exposure products (PREPs) in industry trial testimony. Tob Control. 2006;15(Suppl 4):iv90–7.
76. Carpenter CM, Wayne GF, Connolly GN. The role of sensory perception in the development and targeting of tobacco products. Addiction. 2007;102(1):136–47.
77. Cook BL, Wayne GF, Keithly L, Connolly G. One size does not fit all: how the tobacco industry has altered cigarette design to target consumer groups with specific psychological and psychosocial needs. Addiction. 2003;98(11):1547–61.
78. Caraballo RS, Pederson LL, Gupta N. New tobacco products: do smokers like them? Tob Control. 2006;15(1):39–44.
79. O'Hegarty M, Richter P, Pederson LL. What do adult smokers think about ads and promotional materials for PREPs? Am J Health Behav. 2007;31(5):526–34.
80. Hughes JR, Keely JP, Callas PW. Ever users versus never users of a "less risky" cigarette. Psychol Addict Behav. 2005;19(4):439–42.
81. Hanson K, O'Connor R, Hatsukami D. Measures for assessing subjective effects of potential reduced-exposure products. Cancer Epidemiol Biomarkers Prev. 2009;18(12):3209–24.
82. Rees VW, Kreslake JM, Cummings KM, O'Connor RJ, Hatsukami DK, Parascanola M et al. Assessing consumer responses to potential reduced-exposure tobacco products: a review of tobacco industry and independent research methods. Cancer Epidemiol Biomarkers Prev. 2009;18(12):3225–40.
83. Cappelleri JC, Bushmakin AG, Baker CL, Merikle E, Olufade AO, Gilbert DG. Confirmatory factor analyses and reliability of the modified cigarette evaluation questionnaire. Addict Behav. 2007;32(5):912–23.
84. Cox LS, Tiffany ST, Christen AG. Evaluation of the brief questionnaire of smoking urges (QSU-brief) in laboratory and clinical settings. Nicotine Tob Res. 2001;3(1):7–16.
85. Lüdicke F, Baker G, Magnette J, Picavet P, Weitkunat R. Reduced exposure to harmful and potentially harmful smoke constituents with the tobacco heating system 2.1. Nicotine Tob

Res. 2017;19(2):168–75.
86. Hughes JR, Hatsukami D. Signs and symptoms of tobacco withdrawal. Arch Gen Psychiatry. 1986;43(3):289–94.
87. Liu X, Lugo A, Spizzichino L, Tabuchi T, Pacifici R, Gallus S. Heat-not-burn tobacco products: concerns from the Italian experience. Tob Control. 2018. doi: 10.1136/tobaccocontrol-2017-054054.
88. Miyazaki Y, Tabuchi T. Educational gradients in the use of electronic cigarettes and heat-not-burn tobacco products in Japan. PLoS One. 2018;13(1):e0191008.
89. Calentzopoulos A, Olczak J, King M. Consumer Analyst Group of New York (CAGNY) Conference. New York: Philip Morris International; 2017.
90. Camilleri LC, Calantzopoulos A. Annual meeting of shareholders. Philip Morris International; 2016.
91. Statement on the toxicological evaluation of novel heat-not-burn tobacco products (press release). London: Food Standards Agency; 2017 (https://cot.food.gov.uk/sites/default/files/heat_not_burn_tobacco_statement.pdf, accessed 1 October 2018).
92. King B, Borland R, Abdul-Salaam S, Polzin G, Ashley D, Watson C et al. Divergence between strength indicators in packaging and cigarette engineering: a case study of Marlboro varieties in Australia and the USA. Tob Control. 2010;19(5):398–402.
93. Wayne GF, Connolly GN. Regulatory assessment of brand changes in the commercial tobacco product market. Tob Control. 2009;18(4):302–9.
94. Wu W, Zhang L, Jain RB, Ashley DL, Watson CH. Determination of carcinogenic tobacco-specific nitrosamines in mainstream smoke from US-brand and non-US-brand cigarettes from 14 countries. Nicotine Tob Res. 2005;7(3):443–51.
95. Mitova MI, Campelos PB, Goujon-Ginglinger CG, Maeder S, Mottier N, Rouget EG et al. Comparison of the impact of the tobacco heating system 2.2 and a cigarette on indoor air quality. Regul Toxicol Pharmacol. 2016;80:91–101.
96. Tricker AR, Schorp MK, Urban HJ, Leyden D, Hagedorn HW, Engl J et al. Comparison of environmental tobacco smoke (ETS) concentrations generated by an electrically heated cigarette smoking system and a conventional cigarette. Inhal Toxicol. 2009;21(1):62–77.
97. Protano C, Manigrasso M, Avino P, Sernia S, Vitali M. Second-hand smoke exposure generated by new electronic devices (IQOS(R) and e-cigs) and traditional cigarettes: submicron particle behaviour in human respiratory system. Ann Ig. 2016;28(2):109–12.
98. O'Connell G, Wilkinson P, Burseg KMM, Stotesbury SJ, Pritchard JD. Heated tobacco products create side-stream emissions: implications for regulation. J Environ Anal Chem. 2015;2(5):163.
99. McDonald EA, Popova L, Ling PM. Traversing the triangulum: the intersection of tobacco, legalised marijuana and electronic vaporisers in Denver, Colorado. Tob Control. 2016;25(Suppl 1):i96–102.
100. Lee DC, Crosier BS, Borodovsky JT, Sargent JD, Budney AJ. Online survey characterizing vaporizer use among cannabis users. Drug Alcohol Depend. 2016;159:227–33.
101. Morean ME, Kong G, Camenga DR, Cavallo DA, Krishnan-Sarin S. High school students' use of electronic cigarettes to vaporize cannabis. Pediatrics. 2015;136(4):611–6.
102. Henriksen L, Ribisl KM, Rogers T, Moreland-Russell S, Barker DM, Esquivel NS et al. Standardized tobacco assessment for retail settings (STARS): dissemination and implementation research. Tob Control. 2016;25(Suppl 1):i67–74.

3. Clinical pharmacology of nicotine in electronic nicotine delivery systems

Armando Peruga, Centre of Epidemiology and Health Policies, School of Medicine, Clínica Alemana, University of Desarrollo, Chile

Thomas Eissenberg, Center for the Study of Tobacco Products, Department of Psychology, Virginia Commonwealth University, USA

Contents
- 3.1 Introduction
- 3.2 ENDS operations
- 3.3 Nicotine concentration in e-liquids
- 3.4 Nicotine delivery to ENDS users
- 3.5 Toxicant content of ENDS emissions
    - 3.5.1 Nicotine emissions
    - 3.5.2 Emissions of non-nicotine toxicants
- 3.6 Potential role of ENDS in smoking cessation
- 3.7 Potential health impact of ENDS
    - 3.7.1 Behavioural trajectories associated with use of ENDS
    - 3.7.2 Harm from ENDS and electronic non-nicotine delivery systems
- 3.8 Summary of evidence, research gaps and policy issues derived from the evidence
- 3.9 References

3.1 Introduction

Electronic nicotine delivery systems (ENDS) are a heterogeneous class of products in which an electrically powered coil is used to heat a liquid matrix, or e-liquid, that contains nicotine, solvents (e.g. propylene glycol, vegetable glycerine) and, usually, flavourings. The user inhales the resulting aerosol, which contains variable concentrations of nicotine *(1)*, a dependence-producing central nervous system stimulant. In many countries and certainly in the two largest markets – the European Union and the USA – ENDS are regulated either as generic consumer products or as tobacco products *(2)*.

Products such as ENDS that are marketed to the public and contain drugs that act on the central nervous system, such as nicotine, ideally should have little potential for abuse or dependence for public health reasons. This is true, unless some level of abuse potential is desirable to maintain compliance and support substitution in place of a substance of greater potential abuse and harm. ENDS fall into this category on the basis of claims of a potential role in smoking cessation and reduction.

The purpose of this background paper is to review the literature at the time of writing with some additions after review between March and December 2018 on the nicotine content and nicotine delivery of ENDS and to explore factors that influence the emissions of nicotine and non-nicotine toxicants. In addition, we review the potential role of ENDS in smoking cessation and the prospective population health impact. We also identify some relevant research gaps and make recommendations for policy.

3.2 ENDS operations

Understanding how ENDS operate is useful. Fig. 3.1 is a schematic drawing of a common ENDS configuration. The heating coil is attached to an electrical power source (usually a battery, not shown in the figure) enclosed in a fabric wick that is in turn surrounded by the nicotine-containing e-liquid that saturates the wick. When power is flowing, the coil heats and thus vaporizes some of the e-liquid from the wick. As the user draws air from the mouth-end of the ENDS, the vapour is carried away and re-condenses to form an aerosol, which is inhaled by the user.

Fig. 3.1. Schematic drawing of ENDS operation

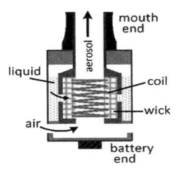

Source: Dr Alan Shihadeh, American University of Beirut, Lebanon.

Several factors influence the amount of nicotine carried by the aerosol, including the electrical power flowing through the ENDS, the inhalation behaviour (or "puff topography") of the user and the amount of nicotine in the e-liquid *(3)*. Electrical power (W) is a function of battery voltage (V) and coil resistance (Ω), such that $W = V^2 / \Omega$. Early ENDS models were powered at ≤ 10 W, but the devices marketed currently are powered at ≥ 250 W *(4, 5)*. Higher power is often achieved with coils with low resistance (e.g. < 1 Ω), application of varying voltage to the coil or a combination.

Puff topography variables include puff number, duration and volume and the interval between puffs (inter-puff interval). User puff topography is highly individual. Experienced ENDS users, however, typically take longer puffs than ENDS-naive cigarette smokers *(6–9)* (see Fig. 3.2 and description below).

Fig. 3.2. Mean plasma nicotine concentrations before and after use of a combusted cigarette and of ENDS

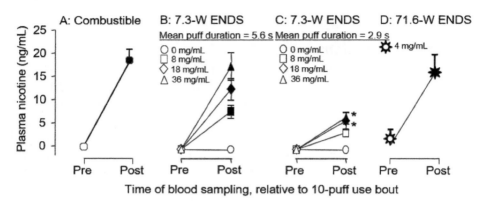

Panel A, N=32 *(8)*; Panel B, N=33 *(8)*; Panel C, N=31 *(8)*; Panel D, N=11 *(4)* (puff topography not available). *Source*: Figure adapted from one published previously *(1)* by adding puff duration data and updating Panel D.

3.3 Nicotine concentration in e-liquids

The nicotine-containing e-liquid used in ENDS comes in prefilled cartridges or refill bottles, depending on the type of device used. The concentration of nicotine in marketed e-liquid can reach 36 mg/mL or more *(1)*, and users can choose from a wide range of concentrations at the point of sale; some manufacturers provide labelling information relevant to the e-liquid. There has been no comprehensive study, however, of the extent to which manufacturers accurately inform consumers of the nicotine concentration in a representative sample of e-liquids, globally or by country. Existing studies give a partial picture based on convenience samples. The proportion of e-liquids that have clear label information on the nicotine content is unknown. Some studies indicate that such information is not always available *(10, 11)* or interpretable *(12)* from the manufacturer's label. Nevertheless, the concentration of nicotine is usually reported on the label as a percentage of total volume or as mg/mL. Table 3.1 lists studies in which the concentration of nicotine was analysed in e-liquids that allegedly contained nicotine and compared with the concentration reported on the manufacturer's label.

Table 3.1. Comparison of labelled and measured concentrations of nicotine in e-liquids with declared nicotine

First author and reference number	Type of e-liquid container	Number of samples		
		Analysed	> ±10% of labelled concentration	> ±25% of labelled concentration
Beauval (13)	Refill bottle	2	0	0
Buettner-Schmidt (14)	Refill bottle	70	36	NA
Cameron (15)	Prefilled cartridge and refill bottle	21	13	7
Cheah (10)	Cartridge	8[a]	8[b]	7[b]
Davis (16)	Refill bottle	81	36	21
El-Hellani (17)	Prefilled cartridge	4	4	4
Etter (18)	Refill bottle	35	4	0
Etter (19)	Refill bottle	34	10	0
Farsalinos (20)	Refill bottle	21	9	0
Goniewiscz (21)	Refill bottle	62	25	7
Kim (22)	Refill bottle	13	7	2
Kirschner (23)	Refill bottle	6	6	4
Kosmider (24)	Refill bottle	9	2	0
Lisko (25)	Refill bottle	29	15	7
Pagano (26)	Prefilled cartridge	4	3	2
Peace (27)	Refill bottle	27	16	7
Rahman (28)	Refill bottle	69	65	53
Raymond (29)	Refill bottle	35	22	22
Trehy (30)	Prefilled cartridge	22	22	19
Trehy (30)	Refill bottle	17	8	6

NA: not available. [a] Number of brands analysed; number of samples analysed not provided. [b] Number of brands in which at least one sample had a nicotine concentration per cartridge above the criterion.

The majority of the studies showed nicotine concentrations below those reported by the manufacturer, and all except one indicated that the nicotine concentrations in some samples were at least 10% below or above that reported on the label of the product, meeting a quality criterion recommended by a United States manufacturers' association (31). In a median of 53% of samples, the nicotine concentration was misreported on the label by at least 10%, and in a median of 26% of samples, the nicotine concentration was misreported by at least 25%.

We know of only three studies of the consistency of nicotine concentration in e-liquids in different batches of the same brand and model of e-liquid. The median variation among production batches was 0.5% in one (19) and 15% (16) and 16% (32) in the other two.

Other studies have shown that some products labelled as not containing nicotine do have measurable nicotine levels. Table 3.2 lists studies in which the concentration of nicotine in e-liquids was analysed and compared with a reported absence of nicotine on the label. Almost half the studies reported that small amounts of nicotine were present in some e-liquids advertised as not containing nicotine. Furthermore, in about 5% of samples of e-liquids allegedly without nicotine, the concentration of nicotine was significant.

Table 3.2. Labelled and measured nicotine concentrations in e-liquids with declared zero nicotine

First author and reference number	Analysed	Samples Nicotine > 0.1 mg/mL	Nicotine > 10 mg/mL	Nicotine concentration in samples containing > 0.1 mg/mL
Beauval (13)	2	0	0	–
Cheah (10)	2	0	0	–
Davis (16)	10	0	0	–
Goniewiscz (21)	28	3	0	0.8–0.9
Kim (22)	20	0	0	–
Lisko (25)	5	0	0	–
Omaiye (33)	125	17	2	0.4–20.4
Raymond (29)	35	6	6	5.7–23.9
Trehy (30)	8	2	2	12.9–24.8/cartridge
Trehy (30)	5	2	2	12–21
Westenberger (34)	5	0	0	–

3.4 Nicotine delivery to ENDS users

The nicotine delivery profile of ENDS may be an important determinant of how effectively the product can substitute for a cigarette for a long-term smoker. Fig. 3.2 demonstrates the influence of the nicotine concentration in e-liquid, user behaviour and device power on the nicotine delivery profile of ENDS relative to a cigarette. Panel A (9) shows the nicotine delivery profile of a cigarette when smokers take 10 puffs with a 30-s inter-puff interval. Panel B shows the nicotine delivery profile of a 7.3-W ENDS loaded with 0, 8, 18 or 36 mg/mL nicotine e-liquid when users took 10 puffs of an average length of 3.6 s at a 30-s inter-puff interval. Clearly, the e-liquid nicotine concentration influences delivery of nicotine to the users' blood. When the 7.3-W ENDS is paired with 36 mg/mL nicotine e-liquid and when users take 10 ~5.6-s puffs, the pairing can match or exceed the nicotine delivery profile of a combusted cigarette (8).

Puff duration is also a factor in ENDS nicotine delivery: Panel C (8) shows the same device and e-liquid nicotine concentration as in Panel B, but the study participants took shorter puffs (2.9 s on average). When the puff duration is shorter and all other device and e-liquid characteristics are constant, less nicotine is delivered. Panel D shows the nicotine delivery profile of higher-powered ENDS devices (mean power, 71.6 W) when users took 10 puffs at a 30-s inter-puff interval (4). When these higher-powered devices were paired with 4 mg/mL nicotine liquid, they approximated the nicotine delivery profile of a combusted cigarette.

Overall, at least in some cases, these data suggest that some ENDS can deliver the same dose of nicotine, at the same rate as a cigarette, to venous blood. Unfortunately, few studies have been conducted to compare the ability of ENDS and cigarettes to deliver nicotine to arterial blood, an important indicator of exposure of the central nervous system to the drug (35). In the only such comparison to date, 10 puffs (30-s inter-puff interval) from a 7.3-W ENDS with 36 mg/mL liquid resulted in a lower mean arterial nicotine concentration (maximum, 12 ng/mL) than 10 puffs (30-s inter-puff interval) from a cigarette (maximum concentration, 27 ng/mL), although the time to peak concentration did not differ *(36)*. The sample was, however, small (four for ENDS; three for cigarettes), and puff duration was not measured. Under the controlled conditions of this study, positron emission tomography imaging showed that this ENDS effectively delivered nicotine to the central nervous system.

While the ENDS used to generate the data for Fig. 3.2 can deliver nicotine as effectively as a cigarette under some conditions, many ENDS cannot *(6, 9, 37–41)*. This heterogeneity in ENDS nicotine delivery is in contrast to regulated nicotine replacement products that deliver nicotine more reliably, although they often achieve lower plasma concentrations at a slower rate. For example, as shown in Panel A in Fig. 3.3 *(42)*, nicotine chewing-gum can take ≥ 30 min to achieve a peak plasma concentration, while Panel C shows that a nicotine patch can take > 2 h *(43, 44)*; other therapeutic products (e.g. nicotine lozenges) also deliver nicotine within this time frame *(43)*. Presumably, ENDS that deliver nicotine to the blood and brain as effectively as a cigarette are more likely to substitute for a cigarette, although this speculation has not been tested empirically, as the ENDS used in clinical trials on the question did not deliver nicotine effectively *(45)*.

Fig. 3.3. Plasma nicotine concentrations before, during and after administration of a single dose of nicotine in several therapeutic forms

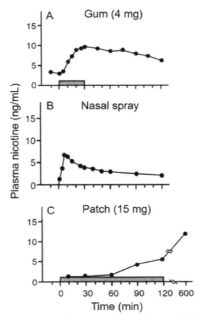

Note: the grey bar indicates duration of product use. *Source*: reference 42. Reprinted with permission from the Massachusetts Medical Society.

3.5 Toxicant content of ENDS emissions

ENDS toxicant emissions are a function of a variety of factors, including device construction, device power, liquid constituents and user behaviour. We review below the literature on ENDS toxicant emissions, beginning with nicotine and then moving to non-nicotine toxicants (for reviews of older literature, see Breland et al. *(1)* and Department of Health and Human Services *(46)*).

3.5.1 Nicotine emissions

The "yield" of nicotine from ENDS is the amount (in mg) of nicotine in the aerosol produced by an ENDS under a specific puffing regimen. Knowing the yield of nicotine from ENDS has been considered important for understanding the pharmacokinetics of nicotine in ENDS users. One review of the literature *(47)* identified seven studies of nicotine yield *(30, 34, 48–52)*; since then, several other studies on this issue have been published *(3, 33, 53, 54)*.

The nicotine yields in these studies were highly variable, depending on the type of ENDS used, the nicotine concentration of the e-liquids and the puffing regime used to obtain the aerosol. Some methodological issues complicate the comparability of studies, including the fact that the ISO methods of machine-smoking ENDS fail to activate some ENDS models. Although the nicotine yields from ENDS in these studies are not fully comparable with those from machine-smoked cigarettes, they are usually much lower than those from cigarettes *(47)*. The literature is, however, limited, for two important reasons. First, nicotine yield does not capture the rate of nicotine emission, which is a measure not only of the amount but also of the speed at which nicotine is made available to the user. The rate of nicotine emission is almost certainly related to the rate of nicotine delivery, and the rate of nicotine delivery is probably a key factor in the capacity of a nicotine-containing product to substitute for cigarettes by providing nicotine that rapidly reaches peak levels in the bloodstream and enters the brain *(55)*. Secondly, ENDS and their e-liquids are so heterogeneous that the results of a study on a particular ENDS are probably not generalizable to another.

To address the first concern, there is growing interest in measuring nicotine "flux", the rate at which nicotine is emitted from ENDS *(56, 57)*. Nicotine flux can be measured (usually reported in μg/s) and can be compared among ENDS and with cigarettes. Those ENDS that mimic the flux of a cigarette may be more likely to substitute well for a cigarette than ENDS that do not. To address the second concern, a physics-based mathematical model has been developed to predict the nicotine flux of any ENDS *(58)* – even those that have not yet been constructed. The model accounts for the time it takes for the coil to heat up after electricity begins flowing and how much the coil cools down between puffs. It also accounts for the various ways in which heat can be transported away from the coil: by the air passing over it, by the latent heat of the e-liquid as it evaporates, by conduction through the metal solder to the body of the device and by radiation to the surroundings. The inputs to the model are the length, diameter, electrical resistance and thermal capacitance of the heater coil; the composition and thermodynamic properties of the e-liquid (including nicotine concentration); puff velocity and duration and inter-puff interval; and the ambient air temperature. In a test of the model, the authors compared its predictions against actual nicotine flux measurements for 100 conditions in which power, puff topography, ENDS type (tank or cartomizer) and liquid composition were varied. The mathematically predicted nicotine flux was highly correlated to measured values ($r = 0.85$, $P < .0001$) *(58)*. In addition, the model accurately predicted the dependence of nicotine flux on device power and nicotine concentration (see Fig. 3.4), the ratio of propylene glycol and vegetable glycerine in the liquid and user puff duration. Fig. 3.4 shows that the higher the electrical power of the device, the lower the e-liquid nicotine concentration required to achieve a given flux. Cigarette flux is 100 µg/s, and the lines depict ENDS nicotine fluxes equivalent to twice, once and half that of a cigarette. Given the relation between ENDS power and liquid nicotine concentration shown in Fig. 3.4, a nicotine flux that is dramatically greater than that of a cigarette can be achieved by pairing a higher-powered ENDS with a higher concentration liquid. The figure does not show that some ENDS are powered well over 100 W *(4, 5)*.

Fig. 3.4. Relation between ENDS power and e-liquid nicotine concentration and effect on nicotine flux

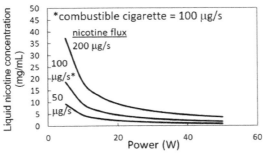

Source: reference 58, reproduced with permission from Dr Alan Shihadeh, American University of Beirut, Lebanon.

Another important issue with regard to ENDS nicotine emissions is the amount of nicotine in e-liquids and aerosols that is present in its more bioavailable, free-base form, as opposed to the less bioavailable protonated form *(17)*. Some studies of nicotine emissions from e-cigarettes have reported nicotine yields without determining whether the methods used resulted in quantification of total nicotine or only one of its forms *(38, 58)*, so that the reported results are difficult to compare or to evaluate with regard to nicotine delivery to the user. In an evaluation of this issue, the free-base nicotine fraction in 19 commercial liquids varied widely (10–90%), and, importantly, the differences were also seen in the aerosol *(17, 59)*, suggesting another factor that probably influences ENDS nicotine delivery to the user. Thus, in addition to measuring nicotine flux, the form of the nicotine in the aerosol should be determined. Overall, as for nicotine delivery to the user, there is considerable variation in nicotine emissions from ENDS, which can be explained and predicted by careful consideration of the many factors that influence it, especially ENDS power, liquid constituents and user behaviour.

3.5.2 Emissions of non-nicotine toxicants

Non-nicotine toxicants in ENDS aerosols are either present in the liquid or formed when the liquid is heated. Those present in the liquid before heating include propylene glycol and vegetable glycerine, which together make up 80–97% of the content of most e-liquids *(60)*, flavourings and other compounds added intentionally and contaminants not added intentionally. Aerosolized propylene glycol is a respiratory irritant *(61–64)* and, when administered intravenously at high doses, can cause potentially fatal lactic acidosis *(65)*. Preclinical work also indicates that vegetable glycerine may be toxic at high doses *(66, 67)*. The health effects of long-term, daily, chronic inhalation of aerosolized propylene glycol and/or vegetable glycerine are unknown. The flavourings used in e-liquids are usually compounds that are added to food, and their effects on the human lung after having been heated and aerosolized are unknown *(68)*. At least three flavourings that have been found in e-liquids and aerosols have raised health concerns: diacetyl (buttery flavour), which causes bronchiolitis obliterans *(69)*; benzaldehyde (fruity flavour), which is cytotoxic and genotoxic *(70)*; and cinnamaldehyde (cinnamon flavour), which is also cytotoxic and genotoxic *(71)* and can cause an inflammatory response in lung cells *(72)*. The contaminants include diethylene glycol, ethylene glycol and ethanol *(73, 74)*. Even if rigorous quality controls are imposed to ensure contaminant-free e-liquids, the uncertain effects of long-term, daily, frequent inhalation of aerosolized propylene glycol and vegetable glycerine and the many chemical flavourings that are often combined in a single liquid pose a potential health threat for ENDS users.

The non-nicotine toxicants formed when the liquid is heated include metals, volatile aldehydes, furans and benzene. In one study of 11 "first-generation" ENDS brands (disposable ENDSs shaped like tobacco cigarettes), three of each brand were puffed for 4.3 s every 5 min for two series of 60 puffs, and the resulting aerosol was analysed for elements, including metals *(75)*. The results revealed substantial variation among brands, but many metals were found in the aerosol generated from most brands, "in some cases at concentrations that were significantly higher than in conventional cigarettes". The authors concluded that most of the elements and metals in ENDS aerosols probably originate from components in the atomizer, such as the filament, solder joints, wick and sheath. These results show how ENDS construction can contribute to the non-nicotine toxicant profile of the aerosol.

In a study of an advanced-generation ENDS with a 1.5-Ω heating element and variable voltage battery (3.3–5.0 V), the aldehyde content of aerosols produced from a variety of liquids (all 6 mg/mL nicotine) was compared after 10 4-s puffs of 91 mL/puff *(76)*. Power was manipulated systematically from 9.1 to 16.6 W. Acetaldehyde, acetone, acrolein and formaldehyde were all present in ENDS aerosols, and aldehyde production increased proportionally as puff volume increased and dramatically when the power was > 11.7 W. The presence of aldehydes in ENDS aerosol is now well documented *(77–79)*, as is the role of device power in forming them: increasing ENDS power from 4.1 to 8.8 W approximately tripled volatile aldehyde emissions *(80–83)*. There also is some suggestion that flavourings contribute to non-nicotine toxicants formed during heating *(84–87)*. For example, heating sweeteners in e-liquids may expose users to furans, a toxic class of compounds. In one study *(88)*, a VaporFi platinum tank ENDS (2.3 Ω) was used to generate aerosol under various conditions, including power (4.2 and 10.8 W), puff duration (4 and 8 s) and sweetener (sorbitol, glucose and sucrose). The per-puff yield of some furans was comparable to values reported for combustible cigarettes, and, again, device power is a factor: increasing power from 4.3 to 10.8 W more than doubled furan emissions. With regard to benzene, increasing ENDS power from 6 to 13 W increased emissions of this carcinogen 100 times *(89)*, although the level remained far below those found in cigarette smoke. The fact that volatile aldehydes, furans and benzene are all formed by thermal degradation of the contents of e-liquids (e.g. propylene glycol, vegetable glycerine, sweeteners), coupled with the fact that increased device power increases the amount of these toxicants in ENDS aerosols, suggests that high-power ENDS are a particular public health concern. To date, most studies of the toxicant profile of ENDS aerosols have been limited to devices powered at 25 W or less (e.g. references *80, 83, 88, 90*), and much of the data reported here may not be relevant to the higher-powered devices common in some locations *(4, 5)*.

3.6 Potential role of ENDS in smoking cessation

Six narrative reviews *(91–96)* and six systematic reviews *(97)*, of which five were meta-analyses *(98–103)*, addressed the role of electronic nicotine and non-nicotine delivery systems (EN&NNDS) in smoking reduction and cessation. Two meta-analyses *(100, 102)* covered studies available up to January 2016.

All five systematic reviews of the quality of the evidence *(97, 98, 100, 102, 103)* concluded that the available studies provide evidence of low to very low certainty, due mainly to the limitations of the cross-sectional and cohort studies included in the reviews and the lack of detail in many of the published articles

Given these limitations, El Dib et al. *(102)*, and Malas et al. *(97)* concluded that no credible inferences could be drawn from their reviews and that the evidence remains inconclusive. Similarly, a review of the systematic reviews concluded that "overall, there is limited evidence that e-cigarettes may be effective aids to promote smoking cessation" *(104)*. The other systematic reviews, however, came to a different conclusion. While Kalkhoran & Glantz *(101)* determined that "as currently used, e-cigarettes were associated with significantly less quitting among smokers", Hartmann-Boyce et al. *(100)* and Rahman et al. *(98)* concluded that use of e-cigarettes is associated with smoking cessation and reduction. Khoudigian et al. *(103)* included only randomized clinical trials. The striking disparity in the conclusions arises from differences in the criteria for selecting eligible studies and the availability of studies at the times at which the reviews were done. Table 3.3 summarizes the studies used in each review.

Table 3.3. Comparison of studies included in reviews of the effectiveness of electronic nicotine and non-nicotine delivery systems as quitting aids

Studies available for review	Franck (91) Sep 2013	Harrell (92) Dec 2013	Rahman (98) May 2014	McRobbie (99) Aug 2014	Lam (93) Mar 2015	Ioakeimidis (94) Jun 2015	Kalkhoran (101) Jul 2015	Hartmann (100) Jan 2016	El Dib (102) Jan 2016	Malas (97) Feb 2016	Khoudigian (103) May 2016
Cohort studies											
Polosa, 2011	✓	✓		✓				✓			
Adkison, 2013		✓					✓			✓	
Caponnetto, 2013b	✓										
Ely, 2013			✓					✓			
Van Staden, 2013			✓					✓			
Vickerman, 2013 *(119)*	✓						✓	✓			
Borderud, 2014 *(123)*						✓	✓	✓			
Choi, 2014			✓				✓	✓			
Etter, 2014		✓	✓					✓			

Study	C1	C2	C3	C4	C5	C6	C7	C8	C9	C10	C11
Oncken, 2015							✓				
Pacifici, 2015							✓				
Pavlov, 2015						✓					
Polosa 2015							✓				
Shi, 2015						✓					
Sutfin, 2015						✓					
Cross-sectional studies											
Siegel, 2011	✓	✓									
Popova, 2013	✓										
Dawkins, 2013 (37)	✓								✓		
Goniewicz, 2013 (32)									✓		
Pokhrel, 2013	✓										
Brown, 2014		✓					✓		✓		
Christensen, 2014							✓		✓		
McQueen, 2015							✓				
Tackett, 2015									✓		
Randomized controlled trials with control group											
Bullen, 2010	✓			✓							✓
Bullen, 2013 (45)	✓	✓	✓	✓	✓	✓	✓	✓	✓	✓	✓
Caponnetto, 2013a	✓	✓	✓	✓	✓	✓		✓	✓		✓
Caponnetto, 2014	✓	✓					✓				
Adriaens, 2014				✓					✓	✓	
Randomized controlled trials without control group											
Hajek, 2015							✓		✓		
Unknown											
Humair, 2014				✓				✓			

The differences in the conclusions do not arise from the evidence provided by the randomized clinical trials. Meta-analysis of the few existing trials showed that ENDS use increases the likelihood of quitting smoking by a factor of two when compared with placebo. Two meta-analyses *(98, 99)* provided an estimated risk ratio of 2.29 (95% CI. 1.05, 4.96) in favour of quitting, one meta-analysis *(102)* gave an estimate of 2.03 (95% CI, 0.94, 4.38) and another *(103)* an estimate of 2.02 (95% CI, 0.97, 4.22). The differences are due to slight variations in the weight attributed to the two randomized clinical trials analysed and treatment of missing data. The different conclusions arise, more specifically, from the conflicting evidence presented by the longitudinal and cross-sectional studies reviewed. Below, we concentrate on the evidence from the longitudinal studies, because it is difficult to interpret the direction of possible associations in cross-sectional studies.

Since the last systematic review, seven new longitudinal studies have been published on the difference in quitting smoking between users and non-users of EN&NNDS *(105–110)*, including an update of a previous one with a longer follow-up *(111)*. Table 3.4 summarizes the findings of longitudinal studies according to sample attributes, characteristics of EN&NNDS products used by participants, measures used to typify ENDS use, criteria for nicotine dependence and abstinence and a summary of the results. It summarizes the seven studies that found a statistically significant positive or negative association between EN&NNDS use and smoking abstinence in systematic reviews. It also summarizes all seven longitudinal studies that were not included in the reviews, for a total of 16 studies. To select the best longitudinal studies for assessing the evidence, we considered that the association between ENDS use and quitting smoking as the outcome of interest should be measured under at least three conditions to obtain valid results:

- Criterion 1: It should be known whether the e-liquid used contains nicotine and the type (electrical power) of device used. Ideally, devices should be classified on the basis of their tested capacity to deliver nicotine, but this might prove difficult in population studies without laboratory testing of the devices used by participants. Otherwise, it is difficult to assess whether the association is linked to the potential role of ENDS as a nicotine replacement aid. We know that some ENDS devices can deliver cigarette-like amounts of nicotine in some instances *(4, 8)*; however, use of ENNDS or ENDS that cannot deliver nicotine because of low power and other factors is still common in the USA *(112)* and many other countries.
- Criterion 2: The analysis must discriminate between people who use ENDS to quit smoking and those who do not. Many use ENDS for reasons other than to quit, including reducing their smoking *(113)*, use indoors when smoking is not allowed or for recreational purposes *(114, 115)*. Conflating ENDS users who do and do not do so for quitting may bias the association towards the null if, as expected, the real effects on smoking cessation are different or even opposite.
- Criterion 3: The measures of ENDS use must be accurate and refined in order to distinguish between established and transient, erratic use to assess the effects of ENDS on population health *(116, 117)*. As ENDS use is a relatively new population behaviour, many people may experiment briefly with EN&NNDS but not adopt an established pattern of use. Comparisons of "ever use" with "never use" of ENDS, for example, might classify as users people who have used an ENDS only once in their lives, while it has been standard practice to consider people smokers if they have smoked at least 100 cigarettes in their lifetime. Conflating experimenters with steadier users may result in the biases described in the previous paragraph.

3. Clinical pharmacology of nicotine in electronic nicotine delivery systems · 273 ·

Table 3.4. Characteristics of longitudinal studies that showed a statistically significant positive or negative association between use of electronic nicotine and non-nicotine delivery systems and smoking abstinence in existing systematic reviews and more recent studies not included in those reviews

Reference no.	Year data collected	Sample					Product		Measures							Results	
		Country and baseline	Age range (years)	No. at baseline	Follow-up	Retention rate at follow-up	EN&NNDS type	Nicotine in e-liquid	EN&NNDS use				Nicotine dependence	Criterion for smoking abstinence	Association between EN&NNDS use and quitting smoking
									Frequency	Quantity	Duration	Reasons			
118	2010	USA. Nationally representative sample of smokers in the general population	≥18	5255	1 year	63%	NM	NM	Ever used vs Never used	NM	NM	NM	M	SR 30 days at follow-up	Ever use of an e-cigarette to quit was associated with less success in quitting than those who had not, adjusted for use of pharmacological help in last attempt, age, sex, race, education, cigarettes smoked per day, nicotine dependence and early smoking initiation.
119	2011–2012	USA Smokers Quit-line callers[a]	≥18	2758	7 months	35%	NM	NM	Ever use vs use for ≥1 month vs use for <1 month	NM	NM	N NU	M NU	SR 30 days at follow-up	Never ENDS users, 31.3% Use <1 month, 16.6% Use >1 month, 21.7% (P <.001). Not adjusted for use of NRT.
120	2011–2012	USA. Current smokers at baseline	18–59	1000	1 year	23%	NM	NM	EN&NNDS users classified as: ever used might use will never	NM	NM	NM	M	SR 1 month at follow-up	Smokers who had ever used e-cigarettes at baseline (not during follow-up) were significantly less likely to be abstinent at follow-up (AOR=0.41; 95% CI=0.186, 0.93) than smokers who reported at baseline they would never use e-cigarettes, adjusted for addiction (time to first cigarette in the morning), age, gender, education, ethnicity, desire to quit smoking, and smoking status. "Have used" and "will never use" include only those respondents with consistent responses at baseline and follow-up.

Ref	Year	Setting/population	Age	N	Duration	%			Exposure		Outcome	Results
121	2011–12	Representative sample of smokers in two United States metropolitan areas	18–65	1374	2 years	51%	NM	NM	Ever used an e-cigarette or not; if so, on how many of the past 30 days Currently used e-cigarettes every day, some days or not at all. If not at all, ever used e-cigarettes "fairly regularly" E-cigarette use: intensive (daily ≥ 1 month); intermittent (≥ once but not daily for ≥ 1 month); no use or at most once or twice.	M	SR 1 month at follow-up	Intensive e-cigarettes users were more than six times as likely to have quit smoking as those who had never used e-cigarettes or used them only once or twice, after adjustment for gender, age group, race/ethnicity and education level as well as baseline smoking level. Intermittent users were three times more likely not to quit than non-users or experimenters, although the association was not statistically significant.
122	2012	USA. Smokers on NRT recruited from a free, publicly available internet cessation programme. Randomized to social network integration or no social network × 2 (access to free NRT, no access)	30–52	3408	3 months	62%	NM	NM	Ever used during follow-up vs never used	NM	SR 1 month at follow-up	The odds of abstinence were lower among smokers who used e-cigarettes to quit than those who did not use e-cigarettes to quit, after adjustment for (measured at baseline) gender, age, race/ethnicity, education, parent, study treatment allocation, Fagerstrom score, number of quit attempts in the past year, cigarettes per day, self-efficacy to quit and stage of change. After further adjustment for use of other quit methods in the past 3 months and number of quit attempts in the past 3 months, the association was not significant.

123	2012–2013	USA In-treatment cancer patients who were current smokers and had received cessation treatment	NR	781	1 year	53%	NM	NM	Selection: 1 < puff Past 30 days at baseline, past 7 days use vs no use at follow-up	NM	NM	NM	M	SR 1 day at follow-up	After adjustment for nicotine dependence, number of past quit attempts and cancer diagnosis, current e-cigarette users were as likely to smoke non-users. Intention-to-treat analysis showed that non-users of e-cigarettes were twice as likely to abstain from smoking as e-cigarette users.
124	2012–2013	Great Britain. Current smokers	≥ 18	1759	1 year	44%	Ciga-like tanks	NM	Daily, less than daily but ≥ 1 weekly less than weekly but ≥ 1 monthly less than monthly not at all	NM	NM	NM	M	SR "I stopped smoking completely in the past year"	In comparison with no e-cigarette use at follow-up: non-daily cigalike users were less likely to quit (P <.001), daily cigalike or non-daily tank users were no more likely to quit (P =.4 and P =.4, respectively), and daily tank users were more likely to quit (P =.0012). Adjusted for age, sex, education level, income, motivation to quit and strength of urge to smoke.
111	2013–2015	Italy. Convenience sample of daily smokers for ≥ 6 months, EN&NNDS users for ≥ 6 months	30–75	932	2 years	69%	NM	NM	EN&NNDS user: inhales ≥ 50 puffs/ week	No. of months of continued EN& NNDS use	NM	NM	NM[a]	NM[b]	SR 30 days at follow-up with CO-tested subsample	Dual use did not improve the chances of smoking cessation at follow-up but reduced smoking.
110	2014	Canada. Smokers enrolled in cessation programme with access to free NRT and behavioural counselling	≥ 16	6526	6 months	–	NM	NM	Any EN&NNDS use in the 3 months before follow-up	NM	NM	NM	M NU	M	SR 7-day point prevalence of abstinence	Any EN&NNDS use was associated with less quitting at 6-month follow-up (AOR=0.50; 0.39, 0.64). Adjusted for heaviness of smoking at baseline and confidence in ability to quit.

#	Year	Population	Age	N	Duration	Retention	Exposure				Outcome	Findings
106	2014–2015	Hong Kong. Young current smokers from Quit line free counselling service not on other cessation programmes	18–25	190	6 months	99%	NM	NM	Ever tried	NM	M	7 days at 6-month follow-up Validated salivary cotinine	EN&NNDS ever users had nonsignificantly lower odds of quitting (AOR, 0.56; 95% CI, 0.24, 1.35) than non-users after adjustment for sex, age, smoking friends, smoking family members, baseline quit attempts and nicotine dependence level. They tried quitting more frequently than non-users.
105	2014–2015	France. Exclusive smokers and dual users	15–85	2057	6 months	695	NM	NM	Use of e-cigarettes: regular use in past 30 days. No use: used e-cigarettes sometimes, rarely or never	NM	NM	SR 7 days at 6-month follow-up	Baseline dual users were not more likely than exclusive tobacco smokers to have quit smoking at follow-up (12.5% versus 9.5%, $P = .18$, AOR, 1.2; 95% CI, 0.8, 1.9) after adjustment for age, sex, intention to quit smoking in the next 6 months, attempt to quit for at least 24 h in the previous 30 days at baseline and heaviness of smoking. They were more likely to reduce their smoking by half and tried to quit.
107	2014	USA. Smokers seeking treatment enrolled in a quitline programme offering counselling and NRT	≥18	5672	NR	NR	NR	NR	Number of days used EN&NNDS in past 30 days Classified as 0, none 1–5 6–20 30, daily	1 month before follow-up	NM	Self-reported 30-day complete abstinence at follow-up	Quit rate of daily EN&NNDS users no different from that of non-users. Quit rate of non-daily users lower than that of non-users. Results were adjusted for age, sex, education level, tobacco use, type of quitline programme and counselling with NRT use.
108	2013–2015	USA. Current cigarette smokers with no current use of EN&NNDS at baseline	≥25	5124	1 year	83%	Reported use of cartridges, refills and tanks	NM	Number of days used EN&NNDS in past 30 days	NM	Time to first cigarette	Self-reported 30-day complete abstinence at follow-up	Daily EN&NNDS users were eight time more likely to quit than non-users when using non-cartridge, refillable tanks. Experimenters were half as likely as non-users to quit. Non-daily users tended to quit less than non-users, but the association was statistically nonsignificant. Less quitting seen mainly among users of cartridge, non-refillable devices.

3. Clinical pharmacology of nicotine in electronic nicotine delivery systems

	Year	Setting/Population	N	Follow-up	Compliance	Intervention	Outcome 1	Outcome 2	Outcome 3	Results
109	2012–2015	USA. Hospital-discharged daily smokers who had received inpatient counselling plus encouragement to use NRT and who planned to stop smoking after discharge ≥ 18	Controls: 677; Intervention: 680	6 months	75% for controls 77% for intervention	NM	Any use of EN&NNDS in past 7 and 30 days	NM	All participants planned to quit smoking	No. of cigarettes per day and time to first cigarette	Cotinine-validated 7-day point prevalence of tobacco abstinence	Proportion of daily smokers motivated to quit who actually quit was slightly higher among EN&NNDS non-users than among users in the 7 days before follow-up, but the difference was statistically nonsignificant.
125	2014–2016	USA. Smokers of ≥5 cigarettes/day for ≥ 1 year, ENDS-naive, not seeking treatment to quit ≥ 18	Control= 22; arm 1 = 25; arm 2 = 21	3 months	Control= 73%; arm 1=76%; arm 2=71%. EMA compliance 58%	Cigalike	Arm 1= 3-week provision of 16 mg/mL nicotine Arm 2= 24 mg/mL nicotine Control = no provision of ENDS. No ENNDs provided but nor precluded from buying them	Number of days used EN&NNDS per week	Number of puffing sessions/day	From baseline to follow-up	CO-verified 7-day abstinence at follow-up	9.5% of users of 24 mg/mL nicotine e-liquid and 4.0% of 16 mg/mL users quit smoking vs 4.6% in controls. Difference was not statistically significant. No analysis by frequency of use of ENDS presented. Motivation to quit interacted, varying by group and over time within groups.

126	2015–2016	Italy. Smokers of ≥ 10 cigarettes/day for ≥ 10 years highly motivated to quit but not quitting or using NRT at baseline. ENDS-naive	≥ 55	70 per arm	Arm 1 = ENDS plus support Placebo = ENNDS plus support Control = no EN&NNDS	eGO 3.3–4.2 V working voltage	Arm 1 = ENDS plus support Placebo = ENNDS plus support Control = no EN&NNDS	Quitting rate higher among EN&NNDS users than in control group; however, there was no difference between ENDS and ENNDS users. Authors indicate that ENDS users had an e-liquid that delivered very little nicotine in the aerosol (0.1 mg/puff).

AOR: adjusted odds ratio; CI: confidence interval; CO: carbon monoxide; EN&NNDS: electronic nicotine and non-nicotine delivery systems; M: measured; NM: not measured; NRT: nicotine replacement therapy; NU: not used in analysis; SR: self-reported; U, used in analysis. [a] The authors did not report whether a caller was abstinent when enrolled. Some of the participants attended multiple calls, and the authors do not indicate whether they were "new" callers. [a] Instead, they measured desire to quit. [b] Instead, they measured strength of urge to smoke.

With these criteria in mind, we find that, of the 14 studies examined,

- only two characterized the type of device used (criterion 1);
- 12 studies did not restrict by or analyse the reasons for use of EN&NNDS, although two included adjustment for or analysis of some variables that could be used as proxies for using EN&NNDS (criterion 2); and
- seven studies compared cessation only between ever and never users of EN&NNDS, three used a crude measure of current use, and six used a more elaborated measure of frequency (criterion 3).

Seven longitudinal studies met at least one of the three criteria; none met all three. The combined evidence from the seven studies suggests that their samples consisted of different subgroups that experienced different or opposing effects of EN&NNDS use on cigarette cessation. Consequently, it could be hypothesized that **some** smokers may successfully quit tobacco use by using **some** types of ENDS frequently or intensively, while others experience no difference or are even prevented from quitting. The findings of these studies are shown in Table 3.5.

Table 3.5. Results of seven longitudinal studies of the efficacy of use of EN&NNDS for quitting tobacco use

Reference	Results by criterion		
	Frequency of use	Type of device used	Use for quitting smoking
Berry (108)	More daily users of EN&NNDS quit smoking than users, especially if they used refillable tank devices; however, the quit rate of experimenters with EN&NNDS and non-daily users of non-refillable cartridge devices was lower than that of non-users.		

EN&NNDS use	AOR[a]	95% CI	P
Non-users	Reference		
Experimenters	0.5	0.26, 1.00	0.05
Current users (not daily)	0.5	0.17, 1.47	0.21
Current users (daily)	7.9	4.45, 13.95	<0.001

EN&NNDS use by type of device	AOR	95% CI	P
Non-users			
Daily users of:			
non-cartridge	10.1	5.4, 10.1	<0.001
refillable	9.1	4.9, 17.0	<0.001
tank	10.2	5.4, 19.4	<0.001
non-tank	3.6	1.04, 12.2	0.04
Experimental or not daily users of:			
cartridge	0.3	0.1, 0.9	0.03
non-refillable	0.25	0.1, 0.9	0.04
non-tank	0.34	0.1, 0.8	0.02

[a] Adjusted for age, sex, race, region, income, frequency, intensity of cigarette use, nicotine dependence and previous quit attempts.

Hitchman (124)	Smokers who were daily tanks users were more likely to quit smoking than non-users of EN&NNDS, while non-daily users of cigalikes were less likely to quit. 	EN&NNDS use	AOR*	95% CI	P	 |---|---|---|---| 	Non-users	Reference			 	Non-daily cigalike	0.35	0.20, 0.60	0.002	 	Daily cigalike	0.74	0.39, 1.42	0.36	 	Non-daily tank	0.7	0.29, 1.68	0.42	 	Daily tank	2.7	1.48, 4.89	0.001	 *Adjusted for gender, age, education, income, motivation to stop smoking, strength of urge to smoke.	Although motivation to quit was measured, no information was collected on whether ENNDS were used specifically to quit.
Biener (121)	Smokers who used EN&NNDS daily were more likely to quit smoking than those who never used EN&NNDS. Non-daily users, however, were less likely to quit than non-users, although the association was not statistically significant. 	EN&NNDS USE	AOR*	95% CI	 	---	---	---	 	Non-users	Reference		 	Current non-daily	0.31	0.04, 2.80	 	Current daily	6.07	1.15, 24.4	 *Adjusted for gender, age, race, education and baseline level of smoking	Motivation to quit with EN&NNDS was measured as two variables: expectation to be smoking in 1 year and plans to quit within 6 months. Smokers who were not daily EN&NNDS users were six times more likely to expect to continue smoking in 1 year than smokers who were non-users. The expectation to continue smoking was similar for daily users and non-users. Smokers who did not use EN&NNDS were less like to have plans to quit smoking than smokers who were EN&NNDS users, but the differences were not statistically significant.
Manzoli (111)	Smokers who used ENDS or ENNDS regularly (inhaled ≥ 50 puffs/week) quit at the same rate as smokers who did not use ENNDS (AOR = 1.25; 95% CI, 0.85, 1.84.) Adjustment for baseline age, gender, body mass index, marital status, educational level, occupation, alcohol use, hypertension, hypercholesterolaemia, diabetes, self-reported health, years of tobacco smoking (former smoking for e-cigarette users), number of tobacco cigarettes smoked per day (or puffs per day for smokers of e-cigarettes only).	

Nowariak (107)	The quitting behaviour of daily and non-daily users of EN&NNDs was different from that of non-users. While the rate of quitting was no different for daily users and non-users, non-daily users quit at a lower rate than non-users. | EN&NNDS use in past 30 days	AOR	95% CI	P
---	---	---	---
Non-users	Reference		
Infrequent (1–5 days)	0.35	0.20, 0.59	< 0.001
Intermediate (6–29 days)	0.50	0.32, 0.80	0.004
Daily (30 days)	1.16	0.71, 1.70	0.453	 Adjusted for age, gender, education, tobacco type, advice from health professional, state programme and medication use.
Rigotti (109)	The quitting rate of smokers was 13.7% among those who had used EN&NNDS in the past 7 days and 17.9% among those who had not used them. The difference in risk was −4.3% (95% CI, −13.6, 5.1), which was not statistically significant. Quitting was therefore slightly but not statistically significantly lower among ENNDS users.
Zawartailo (110)	At the 6-month follow-up, the quit rate of EN&NNDS users (measured crudely as at least once in the past 3 months) who did not use them for smoking cessation was similar to that of non-users (37.6% vs 42.0%; $P = 0.43$), while the quit rate of EN&NNDS users who reported using them to help them quit smoking cigarettes was significantly lower than that of non-users.

AOR: adjusted odds ratio; CI: confidence interval; ENDS: electronic nicotine delivery systems; ENNDS: electronic non-nicotine delivery systems; EN&NNDS: electronic nicotine and non-nicotine delivery systems.

A cross-sectional study by Giovenco et al. *(127)* of current and former smokers who had quit since 2010, as reported in the 2014–2015 National Health Interview in the USA, lends some support to this hypothesis. The prevalence of quitting smoking tripled among daily ENDS users as compared with those who had never used ENDS, in line with the findings of Zhu et al. *(128)*. Interestingly, Giovenco et al. found the opposite effect among non-daily ENDS users and former experimenters, with a prevalence of quitting smoking of 2.6 and 1.5 times less than those who had never used ENDS, respectively. Success or failure in quitting in different subgroups may be influenced by:

- motivation to use EN&NNDS, including for quitting smoking;
- patterns of quantity, frequency and duration of ENDS use;
- technology used, including type of devices and e-liquids;
- type of smoker, including level of nicotine dependence and history of previous successful and unsuccessful quit attempts; and
- the regulatory environment for ENDS and tobacco use *(131-133)*.

Further support for the possibility that some smokers may successfully quit smoking by using ENDS includes the fact that ENDS may be economic substitutes for cigarettes *(134–136)* and the absence of a reversal in the decreasing rate of smoking rate in the two major EN&NNDS markets. Current cigarette smoking among adults in the USA decreased from 20.9% in 2005 to 15.1% in 2015, a 27.7% decrease (P for trend, < 0.05) *(137)*. The decrease includes a significant 1-year drop between 2014 and 2015 of 1.7 percentage points, which coincided with a notable increase in the cessation rate in 2014–2015, attributed by the authors partly to use of EN&NNDS. The results were adjusted for other changes to the policy environment that might affect quit attempts, such as tax increases and the "Tips from former smokers" media campaign of the Centers for Disease Control and Prevention in the USA.

In the United Kingdom, the proportion of current adult ($\geq$ 18 years) smokers in 2016 was 15.8%, the lowest prevalence recorded since the start of the Annual Population Survey in 2010 *(138)*. At the same time, the increase in the use of EN&NNDS in England has been associated with the increasing success of quit attempts *(139)*.

These data in themselves do not prove that use of EN&NNDS by the population is an effective quitting aid. They do show, however, that use of EN&NNDS is at least not changing the trend to a decreasing prevalence of smoking in the United Kingdom.

3.7 Potential health impact of ENDS

As some ENDS may help some smokers to quit, what is their potential health benefit for the population? The overall impact of using ENDS on population health depends primarily on two factors. One is the capacity of ENDS to help prevent smoking, and the other is the relative risk associated with their use in comparison with a defined alternative, such as smoking *(140)*.

3.7.1 Behavioural trajectories associated with use of ENDS

If ENDS prevent smoking, they do not entice nonsmokers into smoking but instead lure smokers into quitting smoking and, ideally, abstaining from nicotine. In other words, whatever the initial status of a person – never, current or former smoker – behavioural paths or trajectories associated with ENDS use must lead away from smoking and ultimately from nicotine dependence. Fig. 3.5 presents the 27 possible paths from an initial state of never, current or former smoker into one of four possible final states: exclusive smoker, exclusive ENDS user, dual user or dual abstainer. The web of trajectories in Fig. 3.5 represents only the behavioural paths between two nicotine products. In reality, it may be complicated by competition among more than two products, be they pharmaceutical, tobacco or consumer products.

Fig. 3.5. Web of trajectories associated with ENDS use

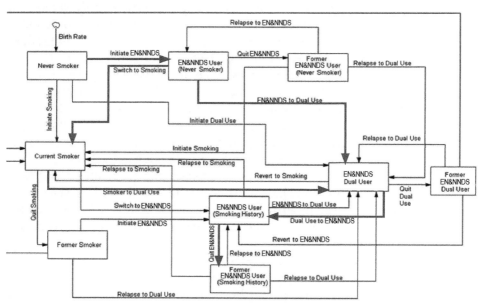

EN&NNDS, electronic nicotine and non-nicotine delivery systems. *Source*: Modified from reference *141*.

The first step in understanding the effect on population health of using ENDS is, therefore, to estimate the probability that people in each initial state will end over time in one of the four final states. The probabilities are context sensitive and therefore cannot be transferred among different cultural and regulatory environments for EN&NNDS and tobacco. Estimating the probabilities is complex, especially in light of the scant empirical evidence for characterizing them. The discussion has focused on the two most relevant combinations of trajectories in which EN&NNDS can play a role for or against health. One is the combination that leads smokers to quit smoking (blue lines in the figure), and the other is that which leads never smokers to smoke (red lines in the figure).

Trajectories that lead smokers to quit smoking

We discussed above the evidence for the role of EN&NNDS in quitting smoking. Contrary to the polarized discussion on whether ENNDS support or dissuade quitting, we concluded that the effects of EN&NNDS use on smoking cessation might depend on individual patterns of use and smoking, attitudes and behaviour, technology and the regulatory environment. The overall usefulness of ENDS for quitting might depend on the predominance of the subgroups for whom ENDS use might have an effect. For example, Giovenco et al. *(127)* showed that daily ENDS users quit smoking 3.2 times more often than never users; however, daily users represented only 5.1% of the sample. Non-daily ENDS users and former attempters, who represented 9.8% and 33.1% of the sample, respectively, however, quit smoking 2.6 and 1.5 times less often than those who had never used ENDS. Overall, the adjusted percentage of the total sample that quit is 26.5% with EN&NNDS and 28.2% without (Table 3.6). Given the predominance of non-daily EN&NNDS users and former experimenters in the population, preventing quitting predominated over promoting quitting among daily users.

Table 3.6. Theoretical impact on the prevalence of population quitting among smokers who use and do not use electronic nicotine and non-nicotine delivery systems (EN&NNDS) by type of user

Type of EN&NNDS user	Prevalence of EN&NNDS use (%)	Rate attributable to EN&NNDS use (%)	Adjusted[a] prevalence of quitting attributable to EN&NNDS use (%)	Prevalence in the absence[b] of EN&NNDS use (%)
Daily	5.1	52.2	4.6	1.4
Non-daily	9.8	12.1	1.1	2.8
Former	33.1	20.2	6.3	9.3
Non-user	51.9	28.2	14.7	14.7
Total	100	--	26.5	28.2

[a] Quit rate adjusted for a prevalence rate for daily and non-daily users, former experimenters and non-users of 3.18, 0.38, 0.67 and 1, respectively. [b] If the whole population were non-users at a quit rate of 28.2%.

Trajectories of never smokers to smoking

Young never smokers who experiment with ENDS are more likely to experiment with smoking later. A meta-analysis *(142)* of three longitudinal studies in the

USA *(143–145)* showed that young people who had used ENDS even once in their lives at baseline were twice as likely to experiment later with smoking than those who had never used ENDS. A more recent meta-analysis *(146)* that included the three previously mentioned studies and six additional ones *(147)* concluded that the likelihood of subsequent smoking initiation by young people who had ever used ENDS was about 3.5 times higher than that of never ENDS users. The authors also reported that using ENDS during the previous 30 days increased the chance of smoking at least once in the next 30 days by four. Two longitudinal studies in the United Kingdom *(148, 149)* showed a similar association between experimental use of ENDS and subsequent experimental smoking. The data available so far do not, however, prove that this evident association is causal or due mostly to ENDS use.

This association is difficult to understand, for several reasons *(150, 151)*. In most of the longitudinal studies, use of these products was measured as at least once in either a lifetime or in the previous 30 days. These recall periods cover a mixture of behaviour in the formative years of young people, including more frequent experimental use of ENDS and smoking, which is tentative and volatile, and also less prevalent established behaviour. It can be assumed that established ENDS use patterns better define the likelihood of future smoking than volatile, tentative ENDS use, such as having a puff once in a while.

Furthermore, there are three theoretical explanations for the association. The first is the "common liability conjecture". According to this theory, ENDS use and smoking are initiated independently of each other because they are the result of a common latent propensity to risky behaviour. Thus, it has been suggested that a large proportion of the young people who try ENDS and then smoke would have tried smoking regardless of the existence of ENDS. The fact that ENDS are used before smoking and not the other way around is due to several factors, including the novelty of ENDS. The second theory is the "renormalization" hypothesis, by which ENDS use is widespread and frequent among young people, and the devices and mannerisms of its use remind them of smoking. The similarity between ENDS use and smoking facilitates the trajectory from one product to the other within a social learning framework. The third theory is the "catalyst" theory, which comprises six hypotheses for initiation of ENDS use: flavour, health, price, role model, concealment and acceptance. Another three hypotheses are proposed to explain the transition to smoking: addiction, accessibility and experience *(152)*. Proving any of these theories will face critical methodological challenges *(153)*. In some longitudinal studies, adjustment has been made for variables to measure common susceptibility traits; however, residual confounding always muddles the association between ENDS use and smoking, and no one has proven beyond doubt which hypothesis or combination best explains the transition from never using nicotine to ENDS use and later to smoking.

The fact that some "never smokers" who experiment with ENDS end up smoking must be reconciled with the fact that the prevalence of current smoking among young people in the two countries with the most prominent ENDS markets continues to decrease. One review *(142)* shows that the prevalence of use of ENDS at least once a month increased quickly in some countries like the USA *(154)* (probably EN&NNDS), while in others such as the United Kingdom the rate among nonsmokers has been stable at very low levels.

3.7.2 Harm from ENDS and electronic non-nicotine delivery systems

Although EN&NNDS may route the population through trajectories in and out of smoking, the overall health impact of use of ENDS depends on the health risks associated with their use. The long-term health effects of EN&NNDS use are still unknown, and determination of such effects with some degree of certainty will require investigations of the health outcomes of large cohorts of well-characterized users who are followed for many years. In the meantime, conclusions about the toxicity of EN&NNDS are based mainly on empirical evidence from chemical and toxicological studies and, to a lesser degree, clinical studies. Reviews of these studies have led various authors to conclude, with more or fewer caveats, that EN&NNDS are not harmless but are generally less dangerous than cigarettes *(155–160)*, especially with regard to death from diseases associated with cigarette use. Efforts have been made to specify and characterize the health risks of EN&NNDS use by type of health condition.

Cancer risk

Ideal combinations of EN&NNDS device power settings, liquid formulation and use should produce an aerosol containing carcinogenic chemicals at a potency < 1% that of tobacco smoke and two orders of magnitude higher than that of a medicinal nicotine inhaler. As shown in Fig. 3.6, however, some products and circumstances can increase the cancer risk of EN&NNDS aerosol considerably, sometimes close to that of tobacco smoke *(161)*. Aerosols with higher carcinogenic potency appear to be formed when the user applies excessive power to the atomizer coil *(76)*. It has been argued that this occurs only under "dry puff" conditions *(162)* – brief situations that are readily detectable by EN&NNDS users. There is no empirical evidence, however, that this is due only to dry puff conditions or, if so, how often such conditions occur.

Fig. 3.6. Carcinogenic potency of formaldehyde and acetaldehyde in aerosol from electronic nicotine and non-nicotine delivery systems and in tobacco smoke, heat-not-burn devices, a nicotine inhaler and ambient air

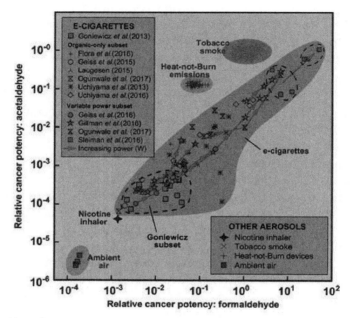

Source: reproduced from reference *161* with permission from BMJ Publishing Group Ltd.

Cardiovascular risk

There is controversy about whether the risk for cardiovascular events associated with use of EN&NNDS is as low as its carcinogenic potential. Some consider that the main cardiovascular risk of ENDS aerosol is due to the toxicity of nicotine, which appears to pose a low short-term cardiovascular risk in healthy users *(163)*. A review of clinical and cell culture studies conducted in 2015–2017 addressed the relation between ENDS use and indicators of risk for cardiovascular disease, including heart rate, blood pressure, and vagal tone; platelet aggregation and adhesion; aortic stiffness and endothelial function; expression of genes for antioxidant defence and immune system function; and indices of oxidative stress. Of the six studies reviewed that showed significant adverse cardiovascular effects, three found that ENDS had less effect on physiological cardiovascular risk indicators than cigarettes, and the other three found that ENDS had the same effect as cigarette smoking. Some studies indicated that these adverse cardiovascular effects are independent of nicotine, although adding nicotine may enhance them *(164)*.

Pulmonary risk

While EN&NNDS aerosol is probably less toxic than tobacco smoke and causes less mortality than cigarettes, the reduction in toxicity in the lung remains unknown for both long-term users who quit smoking and dual users. The authors of a review on the topic concluded that the induction of inflammation by EN&NNDS might differentially affect the risks for lung cancer and chronic obstructive pulmonary disease *(165)*. Thus, the most recent empirical evidence suggests that EN&NNDS aerosol is less toxic than cigarette smoke; however, there are no empirical data to quantify the relative risks of exposure to EN&NNDS aerosol and tobacco smoke.

Several efforts have been made to model the potential population impact of EN&NNDS *(166–168)*; however, the results are only as good as the data put into the model. Given the paucity of data, it is unclear which should be included in calculating the benefits of ENDS in worst- and best-case scenarios *(169, 170)*, especially for variables such as the efficacy of ENDS in helping people quit smoking and their safety relative to cigarettes.

Quantifying the effects of ENDS use on the health of the population is highly complex, as many variables must be taken into account. The available evidence indicates a possible positive effect of ENDS on population health, particularly if appropriate ENDS regulation is enacted to maximize their benefits and minimize their risks.

3.8 Summary of evidence, research gaps and policy issues derived from the evidence

ENDS are a heterogeneous class of products, with various profiles of nicotine and non-nicotine toxicants, which depend on factors including their construction, power, liquid constituents, nicotine concentration and user behaviour. The amount of nicotine delivered can range from none to doses that exceed those delivered by tobacco cigarettes in the same number of puffs. Nicotine from ENDS reaches users' blood faster than from most types of nicotine replacement therapy (NRT), and, at least with some ENDS, at higher concentrations. ENDS could be effective in cessation for some smokers under some circumstances, while, for other smokers, in different circumstances, it might have the opposite effect. Whether an ENDS has beneficial or detrimental effects on smoking cessation appears to depend on the technology, the motivation and consumer behaviour of the ENDS user, the type of smoker who seeks ENDS use and the regulatory environment for ENDS and tobacco use.

Translating the evidence into a potential role of EN&NNDS in smoking cessation is difficult. The evidence does not allow a blanket policy recommendation for or against general use of ENDS and ENNDS as cessation aids. Nevertheless, it points to four areas for regulatory consideration by policy-makers.

The concept of nicotine flux in ENDS regulation: regulators who wish to maximize the potential of the ENDS technology for nicotine substitution should consider the rate at which nicotine is emitted (i.e. nicotine flux) as a primary factor in their decision. In practical terms, factors that influence nicotine flux should not be regulated in isolation. ENDS nicotine flux can be modelled mathematically for product standards for regulatory purposes, although such standards should also be based on a clinical evaluation (i.e. effects in humans who are and are not ENDS users).

The relation between nicotine flux and toxicant profile: a corollary to the above is that the conditions under which different nicotine fluxes are obtained may affect the toxicant profile, because some of the same factors that increase the nicotine flux, such as power, also increase the concentrations of some toxicants in the aerosol, such as aldehydes. Therefore, regulators might consider how the manufacturers and the government should inform users of the balance between creating an adequate nicotine flux and the associated toxicant delivery.

Nicotine e-liquid concentration: despite some industry guidelines on labelling nicotine concentrations, the labels on many e-liquids do not indicate the concentration, are difficult to interpret or, most often, do not provide accurate information. Depriving ENDS users of accurate information on the nicotine concentration in e-liquids denies them important information for controlling their self-administration of nicotine.

Labelling and quality control for ENDS devices and e-liquids: the labels on all e-liquids should display the total amount of nicotine per receptacle, the ratio of free-base to protonated nicotine and the liquid concentration in mg/mL, visibly and understandably; otherwise, they should indicate that the e-liquids do not contain nicotine at a concentration above, for example, 0.1 mg/mL. Quality control must be used to ensure the veracity of labelling information and conformity to production standards.

Although the topic is not reviewed in this paper, there is conclusive evidence that exposure to nicotine in e-liquids other than through aerosol inhalation can harm health, sometimes fatally *(171)*. In order to avoid accidental exposure to nicotine, regulators should consider requiring child-resistant containers for all e-liquid receptacles.

The development of adequate policies and regulations on the ENDS issues described in this paper would benefit from disclosure requirements for manufacturers and effective, organized, systematic national surveillance. Key disclosure data to be requested from manufacturers include the voltage, resistance and power of marketed devices and the e-liquid constituents. In addition, monitoring should be conducted to determine consumer behaviour towards ENDS, such as who uses them, for what purpose, what and how products are used and the frequency of use.

Table 3.7 summarizes the evidence on the delivery of nicotine by ENDS, their effect on smoking cessation and their prospective impact on population health. The table also lists gaps in research and policy issues for each element of the evidence.

Table 3.7. Summary of evidence, research gaps and policy issues

Topic	Evidence	Research gaps	Policy and regulatory issues
Technology	ENDS differ widely in device technology and e-liquid components. Although the technological characteristics of ENDS that govern the delivery of nicotine are known, further research is needed to generalize the findings to the whole class of ENDS, beyond individual products.		The concept of nicotine flux in ENDS regulation As nicotine flux is the primary determinant of the capacity of ENDS to substitute for nicotine from cigarettes, regulators should consider this factor in endeavours to maximize the nicotine substitution potential of ENDS technology.
Effective nicotine delivery	A nicotine flux similar to that of cigarettes with regard to the levels and speed of nicotine delivery can be produced. The flux is influenced by: • the voltage applied to a coil of a given resistance: the higher the power, the higher the concentration of nicotine in the aerosol in a range of values between a threshold and a ceiling; • the concentration of nicotine in the e-liquid: the higher the concentration in the e-liquid the higher concentration in aerosol at a given power value; and • the puffing behaviour of the user: the longer the puff, the more nicotine is delivered.	Programmatic and transdisciplinary research is required, combining aerosol research, analytical chemistry and clinical laboratory methods to complete the mathematical model for predicting nicotine flux and to begin to describe similar models for predicting non-nicotine toxicants to the extent possible. The relation between nicotine flux and smoking cessation must be characterized at population level. Characterizing the relation between nicotine flux and potential abuse might help to explain and prevent initiation of cigarette smoking by non-tobacco users via ENDS.	Regulation of only one of the factors that influence nicotine flux (e.g. concentration in e-liquids) might result in changes to other factors by ENDS users (e.g. increased power). Such changes may or may not increase nicotine flux but may increase health risks (e.g. by increasing volatile aldehyde emissions at higher power values). ENDS that cannot deliver nicotine at a speed and concentration similar to those of cigarettes should, at a minimum, bear a warning that they cannot assist in smoking cessation.
Relation between toxicant profile of ENDS emissions and nicotine flux	Some of the factors that increase the nicotine flux, such as power, also increase the concentrations of some toxicants in the aerosol, such as aldehydes.	Further research is needed to characterize the association between the toxicant profile of ENDS emissions and the factors that influence nicotine flux, under a variety of conditions, such as: e-liquid composition: types of solvents and flavourings. Many flavourings used in ENDS liquids are intended to be consumed orally and have not been tested for safety after heating and inhalation. device construction: including general design, metals used, wicking materials.	Toxicant profile information Regulators might consider how to make users aware of the factors that influence the balance between creating an adequate nicotine flux and the associated toxicant delivery.

	Range of nicotine concentration in e-liquids	Regulation of a minimum and a maximum amount and concentration of nicotine in e-liquids should ensure a balance between:
		ensuring a sufficient concentration to reach an adequate flux for nicotine replacement; and
		the risk of accidental exposure to e-liquid containing nicotine.
		The total amount of nicotine per e-liquid container should be small enough to avoid the risk of lethal or serious cases of accidental nicotine poisoning if packaging safeguards fail.
		The nicotine concentration should be high enough to provide an adequate nicotine flux and to limit production of toxicants when the coil is heated too much to obtain more nicotine aerosol.
	Labelling and quality control of ENDS devices and e-liquids	
	Consumers must have accurate, reliable information on product design characteristics and e-liquid ingredients, including:	
		the ingredients listed comprise all liquid constituents (i.e. no contaminants); and
		all e-liquid containers are labelled to provide accurate information on the amount of nicotine, the ratio of free-base to protonated nicotine and the concentration.
		E-liquid contaminants known to pose a severe risk should be banned (e.g., diethylene glycol, diacetyl).
		Quality control must be used ensure the veracity of labelling information and implementation of production standards.

Concentration of nicotine in e-liquid

Despite some industry guidelines on labelling of e-liquids for nicotine concentration, many do not carry such a label, it is difficult to interpret or, most often, does not provide accurate information.

Depriving ENDS users of accurate information on the nicotine concentration in e-liquids denies them adequate information for deciding whether to self-administer nicotine.

A major research gap is how to communicate this information in a manner that is useful to the consumer and increases the chance that ENDS will be used to increase cessation rates. The same applies to ENDS power. Some ENDS users may not realize that they are using an ENDS that cannot deliver nicotine effectively, no matter what strength of nicotine liquid it contains.

Effectiveness of ENDS as a smoking cessation aid	ENDS can be effective for cessation for some smokers under some circumstances. It may have the opposite effect for other smokers under different circumstances. The conditions that appear to affect the potential of ENDS as a smoking cessation aid include: the appropriate combination of ENDS device and e-liquid to deliver nicotine at levels and speed similar to those of cigarettes; use of ENDS for quitting smoking with a minimal pattern of quantity, frequency and duration of use; the type of smoker, including level of nicotine dependence and history of previous successful and unsuccessful attempts to quit; and the regulatory environment of ENDS and tobacco use.	Better understanding is needed of the circumstances in which ENDS use can promote or be detrimental to smoking cessation, including: features of ENDS that facilitate cessation: the nicotine delivery profiles most strongly associated with quitting; whether flavours are necessary and, if so, which and how many will maximize quitting; and which user behaviour and use frequencies are most strongly associated with quitting and avoiding relapse to cigarettes; the subpopulations of smokers in whom ENDS are likely to be effective for cessation and those in whom they are not; the features of ENDS that maintain dual cigarette and ENDS use and how they could be manipulated to encourage cigarette cessation; andhe features of ENDS that maintain dual cigarette and ENDS use and how they could be manipulated to encourage cigarette cessation; and how to help long-term ENDS users to cease ENDS use, should they desire that outcome.	Regulators should be aware that ENDS use may have opposite effects on smoking cessation. Without further research, however, it is not possible to recommend policies to maximize their potential to help quit smoking and to minimize their detrimental effects on cessation.

Potential health impact of ENDS	The overall population health effects of using ENDS depends primarily on: – the capacity of ENDS to lead smokers away from smoking, while dissuading never smokers from starting to smoke and ex-smokers from relapsing; and – the health risks associated with their use relative to defined alternatives such as never using ENDS or smoking.	The gaps in research on the effectiveness of ENDS as smoking cessation aid are described above. Further research is needed on the long-term health consequences of ENDS use in comparison with non-use and with cigarette smoking in: smokers, including adults, children and pregnant women; and nonsmokers. In some cases, in-vitro (e.g. cell preparations) or in-vivo (e.g. animal models) methods may be appropriate for addressing these questions or to guide subsequent clinical investigation. For human studies, any diseases associated with ENDS use may not necessarily be those associated with cigarette smoking. Thus, particular attention must be paid to disease states and biomarkers of disease that might be associated with ENDS use, with an initial focus on known ENDS emissions that include propylene glycol, vegetable glycerine, flavourings, sweeteners and by-products of these liquid constituents that are produced when they are heated.	Development of adequate policies and regulations on the ENDS issues described in this paper would benefit from effective, organized national and global surveillance of the types of ENDS marketed and their use. More specifically, information is required on who is using them, for what purpose (including to deliver other drugs of abuse) and the products being used (including measures of voltage, resistance, power, liquid constituents, use frequency).

3.9 References

1. Breland A, Soule E, Lopez A, Ramôa C, El-Hellani A, Eissenberg T et al. Electronic cigarettes: what are they and what do they do? Ann N Y Acad Sci. 2016;1394:5–30.
2. Kennedy RD, Awopegba A, De León E, Cohen JE. Global approaches to regulating electronic cigarettes. Tob Control 2016;26:440–5.
3. Talih S, Balhas Z, Eissenberg T, Salman R, Karaoghlanian N, El Hellani H et al. Effects of user puff topography, device voltage, and liquid nicotine concentration on electronic cigarette nicotine yield: measurements and model predictions. Nicotine Tob Res. 2014;17:150–7.
4. Wagener T, Floyd E, Stepanov I, Driskill LM, Frank SG, Meier E et al. Have combustible cigarettes met their match? The nicotine delivery profiles and harmful constituent exposures of second-generation and third-generation electronic cigarette users. Tob Control. 2016;26:e23–8.
5. Rudy A, Leventhal A, Goldenson NI, Eissenberg T. Assessing electronic cigarette effects and regulatory impact: challenges with user self-reported device power. Drug Alcohol Depend. 2017;179:337–40.
6. Farsalinos K, Spyrou A, Stefopoulos C, Tsimopoulou K, Kourkoveli P, Tsiapris D et al. Nicotine absorption from electronic cigarette use: comparison between experienced consumers (vapers) and naïve users (smokers). Sci Rep. 2015;5:11269.
7. Hua M, Yip H, Talbot P. Mining data on usage of electronic nicotine delivery systems (ENDS) from YouTube videos. Tob Control. 2011;22:103–6.
8. Hiler M, Breland A, Spindle T, Maloney S, Lipato T, Karaoghlanian N et al. Electronic cigarette user plasma nicotine concentration, puff topography, heart rate, and subjective effects: the influence of liquid nicotine concentration and user experience. Exp Clin Psychopharmacol. 2017;25(5):380–92.
9. Vansickel A, Cobb C, Weaver M, Eissenberg T. A clinical laboratory model for evaluating the acute effects of electronic "cigarettes": nicotine delivery profile and cardiovascular and subjective effects. Cancer Epidemiol Biomarkers Prev. 2010;19:1945–53.
10. Cheah N, Chong N, Tan J, Morsed FA, Yee SK. Electronic nicotine delivery systems: regulatory and safety challenges: Singapore perspective. Tob Control. 2012;23:119–25.
11. Chaudhry IW, Leigh NJ, Smith DM, O'Connor RJ, Goniewicz ML. Labeling information on electronic nicotine delivery systems. Tob Regul Sci. 2017;3:3–9.
12. Buonocore F, Marques Gomes ACN, Nabhani-Gebara S, Barton SJ, Calabrese G. Labelling of electronic cigarettes: regulations and current practice. Tob Control. 2016;26:46–52.
13. Beauval N, Antherieu S, Soyez M, Gengler N, Grova N, Howsam M et al. Chemical evaluation of electronic cigarettes: multicomponent analysis of liquid refills and their corresponding aerosols. J Anal Toxicol. 2017;41:670–8.
14. Buettner-Schmidt K, Miller DR, Balasubramanian N. Electronic cigarette refill liquids: child-resistant packaging, nicotine content, and sales to minors. J Pediatric Nurs. 2016;31:373–9.
15. Cameron JM, Howell DN, White JR, Andrenyak DM, Layton ME, Roll JM. Variable and potentially fatal amounts of nicotine in e-cigarette nicotine solutions: Table 1. Tob Control 2013;23:77–8.
16. Davis B, Dang M, Kim J, Talbot P. Nicotine concentrations in electronic cigarette refill and do-it-yourself fluids. Nicotine Tob Res. 2014;17:134–41.
17. El-Hellani A, El-Hage R, Baalbaki R, Salman R, Talih S, Shihadeh A et al. Free-base and protonated nicotine in electronic cigarette liquids and aerosols. Chem Res Toxicol. 2015;28:1532–7.
18. Etter J, Zäther E, Svensson S. Analysis of refill liquids for electronic cigarettes. Addiction. 2013;108:1671–9.
19. Etter J, Bugey A. E-cigarette liquids: constancy of content across batches and accuracy of labeling. Addictive Behav. 2017;73:137–43.
20. Farsalinos KE, Gillman IG, Melvin MS, Paolantonio AR, Gardow WJ, Humphries KE et al. Nicotine

levels and presence of selected tobacco-derived toxins in tobacco flavoured electronic cigarette refill liquids. Int J Environ Res Public Health. 2015;12:3439–52.
21. Goniewicz ML, Gupta R, Lee YH, Reinhardt S, Kim S, Kim B et al. Nicotine levels in electronic cigarette refill solutions: a comparative analysis of products from the US, Korea, and Poland. Int J Drug Policy. 2015;26:583–8.
22. Kim S, Goniewicz M, Yu S, Kim B, Gupta R. Variations in label information and nicotine levels in electronic cigarette refill liquids in South Korea: regulation challenges. Int J Environ Res Public Health. 2015;12:4859–68.
23. Kirschner RI, Gerona R, Jacobitz KL. Nicotine content of liquid for electronic cigarettes. Clin Toxicol. 2013;51:684–4.
24. Kosmider L, Sobczak A, Szołtysek-Bołdys I, Prokopowicz A, Skórka A, Abdulafeez O et al. Assessment of nicotine concentration in electronic nicotine delivery system (ENDS) liquids and precision of dosing to aerosol. Przegl Lek. 2015;72:500–4.
25. Lisko JG, Tran H, Stanfill SB, Blount BC, Watson CH. Chemical composition and evaluation of nicotine, tobacco alkaloids, pH, and selected flavors in e-cigarette cartridges and refill solutions. Nicotine Tob Res. 2015;17:1270–8.
26. Pagano T, DiFrancesco AG, Smith SB, George J, Wink G, Rahman I et al. Determination of nicotine content and delivery in disposable electronic cigarettes available in the United States by gas chromatography–mass spectrometry. Nicotine Tob Res. 2015;18:700–7.
27. Peace MR, Baird TR, Smith N, Wolf CE, Poklis JL, Poklis A. Concentration of nicotine and glycols in 27 electronic cigarette formulations. J Anal Toxicol. 2016;40:403–7.
28. Rahman A, Nik Mohamed MH, Mahmood S. Nicotine estimations in electronic cigarette e-liquids among Malaysian marketed samples. Anal Chem Lett. 2018;8:54–62.
29. Raymond BH, Collette-Merrill K, Harrison RG, Jarvis S, Rasmussen RJ. The nicotine content of a sample of e-cigarette liquid manufactured in the United States. J Addict Med. 2018;12:127–31.
30. Trehy ML, Ye W, Hadwiger ME, Moore TW, Allgire JF, Woodruff JT et al. Analysis of electronic cigarette cartridges, refill solutions, and smoke for nicotine and nicotine related impurities. J Liquid Chromatogr Relat Technol. 2011;34:1442–58.
31. E-liquid manufacturing standards 2017 version 2.3.2. Mooresville (NC): American E-liquid Manufacturing Standards Association; 2017 (http://www.aemsa.org/standards/, accessed 19 September 2017).
32. Goniewicz ML, Hajek P, McRobbie H. Nicotine content of electronic cigarettes, its release in vapour and its consistency across batches: regulatory implications. Addiction. 2013;109:500–7.
33. Omaiye E, Cordova I, Davis B, Talbot P. Counterfeit electronic cigarette products with mislabeled nicotine concentrations. Tob Regul Sci. 2017;3:347–57.
34. Westenberger B. Evaluation of e-cigarettes. St Louis (MO): Center for Drug Evaluation and Research, Office of Pharmaceutical Science, Office of Testing and Research; 2009 (https://www.fda.gov/AboutFDA/CentersOffices/OfficeofMedicalProductsandTobacco/CDER/ucm088761.htm, accessed 1 October 2018).
35. Henningfield J, London E, Benowitz N. Arterial–venous differences in plasma concentrations of nicotine after cigarette smoking. JAMA. 1990;263:2049–50.
36. Baldassarri SR, Hillmer AT, Anderson JM, Jatlow P, Nabulsi N, Labaree D et al. Use of electronic cigarettes leads to significant beta2-nicotinic acetylcholine receptor occupancy: evidence from a PET imaging study. Nicotine Tob Res. 2018;20(4):425–33.
37. Dawkins L, Corcoran O. Acute electronic cigarette use: nicotine delivery and subjective effects in regular users. Psychopharmacology. 2013;231:401–7.
38. Hajek P, Goniewicz ML, Phillips A, Myers Smith K, West O, McRobbie H. Nicotine intake from electronic cigarettes on initial use and after 4 weeks of regular use. Nicotine Tob Res. 2014;17:175–9.
39. Nides MA, Leischow SJ, Bhatter M, Simmons M. Nicotine blood levels and short-term smoking

reduction with an electronic nicotine delivery system. Am J Health Behav. 2014;38:265–74.
40. St Helen G, Havel C, Dempsey DA, Jacob P 3rd, Benowitz NL. Nicotine delivery, retention and pharmacokinetics from various electronic cigarettes. Addiction. 2015;111:535–44.
41. Yan XS, D'Ruiz C. Effects of using electronic cigarettes on nicotine delivery and cardiovascular function in comparison with regular cigarettes. Regul Toxicol Pharmacol 2015;71:24–34.
42. Henningfield JE. Nicotine medications for smoking cessation. N Engl J Med. 1995;333:1196–203.
43. Choi JH, Dresler CM, Norton MR, Strahs KR. Pharmacokinetics of a nicotine polacrilex lozenge. Nicotine Tob Res. 2003;5:635–44.
44. Evans SE, Blank M, Sams C, Weaver MF, Eissenberg T. Transdermal nicotine-induced tobacco abstinence symptom suppression: nicotine dose and smokers' gender. Exp Clin Psychopharmacol. 2006;14:121–35.
45. Bullen C, Howe C, Laugesen M, McRobbie H, Parag V, Williman J et al. Electronic cigarettes for smoking cessation: a randomised controlled trial. Lancet 2013;382:1629–37.
46. E-Cigarette use among youth and young adults. A report of the Surgeon General. Atlanta (GA): Department of Health and Human Services, National Center for Chronic Disease Prevention and Health Promotion, Office on Smoking and Health; 2016.
47. Schroeder MJ, Hoffman AC. Electronic cigarettes and nicotine clinical pharmacology. Tob Control 2014;23:ii30–5.
48. Cobb NK, Byron MJ, Abrams DB, Shields PG. Novel nicotine delivery systems and public health: the rise of the "e-cigarette". Am J Public Health. 2010;100:2340–2.
49. McAuley TR, Hopke PK, Zhao J, Babaian S. Comparison of the effects of e-cigarette vapor and cigarette smoke on indoor air quality. Inhal Toxicol. 2012;24:850–7.
50. Pellegrino RM, Tinghino B, Mangiaracina G, Marani A, Vitali M, Protano C et al. Electronic cigarettes: an evaluation of exposure to chemicals and fine particulate matter (PM). Ann Ig. 2012;24:79–88.
51. Goniewicz ML, Kuma T, Gawron M, Knysak J, Kosmider L. Nicotine levels in electronic cigarettes. Nicotine Tob Res. 2012;15:158–66.
52. Farsalinos KE, Romagna G, Tsiapras D, Kyrzopoulos S, Voudris V. Evaluation of electronic cigarette use (vaping) topography and estimation of liquid consumption: implications for research protocol standards definition and for public health authorities' regulation. Int J Environ Res Public Health 2013;10:2500–14.
53. Laugesen M. Nicotine and toxicant yield ratings of electronic cigarette brands in New Zealand. NZ Med J. 2015;128:77–82.
54. Tayyarah R, Long GA. Comparison of select analytes in aerosol from e-cigarettes with smoke from conventional cigarettes and with ambient air. Regul Toxicol Pharmacol. 2014;70:704–10.
55. Henningfield JE, Keenan RM. Nicotine delivery kinetics and abuse liability. J Consult Clin Psychol 1993;61:743–50.
56. Shihadeh A, Eissenberg T. Electronic cigarette effectiveness and abuse liability: predicting and regulating nicotine flux. Nicotine Tob Res. 2014;17:158–62.
57. Eissenberg T, Shihadeh A. Nicotine flux: a potentially important tool for regulating electronic cigarettes. Nicotine Tob Res. 2014;17:165–7.
58. Talih S, Balhas Z, Salman R, El-Hage R, Karaoghlanian N, El-Hellani A et al. Transport phenomena governing nicotine emissions from electronic cigarettes: model formulation and experimental investigation. Aerosol Sci Technol. 2016;51:1–11.
59. Stepanov I, Fujioka N. Bringing attention to e-cigarette pH as an important element for research and regulation. Tob Control. 2014;24:413–4.
60. Han S, Chen H, Zhang X, Liu T, Fu Y. Levels of selected groups of compounds in refill solutions for electronic cigarettes. Nicotine Tob Res. 2015;18:708–14.
61. Burr GA, Van Gilder TJ, Trout BD, Wilcox TC, Driscoll R. Health hazard evaluation report HETA

90-0355-2449, Actors' Equity Association/The League of American Theatres and Producers, Inc. Washington (DC): National Institute for Occupational Safety and Health, Hazard Evaluations and Technical Assistance Branch Division of Surveillance, Hazard Evaluations, and Field Studies; 1994 (http://www.cdc.gov/niosh/hhe/reports/pdfs/1990-0355-2449.pdf, accessed 1 October 2018).
62. Moline JM, Golden AL, Highland JH, Wilmarth KR, Kao AS. Health effects evaluation of theatrical smoke, haze, and pyrotechnics. Report to Equity-League Pension and Health Trust Funds, 2000.
63. Varughese S, Teschke K, Brauer M, Chow Y, van Netten C, Kennedy SM. Effects of theatrical smokes and fogs on respiratory health in the entertainment industry. Am J Ind Med. 2005;47:411–8.
64. Wieslander G. Experimental exposure to propylene glycol mist in aviation emergency training: acute ocular and respiratory effects. Occup Environ Med. 2001;58:649–55.
65. Neale BW, Mesler EL, Young M, Rebuck JA, Weise WJ. Propylene glycol-induced lactic acidosis in a patient with normal renal function: a proposed mechanism and monitoring recommendations. Ann Pharmacother. 2005;39:1732–6.
66. Deichmann W. Glycerol:–effects upon rabbits and rats–. Am Ind Hyg Assoc Q 1941;2:5–6.
67. Hine CH, Anderson HH, Moon HO, Dunlop MK, Morse MS. Comparative toxicity of synthetic and natural glycerin. Arch Ind Hyg Occup Med. 1953;7:282–91.
68. Safety assessment and regulatory authority to use flavors: focus on electronic nicotine delivery systems and flavored tobacco products. Washington (DC): Federal Emergency Management Agency; 2013 (https://www.femaflavor.org/safety-assessment-and-regulatory-authority-use-flavors-focus-electronic-nicotine-delivery-systems, accessed 18 Oct 2017).
69. Farsalinos KE, Kistler KA, Gillman G, Voudris V. Evaluation of electronic cigarette liquids and aerosol for the presence of selected inhalation toxins. Nicotine Tob Res. 2014;17:168–74.
70. Kosmider L, Sobczak A, Prokopowicz A, Kurek J, Zaciera M, Knysak J et al. Cherry-flavoured electronic cigarettes expose users to the inhalation irritant, benzaldehyde. Thorax. 2016;71:376–7.
71. Behar RZ, Luo W, Lin SC, Wang Y, Valle J, Pankow JE et al. Distribution, quantification and toxicity of cinnamaldehyde in electronic cigarette refill fluids and aerosols. Tob Control 2016;25:ii94–102.
72. Gerloff J, Sundar IK, Freter R, Sekera ER, Friedman AE, Robinson R et al. Inflammatory response and barrier dysfunction by different e-cigarette flavoring chemicals identified by gas chromatography–mass spectrometry in e-liquids and e-vapors on human lung epithelial cells and fibroblasts. Appl In Vitro Toxicol. 2017;3:28–40.
73. Hutzler C, Paschke M, Kruschinski S, Henkler F, Hahn J, Luch A. Chemical hazards present in liquids and vapors of electronic cigarettes. Arch Toxicol. 2014;88:1295–308.
74. Varlet V, Farsalinos K, Augsburger M, Thomas A, Etter JF. Toxicity assessment of refill liquids for electronic cigarettes. Int J Environ Res Public Health. 2015;12:4796–815.
75. Williams M, Bozhilov K, Ghai S, Talbot P. Elements including metals in the atomizer and aerosol of disposable electronic cigarettes and electronic hookahs. PLoS One. 2017;12:e0175430.
76. Ogunwale MA, Li M, Ramakrishnam Raju MV, Chen Y, Nantz MH, Conklin DJ et al. Aldehyde detection in electronic cigarette aerosols. ACS Omega 2017;2:1207–14.
77. Herrington JS, Myers C. Electronic cigarette solutions and resultant aerosol profiles. J Chromatogr A 2015;1418:192–9.
78. Talih S, Balhas Z, Salman R, Karaoghlanian N, Shihadeh A. "Direct dripping": a high-temperature, high-formaldehyde emission electronic cigarette use method. Nicotine Tob Res. 2015;18:453–9.
79. Bekki K, Uchiyama S, Ohta K, Inaba Y, Nakagome H, Kunugita N. Carbonyl compounds generated from electronic cigarettes. Int J Environ Res Public Health. 2014;11:11192–200.

80. Sleiman M, Logue JM, Montesinos VN, Russell ML, Litter MI, Gundel LA et al. Emissions from electronic cigarettes: key parameters affecting the release of harmful chemicals. Environ Sci Technol. 2016;50:9644–51.
81. Wang P, Chen W, Liao J, Matsuo T, Ito K, Fowles J et al. A device-independent evaluation of carbonyl emissions from heated electronic cigarette solvents. PLoS One. 2017;12:e0169811.
82. Havel CM, Benowitz NL, Jacob P 3rd, St Helen G. An electronic cigarette vaping machine for the characterization of aerosol delivery and composition. Nicotine Tob Res. 2017;19(10):1224–31.
83. Geiss O, Bianchi I, Barrero-Moreno J. Correlation of volatile carbonyl yields emitted by e-cigarettes with the temperature of the heating coil and the perceived sensorial quality of the generated vapours. Int J Hyg Environ Health. 2016;219:268–77.
84. Khlystov A, Samburova V. Flavoring compounds dominate toxic aldehyde production during e-cigarette vaping. Environ Sci Technol. 2016;50:13080–5.
85. Farsalinos K, Gillman G, Kistler K, Yannovits N. Comment on "Flavoring compounds dominate toxic aldehyde production during e-cigarette vaping". Environ Sci Technol. 2017;51:2491–2.
86. Khlystov A, Samburova V. Response to comment on "Flavoring compounds dominate toxic aldehyde production during e-cigarette vaping". Environ Sci Technol. 2017;51:2493–4.
87. Tierney P, Karpinski C, Brown J, Luo W, Pankow JF. Flavour chemicals in electronic cigarette fluids. Tob Control. 2015;25:e10–5.
88. Soussy S, EL-Hellani A, Baalbaki R, Salman R, Shihadeh A, Saliba NA. Detection of 5-hydroxymethylfurfural and furfural in the aerosol of electronic cigarettes. Tob Control. 2016;25:ii88–93.
89. Pankow JF, Kim K, McWhirter KJ, Luo W, Escobedo JO, Strongin RM et al. Benzene formation in electronic cigarettes. PLoS One. 2017;12:e0173055.
90. Gillman I, Kistler K, Stewart E, Paolantonio AR. Effect of variable power levels on the yield of total aerosol mass and formation of aldehydes in e-cigarette aerosols. Regul Toxicol Pharmacol. 2016;75:58–65.
91. Franck C, Budlovsky T, Windle S, Filion KB, Eisenberg MJ. Electronic cigarettes in North America: history, use, and implications for smoking cessation. Circulation. 2014;129:1945–52.
92. Harrell PT, Simmons VN, Correa JB, Padhya TA, Brandon TH. Electronic nicotine delivery systems ("e-cigarettes"): review of safety and smoking cessation efficacy. Otolaryngol Head Neck Surg. 2014;151:381–93.
93. Lam C, West A. Are electronic nicotine delivery systems an effective smoking cessation tool? Can J Resp Ther. 2015;51:93–8.
94. Ioakeimidis N, Vlachopoulos C, Tousoulis D. Efficacy and safety of electronic cigarettes for smoking cessation: a critical approach. Hell J Cardiol. 2016;57:1–6.
95. Waghel RC, Battise DM, Ducker ML. Effectiveness of electronic cigarettes as a tool for smoking cessation or reduction. J Pharm Technol. 2014;31:8–12.
96. Orr KK, Asal NJ. Efficacy of electronic cigarettes for smoking cessation. Ann Pharmacother 2014;48:1502–6.
97. Malas M, van der Tempel J, Schwartz R, Mimichiello A, Lightfoot C, Noormohamed A et al. Electronic cigarettes for smoking cessation: a systematic review. Nicotine Tob Res. 2016;18:1926–36.
98. Rahman MA, Hann N, Wilson A, Mnatzganian G, Worrell-Carter L. E-cigarettes and smoking cessation: evidence from a systematic review and meta-analysis. PLoS One. 2015;10:e0122544.
99. McRobbie H, Bullen C, Hartmann-Boyce J, Hajek P. Electronic cigarettes for smoking cessation and reduction. Cochrane Database Syst Rev. 2014;(12):CD010216.
100. Hartmann-Boyce J, McRobbie H, Bullen C, Begh R, Stead LF, Hajek P. Electronic cigarettes for smoking cessation. Cochrane Database Syst Rev. 2016;9:CD010216.
101. Kalkhoran S, Glantz SA. E-cigarettes and smoking cessation in real-world and clinical settings: a systematic review and meta-analysis. Lancet Resp Med. 2016;4:116–28.
102. El Dib R, Suzumura EA, Akl EA, Gomaa H, Agarwal A, Chang Y et al. Electronic nicotine delivery

systems and/or electronic non-nicotine delivery systems for tobacco smoking cessation or reduction: a systematic review and meta-analysis. BMJ Open. 2017;7:e012680.
103. Khoudigian S, Devji T, Lytvyn L, Campbell K, Hopkins R, O'Reilly D. The efficacy and short-term effects of electronic cigarettes as a method for smoking cessation: a systematic review and a meta-analysis. Int J Public Health. 2016;61:257–67.
104. Stratton K, Kwan LY, Eaton DL, editors. Smoking cessation among adults. Chapter 17. In: Public health consequences of e-cigarettes. Washington (DC): National Academies Press;2018;541–88 (http://nationalacademies.org/hmd/Reports/2018/public-health-consequences-of-e-cigarettes.aspx, accessed 20 April 2018).
105. Pasquereau A, Guignard R, Andler R, Nguyen-Thanh V. Electronic cigarettes, quit attempts and smoking cessation: a 6-month follow-up. Addiction. 2017;112:1620–8.
106. Wang M, Li W, Wu Y, Lam TH, Chan SS. Electronic cigarette use is not associated with quitting of conventional cigarettes in youth smokers. Pediatric Res. 2017;82:14–8.
107. Subialka Nowariak EN, Lien RK, Boyle RG, Amato MS, Beebe LA. E-cigarette use among treatment-seeking smokers: moderation of abstinence by use frequency. Addict Behav 2018;77:137–42.
108. Berry KM, Reynolds LM, Collins JM, Siegel MB, Fetterman JL, Hamburg NM et al. E-cigarette initiation and associated changes in smoking cessation and reduction: the Population Assessment of Tobacco and Health Study, 2013–2015. Tob Control 2018; doi:10.1136/tobaccocontrol-2017-054108.
109. Rigotti NA, Chang Y, Tindle HA, Kalkhoran SM, Levy DE, Regan S et al. Association of e-cigarette use with smoking cessation among smokers who plan to quit after a hospitalization. Ann Internal Med. 2018;168(9):613–20..
110. Zawertailo L, Pavlov D, Ivanova A, Ng G, Baliunas D, Selby P. Concurrent e-cigarette use during tobacco dependence treatment in primary care settings: association with smoking cessation at three and six months. Nicotine Tob Res. 2016;19:183–9.
111. Manzoli L, Flacco ME, Ferrante M, La Vecchia C, Siliquini R, Ricciardi W et al. Cohort study of electronic cigarette use: effectiveness and safety at 24 months. Tob Control 2017;26:284–92.
112. Barrington-Trimis JL, Gibson LA, Halpern-Felsher B, Harrell MB, Kong G, Krishnan-Sarin S et al. Type of e-cigarette device used among adolescents and young adults: findings from a pooled analysis of 8 studies of 2,166 vapers. Nicotine Tob Res. 2018;20(2):271–4.
113. Hummel K, Hoving C, Nagelhout GE, de Vries H, van den Putte B, Candel MJJM et al. Prevalence and reasons for use of electronic cigarettes among smokers: findings from the International Tobacco Control (ITC) Netherlands Survey. Int J Drug Policy. 2014;26(6):601–8.
114. Attitudes of Europeans towards tobacco and electronic cigarettes (Special Eurobarometer 458). Brussels: European Union; 2017.
115. Ayers JW, Leas EC, Allem JP, Benton A, Dredze M, Althouse BM et al. Why do people use electronic nicotine delivery systems (electronic cigarettes)? A content analysis of Twitter, 2012–2015. PLoS One. 2017;12(3):e0170702.
116. Pearson JL, Hitchman SC, Brose LS, Bauld L, Glasser AM, Vilanti AC et al. Recommended core items to assess e-cigarette use in population-based surveys. Tob Control. 2018;27(3):341–6.
117. Amato MS, Boyle RG, Levy D. How to define e-cigarette prevalence? Finding clues in the use frequency distribution. Tob Control. 2015;25:e24–9.
118. Shi Y, Pierce J, White M, Vijayaraghavan M, Compton W, Conway K et al. E-cigarette use and smoking reduction or cessation in the 2010/2011 TUS-CPS longitudinal cohort. BMC Public Health. 2016;16(1):1105.
119. Vickerman K, Carpenter K, Altman T, Nash CM, Zbikowski SM. Use of electronic cigarettes among state tobacco cessation quitline callers. Nicotine Tobacco Res. 2013;15:1787–791.
120. Al-Delaimy W, Myers M, Leas E, Strong DR, Hofstetter CR. E-Cigarette use in the past and quit-

ting behavior in the future: a population-based study. Am J Public Health. 2015;105:1213–9.
121. Biener L, Hargraves JL. A longitudinal study of electronic cigarette use among a population-based sample of adult smokers: association with smoking cessation and motivation to quit. Nicotine Tob Res. 2014;17:127–33.
122. Pearson J, Stanton C, Cha S, Niaura RS, Luta G, Graham AL. E-Cigarettes and smoking cessation: insights and cautions from a secondary analysis of data from a study of online treatment-seeking smokers. Nicotine Tob Res. 2014;17:1219–27.
123. Borderud S, Li Y, Burkhalter J, Sheffer CE, Ostroff JS. Electronic cigarette use among patients with cancer: Characteristics of electronic cigarette users and their smoking cessation outcomes. Cancer. 2014;120:3527–35.
124. Hitchman SC, Brose LS, Brown J, Robson D, McNeill A. Associations between e-cigarette type, frequency of use, and quitting smoking: findings from a longitudinal online panel survey in Great Britain. Nicotine Tob Res. 2015;17:1187–94.
125. Carpenter M, Heckman B, Wahlquist A, Wagener TL, Goniewicz ML, Gray KM et al. A naturalistic, randomized pilot trial of e-cigarettes: uptake, exposure, and behavioral effects. Cancer Epidemiol Biomarkers Prev. 2017;26:1795–803.
126. Masiero M, Lucchiari C, Mazzocco K, Veronesi G, Maisonneuve P, Jemos C et al. E-Cigarettes may support smokers with high smoking-related risk awareness to stop smoking in the short run: preliminary results by randomized controlled trial. Nicotine Tob Res. 2018. doi:10.1093/ntr/nty175.
127. Giovenco DP, Delnevo CD. Prevalence of population smoking cessation by electronic cigarette use status in a national sample of recent smokers. Addict Behav. 2017;76:129–34.
128. Zhu S, Zhuang Y, Wong S, Cummins SE, Tedeschi GJ. E-cigarette use and associated changes in population smoking cessation: evidence from US current population surveys. BMJ. 2017;358:j3262.
129. Cho HJ, Dutra LM, Glantz SA. Differences in adolescent e-cigarette and cigarette prevalence in two policy environments: South Korea and the United States. Nicotine Tob Res. 2018;20(8):949–53.
130. Benmarhnia T, Leas E, Hendrickson E, Trinidad DR, Strong DR, Pierce JP. The potential influence of regulatory environment for e-cigarettes on the effectiveness of e-cigarettes for smoking cessation: different reasons to temper the conclusions from inadequate data. Nicotine Tob Res. 2018;20(5):659.
131. Yong HH, Hitchman S, Cummings K, Borland R, Gravely SML, McNeill A et al. Does the regulatory environment for e-cigarettes influence the effectiveness of e-cigarettes for smoking cessation? Longitudinal findings from the ITC four country survey. Nicotine Tob Res. 2017;19(11):1268–76.
132. Yong HH, Borland R, Balmford J, McNeill A, Hitchman S, Driezen P et al. Trends in e-cigarette awareness, trial, and use under the different regulatory environments of Australia and the United Kingdom. Nicotine Tob Res. 2015;17:1203–11.
133. Yong HH, Borland R, Hitchman SCB, Cummings KM, Gravely SML, McNeill AD et al. Response to letter to the Editor by Benmarhnia T, Leas E, Hendrickson E, Trinidad D, Strong D, Pierce J. The potential influence of regulatory environment for e-cigarettes on the effectiveness of e-cigarettes for smoking cessation: different reasons to temper the conclusions from inadequate data. Nicotine Tob Res. 2017;20(5):660–1.
134. Grace RC, Kivell BM, Laugesen M. Estimating cross-price elasticity of e-cigarettes using a simulated demand procedure. Nicotine Tob Res. 2014;17:592–8.
135. Johnson MW, Johnson PS, Rass O, Pacek LR. Behavioral economic substitutability of e-cigarettes, tobacco cigarettes, and nicotine gum. J Psychopharmacol. 2017;31:851–60.
136. Snider SE, Cummings KM, Bickel WK. Behavioral economic substitution between conventional cigarettes and e-cigarettes differs as a function of the frequency of e-cigarette use. Drug Al-

cohol Depend. 2017;177:14–22.
137. Jamal A, King BA, Neff LJ, Whitmill J, Babb SD, Graffunder CM. Current cigarette smoking among adults – United States, 2005–2015. Morbid Mortal Wkly Rep. 2016;65:1205–11.
138. Adult smoking habits in the UK: 2016. Cigarette smoking among adults including the proportion of people who smoke, their demographic breakdowns, changes over time, and e-cigarettes. Lonson: Office for National Statistics; 2017 (https://www.ons.gov.uk/peoplepopulationandcommunity/healthandsocialcare/healthandlifeexpectancies/bulletins/adultsmokinghabitsingreatbritain/2016#toc, accessed October 2018).
139. Beard E, West R, Michie S, Brown J. Association between electronic cigarette use and changes in quit attempts, success of quit attempts, use of smoking cessation pharmacotherapy, and use of stop smoking services in England: time series analysis of population trends. BMJ 2016;354:i4645.
140. McRobbie H. Modelling the population health effects of e-cigarettes use: current data can help guide future policy decisions. Nicotine Tob Res. 2016;19:131–2.
141. Hill A, Camacho OM. A system dynamics modelling approach to assess the impact of launching a new nicotine product on population health outcomes. Regul Toxicol Pharmacol 2017;86:265–78.
142. Yoong S, Tzelepis F, Wiggers JH, Oldmeadow C, Chai K, Paul CL et al. Prevalence of smoking-proxy electronic inhaling system (SEIS) use and its association with tobacco initiation in youths: a systematic review. Geneva: World Health Organization; 2015 (http://who.int/tobacco/industry/product_regulation/BackgroundPapersENDS2_4November.pdf?ua=1, accessed 3 October 2017).
143. Leventhal AM, Strong DR, Kirkpatrick MG, Unger JB, Sussman S, Riggs NR et al. Association of electronic cigarette use with initiation of combustible tobacco product smoking in early adolescence. JAMA. 2015;314:700.
144. Primack B, Soneji S, Stoolmiller M, Fine MJ, Sargent JD. Progression to traditional cigarette smoking after electronic cigarette use among US adolescents and young adults. JAMA Pediatrics. 2015;169:1018–23.
145. Wills T, Knight R, Sargent J, Gibbons FX, Pagano I, Williams RJ. Longitudinal study of e-cigarette use and onset of cigarette smoking among high school students in Hawaii. Tob Control 2016;26:34–9.
146. Soneji S, Barrington-Trimis J, Wills T, Leventhal AM, Unger JB, Gibson LA et al. Association between initial use of e-cigarettes and subsequent cigarette smoking among adolescents and young adults. JAMA Pediatrics. 2017;171:788.
147. Miech R, Patrick M, O'Malley P et al. E-cigarette use as a predictor of cigarette smoking: results from a 1-year follow-up of a national sample of 12th grade students. Tobacco Control 2017. doi:10.1136/tobaccocontrol-2016-053291.
148. Best C, Haseen F, Currie D, Ozakinci G, MacKintosh AM, Stead M et al. Relationship between trying an electronic cigarette and subsequent cigarette experimentation in Scottish adolescents: a cohort study. Tob Control. 2017. doi:10.1136/tobaccocontrol-2017-053691.
149. Conner M, Grogan S, Simms-Ellis R, Flett K, Sykes-Muskett B, Cowap L et al. Do electronic cigarettes increase cigarette smoking in UK adolescents? Evidence from a 12-month prospective study. Tob Control. 2017. doi:10.1136/tobaccocontrol-2016-053539.
150. Kozlowski LT, Warner KE. Adolescents and e-cigarettes: Objects of concern may appear larger than they are. Drug Alcohol Depend. 2017;174:209–14.
151. Gartner CE. E-cigarettes and youth smoking: be alert but not alarmed. Tob Control 2017. doi:10.1136/tobaccocontrol-2017-054002.
152. Schneider S, Diehl K. Vaping as a catalyst for smoking? An initial model on the initiation of electronic cigarette use and the transition to tobacco smoking among adolescents. Nicotine Tob Res 2015;18:647–53.

153. Lee PN. Appropriate and inappropriate methods for investigating the "gateway" hypothesis, with a review of the evidence linking prior snus use to later cigarette smoking. Harm Reduction J. 2015;12:8.
154. Jamal A, Gentzke A, Hu SS, Cullen KA, Apelberg BJ, Homa DM et al. Tobacco use among middle and high school students – United States, 2011–2016. Morb Mortal Wkly Rep. 2017;66:597–603.
155. Dinakar C, O'Connor GT. The health effects of electronic cigarettes. N Engl J Med. 2016;375:1372–81.
156. Hua M, Talbot P. Potential health effects of electronic cigarettes: a systematic review of case reports. Prev Med Rep. 2016;4:169–78.
157. Farsalinos KE, Polosa R. Safety evaluation and risk assessment of electronic cigarettes as tobacco cigarette substitutes: a systematic review. Ther Adv Drug Saf. 2014;5:67–86.
158. Burstyn I. Peering through the mist: systematic review of what the chemistry of contaminants in electronic cigarettes tells us about health risks. BMC Public Health. 2014;14:18.
159. Pisinger C. A systematic review of health effects of electronic cigarettes. Geneva: World Health Organization; 2017 (http://www.who.int/tobacco/industry/product_regulation/Background PapersENDS3_4November-.pdf, accessed 30 September 2017).
160. Glasser AM, Collins L, Pearson JL, Abudayyeh H, Niaura RS, Abrams DB et al. Overview of electronic nicotine delivery systems: a systematic review. Am J Prev Med. 2017;52:e33–66.
161. Stephens WE. Comparing the cancer potencies of emissions from vapourised nicotine products including e-cigarettes with those of tobacco smoke. Tob Control. 2017. doi:10.1136/tobaccocontrol-2017-053808.
162. Farsalinos KE, Voudris V, Poulas K. E-cigarettes generate high levels of aldehydes only in "dry puff" conditions. Addiction. 2015;110:1352–6.
163. Benowitz NL, Burbank AD. Cardiovascular toxicity of nicotine: implications for electronic cigarette use. Trends Cardiovasc Med. 2016;26:515–23.
164. Shields PG, Berman M, Brasky TM, Freudenheim JL, Mathe E, McElroy JP et al. A review of pulmonary toxicity of electronic cigarettes in the context of smoking: a focus on inflammation. Cancer Epidemiol Biomarkers Prev. 2017;26:1175–91.
165. Levy DT, Borland R, Villanti AC, Niaura R, Yuan Z, Zhang Y et al. The application of a decision–theoretic model to estimate the public health impact of vaporized nicotine product initiation in the United States. Nicotine Tob Res 2016;19:149–59.
166. Kalkhoran S, Glantz SA. Modeling the health effects of expanding e-cigarette sales in the United States and United Kingdom. JAMA Internal Med. 2015;175:1671.
167. Levy D, Borland R, Lindblom E, Goniewicz ML, Meza R, Holford TR et al. Potential deaths averted in USA by replacing cigarettes with e-cigarettes. Tob Control 2017. doi:10.1136/tobaccocontrol-2017-053759.
168. Soneji S, Sung H, Primack B, Pierce JP, Sargent J. Problematic assessment of the impact of vaporized nicotine product initiation in the United States. Nicotine Tob Res 2016;19:264–5.
169. Glantz SA. Need for examination of broader range of risks when predicting the effects of new tobacco products. Nicotine Tob Res. 2016;19:266–7.
170. Levy DT, Borland R, Fong GT, Vilanti AC, Niaura R, Meza R et al. Developing consistent and transparent models of e-cigarette use: reply to Glantz and Soneji et al. Nicotine Tob Res. 2016;19:268–70.
171. Stratton K, Kwan LY, Eaton DL, editors. Injuries and poisonings. Chapter 14. In: Public health consequences of e-cigarettes. Washington (DC): National Academies Press;2018;541–88 (http://nationalacademies.org/hmd/Reports/2018/public-health-consequences-of-e-cigarettes.aspx, accessed 20 April 2018).

4. A global nicotine reduction strategy: state of the science

Geoffrey Ferris Wayne, Portland, OR, USA

Eric Donny, Department of Physiology and Pharmacology, Wake Forest School of Medicine, Winston-Salem (NC), USA

Kurt M. Ribisl, Health Behavior, Gillings School of Global Public Health, University of North Carolina School of Medicine, Chapel Hill (NC), USA

Contents
4.1 Background
4.2 Individual outcomes of nicotine reduction
    4.2.1 Behavioural compensation and exposure to toxicants
    4.2.2 Threshold for establishing or maintaining nicotine addiction
    4.2.3 Tobacco cessation after use of VLNC cigarettes
    4.2.4 Non-compliance in clinical trials
    4.2.5 Adverse health effects and vulnerable populations
    4.2.6 Summary
4.3 Population impact of nicotine reduction
    4.3.1 VLNC cigarettes as replacements for regular cigarettes
    4.3.2 Substitution for VLNC cigarettes with alternative tobacco products
    4.3.3 The black market
    4.3.4 Manipulation of VLNC cigarettes
    4.3.5 Beliefs and attitudes regarding VLNC cigarettes and nicotine reduction
    4.3.6 Summary
4.4 Regulatory approaches to nicotine reduction
    4.4.1 Feasibility of nicotine reduction and potential challenges
    4.4.2 Prerequisites for successful implementation of a nicotine reduction policy
    4.4.3 Strategies for implementation of a nicotine reduction policy
    4.4.4 Philosophical objections to nicotine reduction
    4.4.5 Summary
4.5 Research questions
4.6 Policy recommendations
4.7 References

4.1 Background

In 2015, TobReg published an advisory note in support of policy to reduce the nicotine content of the tobacco used in cigarettes to a level below that necessary to develop or maintain addiction *(1)*. Policy to reduce nicotine could have a significant effect on public health by minimizing progression from experimental cigarette use to dependence, reducing duration of use, facilitating quitting, reducing the prevalence of smoking among addicted smokers and encouraging a transition to less harmful products for those who want or need nicotine from other sources *(1, 2)*. About one billion people in the world smoke, and global consumption of cigarettes

4. A global nicotine reduction strategy: state of the science

and other combusted tobacco products (including cigarillos and roll-your-own tobacco) continues to rise *(3)*. Article 9 of the WHO FCTC authorizes Parties to regulate the content and emissions of tobacco products, including nicotine *(4, 5)*. There is clear evidence that reducing the nicotine content of cigarettes to a very low level can reduce dependence on cigarettes *(1, 6, 7)*.

The WHO advisory note described the potential health outcomes of a nicotine reduction policy, while observing that many research projects, including clinical trials of very low nicotine content (VLNC) cigarettes, were still under way. The advisory note identified a number of open questions on possible health outcomes, including:

- the use and effects of VLNC cigarettes in non-smoking adolescents and adults and non-dependent smokers;
- the use and effects of VLNC cigarettes in vulnerable populations such as those with moderate or severe mental illness and pregnant women;
- the use and effects of VLNC cigarettes with other forms of nicotine or other drugs; and
- the effects of long-term use of VLNC cigarettes.

Critical discussion since publication of the advisory note focused on the open questions listed above *(8–10)*, the difficulty of drawing conclusions about the outcomes of restricted clinical trials for the open market *(8, 9, 11)* and potential unintended effects, such as the belief that VLNC products are less toxic than regular cigarettes *(11–13)*. The advisory note did not provide a detailed proposal of how a nicotine reduction policy would be enacted, and critical discussion has highlighted practical challenges to implementation of a nicotine reduction policy *(9, 14)*.

This section updates the fast-growing science on nicotine reduction and use of VLNC products and addresses the questions about a nicotine reduction policy raised in both the 2015 advisory note and the critical discussion that followed. More than 100 relevant studies have been published since the background paper on nicotine reduction was presented to TobReg at its seventh meeting in December 2013, when discussions on nicotine reduction began. In line with decision FCTC/COP7(14) of the seventh session of the COP to the WHO FCTC, the Convention Secretariat and WHO convened a face-to-face meeting of experts in many disciplines on measures for reducing the addictiveness of tobacco (15–16 May 2018, Berlin, Germany). Various clinical trials remained under way as of September 2017, at the time this section was written (19 September 2017; see Table 4.1), and the scope and relevant preliminary outcomes of these trials are discussed below. The plan of the FDA to pursue a nicotine reduction strategy *(15)* is also discussed.

Table 4.1. Clinical trials of very-low-nicotine content (VLNC) cigarettes under way or still being analysed at 19 September 2017

Principal investigator(s)	Clinical trial number	Purpose	Key feature	Sample size	Estimated completion date
Cassidy	NCT02587312	To determine how reducing the level of nicotine in cigarettes affects adolescent smoking behaviour.	Adolescents	90	May 2019
Cinciripini	NCT02964182	To assess the effects of different nicotine levels in cigarettes and electronic cigarettes.	Combination with e-cigarettes	480	November 2020
Colby	NCT03194256	To determine how the nicotine content of cigarettes and the nicotine concentration and flavours in e-liquids influence responses to and use of these products by adolescent smokers.	Laboratory-based; adolescents; e-cigarettes as alternative	120	February 2022
Donny	NCT02301325	To evaluate the impact of low-nicotine cigarettes with and without transdermal nicotine.	Transdermal nicotine	240	April 2017
Donny	NCT03185546	To evaluate the effects of VLNC cigarettes, e-cigarettes with different nicotine contents and e-cigarette flavourings on smoking.	Access to e-cigarettes with varying nicotine content and flavours	480	September 2021
Drobes	NCT02796391	To determine the impact of gradual vs immediate reduction of nicotine content in cigarettes, in combination with targeted behavioural treatment, on smoking cessation and intermediate outcomes.	Cessation	220	April 2021
Foulds, Evins	NCT01928758	To evaluate the effect of progressive nicotine reduction in cigarettes on smoking behaviour, exposure to toxins and psychiatric symptoms in smokers with comorbid mood and/or anxiety disorders.	Affective disorders	280	October 2018
Ganz	NCT01964807	To test products that provide a wide range of concentrations of nicotine, particles and other cardiovascular toxins to determine how the components associated with tobacco use adversely affect cardiovascular risk.	Laboratory-based; cardiovascular effects	90	September 2018
Hatsukami	NCT02139930	To compare two approaches to reducing levels of nicotine in cigarettes: an immediate vs a gradual reduction.	Gradual vs immediate	1250	March 2017
Hatsukami	NCT03272685	To determine the effect of VLNC cigarettes in the complex tobacco and nicotine product marketplace.	Access to wide range of alternative products	700	December 2022
Higgins, Heil	NCT02250534	To determine the effect of extended exposure to cigarettes with various nicotine contents in disadvantaged women.	Disadvantaged women	282	October 2019
Higgins, Sigmon	NCT02250664	To determine the effect of extended exposure to cigarettes with various nicotine contents in opioid abusers.	Opioid abusers	282	October 2019
Klemperer	NCT03060083	To examine two strategies by randomizing smokers to (i) switch to VLNC cigarettes or (ii) reduce the number of cigarettes smoked per day. All smokers use a nicotine patch to help them reduce their nicotine intake.	Compare nicotine reduction to cigarette reduction	74	November 2017

Koffarnus, Bickel	NCT02951143	To assess how smokers purchase and consume reduced-nicotine cigarettes.	Laboratory-based; cost	232	July 2020
Kollins, McClernon	NCT02599571	To investigate the effects of different nicotine levels in cigarettes in individuals with attention deficit hyperactivity disorder.	Attention deficit hyperactivity disorder	350	June 2020
McClernon	NCT02989038	To evaluate reactions and choices to self-administer cigarette smoke with various nicotine contents among low-frequency, non-dependent smokers aged 18–25 years.	Young adult light smokers	90	April 2019
Muscat, Horn	NCT01928719	To determine whether progressively lowering the nicotine content of cigarettes reduces or eliminates nicotine dependence in smokers of low socioeconomic status.	Low socioeconomic status	400	October 2018
Oncken	NCT02048852	To determine the effect of reducing the nicotine or menthol content of cigarettes or both in women of reproductive age.	Menthol; women of reproductive age	320	December 2018
Oncken, Dornelas	NCT02592772	To observe the effect of reducing the nicotine or menthol content of cigarettes or both in men.	Menthol; men; gender differences	57	December 2018
Peters	NCT02990455	To determine whether smokers with and without current alcohol use disorder respond to reduced-nicotine cigarettes by increasing their alcohol consumption or exposure to smoke.	Alcohol use	90	November 2019
Richie	NCT02415270	To determine the short-term effects of switching to tobacco products that deliver low levels of nicotine or reactive oxygen or nitrogen species on smoking behaviour and biomarkers of exposure to tobacco smoke and oxidative stress.	Biomarkers of exposure and harm	70	May 2017
Rohsenow	NCT01989507	To determine the effect of VLNC cigarettes in smokers with substance use disorders currently or in the past year.	Substance abuse	250	May 2019
Rose	NCT02870218	To assess use of cigarettes with 0.4, 1.4, 2.5, 5.6 or 16.9 mg nicotine in heavy, long-time smokers and light smokers with a shorter smoking history (< 10 cigarettes/day for < 10 years). Participants have free access to nicotine-containing e-cigarettes (15 mg/mL) throughout the 12-week study.	Dose-effect relation; heavy and light smokers; access to e-cigarettes	320	June 2019
Shiffman	NCT02228824	To investigate the effects of cigarettes with different nicotine levels in non-daily smokers.	Non-daily smokers	312	July 2017
Strasser	NCT01898507	To examine the effects of smoking low-nicotine cigarettes in groups of smokers with different nicotine metabolism.	Nicotine metabolism	210	July 2018
Tidey	NCT02019459	To examine whether reducing the nicotine content of cigarettes to non-addictive levels reduces smoking in smokers with schizophrenia.	Schizophrenia	80	August 2018
Tidey	NCT02232737	To determine the effect of extended exposure to cigarettes with various nicotine contents in people with current affective disorders.	Affective disorders	282	October 2019

Source: Data from clinicaltrials.gov and personal communication (EC Donny). Studies found at Clinicaltrials.gov on or before 19 September 2017 were identified with the search terms "low nicotine" and "reduced nicotine" and listed as not yet recruiting, recruiting, enrolling by invitation and active not recruiting. Studies with fewer than 50 participants or designs that did not address the impact of nicotine reduction relative to usual brand or normal nicotine content controls were omitted. A previous version of the table was published by Donny et al. (*16*). Listed by estimated date of completion.

4.2 Individual outcomes of nicotine reduction

Most estimates of the impact of nicotine reduction on individuals are from randomized clinical trials in which participants are given normal or VLNC cigarettes for an extended time, usually under double-blind conditions (see Table 4.1). Laboratory assessments that allow relatively rapid assessment of issues such as abuse liability and compensatory smoking have provided useful complementary information about nicotine reduction. In both cases, the cigarettes used contain less nicotine in the filler, generally as a result of genetic engineering (e.g. Quest brands; Spectrum cigarettes made available by the United States National Institute on Drug Abuse). Although nicotine can be reduced by other methods (e.g. nicotine extraction, as in Philip Morris' Next cigarette in the 1990s), relatively little clinical research has been done on cigarettes made by these methods because of practical issues such as cost, taste and availability. Furthermore, to our knowledge, no studies have been conducted to assess the effects of nicotine reduction in other forms of combusted tobacco (e.g. roll-your-own products, little cigars, cigarillos). The results of research to date cannot predict the full effects of a policy of nicotine reduction in the real world, although this is important. Hence, post-marketing surveillance (see e.g. International Tobacco Control project: http://www.itcproject.org/) would be an important component of any regulatory action on nicotine reduction.

4.2.1 Behavioural compensation and exposure to toxicants

The toxicants emitted from VLNC cigarettes are generally similar to those from regular cigarettes *(17)*; consequently, their effect on health depends largely on changes in smoking behaviour. Randomized clinical trials confirm earlier suggestions *(1, 18, 19)* that nicotine reduction reduces the number of cigarettes smoked per day from the number of both the usual brand and control cigarettes with a normal nicotine content *(6, 20)*.[4] Within 6 weeks of use, participants typically reported smoking 25–40% fewer cigarettes than controls *(6, 19–21)*.[1]

The common concern that smokers will compensate for the lack of nicotine by smoking more intensely is not supported by the evidence. Compensatory smoking is often observed when the nicotine content of the filler is only modestly reduced *(20, 22)*[1] and when the nicotine yield of cigarettes is reduced by ventilation but the filler remains unchanged (often called "light" cigarettes) *(23)*. Compensatory smoking does not, however, appear to be induced by cigarettes with ≤b2.4 mg nicotine per g of tobacco (as compared with a typical range of 10–20 mg/g) *(6, 18, 19, 22)*, possibly because the noxious components of smoke limit extreme compensatory behaviour, and more modest compensatory behaviour is ineffective in maintaining exposure to nicotine, given the large reduction in nicotine content. Instead, many studies found a reduction in exposure to the

---

[4] EC Donny, unpublished observation.

toxicants that result from smoking. For example, participants who switched to VLNC cigarettes had similar or lower expired CO, total puff volume per cigarette, mouth-level exposure to smoke constituents *(24)* and biomarkers of exposure to toxicants *(6, 18, 19, 21, 22, 25–27)*.

4.2.2 Threshold for establishing or maintaining nicotine addiction

An important question raised in the 2015 WHO TobReg advisory note is the threshold for establishing or maintaining nicotine addiction. It is not known whether there is a single threshold for different measures of addiction and different populations of smokers; however, data suggest that nicotine should be reduced to a maximum of 2.4 mg/g of tobacco and that decreasing the content to 0.4 mg/g of tobacco may have additional benefits. Although most smokers cannot discriminate between VLNC cigarettes with 2.4 and 0.4 mg/g, some can, suggesting that reducing the nicotine content below 2.4 mg/g might affect more smokers *(28)*. Furthermore, Donny et al. *(6)* found that, although the number of cigarettes smoked per day and self-reported craving after abstinence were significantly reduced at ≤f2.4 mg/g, composite measures of nicotine dependence were significantly reduced only at 1.3 mg/g (one measure) and 0.4 mg/g (multiple measures), and the number of quit attempts during the follow-up period was significantly increased only at 0.4 mg/g. Interestingly, research on rats also suggested that a reduction of 85–90% or more (relative to doses that reinforce dependence in most rats) is required to decrease intravenous self-administration of nicotine reliably *(29)*.

It is not known whether reducing nicotine to ≤t0.4 mg/g would also limit the development of addiction in adolescent smokers. Although some studies suggested greater reinforcing effects of nicotine in adolescent rats *(30, 31)*, other research suggests that the threshold dose required for animals to learn to self-administer nicotine is similar to or higher than the dose required to maintain behaviour *(32–34)*. The latter studies indicate that a product standard based on evidence for adult smokers would be expected also to limit smoking in tobacco-naive young people. Furthermore, a secondary analysis of the data of Donny et al. *(6)* showed that nicotine reduction resulted in more rapid decreases in indices of abuse liability (smoking satisfaction, psychological reward and enjoyment of respiratory tract sensations) in 18–24-year-old smokers than in adults aged ≥ 25 years *(35)*. A study of adolescent smokers' responses to use of single cigarettes with different nicotine contents *(36)* found that reduced-nicotine cigarettes had fewer positive subjective effects but no significant effect on withdrawal or exposure. Additional studies of adolescent and young adult smokers were under way as of September 2017 (see Table 4.1). Nevertheless, the possibility that adolescents (and other populations) are more sensitive to nicotine is one reason for considering product standards with a nicotine concentration well below 2.4 mg/g.

4.2.3 Tobacco cessation after use of VLNC cigarettes

Whether reductions in the number of cigarettes per day, exposure to nicotine and nicotine dependence increase cessation is less clear. Research with participants who were intending to quit smoking suggest that use of VLNC cigarettes can increase the rate of cessation *(19, 37, 38)*, whereas studies with participants who were not currently interested in quitting found persistence of smoking after weeks or even months of VLNC cigarette use *(6, 22, 26, 39)*. Most of the studies, however, were not designed to address cessation, were not powerful enough to determine cessation outcomes, assessed behaviour for only a relatively short time, did not include giving participants alternative sources of nicotine and were conducted with commercially available cigarettes with a normal nicotine content. Two studies are important in this regard. In a 10-site clinical trial of 1250 smokers who were not intending to quit, participants who were randomized to receive VLNC cigarettes for 20 weeks self-reported more days of abstinence (10.99) than those assigned to cigarettes with a normal nicotine content (3.1; $P < .0001$) and were more likely to be abstinent (biochemically confirmed) at the end of the trial *(20)*. In an exploratory trial, Hatsukami et al. *(40)* found that VLNC cigarettes increased the use of alternative nicotine products, including ENDS, cigars, cigarillos and NRT and that the more non-combusted products were used, the greater the number of days of abstinence from smoking. The potential interactive effect of nicotine reduction and alternative products has been discussed by several researchers *(41, 42)*, was the topic of an announcement by the FDA *(15)* (discussed further below) and was the objective of several clinical trials that were under way in September 2017 (see Table 4.1).

4.2.4 Non-compliance in clinical trials

Since the 2015 WHO TobReg advisory note, it has become clear that participants in most studies of VLNC cigarettes continue to use non-study cigarettes, despite instructions to the contrary and the provision of free study cigarettes. For example, 75–80% of participants randomized to VLNC study cigarettes in the study by Donny et al. *(6)* were at least partially non-compliant *(43)*. Smokers who were young and heavily dependent and reported that VLNC cigarettes were dissatisfying were more likely to be non-compliant. Most smokers had clearly reduced exposure to nicotine, suggesting that they had replaced many of their daily cigarettes with VLNC cigarettes but still used some regular-nicotine cigarettes. The most commonly reported context for non-compliance is the first cigarette of the day, underscoring the fact that non-compliance is due to nicotine dependence.[5] To address this limitation, statistical approaches have been used in which data are weighted according to the probability of compliance. This

---

5 EC Donny, unpublished observation.

indicated similar effects on the number of cigarettes per day and dependence as non-weighted data *(44)*.[6] Other studies of incentivized compliance resulted in somewhat less non-compliance (40–50%) (based on urinary biomarkers) and replicated the reported effects of VLNC cigarettes on smoking behaviour and dependence, with little evidence of negative consequences beyond mild, transient withdrawal symptoms.[1,2] Nevertheless, non-compliance may attenuate both positive (e.g. abstinence; see *45*) and negative effects (e.g. withdrawal symptoms) of nicotine reduction. The extent of non-compliance despite the provision of free VLNC cigarettes and incentives for compliance indicates that some smokers are likely to seek alternative sources of nicotine, whether on the black market, by hoarding, by product tampering or in non-cigarette products.

4.2.5 Adverse health effects and vulnerable populations

Analyses of the potential unintended consequences of nicotine reduction have shown few effects on the health of participants. VLNC cigarette use is associated with either no change in or less expired CO *(6, 18, 19, 21)*.[7] VLNC cigarette use may temporarily increase the occurrence of non-serious adverse events related to withdrawal symptoms *(19, 46)*[1] and attention deficit *(47, 48)*, although few deficits are observed after the first week or so.[8] Smoking of regular cigarettes suppresses body weight gain, raising the possibility that nicotine reduction could have unique adverse effects on obese smokers *(49, 50)*. It is important to note that unintended consequences could be masked by the use of regular commercial cigarettes, as described above. Thus, for example, an association between VLNC cigarettes use and weight gain in a clinical trial was obscured by participant non-compliance *(49)*.

Most clinical trials of VLNC cigarettes have involved relatively healthy daily smokers, generally with the exclusion of individuals such as those with serious mental illness or those who did not smoke every day. The positive and/or negative effects of nicotine reduction might vary among subpopulations, and several clinical trials were under way or were not published as of September 2017 on subpopulations who might be at risk (see Table 4.1). To date, the best data on the potential effects on different subpopulations of smokers are from secondary analyses of previous trials (commonly 6) and from laboratory studies. Bandiera et al. *(51)* reported greater increases in the number of cigarettes smoked per day in a relatively small subgroup of 24 highly dependent smokers in the trial of gradual nicotine reduction by Benowitz *(52)*; however, analysis of highly dependent smokers in the trial of abrupt reduction by Donny et al. *(6)* indicated no effect of dependence on the

---

6   DK Hatsukami, personal communication.
7   DK Hatsukami, personal communication.
8   Ribisl KM, Hatsukami D, Johnson, Huang J, Williams R, Donny E. Strategies to prevent illicit trade for very low nicotine cigarettes (unpublished information).

number of cigarettes smoked per day or other measures of possible compensatory smoking after randomization to VLNC cigarettes.[2] In addition, highly dependent smokers showed the greatest reduction in dependence as a consequence of nicotine reduction.[2] As noted above, studies often exclude non-daily smokers; however, a completed trial with non-daily smokers showed a significant (> 50%) decrease in the number of cigarettes per day as a consequence of nicotine reduction *(53)*.

Participants with a history of using alcohol *(54)* or cannabis *(55)* had reductions in smoking similar to that of smokers without such a history when randomized to VLNC cigarettes, with no increase in alcohol or cannabis use, although nicotine deprivation might increase the motivation of some individuals to drink *(45, 56)*. Tidey et al. *(57)* found that having depressive symptoms at baseline did not moderate the effects of nicotine reduction on the number of cigarettes smoked or measures of nicotine dependence and that randomization to VLNC cigarettes of smokers with severe depressive symptoms at baseline resulted in lower levels of depression than controls by the end of the trial. These findings are consistent with the results of a broader review of the potential impact of nicotine reduction on smokers with affective disorders *(58)*. An intensive laboratory study of smokers in three vulnerable populations (with affective disorders, with opioid dependence and socioeconomically disadvantaged women of childbearing age) found that reducing nicotine levels in a dose-dependent fashion decreased the reinforcing effects of cigarettes similarly in these populations *(59)*. Other trials under way as of September 2017 may provide further information about the potential unintended consequences of extended use of VLNC cigarettes in vulnerable populations; however, any negative effects must be weighed against the potential benefits, given the disproportionate harm of smoking to these populations *(60–64)*.

4.2.6 Summary

- Use of VLNC cigarettes in place of regular cigarettes reduces the number of cigarettes smoked per day and results in similar or less exposure to toxicants in clinical trials.
- Compensatory smoking is observed when the nicotine content of tobacco filler is modestly reduced but not with VLNC cigarettes.
- Less cigarette use is observed at a nicotine content of ≤ 2.4 mg/g. Less nicotine dependence and more quit attempts have been most reliably observed at levels of 0.4 mg/g, although this observation must be confirmed in long-term studies. Reducing nicotine to ≤ 0.4 mg/g may benefit the broadest population of current and potential smokers.
- Most studies were not designed to address cessation; however, most of the evidence suggests that use of VLNC cigarettes is likely to increase abstinence, even among smokers who are not intending to quit.

- Although the data are limited, reducing the nicotine content of cigarettes is likely to reduce smoking among adolescents.
- Use of VLNC cigarettes increases the use of alternative nicotine products such as ENDS, cigars, cigarillos and NRT in place of regular cigarettes. This conclusion warrants consideration of how nicotine reduction should be applied to nicotine products other than regular cigarettes.
- Significant non-compliance in clinical trials indicates that some smokers experiencing nicotine reduction are likely to seek alternative sources of nicotine.
- No significant adverse health effects have been identified with use of VLNC cigarettes in place of regular cigarettes.
- Reduced nicotine levels decrease the reinforcing effects of regular cigarettes similarly in a number of potentially vulnerable populations, including smokers with affective disorders, those who use alcohol and other substances and women of childbearing age of low socioeconomic status.

4.3   Population impact of nicotine reduction

Although clinical trials provide mounting evidence on the effects of VLNC cigarettes on individual behaviour and health when they are used instead of regular cigarettes, estimation of the public health impact of a nicotine reduction policy requires projection of these findings to the broader population *(9)* under market conditions, which may differ significantly from clinical trial settings, and will be affected by industry promotional activity *(9, 11)*. For example, observational studies indicate a smaller reduction in cigarette consumption with use of NRT than would be predicted by clinical trials, possibly because of underuse of the products in real situations *(65)*. Compliance with medication is probably better in a clinical setting than in real situations, whereas the opposite may be true in the case of clinical trials with VLNC cigarettes because of the commercial availability of regular nicotine tobacco products outside the trial *(44)*.

In the absence of population-based evidence on nicotine reduction, the WHO advisory note concluded that data from clinical studies and studies in experimental animals strongly suggest reduced risks at population level *(1, 2)*. Questions remained, however, about the potential role of illicit sales, potential product manipulation by either tobacco companies or smokers, the effects of a nicotine reduction policy on health beliefs, experimentation and potential use of other highly toxic combusted tobacco products (e.g. cigarillos or roll-your-own tobacco), either as substitutes for or in combination with VLNC cigarettes in the absence of policy measures to discourage their use *(1)* and the impacts of tobacco industry marketing that promotes use of VLNC cigarettes as part of a programme to counter further constraints on their other (regular) products. In this section, we review the current evidence on the probable population response and health impact of nicotine reduction.

4.3.1 VLNC cigarettes as replacements for regular cigarettes

The popularity of regular cigarettes is supported by their efficacy as a behavioural reinforcer *(66)*. Nicotine is absorbed more rapidly from regular cigarettes than from other nicotine-containing products, even when systemic exposure to nicotine is similar (although the technology used in alternative products such as ENDS is developing rapidly) *(66, 67)*. Cigarettes also provide a rich network of behavioural and sensory stimuli associated with nicotine delivery that further contributes to the strong conditioned rewarding effects of smoking *(1, 68, 69)* and to the ability of VLNC cigarettes to reinforce behaviour *(1, 70, 71)*. Smokers' consumption of regular cigarettes is correspondingly less sensitive to increases in price *(72–75)*, indicating greater abuse liability than with other forms of tobacco or nicotine, including *snus (76, 77)*, little cigars *(75, 78)*, large cigars *(75)*, loose smoking tobacco *(75, 79)*, nicotine chewing-gum *(77, 80)* and e-cigarettes *(80–83)*.

Despite the high abuse liability of regular nicotine cigarettes, extensive research shows that increased cost can promote quitting by current users, deter initiation by potential users and reduce tobacco use by continuing users *(72, 74, 84, 85)*. Sensitivity to cigarette price is greatest among low-income smokers *(86, 87)* and is influenced by environmental factors that include tobacco control and public education campaigns *(88)*. Smokers of high-equity or "premium" cigarette brands are less likely to quit, indicating the importance of marketing and brand-consumer relations *(89)*. Subjective characteristics, such as flavour and acceptability, also play a role in product use *(90–92)*, and perceived negative characteristics (e.g. of *snus* or chewing-gum) outweigh potential reinforcing effects, resulting in limited demand for some tobacco substitutes regardless of price *(77)*.

VLNC cigarettes can substitute at least partly for regular cigarettes *(1, 93–95)*, and laboratory studies suggest that they are a more reinforcing alternative than nicotine chewing-gum *(96)*. Use of VLNC cigarettes in place of regular cigarettes decreases cigarette demand *(6, 94)*. As the price breakpoint (the point at which purchase is no longer sustained) is higher for regular cigarettes than for VLNC cigarettes, smokers would quit at relatively lower prices if VLNC cigarettes replaced regular ones, and a subset of smokers have no demand for VLNC cigarettes, regardless of price *(95)*. This is consistent with self-reports from as many as half of participants who use VLNC cigarettes that they would stop smoking within 1 year if these were the only cigarettes available for purchase *(94)*. Smokers of mentholated cigarettes reported a similar intention to quit when confronted with a potential ban on menthol *(97)*. Such sensitivity to cost suggests that a nicotine reduction policy could increase the effectiveness of other tobacco control approaches such as raised price, smoke-free environments and media campaigns. Conversely, marketing and other social approaches could be used to support continued demand if the nicotine content of cigarettes were reduced.

As indicated in clinical trials (section 4.2.4), many smokers of VLNC cigarettes supplement their cigarette consumption outside of the study parameters. In a small pilot trial in New Zealand, a large price differential based on nicotine content prompted smokers to reduce the number of but not to eliminate their usual cigarettes and to replace them with VLNC cigarettes *(93)*. This suggests that, in a market in which both VLNC and regular cigarettes are available, most smokers who continue to smoke are unlikely to use VLNC cigarettes exclusively, even with a significant differential in price *(94, 95)*.

Few data are available on young and tobacco-naive users, although clinical trials with adolescents were under way as of September 2017. Meanwhile, the implications of studies of price and other abuse liability is that VLNC cigarettes are less likely than regular nicotine cigarettes to encourage the transition from experimentation to regular use. A possible, although not yet demonstrated outcome of nicotine reduction is that it may prevent some (possibly significant) proportion of children and adolescents who experiment with the most toxic (i.e. combusted) forms of nicotine from becoming long-term users of those products. In contrast, promotion may directly or indirectly encourage young people to take up VLNC products by convincing them that they are acceptable from a health perspective.

4.3.2 Substitution of VLNC cigarettes with alternative tobacco products

Tobacco products differ widely in terms of price, availability, marketing, social acceptability and perceived risk, as well as nicotine delivery and sensory and non-pharmacological characteristics. The balance of these environmental and product factors can result in significant population changes in tobacco product use. For example, studies in many countries have found a shift from manufactured to roll-your-own cigarettes, for reasons that include price and lower perceived risk *(98–100)*, even though roll-your-own products are inferior on subjective measures *(79)*. Concurrent use of several tobacco products is increasingly common *(101, 102)*, including the use of tobacco with e-cigarettes and other ENDS such as HTPs *(103, 104)*.

Although the prices of other combusted tobacco products are more sensitive than those of cigarettes, most are strong substitutes for regular cigarettes *(75)*. Some cigarette smokers maintain their tobacco consumption when cigarette prices rise by changing from regular cigarettes to little cigars *(78, 105)*, and interest is increasing in other substitutes (ENDS, *snus*, large cigars, roll-your-own) *(106)*. Large proportions of current smokers do not accurately differentiate regular cigarettes, little cigars, cigarillos, roll-your-own tobacco and cigars when presented with images of these products and their packaging, indicating the possibility of substitutions among relatively similar products *(107)*. The perceptions may be exaggerated by manufacturing and/or marketing by the tobacco industry, which may mimic or emphasize similarities in appearance or function among combusted or alternative products (e.g. ENDS and HTPs).

The presence of product substitutes can determine interactions among multiple products. For example, in a behavioural study in 2016, *snus* became a significant substitute for cigarettes once cigarillos had been removed as an option, and participants were less likely to break the study protocol by purchasing other products when cigarillos were available *(108)*. Findings on the substitution patterns of users of smokeless tobacco (moist snuff, *snus*, chewing tobacco) and cigarettes are mixed *(75, 92, 109–112)*: smokers, particularly in the USA, showed poor acceptance of smokeless products *(90, 92, 113–115)*. In Sweden, although lower taxes on *snus* contributed to the switch of many male cigarette smokers to *snus*, women did not switch to the same extent, illustrating the significant role of social and demographic factors *(114, 116)*.

ENDS are potential substitutes for regular cigarettes *(75, 108, 117–120)*, and, although ENDS are a heterogeneous class of products (see section 3), at least some are more effective in decreasing regular cigarette use than NRT *(119, 121–123)*, partly because of similar subjective and sensorimotor characteristics *(69)*. It has been reported ENDS are better accepted, provide more satisfaction and can better reduce craving, negative affect and stress in smokers than NRT *(122, 124)*, although they are rated as less satisfying than regular cigarettes *(42)*. ENDS remain highly sensitive to price *(81, 118)* and were found to be more expensive than combusted products in 45 countries *(125)*. Users of both cigarettes and e-cigarettes more successfully maintain reduced cigarette consumption in the short term than those who smoke only cigarettes *(126)*. ENDS are generally perceived to pose a lower risk than regular cigarettes, although it is a misperception that ENDS pose an equal or greater health risk than cigarettes *(127)*.

Use of ENDS by adults is concentrated primarily among current and former smokers *(104, 128–137)*. Studies on young people have had mixed results, some indicating a high level of interest and/or experimentation by never smokers *(133, 138, 139)* and others reporting concurrent use of ENDS and other tobacco products by current smokers *(130–132, 133, 140–145)*. Some countries have reported dramatic increases in ENDS use among young people; current use exceeded 20% among high-school-aged young people in the USA in 2018 *(146)*. The association between ENDS use and cigarette smoking has led to concern that ENDS might act as a gateway to the use of other tobacco products, specifically regular cigarettes, particularly among adolescents and young adults *(137, 147–*

154), although confounding factors make it difficult to interpret the study results *(155–159)*. Further substantial concern has, however, been expressed recently in the USA about the "epidemic of youth e-cigarette use" *(160)* and the possible involvement of e-cigarettes in encouraging young people to smoke conventional products *(161)*. ENDS are less likely to displace cigarette smoking among people who want or need to quit as long as regular nicotine cigarettes are widely available. A reduction of the nicotine level in combusted and other tobacco products might, however, increase substitution with ENDS *(40)*.

4.3.3 The black market

If the only combusted tobacco products legally available contained very low nicotine levels, there would potentially be illicit trade in cigarettes with regular nicotine levels. As no jurisdiction has yet restricted cigarettes with regular nicotine, the information about illicit trade pertains to products that have been banned or on which taxes have been raised, and consumers have sought cheaper, lower-tax products.

Illicit trade can be conducted at retail locations, by street sellers and via the internet. Banning of some tobacco products in some jurisdictions (e.g. menthol and other flavoured cigarettes in Canada and clove and other flavoured cigarettes in the USA) has not created a major black market, and most illicit trade is conducted to avoid cigarette excise taxes. The National Research Council in the USA estimated that illicit sales accounted for 8.5–21% of the total cigarette market globally and for 11.6% of cigarette consumption in 84 countries in 2007 *(162)*. The volume of illicit trade depends on factors that include the magnitude of the price (or tax) differential, the ease of accessing products and the efficiency of tax administration.[9] In jurisdictions with strong tax administration, illicit trade is modest. The practices of these jurisdictions include strong "track-and-trace" systems to follow products from the manufacturer to the distributor and the retailer, strong customs control and rigorous tax stamping. Although thermal onion-skin stamps are used in many countries, encrypted tax stamps are used in California and Massachusetts, USA, as they are more durable and can include encoded information about the supply chain *(163)*. Many countries do not have strong track-and-trace systems to minimize illicit trade, and these should be instituted to minimize illicit trade in cigarettes containing normal nicotine levels.

Another means of reducing illicit trade is to implement strong policies, including a penalty structure, the likelihood of detection and provisions for adequate staff for surveillance and enforcement. To reduce the likelihood of a black market, policies could address the manufacture, sales, purchase, use and/or possession of contraband products with a normal nicotine level.

---

9   Ribisl KM, Hatsukami D, Johnson, Huang J, Williams R, Donny E. Strategies to prevent illicit trade for very low nicotine cigarettes (unpublished information).

Manufacturers should be prohibited from making regular nicotine cigarettes for domestic distribution, and importers should be prohibited from importing such products. As for many other illicit products, the strongest penalties and the most enforcement should be imposed on the manufacturers and sellers of the products, such as retailers and internet vendors, rather than the smokers who possess them.

In the same way that taxation agencies check compliance by visiting manufacturing plants and retailers, an inspections programme should exist that includes testing of purchases to ensure that products with regular nicotine are not circulated through the retail supply chain. Policies to restrict sales from internet vendors are also required. In the USA, high rates of cigarette sales to minors and violation of tax policies led to new restrictions on payment options (e.g. Visa, Mastercard) and shipping options (UPS, FedEx), which drastically reduced the number of internet cigarette vendors and traffic to these websites *(164)*. Similar policies should be enacted to restrict sales into countries that have low-nicotine products. Australia, for instance, has some of the highest tobacco excise taxes in the world, yet has successfully blocked nearly all websites selling regular cigarettes. In addition, enforcement agencies should make routine test purchases to ensure that internet vendors do not sell regular nicotine products in countries in which they are restricted.

Illicit trade can also be reduced by the widespread availability of substitute products that appeal to adult consumers. As discussed earlier, these products may include ENDS or *snus* containing nicotine. If these products are banned or restricted, as they are in some countries, there is likely to be greater demand for cigarettes on the black market.

4.3.4 Manipulation of VLNC cigarettes

Research on VLNC cigarettes has been conducted with experimental cigarettes that have chemical and physical properties sufficiently similar to those of commercial cigarettes that they are considered acceptable for use in studies of behaviour *(17)*. Experimental products do not necessarily mimic the complex product designs and are not subject to the marketing tactics used to increase the appeal of commercial cigarettes and other tobacco products *(89, 165)*. In an open market, branding may alter expectations or perceptions of VLNC cigarettes, with subsequent behavioural effects *(89)*. Companies are likely to use heavy promotion both to use VLNC cigarettes to deflect attention from evidence-based measures to reduce smoking and as a means of maintaining use of current products. The physical or chemical parameters of cigarette construction could also be manipulated to alter the formulation of VLNC cigarettes *(1)*.

A number of studies have addressed the behavioural effects of non-nicotine constituents present in tobacco to determine whether they add to or interact with the abuse liability of nicotine. Some studies of self-administration by experimental animals indicated greater reinforcing effects than nicotine alone *(166–169)*, while others found no additive or synergistic effects *(170–172)*. The role of specific tobacco compounds and their interaction with nicotine remains poorly understood *(1, 7, 173)*. Nornicotine, norharman and acetaldehyde have been shown to support self-administration independently of tobacco *(1, 174–176)* but only at doses substantially higher than those in cigarette smoke *(7)*. Monoamine inhibitors increase self-administration of low doses of nicotine by rats, although reinforcement is minimal if nicotine is further reduced *(177)*. Menthol appears to strengthen the reinforcing mechanism of nicotine and thus promote nicotine consumption and tobacco smoking, even at reduced levels of nicotine *(178, 179)*, although this has not been demonstrated at the levels produced by VLNC cigarettes. Other research has been conducted on the development of nicotine analogues and their potential as substitutes *(180)*. These and other non-nicotine constituents may play a greater role as determinants of behavioural reinforcement in the context of the low doses of nicotine available in VLNC cigarettes *(171, 181)*.

Another possible response to a nicotine reduction policy is manipulation of products by cigarette smokers *(2)*. In Malaysia, banning of the sale of nicotine-containing e-liquid in ENDS shops resulted in not only a black-market supply of nicotine-containing e-liquid but also a significant increase in home-made nicotine-containing e-liquid *(182)*. It is not known whether adding nicotine to VLNC cigarettes or other direct consumer manipulation of products would yield an appealing, acceptable alternative. Further types of commercial product manipulation that have not yet been studied include changes to the protonation of nicotine or the size distribution of aerosol particles, which determine the deposition and absorption of nicotine and other constituents *(1)*.

4.3.5 Beliefs and attitudes regarding VLNC cigarettes and nicotine reduction

Many cigarette smokers have inaccurate beliefs about nicotine. Most participants in many studies misattributed nicotine as the primary cause of cancers and smoking-related morbidity *(1, 12, 183)*. Beliefs about use of VLNC cigarettes reflect these misconceptions, as smokers perceive them as significantly less harmful than regular cigarettes *(184, 185)*, potentially reinforcing interest in and use of these cigarettes to a greater extent than would be expected on the basis of subjective ratings *(185, 186)*. Although nicotine is a precursor of carcinogenic TSNAs, and reduction of nicotine might decrease the levels of TSNAs in cigarette smoke, there is no evidence that VLNC cigarettes are less toxic. Health communication strategies are necessary to educate smokers about their relative toxicity and harm. Inaccurate beliefs about risk are not limited to VLNC cigarettes, as substantial proportions of smokers are unable to accurately assess the risks associated with a range of tobacco product categories *(187, 188)*. A policy targeting nicotine might add further confusion about its relative harm, which could result in more negative perceptions of NRT, ENDS or other alternative nicotine delivery systems than of more toxic tobacco products. Clear communication that the purpose of a nicotine reduction policy is to reduce dependence on the most toxic (combusted) tobacco products and education about the role of nicotine in addiction are therefore critical. The announcement by the FDA *(15)*, which identified a continuum of risk across tobacco products, is an example of a health message that is sensitive to the role of nicotine in use versus harm.

Expectations about nicotine content influence smokers' subjective response to regular and VLNC cigarettes *(189, 190)*, including responses of the insula to craving and learning, as measured by functional magnetic resonance imaging *(191, 192)*. Sensory responses to VLNC cigarettes, specifically taste and strength ratings, moderate associations between beliefs and daily cigarette consumption, with higher subjective ratings (on scales of strength, taste, harshness and mildness), and beliefs about reduced harm support increased consumption *(185)*. Expectations may also influence brand choice and smoking behaviour; for example, positive expectations of ENDS users are associated with a greater likelihood of quitting smoking but less likelihood of intention to quit ENDS use *(193)*.

The WHO advisory note in 2015 reported strong public support for nicotine reduction *(1, 194)*, and support has been demonstrated in population-based surveys in New Zealand and the USA *(186, 195)*, which may reflect beliefs about the disease risk associated with nicotine *(186)*. Both surveys found about 80% support, mainly among smokers but also among nonsmokers and various ethnic minority populations *(186, 195)*. Consistently less support is shown for a ban on menthol than for regulation of nicotine *(179, 186)*. Trial surveys differ from population surveys by providing an experiential basis for expectations. In an interview study in which New Zealand smokers were first given VLNC cigarettes to smoke, smokers were less interested in mandated nicotine reduction but supported their sale at a much cheaper price concurrently with regular cigarettes *(196)*. In a trial in the USA, however, smokers who used VLNC cigarettes for 6 weeks and who judged that their experimental cigarettes contained low or very low nicotine were approximately twice as likely to support as to oppose a regulated reduction in nicotine content *(197)*.

4.3.6 Summary

- VLNC cigarettes are a partial but incomplete substitute for regular cigarettes. A subset of current smokers would probably give up smoking rather than switch to VLNC cigarettes if current high-nicotine cigarettes were no longer legally marketed. Most smokers who continue to smoke are unlikely to use VLNC cigarettes exclusively.

- Countries should put mechanisms in place to ensure strong tax administration and consider other enforcement measures to limit illicit trade, restrict sales of regular nicotine content cigarettes into countries requiring low-nicotine content cigarettes and restrict websites selling regular nicotine content cigarettes.
- Research was under way as of September 2017 on adolescents' response to low-nicotine products. It is anticipated that nicotine reduction will prevent a significant portion of children and adolescents who experiment with combusted tobacco products from becoming long-term users of those products.
- Substitution among combusted tobacco products is common. A nicotine reduction policy might have to be applied to all combusted tobacco products rather than only to regular cigarettes to ensure a shift in population use from the most toxic and addictive products.
- Reducing the nicotine content of combusted tobacco products might increase substitution by ENDS and possibly HTPs, which are marketed and promoted by manufacturers as reduced risk products and safer alternatives to regular cigarettes. Independent research to substantiate industry research and claims, appropriate health communication, effective surveillance and regulatory oversight of these products are necessary.
- Strong policies and enforcement systems can help reduce the supply of illicit cigarettes. Policies to reduce demand (e.g. treatment services, substitute non-combusted products) may also reduce illicit trade, although the evidence is limited.
- Smokers perceive VLNC cigarettes as less harmful than regular cigarettes, which might support greater interest in and use of VLNC cigarettes than would be expected from subjective ratings. Health communication can address misperceptions of risk.
- Physical or chemical parameters of cigarette construction could be manipulated by manufacturers or smokers to alter the formulation and reinforcing effects of VLNC cigarettes, and the effects might not readily be anticipated.
- Support for nicotine reduction is reported from both population-based and trial surveys.
- Overall, the research suggests that a nicotine reduction policy, as part of a comprehensive approach and implemented with appropriate safeguards, could be an effective method for reducing the population prevalence of smoking.

4.4 Regulatory approaches to nicotine reduction

Much of the concern about a nicotine reduction policy is based not on evidence of the individual effects of use of VLNC cigarettes but on the difficulties in successful implementation of such a policy *(14)* and potential unanticipated consequences of implementation, such as continued availability of toxic forms of nicotine on the black market and/or failure to adequately support less toxic forms of nicotine as alternatives to cigarettes *(8, 10, 11, 13)*. These real challenges are difficult to model in advance of regulatory action *(11, 14)*. One proposed solution is to monitor the impact of implementation in one or more communities that adopted reduced nicotine early on *(11, 198)*. New Zealand has been discussed as a potentially ideal site for nicotine reduction, partly because of strong public support *(194)*, continuing discussion of a policy of nicotine reduction *(95, 198, 199)* and a geographical situation that limits the likelihood of illicit sales *(198)*. In July 2017, the FDA announced that nicotine reduction was part of a comprehensive multi-year plan to reduce the health effects of tobacco use *(15)*. In this section, we consider the feasibility of nicotine reduction, various regulatory approaches, potential challenges to implementation and broader philosophical arguments against nicotine reduction.

4.4.1 Feasibility of nicotine reduction and potential challenges

Commercial production of VLNC cigarettes that meet the product standard of om0.4 mg of nicotine per gram of tobacco without measurably increasing the toxicity of the product has been clearly demonstrated *(1, 17, 198)*. The probable substitution for VLNC cigarettes of other combusted tobacco products (such as little cigars and roll-your-own tobacco) that approximate regular cigarettes in appearance and nicotine content indicates that reduced nicotine product standards should be extended to all combusted tobacco products. As noted in section 4.3.4, changes to VLNC products by manufacturers in response to a nicotine reduction policy could result in products with higher levels of pharmacologically relevant compounds that might impact behaviour. Products must be monitored to ensure successful policy implementation *(1, 93)* and to verify that non-nicotine compounds such as monoamine oxidase inhibitors and nicotine analogues do not change the reinforcement threshold of VLNC cigarettes *(177)*.

As noted in both the WHO advisory note *(1)* and in subsequent discussions *(8, 10)*, a nicotine reduction policy is unlikely to be successful without support from smokers of regular cigarettes, particularly those who are dependent on nicotine. While strong public acceptance has been found in population surveys (section 4.3.5), at least some smokers will consider the subjective trade-offs and policy unacceptable *(200)*. Informing the public that nicotine reduction is part of a long-term strategy to phase out addiction to the most highly toxic tobacco products and to support the availability of less toxic alternatives, including easy access to smoking cessation medication and behavioural treatment, will be necessary to maintain support among smokers and to provide time for smokers to switch from regular cigarettes before the policy is enacted *(1, 198)*.

Political and legal challenges to nicotine reduction will be substantial, given that nicotine is a central feature of all successful tobacco products, as will tobacco industry efforts to undermine and misuse such policy *(9, 14)*. One strategy that should be considered would be to prohibit the manufacture and sale of cigarettes with regular nicotine without criminalizing possession or use. This would shift the burden of compliance from individuals, and continued consumer demand could be moderated by education, treatment services and the availability of legal alternatives. Although countries that are considering nicotine reduction will recognize the potential threat to tax revenue, significant savings in health care would be expected. Ultimately, the net economic impact will be specific to each country and should be evaluated, as it will be an important determinant of the viability of a policy, even if the public health impact is clear.

4.4.2 Prerequisites for successful implementation of a nicotine reduction policy

The WHO advisory note *(1)* listed a number of preconditions that were considered necessary for a successful nicotine reduction policy:

- comprehensive regulation of all nicotine- and tobacco-containing products;
- other comprehensive tobacco control (increased taxes, smoking bans, graphic warning labels, plain packaging);
- continuing communication of the health risks of smoking to both the general population and health professionals;
- availability of affordable treatment and alternative forms of nicotine to reduce withdrawal symptoms in dependent smokers; and
- capacity for market surveillance and product testing.

The inaccurate belief that nicotine is the main harmful constituent of cigarettes could decrease the sense of urgency of current smokers to quit *(12, 183)*. Although this unintended consequence is unlikely to outweigh the potential benefits of nicotine reduction, it should be addressed by clear communication *(185, 198)*. Further research and continuous evaluation will be required to ensure appropriate messages (e.g. accounting for differences among populations) without denying the harmful effects of nicotine *(201)*. It must also be recognized that effective communication may not be possible in an environment where governments underfund evidence-based communication about the harms of smoking and where the bulk of communication still comes from tobacco interests and their allies. Ultimately, subjective effects rather than the perceived harmfulness of cigarettes may best predict long-term smoking behaviour *(184)*.

It has been speculated that regulatory environments in which principles of harm reduction are included with other elements of tobacco control would provide more favourable conditions for a nicotine reduction strategy *(1, 10, 11, 198)*. This might entail readily available, consumer-acceptable non-combusted forms of nicotine that could be used to manage withdrawal symptoms and provide smokers with a viable alternative to regular cigarettes, and it would demonstrate that reduction of the nicotine in combusted tobacco is not prohibition of nicotine *(41)*. This will require further consideration in a real-world context in which there is already concern about the use and promotion of some alternative nicotine products.

While a nicotine reduction approach could substantially improve public health, it also offers considerable scope for tobacco companies to mislead consumers, both directly and indirectly. For example, they might use lower-nicotine products to promote similarly branded products, to imply the safety of both new and established products and to present themselves as policy advisers. Given the industry's long history of abusing health policy initiatives for promotional purposes and public relations, any enacted regulations should preclude inappropriate or misleading promotion by tobacco companies, directly or indirectly, with strong monitoring and enforcement capacity. Further, regulations should ensure that information and advice on the health aspects of VLNC cigarettes or other tobacco products come from health authorities, not tobacco companies or their agents. It is recommended that an international approach be found for regulation of communication strategies and promotion of VLNC, with commitments from all relevant Parties.

The WHO advisory note *(1)* raised the issue of the potential health risks introduced by alternative forms of tobacco or nicotine, which have also been discussed critically elsewhere (see section 4.3.2). Many of the concerns about the promotion of ENDS and other non-combusted tobacco product alternatives are related to the possibility that they would support cigarette initiation or use; however, these concerns should become less relevant in the context of cigarettes with low addiction potential *(41)*. Likewise, concurrent use of two or more products, which could potentially support continued use of VLNC cigarettes, would be less of a problem, because these cigarettes would be less desirable and alternative products would be more satisfying in comparison *(41)*. Evaluation and monitoring of the use and health effects of other products, including less toxic alternatives, and patterns of concurrent use of several products would nonetheless be a critical component of a nicotine reduction policy *(1, 2, 9)*. Constraints on marketing and on the availability of ENDS and other alternative nicotine products would be appropriate to limit experimentation by nonsmokers, especially young people *(10, 41)*. Health agencies should clearly and accurately communicate to the public the potential health risks of alternative products, including ENDS and HTPs, and should be careful not to encourage addiction to alternative products.

A nicotine reduction strategy should be part of a comprehensive tobacco control strategy that complements other evidence-based policies, such as increased taxes, sustained and adequately funded public education campaigns, restrictions on smoking in public, further product regulation as required, bans on tobacco industry marketing and public relations, the availability of low-cost NRT and behavioural cessation treatment (e.g. telephone counselling lines) and special programmes for disadvantaged groups *(95)*. The regulatory considerations reported by the FDA in its announcement of nicotine reduction *(15)* were continuing regulation of combusted cigarettes and the characteristics related to toxicity and the attractiveness of combusted products; the use, availability and health effects of ENDS and other alternatives to cigarettes, such as medicinal nicotine products; and regulation of health claims and other forms of communication of risk.

4.4.3 Strategies for implementation of a nicotine reduction policy

The WHO advisory note *(1)* addressed the question of a gradual versus an immediate reduction in nicotine content, noting that an immediate reduction is preferable insofar as it is less likely to introduce such unintended consequences as a short-term increase in compensatory smoking by dependent smokers *(93)*. A 10-site, double-blind study showed that immediate, not gradual, reductions in nicotine content decreased the number of cigarettes smoked per day and biomarkers of exposure in a 20-week intervention, and measures of craving and dependence were significantly lower at week 20 in the group that experienced an immediate reduction *(20)*. Although these data suggest potential benefits of an immediate reduction, it is important to note that, consistent with other reports *(202)*, more participants drop out of studies when the reduction is immediate, suggesting that immediate reductions may make it more difficult for smokers to adjust, particularly in the absence of alternative sources of nicotine or smoking cessation treatment. As smokers will continue to have access to regular cigarettes until the supplies have been exhausted, more gradual introduction of VLNC cigarettes is the probable practical outcome, even in an immediate reduction policy *(40)*. Concern about the potential discomfort of dependent smokers of regular cigarettes faced with immediate reduction emphasizes the need for readily available alternative forms of nicotine and/or medication *(40, 198)*.

Population responses to nicotine reduction will partly reflect contextual differences. For example, reducing nicotine in Sweden would be likely to result in the adoption of *snus*, whereas *snus* is banned in most other countries in the European Union *(11)*. Thus, strategies for reducing nicotine must be sensitive to differences in context, including other regulation and population preferences.

Some countries have restricted or banned the sale and importation of ENDS (e.g. Brazil, Singapore, Thailand, Uruguay and Venezuela). Such a ban would not necessarily preclude a nicotine reduction policy for combusted tobacco products; however, it might raise additional challenges. Care must be taken to ensure the availability of alternative sources of nicotine, such as medicinal nicotine, access to tobacco dependence treatment and the capacity or infrastructure necessary to address a potentially stronger illicit market.

4.4.4 Philosophical objections to nicotine reduction

Some discussions of nicotine reduction represent the policy as a de-facto prohibition of cigarettes *(2, 10, 13, 14)*. In this viewpoint, nicotine reduction differs significantly from other tobacco control measures such as plain packaging, graphic warnings or flavour bans, in that it fundamentally alters the primary function of the product, which is to deliver nicotine *(14)*. Nevertheless, smokers use VLNC cigarettes even when no other products are available, although they may be less likely to persist in their use over time *(196)*. Further, unlike in a prohibition strategy, smokers would still have access to nicotine but in potentially less harmful delivery systems. The availability of other forms of nicotine would also minimize the efforts of dependent smokers to seek nicotine through illegal means *(198)*.

A second objection to nicotine reduction is that it is fundamentally a "top–down" or paternalistic means for reducing the health risks of tobacco *(9, 10)*. Limiting the sale of cigarettes to those with a VLNC intrudes on personal choice. Some argue, however, that support for ENDS and other alternative sources of nicotine could attract smokers away from cigarettes without imposing invasive regulations *(11, 13)*. The advantages of avoiding a nicotine reduction policy include the elimination of legal or political challenges to policy, a potential black market and opposition from smokers *(9)*. Questions remain about the long-term health consequences of alternative products, including ENDS and HTPs, the use, function and emissions of which continue to evolve rapidly. Ultimately, some combination of policies may be necessary to reduce the desirability of combusted products relative to alternatives; this might include increased communication of relative risk, increased availability of lower-risk products, an increased price differential between low- and high-risk products or nicotine product standards that reduce dependence on the most toxic products.

4.4.5 Summary

- The feasibility of producing commercial VLNC cigarettes with subjective characteristics similar to those of regular cigarettes but with reduced potential for dependence has now been reasonably established. The toxicity of these cigarettes remains high, however, due to combustion and inhalation of tobacco.

- The physical characteristics and construction of VLNC cigarettes used in clinical trials are relatively uniform; in a commercial market of VLNC cigarettes, product characteristics other than nicotine might diverge significantly. Some differences, such as the levels of non-nicotine compounds with pharmacological effects, might have to be carefully monitored, or restrictions might have to be imposed to limit product changes.
- Public education campaigns should communicate nicotine reduction as part of a long-term strategy, including a comprehensive tobacco control programme, to phase out addiction to the most highly toxic tobacco products. If VLNC cigarettes are sold concurrently with regular cigarettes, without substantive policy interventions or education campaigns, their use is likely to remain low.
- The preconditions for a successful nicotine reduction strategy identified in the WHO advisory note remain valid, namely: comprehensive regulation of all tobacco- and nicotine-containing products, capacity for surveillance and product testing, commitment to continuing public education and the availability of and access to tried and tested forms of treatment, as approved by national authorities. Inappropriate or misleading promotion by tobacco product manufacturers and manufacturers of alternative products should be prohibited.
- An immediate reduction would have clear advantages over a gradual reduction in nicotine content but might require a comprehensive approach in order to reduce unintended consequences.
- The intent of reducing the nicotine content of regular cigarettes is not to prohibit their sale and use but to reduce their addictiveness, so that people can choose whether or not to continue using an extremely toxic product.

4.5 Research questions

- Research to date has focused exclusively on reduced-nicotine cigarettes. Future studies should address the potential effects of reduced nicotine in other combusted tobacco products (e.g. roll-your-own tobacco, little cigars, cigarillos, waterpipe tobacco) and the benefits and drawbacks of reducing nicotine in non-combusted tobacco products.
- The positive and negative impacts of VLNC cigarettes on vulnerable populations, including women of reproductive age, should continue to be assessed. Research on the potential uptake and use of VLNC cigarettes by adolescents and tobacco-naive users would be particu-

larly informative; this may have to be assessed primarily after the implementation of product standards through surveillance.
- Further research should be conducted on concomitant use of VLNC cigarettes, ENDS and other alternative tobacco and nicotine products, by both dependent and non-dependent smokers and tobacco-naive users.
- Studies conducted to date do not show that use of VLNC cigarettes increases the use of alcohol or dependence on other substances (such as cannabis), but continued assessment is warranted.
- VLNC products with altered physical or chemical characteristics designed to replace or enhance the pharmacological effects of nicotine have not been developed or tested to date.
- Research on the impact of specific public communication strategies would help to maximize the benefit of a policy, minimize misunderstanding of the impact of the change in cigarette addictiveness on its toxicity and inform those who are dependent on nicotine about appropriate alternatives.
- Research on the nature of tobacco industry abuse of nicotine reduction strategies for continued promotion of current products and for undermining evidence-based tobacco control policies should be conducted. Research is also needed on means of regulating such activity nationally and internationally.
- Few data are available for estimating the extent of illicit sales or of product tampering and related factors that could mitigate the effects of nicotine reduction.

4.6 Policy recommendations

- A nicotine reduction policy has considerable promise as a means to accelerate a reduction in use of the most toxic tobacco products and encourage smoking cessation. Such a policy should be part of a comprehensive approach and might be most effective when coordinated with other relevant policies designed to support cessation and reduce smoking.
- Substitution among combusted tobacco products is common, and the availability of high-nicotine combusted products that are similar in use and appearance to regular cigarettes is likely to mitigate the potential public health benefit of nicotine reduction. Therefore, a nicotine reduction policy might have to be applied to all combusted

products rather than only to regular cigarettes to support a shift in population use from the most toxic and addictive products.
- In view of the reduced exposure to nicotine and decreased tobacco consumption demonstrated in many clinical and behavioural studies, there is insufficient evidence that a significant population of non-smoking adolescents or adults who would not otherwise have smoked will adopt and maintain use of VLNC cigarettes. To minimize any unintended consequences in these populations, the health risks of both VLNC cigarettes and alternative forms of nicotine must be clearly communicated, including the fact that VLNC cigarettes are not safer than regular cigarettes.
- As indicated in the 2015 WHO advisory note, a nicotine reduction policy should not be considered to be supplanting a comprehensive tobacco control policy, which should include increased taxes, sustained, adequately funded public education, smoking bans, graphic warning labels, plain packaging and other forms of product regulation. A coordinated policy for all tobacco- and nicotine-containing products is essential, as is access to treatment for tobacco dependence.
- Capacity to conduct continuous population surveillance, monitoring and testing of products and enforcement of product standards is critical to support a nicotine reduction policy. Monitoring should also include industry promotional and other activities as well as product availability and sales.

4.7 References

1. WHO Study Group on Tobacco Product Regulation. Advisory note: global nicotine reduction strategy (WHO Technical Report Series No. 989). Geneva: World Health Organization; 2015 (http://apps.who.int/iris/bitstream/10665/189651/1/9789241509329_eng.pdf, accessed 15 May 2019).
2. Hatsukami DK, Zaatari G, Donny E. The case for the WHO advisory note, global nicotine reduction strategy. Tob Control. 2017;26(e1):e29–30.
3. Eriksen M, Mackay J, Schluger N, Gomeshtapeh FI, Drope J, editors. The tobacco atlas, 5th edition, Atlanta (GA), American Cancer Society; 2015.
4. WHO Framework Convention on Tobacco Control. Geneva: World Health Organization; 2003 (http://www.who.int/fctc/en/, accessed 15 May 2019).
5. Report on the scientific basis of tobacco product regulation: fourth report of a WHO study group (WHO Technical Report Series No. 967). Geneva: World Health Organization; 2012 (http://www.who.int/tobacco/publications/prod_regulation/trs_967/en/index.html, accessed 15 May 2019).
6. Donny EC, Denlinger RL, Tidey JW, Koopmeiners JS, Benowitz NL, Vandrey RG et al. Randomized trial of reduced-nicotine standards for cigarettes. N Engl J Med. 2015;373(14):1340–9.
7. Smith TT, Rupprecht LE, Denlinger-Apte RL, Weeks JJ, Panas RS, Donny EC et al. Animal research

on nicotine reduction: current evidence and research gaps. Nicotine Tob Res. 2017;19(9):1005–15.
8. Kozlowski LT. Commentary on Nardone et al. (2016): satisfaction, dissatisfaction and complicating the nicotine-reduction strategy with more nicotine. Addiction. 2016;111(12):2217–8.
9. Kozlowski LT. Cigarette prohibition and the need for more prior testing of the WHO TobReg's global nicotine-reduction strategy. Tob Control. 2017;26(e1):e31–4.
10. Borland R. Paying more attention to the "elephant in the room". Tob Control. 2017;26(e1):e35–6.
11. Kozlowski LT. Let actual markets help assess the worth of optional very-low-nicotine cigarettes before deciding on mandatory regulations. Addiction. 2017;112(1):3–5.
12. Pacek LR, Rass O, Johnson MW. Knowledge about nicotine among HIV-positive smokers: implications for tobacco regulatory science policy. Addict Behav. 2017;65:81–6.
13. Kozlowski LT, Abrams DB. Obsolete tobacco control themes can be hazardous to public health: the need for updating views on absolute product risks and harm reduction. BMC Public Health. 2016;16:432.
14. Kozlowski LT. Prospects for a nicotine-reduction strategy in the cigarette endgame: alternative tobacco harm reduction scenarios. Int J Drug Policy. 2015;26(6):543–7.
15. Gottlieb S, Zeller M. A nicotine-focused framework for public health. N Engl J Med. 2017;377(12):1111–4.
16. Donny EC, Walker N, Hatsukami DK, Bullen C. Reducing the nicotine content of combusted tobacco products sold in New Zealand. Tob Control. 2017;26:e37–42.
17. Richter P, Steven PR, Bravo R, Lisko JG, Damian M, Gonzalez-Jimenez N et al. Characterization of Spectrum variable nicotine research cigarettes. Tob Regul Sci. 2016;2(2):94–105.
18. Donny EC, Houtsmuller E, Stitzer ML. Smoking in the absence of nicotine: behavioral, subjective and physiological effects over 11 days. Addiction. 2007;102(2):324–34.
19. Hatsukami DK, Kotlyar M, Hertsgaard LA, Zhang Y, Carmella SG, Jensen JA et al. Reduced nicotine content cigarettes: effects on toxicant exposure, dependence and cessation. Addiction. 2010;105(2):343–55.
20. Hatsukami DK, Luo X, Jensen JA, al'Absi M, Allen SS, Carmella SG et al. Effect of immediate vs gradual reduction in nicotine content of cigarettes on biomarkers of smoke exposure: a randomized clinical trial. JAMA. 2018;320(9):880–91.
21. Hatsukami DK, Hertsgaard LA, Vogel RI, Jensen JA, Murphy SE, Hecht SS et al. Reduced nicotine content cigarettes and nicotine patch. Cancer Epidemiol Biomarkers Prev. 2013;22(6):1015–24.
22. Hammond D, O'Connor RJ. Reduced nicotine cigarettes: smoking behavior and biomarkers of exposure among smokers not intending to quit. Cancer Epidemiol Biomarkers Prev. 2014;23(10):2032–40.
23. Kozlowski LT, O'Connor RJ. Cigarette filter ventilation is a defective design because of misleading taste, bigger puffs, and blocked vents. Tob Control. 2002;11(Suppl 1):I40–50.
24. Ding YS, Ward J, Hammond D, Watson CH. Mouth-level intake of benzo[a]pyrene from reduced nicotine cigarettes. Int J Environ Res Public Health. 2014;11(11):11898–914.
25. Benowitz NL, Jacob P 3rd, Herrera B. Nicotine intake and dose response when smoking reduced-nicotine content cigarettes. Clin Pharmacol Ther. 2006;80(6):703–14.
26. Benowitz NL, Nardone N, Dains KM, Hall SM, Stewart S, Dempsey D et al. Effect of reducing the nicotine content of cigarettes on cigarette smoking behavior and tobacco smoke toxicant exposure: 2-year follow up. Addiction. 2015;110(10):1667–75.
27. Mercincavage M, Souprountchouk V, Tang KZ, Dumont RL, Wileyto EP, Carmella SG et al. A randomized controlled trial of progressively reduced nicotine content cigarettes on smoking behaviors, biomarkers of exposure, and subjective ratings. Cancer Epidemiol Biomarkers Prev. 2016;25(7):1125–33.
28. Perkins KA, Kunkle N, Karelitz JL, Perkins KA, Kunkle N, Karelitz JL. Preliminary test of cigarette nicotine discrimination threshold in non-dependent versus dependent smokers. Drug Alcohol

Depend. 2017;175:36–41.
29. Smith TT, Levin ME, Schassburger RL, Buffalari DM, Sved AF, Donny EC. Gradual and immediate nicotine reduction result in similar low-dose nicotine self-administration. Nicotine Tob Res. 2013;15(11):1918–25.
30. Chen H, Matta SG, Sharp BM. Acquisition of nicotine self-administration in adolescent rats given prolonged access to the drug. Neuropsychopharmacology. 2007;32(3):700–9.
31. Levin ED, Lawrence SS, Petro A, Horton K, Rezvani AH, Seidler FJ et al. Adolescent vs. adult-onset nicotine self-administration in male rats: duration of effect and differential nicotinic receptor correlates. Neurotoxicol Teratol. 2007;29(4):458–65.
32. Donny EC, Hatsukami DK, Benowitz NL, Sved AF, Tidey JW, Cassidy RN. Reduced nicotine product standards for combustible tobacco: building an empirical basis for effective regulation. Prev Med. 2014;68:17–22.
33. Smith TT, Schassburger RL, Buffalari DM, Sved AF, Donny EC. Low-dose nicotine self-administration is reduced in adult male rats naive to high doses of nicotine: implications for nicotine product standards. Exp Clin Psychopharmacol. 2014;22(5):453–9.
34. Schassburger RL, Pitzer EM, Smith TT, Rupprecht LE, Thiels E, Donny EC et al. Adolescent rats self-administer less nicotine than adults at low doses. Nicotine Tob Res. 2016;18(9):1861–8.
35. Cassidy RN, Tidey JW, Cao Q, Colby SM, McClernon FJ, Koopmeiners JS et al. Age moderates smokers' subjective response to very low nicotine content cigarettes: evidence from a randomized controlled trial. Nicotine Tob Res. 2018; doi:10.1093/ntr/nty079.
36. Jackson KM, Cioe PA, Krishnan-Sarin S, Hatsukami D. Adolescent smokers' response to reducing the nicotine content of cigarettes: acute effects on withdrawal symptoms and subjective evaluations. Drug Alcohol Depend. 2018;188:153–60.
37. Walker N, Howe C, Bullen C, Grigg M, Glover M, McRobbie H et al. The combined effect of very low nicotine content cigarettes, used as an adjunct to usual Quitline care (nicotine replacement therapy and behavioural support), on smoking cessation: a randomized controlled trial. Addiction. 2012;107(10):1857–67.
38. McRobbie H, Przulj D, Smith KM, Cornwall D. Complementing the standard multicomponent treatment for smokers with denicotinized cigarettes: a randomized trial. Nicotine Tob Res. 2016;18(5):1134–41.
39. Donny EC, Jones M. Prolonged exposure to denicotinized cigarettes with or without transdermal nicotine. Drug Alcohol Depend. 2009;104(1–2):23–33.
40. Hatsukami DK, Luo X, Dick L, Kangkum M, Allen SS, Murphy SE et al. Reduced nicotine content cigarettes and use of alternative nicotine products: exploratory trial. Addiction. 2017;112(1):156–67.
41. Benowitz NL, Donny EC, Hatsukami DK. Reduced nicotine content cigarettes, e-cigarettes and the cigarette end game. Addiction. 2017;112(1):6–7.
42. Pechacek TF, Nayak P, Gregory KR, Weaver SR, Eriksen MP. The potential that electronic nicotine delivery systems can be a disruptive technology: results from a national survey. Nicotine Tob Res. 2016;18(10):1989–97.
43. Nardone N, Donny EC, Hatsukami DK, Koopmeiners JS, Murphy SE, Strasser AA et al. Estimations and predictors of non-compliance in switchers to reduced nicotine content cigarettes. Addiction. 2016;111(12):2208–16.
44. Boatman JA, Vock DM, Koopmeiners JS, Donny EC. Estimating causal effects from a randomized clinical trial when noncompliance is measured with error. Biostatistics. 2018;19(1):103–18.
45. Dermody SS, Donny EC, Hertsgaard LA, Hatsukami DK. Greater reductions in nicotine exposure while smoking very low nicotine content cigarettes predict smoking cessation. Tob Control. 2015;24(6):536–9.
46. Dermody SS, McClernon FJ, Benowitz N, Luo X, Tidey JW, Smith TT et al. Effects of reduced nicotine content cigarettes on individual withdrawal symptoms over time and during abstinence.

Exp Clin Psychopharmacol. 2018;26(3):223–32.
47. Faulkner P, Ghahremani DG, Tyndale RF, Cox CM, Kazanjian AS, Paterson N et al. Reduced-nicotine cigarettes in young smokers: impact of nicotine metabolism on nicotine dose effects. Neuropsychopharmacology. 2017;42(8):1610–8.
48. Keith DR, Kurti AN, Davis DR, Zvorsky IA, Higgins ST. A review of the effects of very low nicotine content cigarettes on behavioral and cognitive performance. Prev Med. 2017;104:100–16.
49. Rupprecht LE, Koopmeiners JS, Dermody SS, Oliver JA, al'Absi M, Benowitz NL et al. Reducing nicotine exposure results in weight gain in smokers randomised to very low nicotine content cigarettes. Tob Control. 2017;26(e1):e43–8.
50. Rupprecht LE, Donny EC, Sved AF. Obese smokers as a potential subpopulation of risk in tobacco reduction policy. Yale J Biol Med. 2015;88(3):289–94.
51. Bandiera FC, Ross KC, Taghavi S, Delucchi K, Tyndale RF, Benowitz NL. Nicotine dependence, nicotine metabolism, and the extent of compensation in response to reduced nicotine content cigarettes. Nicotine Tob Res. 2015;17(9):1167–72.
52. Benowitz NL, Dains KM, Hall SM, Stewart S, Wilson M, Dempsey D et al. Smoking behavior and exposure to tobacco toxicants during 6 months of smoking progressively reduced nicotine content cigarettes. Cancer Epidemiol Biomarkers Prev. 2012;21(5):761–9.
53. Shiffman S, Kurland BF, Scholl SM, Mao JM. Nondaily smokers' changes in cigarette consumption with very low-nicotine-content cigarettes: a randomized double-blind clinical trial. JAMA Psychiatry. 2018;75(10):995–1002.
54. Dermody SS, Tidey JW, Denlinger RL, Pacek LR, al'Absi M, Drobes DJ. The impact of smoking very low nicotine content cigarettes on alcohol use. Alcohol Clin Exp Res. 2016;40(3):606–15.
55. Pacek LR, Vandrey R, Dermody SS, Denlinger-Apte RL, Lemieux A, Tidey JW et al. Evaluation of a reduced nicotine product standard: moderating effects of and impact on cannabis use. Drug Alcohol Depend. 2016;167:228–32.
56. Cohn A, Ehlke S, Cobb CO. Relationship of nicotine deprivation and indices of alcohol use behavior to implicit alcohol and cigarette approach cognitions in smokers. Addict Behav. 2017;67:58–65.
57. Tidey JW, Pacek LR, Koopmeiners JS, Vandrey R, Nardone N, Drobes DJ et al. Effects of 6-week use of reduced-nicotine content cigarettes in smokers with and without elevated depressive symptoms. Nicotine Tob Res. 2017;19(1):59–67.
58. Gaalema DE, Miller ME, Tidey JW. Predicted impact of nicotine reduction on smokers with affective disorders. Tob Regul Sci. 2015;1(2):154–65.
59. Higgins ST, Heil SH, Sigmon SC, Tidey JW, Gaalema DE, Stitzer ML et al. Response to varying the nicotine content of cigarettes in vulnerable populations: an initial experimental examination of acute effects. Psychopharmacology. 2017;234(1):89–98.
60. Lawrence D, Mitrou F, Zubrick SR. Smoking and mental illness: results from population surveys in Australia and the United States. BMC Public Health. 2009;9:285.
61. Lasser K1, Boyd JW, Woolhandler S, Himmelstein DU, McCormick D, Bor DH. Smoking and mental illness: a population-based prevalence study. JAMA. 2000;284(20):2606–10.
62. Smith PH, Mazure CM, McKee SA. Smoking and mental illness in the US population. Tob Control. 2014;23(e2):e147–53.
63. Adler NE, Glymour MM, Fielding J. Addressing social determinants of health and health inequalities. JAMA. 2016;316(16):1641–2.
64. Das S, Prochaska JJ. Innovative approaches to support smoking cessation for individuals with mental illness and co-occurring substance use disorders. Expert Rev Respir Med. 2017;11(10):841–50.
65. Beard E, Bruguera C, McNeill A, Brown J, West R. Association of amount and duration of NRT use in smokers with cigarette consumption and motivation to stop smoking: a national survey of smokers in England. Addict Behav. 2015;40:33–8.

66. Benowitz NL, Hukkanen J, Jacob P III. Nicotine chemistry, metabolism, kinetics and biomarkers. Handb Exp Pharmacol. 2009;192:29–60.
67. Digard H, Proctor C, Kulasekaran A, Malmqvist U, Richter A. Determination of nicotine absorption from multiple tobacco products and nicotine gum. Nicotine Tob Res. 2013;15(1):255–61.
68. Caggiula AR, Donny EC, Palmatier MI, Liu X, Chaudhri N, Sved AF. The role of nicotine in smoking: a dual-reinforcement model. Nebr Symp Motiv. 2009;55:91–109.
69. Van Heel M, Van Gucht D, Vanbrabant K, Baeyens F. The importance of conditioned stimuli in cigarette and e-cigarette craving reduction by e-cigarettes. Int J Environ Res Public Health. 2017;14(2):193.
70. Rose JE. Nicotine and nonnicotine factors in cigarette addiction. Psychopharmacology. 2006;184(3–4):274–85.
71. Perkins K, Sayette M, Conklin C, Caggiula A. Placebo effects of tobacco smoking and other nicotine intake. Nicotine Tob Res. 2003;5(5):695–709.
72. Chaloupka FJ, Warner KE. The economics of smoking. In: Culyer AJ, Newhouse JP, editors. Handbook of health economics. Oxford: Elsevier Science; 2000:1539–627.
73. MacKillop J, Murphy JG, Ray LA, Eisenberg DT, Lisman SA, Lum JK et al. Further validation of a cigarette purchase task for assessing the relative reinforcing efficacy of nicotine in college smokers. Exp Clin Psychopharmacol. 2008;16(1):57–65.
74. Guindon GE, Paraje GR, Chaloupka FJ. The impact of prices and taxes on the use of tobacco products in Latin America and the Caribbean. Am J Public Health. 2015;105(3):e9–19.
75. Zheng Y, Zhen C, Dench D, Nonnemaker JM. US demand for tobacco products in a system framework. Health Econ. 2017;26(8):1067–86.
76. O'Connor RJ, June KM, Bansal-Travers M, Rousu MC, Thrasher JF, Hyland A et al. Estimating demand for alternatives to cigarettes with online purchase tasks. Am J Health Behav. 2014;38(1):103–13.
77. Stein JS, Wilson AG, Koffarnus MN, Judd MC, Bickel WK. Naturalistic assessment of demand for cigarettes, snus, and nicotine gum. Psychopharmacology. 2017;234(2):245–54.
78. Gammon DG, Loomis BR, Dench DL, King BA, Fulmer EB, Rogers T. Effect of price changes in little cigars and cigarettes on little cigar sales: USA, Q4 2011–Q4 2013. Tob Control. 2016;25(5):538–44.
79. Tait P, Rutherford P, Saunders C. Do consumers of manufactured cigarettes respond differently to price changes compared with their roll-your-own counterparts? Evidence from New Zealand. Tob Control. 2015;24(3):285–9.
80. Etter JF, Eissenberg T. Dependence levels in users of electronic cigarettes, nicotine gums and tobacco cigarettes. Drug Alcohol Depend. 2015;147:68–75.
81. Huang J, Tauras J, Chaloupka FJ. The impact of price and tobacco control policies on the demand for electronic nicotine delivery systems. Tob Control. 2014;23(Suppl 3):iii41–7.
82. McPherson S, Howell D, Lewis J, Barbosa-Leiker C, Bertotti Metoyer P, Roll J. Self-reported smoking effects and comparative value between cigarettes and high dose e-cigarettes in nicotine-dependent cigarette smokers. Behav Pharmacol. 2016;27(2–3 Spec Issue):301–7.
83. Liu G, Wasserman E, Kong L, Foulds J. A comparison of nicotine dependence among exclusive e-cigarette and cigarette users in the PATH study. Prev Med. 2017;104:86–91.
84. National Cancer Institute, World Health Organization. The economics of tobacco and tobacco control (NCI Tobacco Control Monograph 21). Bethesda (MD): Department of Health and Human Services; Geneva: World Health Organization; 2016.
85. Chaloupka FJ, Yurekli A, Fong GT. Tobacco taxes as a tobacco control strategy. Tob Control. 2012;21(2):172–80.
86. Choi SE. Are lower income smokers more price sensitive? The evidence from Korean cigarette tax increases. Tob Control. 2016;25(2):141–6.
87. Koffarnus MN, Wilson AG, Bickel WK. Effects of experimental income on demand for potentially

real cigarettes. Nicotine Tob Res. 2015;17(3):292–8.
88. Yang T, Peng S, Yu L, Jiang S, Stroub WB, Cottrell RR et al. Chinese smokers' behavioral response toward cigarette price: individual and regional correlates. Tob Induc Dis. 2016;14:13.
89. Lewis M, Wang Y, Cahn Z, Berg CJ. An exploratory analysis of cigarette price premium, market share and consumer loyalty in relation to continued consumption versus cessation in a national US panel. BMJ Open. 2015;5(11):e008796.
90. Biener L, Roman AM, McInerney SA, Bolcic-Jankovic D, Hatsukami DK, Loukas A et al. Snus use and rejection in the USA. Tob Control. 2016;25(4):386–92.
91. Nonnemaker J, Kim AE, Lee YO, MacMonegle A. Quantifying how smokers value attributes of electronic cigarettes. Tob Control. 2016;25(e1):e37–43.
92. Meier E, Burris JL, Wahlquist A, Garrett-Mayer E, Gray KM, Alberg AJ et al. Perceptions of snus among US adult smokers given free product. Nicotine Tob Res. 2017;20(1):22–9.
93. Walker N, Fraser T, Howe C, Laugesen M, Truman P, Parag V et al. Abrupt nicotine reduction as an endgame policy: a randomised trial. Tob Control. 2015;24(e4):e251–7.
94. Smith TT, Cassidy RN, Tidey JW, Luo X, Le CT, Hatsukami DK et al. Impact of smoking reduced nicotine content cigarettes on sensitivity to cigarette price: further results from a multi-site clinical trial. Addiction. 2017;112(2):349–59.
95. Tucker MR, Laugesen M, Grace RC. Estimating demand and cross-price elasticity for very low nicotine content (VLNC) cigarettes using a simulated demand task. Nicotine Tob Res. 2018;20(4):528.
96. Johnson MW, Bickel WK, Kirshenbaum AP. Substitutes for tobacco smoking: a behavioral economic analysis of nicotine gum, denicotinized cigarettes, and nicotine-containing cigarettes. Drug Alcohol Depend. 2004;74(3):253–64.
97. O'Connor RJ, Bansal-Travers M, Carter LP, Cummings KM. What would menthol smokers do if menthol in cigarettes were banned? Behavioral intentions and simulated demand. Addiction. 2012;107(7):1330–8.
98. Sureda X, Fu M, Martínez-Sánchez JM, Martínez C, Ballbé M, Pérez-Ortuño R et al. Manufactured and roll-your-own cigarettes: a changing pattern of smoking in Barcelona, Spain. Environ Res. 2017;155:167–74.
99. Brown AK, Nagelhout GE, van den Putte B, Willemsen MC, Mons U, Guignard R et al. Trends and socioeconomic differences in roll-your-own tobacco use: findings from the ITC Europe surveys. Tob Control. 2015;24(Suppl 3):iii11–6.
100. Ayo-Yusuf OA, Olutola BG. "Roll-your-own" cigarette smoking in South Africa between 2007 and 2010. BMC Public Health. 2013;13:597.
101. Agaku IT, Filippidis FT, Vardavas CI, Odukoya OO, Awopegba AJ, Ayo-Yusuf OA et al. Poly-tobacco use among adults in 44 countries during 2008–2012: evidence for an integrative and comprehensive approach in tobacco control. Drug Alcohol Depend. 2014;139:60–70.
102. Lee YO, Hebert CJ, Nonnemaker JM, Kim AE. Multiple tobacco product use among adults in the United States: cigarettes, cigars, electronic cigarettes, hookah, smokeless tobacco, and snus. Prev Med. 2014;62:14–9.
103. Gravely S, Fong GT, Cummings KM, Yan M, Quah AC, Borland R et al. Awareness, trial, and current use of electronic cigarettes in 10 countries: findings from the ITC project. Int J Environ Res Public Health. 2014;11(11):11691–704.
104. Palipudi KM, Mbulo L, Morton J, Mbulo L, Bunnell R, Blutcher-Nelson G et al. Awareness and current use of electronic cigarettes in Indonesia, Malaysia, Qatar, and Greece: findings from 2011–2013 global adult tobacco surveys. Nicotine Tob Res. 2016;18(4):501–7.
105. Delnevo CD, Giovenco DP, Miller Lo EJ. Changes in the mass-merchandise cigar market since the Tobacco Control Act. Tob Regul Sci. 2017;3(2 Suppl 1):S8–16.
106. Jo CL, Williams RS, Ribisl KM. Tobacco products sold by Internet vendors following restrictions on flavors and light descriptors. Nicotine Tob Res. 2015;17(3):344–9.

107. Casseus M, Garmon J, Hrywna M, Delnevo CD. Cigarette smokers' classification of tobacco products. Tob Control. 2016;25(6):628–30.
108. Quisenberry AJ, Koffarnus MN, Hatz LE, Epstein LH, Bickel WK. The experimental tobacco marketplace I: Substitutability as a function of the price of conventional cigarettes. Nicotine Tob Res. 2016;18(7):1642–8.
109. Dave D, Saffer H. Demand for smokeless tobacco: role of advertising. J Health Econ. 2013;32(4):682–97.
110. Adams S, Cotti C, Fuhrmann D. Smokeless tobacco use following smoking bans in bars. South Econ J. 2013;80(1):162–80.
111. Tam J, Day HR, Rostron BL, Apelberg BJ. A systematic review of transitions between cigarette and smokeless tobacco product use in the United States. BMC Public Health. 2015;15:258.
112. Lund KE, Vedøy TF, Bauld L. Do never smokers make up an increasing share of snus users as cigarette smoking declines? Changes in smoking status among male snus users in Norway 2003–15. Addiction. 2017;112(2):340–8.
113. Hatsukami DK, Severson H, Anderson A, Vogel RI, Jensen J, Broadbent B et al. Randomised clinical trial of snus versus medicinal nicotine among smokers interested in product switching. Tob Control. 2016;25(3):267–74.
114. Allen A, Vogel RI, Meier E, Anderson A, Jensen J, Severson HH et al. Gender differences in snus versus nicotine gum for cigarette avoidance among a sample of US smokers. Drug Alcohol Depend. 2016;168:8–12.
115. Berman ML, Bickel WK, Harris AC, LeSage MG, O'Connor RJ, Stepanov I et al. Consortium on methods evaluating tobacco: research tools to inform FDA regulation of snus. Nicotine Tob Res. 2018;20(11):1292–300.
116. Chaloupka FJ, Sweanor D, Warner KE. Differential taxes for differential risks – toward reduced harm from nicotine-yielding products. N Engl J Med. 2015;373(7):594–7.
117. Grace RC, Kivell BM, Laugesen M. Estimating cross-price elasticity of e-cigarettes using a simulated demand procedure. Nicotine Tob Res. 2015;17(5):592–8.
118. Stoklosa M, Drope J, Chaloupka FJ. Prices and e-cigarette demand: evidence from the European Union. Nicotine Tob Res. 2016;18(10):1973–80.
119. Johnson MW, Johnson PS, Rass O, Pacek LR. Behavioral economic substitutability of e-cigarettes, tobacco cigarettes, and nicotine gum. J Psychopharmacol. 201731(7);851–60.
120. Snider SE, Cummings KM, Bickel WK. Behavioral economic substitution between conventional cigarettes and e-cigarettes differs as a function of the frequency of e-cigarette use. Drug Alcohol Depend. 2017;177:14–22.
121. Beard E, West R, Michie S, Brown J. Association between electronic cigarette use and changes in quit attempts, success of quit attempts, use of smoking cessation pharmacotherapy, and use of stop smoking services in England: time series analysis of population trends. BMJ. 2016;354:i4645.
122. O'Brien B, Knight-West O, Walker N, Parag V, Bullen C. E-cigarettes versus NRT for smoking reduction or cessation in people with mental illness: secondary analysis of data from the ASCEND trial. Tob Induc Dis. 2015;13(1):5.
123. Brown J, Beard E, Kotz D, Michie S, West R. Real-world effectiveness of e-cigarettes when used to aid smoking cessation: a cross-sectional population study. Addiction. 2014;109(9):1531–40.
124. Harrell PT, Marquinez NS, Correa JB, Meltzer LR, Unrod M, Sutton SK et al. Expectancies for cigarettes, e-cigarettes, and nicotine replacement therapies among e-cigarette users (aka vapers). Nicotine Tob Res. 2015;17(2):193–200.
125. Liber AC, Drope JM, Stoklosa M. Combustible cigarettes cost less to use than e-cigarettes: global evidence and tax policy implications. Tob Control. 2017;26(2):158–63.
126. Jorenby DE, Smith SS, Fiore MC, Baker TB. Nicotine levels, withdrawal symptoms, and smoking

reduction success in real world use: a comparison of cigarette smokers and dual users of both cigarettes and E-cigarettes. Drug Alcohol Depend. 2017;170:93–101.
127. Majeed BA, Weaver SR, Gregory KR, Whitney CF, Slovic P, Pechacek TF et al. Changing perceptions of harm of e-cigarettes among U.S. adults, 2012–2015. Am J Prev Med. 2017;52(3):331–8.
128. Chou SP, Saha TD, Zhang H, Ruan WJ, Huang B, Grant BF et al. Prevalence, correlates, comorbidity and treatment of electronic nicotine delivery system use in the United States. Drug Alcohol Depend. 2017;178:296–301.
129. Lee JA, Kim SH, Cho HJ. Electronic cigarette use among Korean adults. Int J Public Health. 2016;61(2):151–7.
130. Wang M, Wang JW, Cao SS, Wang HQ, Hu RY. Cigarette smoking and electronic cigarettes use: a meta-analysis. Int J Environ Res Public Health. 2016;13(1):120.
131. Carroll Chapman SL, Wu LT. E-cigarette prevalence and correlates of use among adolescents versus adults: a review and comparison. J Psychiatr Res. 2014;54:43–54.
132. Shiplo S, Czoli CD, Hammond D. E-cigarette use in Canada: prevalence and patterns of use in a regulated market. BMJ Open. 2015;5(8):e007971.
133. Reid JL, Rynard VL, Czoli CD, Hammond D. Who is using e-cigarettes in Canada? Nationally representative data on the prevalence of e-cigarette use among Canadians. Prev Med. 2015;81:180–3.
134. Kilibarda B, Mravcik V, Martens MS. E-cigarette use among Serbian adults: prevalence and user characteristics. Int J Public Health. 2016;61(2):167–75.
135. Vardavas CI, Filippidis FT, Agaku IT. Determinants and prevalence of e-cigarette use throughout the European Union: a secondary analysis of 26 566 youth and adults from 27 Countries. Tob Control. 2015;24(5):442–8.
136. Farsalinos KE, Poulas K, Voudris V, Le Houezec J. Electronic cigarette use in the European Union: analysis of a representative sample of 27 460 Europeans from 28 countries. Addiction. 2016;111(11):2032–40.
137. E-cigarette use among youth and young adults: a report of the Surgeon General. Washington (DC): Department of Health and Human Services; 2016 (http://www.surgeongeneral.gov/library/2016ecigarettes/index.html#execsumm, accessed 15 May 2019).
138. Rennie LJ, Bazillier-Bruneau C, Rouëssé J. Harm reduction or harm introduction? Prevalence and correlates of E-cigarette use among French adolescents. J Adolesc Health. 2016;58(4):440–5.
139. Mays D, Smith C, Johnson AC, Tercyak KP, Niaura RS. An experimental study of the effects of electronic cigarette warnings on young adult nonsmokers' perceptions and behavioral intentions. Tob Induc Dis. 2016;14:17.
140. Jiang N, Wang MP, Ho SY, Leung LT, Lam TH. Electronic cigarette use among adolescents: a cross-sectional study in Hong Kong. BMC Public Health. 2016;16:202.
141. Lee JA, Lee S, Cho HJ. The relation between frequency of e-cigarette use and frequency and intensity of cigarette smoking among South Korean adolescents. Int J Environ Res Public Health. 2017;14(3):305.
142. Chang HC, Tsai YW, Shiu MN, Wang YT, Chang PY. Elucidating challenges that electronic cigarettes pose to tobacco control in Asia: a population-based national survey in Taiwan. BMJ Open. 2017;7(3):e014263.
143. White J, Li J, Newcombe R, Walton D. Tripling use of electronic cigarettes among New Zealand adolescents between 2012 and 2014. J Adolesc Health. 2015;56(5):522–8.
144. Chaffee BW, Couch ET, Gansky SA. Trends in characteristics and multi-product use among adolescents who use electronic cigarettes, United States 2011–2015. PLoS One. 2017;12(5):e0177073.
145. Fotiou A, Kanavou E, Stavrou M, Richardson C, Kokkevi A. Prevalence and correlates of electronic cigarette use among adolescents in Greece: a preliminary cross-sectional analysis of nationwide survey data. Addict Behav. 2015;51:88–92.

146. Cullen KA, Ambrose BK, Gentzke AS, Apelberg BJ, Jamal A, King BA. Notes from the field: use of electronic cigarettes and any tobacco product among middle and high school students – United States, 2011–2018. Morb Mortal Wkly Rep. 2018;67:1276–7.
147. Chatterjee K, Alzghoul B, Innabi A, Meena N. Is vaping a gateway to smoking: a review of the longitudinal studies. Int J Adolesc Med Health. 2016;30(3). doi: 10.1515/ijamh-2016-0033.
148. Lanza ST, Russell MA, Braymiller JL. Emergence of electronic cigarette use in US adolescents and the link to traditional cigarette use. Addict Behav. 2017;67:38–43.
149. Schneider S, Diehl K. Vaping as a catalyst for smoking? An initial model on the initiation of electronic cigarette use and the transition to tobacco smoking among adolescents. Nicotine Tob Res. 2016;18(5):647–53.
150. Zhong J, Cao S, Gong W, Fei F, Wang M. Electronic cigarettes use and intention to cigarette smoking among never-smoking adolescents and young adults: a meta-analysis. Int J Environ Res Public Health. 2016;13(5):465.
151. Barrington-Trimis JL, Urman R, Berhane K, Unger JB, Cruz TB, Pentz MA et al. E-cigarettes and future cigarette use. Pediatrics. 2016;138(1). doi: 10.1542/peds.2016-0379.
152. Wills TA, Sargent JD, Knight R, Pagano I, Gibbons FX. E-cigarette use and willingness to smoke: a sample of adolescent non-smokers. Tob Control. 2016;25(e1):e52–9.
153. Conner M, Grogan S, Simms-Ellis R, Flett K, Sykes-Muskett B, Cowap L et al. Do electronic cigarettes increase cigarette smoking in UK adolescents? Evidence from a 12-month prospective study. Tob Control. 2017;27(4). doi: 10.1136/tobaccocontrol-2016-053539.
154. Soneji S, Barrington-Trimis JL, Wills TA, Leventhal AM, Unger JB, Gibson LA et al. Association between initial use of e-cigarettes and subsequent cigarette smoking among adolescents and young adults: a systematic review and meta-analysis. JAMA Pediatr. 2017;171(8):788–97.
155. Etter JF. Gateway effects and electronic cigarettes. Addiction. 2018;113(10):1776–83.
156. Franck C, Filion KB, Kimmelman J, Grad R, Eisenberg MJ. Ethical considerations of e-cigarette use for tobacco harm reduction. Respir Res. 2016;17(1):53.
157. Meier EM, Tackett AP, Miller MB, Grant DM, Wagener TL. Which nicotine products are gateways to regular use? First-tried tobacco and current use in college students. Am J Prev Med. 2015;48(1 Suppl 1):S86–93.
158. Polosa R, Russell C, Nitzkin J, Farsalinos KE. A critique of the US Surgeon General's conclusions regarding e-cigarette use among youth and young adults in the United States of America. Harm Reduct J. 2017;14(1):61.
159. Kozlowski LT, Warner KE. Adolescents and e-cigarettes: objects of concern may appear larger than they are. Drug Alcohol Depend. 2017;174:209–14.
160. Surgeon General's advisory on e-cigarette use among youth. Atlanta (GA): Office of the Surgeon General, Office on Smoking and Health; 2018 (https://e-cigarettes.surgeongeneral.gov/documents/surgeon-generals-advisory-on-e-cigarette-use-among-youth-2018.pdf, accessed 1 January 2019).
161. Azar AM, Gottlieb S. Opinion. We cannot let e-cigarettes become an on-ramp for teenage addiction. The Washington Post, 11 October 2018 (https://www.washingtonpost.com/opinions/we-cannot-let-e-cigarettes-become-an-on-ramp-for-teenage-addiction/2018/10/11/55ce424e-ccc6-11e8-a360-85875bac0b1f_story.html?utm_term=.3b243a1a45f5, accessed 1 January 2019).
162. National Research Council. Understanding the US illicit tobacco market: characteristics, policy context, and lessons from international experiences. Washington (DC): National Academies Press; 2015 (https://doi.org/10.17226/19016).
163. Lee JGL, Golden SD, Ribisl KM. Limited indications of tax stamp discordance and counterfeiting on cigarette packs purchased in tobacco retailers, 97 counties, USA, 2012. Prev Med Rep. 2017;8:148–52.
164. Ribisl KM, Williams R, Gizlice Z, Herring AH. Effectiveness of state and federal government

agreements with major credit card and shipping companies to block illegal Internet cigarette sales. PLoS One. 2011;6(2):e16745.
165. Hatsukami DK, Heishman SJ, Vogel RI, Denlinger RL, Roper-Batker AN, Mackowick KM et al. Dose–response effects of Spectrum research cigarettes. Nicotine Tob Res. 2013;15(6):1113–21.
166. Wiley JL, Marusich JA, Thomas BF, Jackson KJ. Determination of behaviorally effective tobacco constituent doses in rats. Nicotine Tob Res. 2015;17(3):368–71.
167. Grizzell JA, Echeverria V. New insights into the mechanisms of action of cotinine and its distinctive effects from nicotine. Neurochem Res. 2015;40(10):2032–46.
168. Brennan KA, Laugesen M, Truman P. Whole tobacco smoke extracts to model tobacco dependence in animals. Neurosci Biobehav Rev. 2014;47:53–69.
169. Costello MR, Reynaga DD, Mojica CY, Zaveri NT, Belluzzi JD, Leslie FM. Comparison of the reinforcing properties of nicotine and cigarette smoke extract in rats. Neuropsychopharmacology. 2014;39(8):1843–51.
170. Gellner CA, Belluzzi JD, Leslie FM. Self-administration of nicotine and cigarette smoke extract in adolescent and adult rats. Neuropharmacology. 2016;109:247–53.
171. Smith TT, Schaff MB, Rupprecht LE, Schassburger RL, Buffalari DM, Murphy SE et al. Effects of MAO inhibition and a combination of minor alkaloids, β-carbolines, and acetaldehyde on nicotine self-administration in adult male rats. Drug Alcohol Depend. 2015;155:243–52.
172. Harris AC, Tally L, Schmidt CE, Muelken P, Stepanov I, Saha S et al. Animal models to assess the abuse liability of tobacco products: effects of smokeless tobacco extracts on intracranial self-stimulation. Drug Alcohol Depend. 2015;147:60–7.
173. Truman P, Grounds P, Brennan KA. Monoamine oxidase inhibitory activity in tobacco particulate matter: Are harman and norharman the only physiologically relevant inhibitors? Neurotoxicology. 2017;59:22–6.
174. Arnold MM, Loughlin SE, Belluzzi JD, Leslie FM. Reinforcing and neural activating effects of norharmane, a non-nicotine tobacco constituent, alone and in combination with nicotine. Neuropharmacology. 2014;85:293–304.
175. Caine SB, Collins GT, Thomsen M, Wright C, Lanier RK, Mello NK. Nicotine-like behavioral effects of the minor tobacco alkaloids nornicotine, anabasine, and anatabine in male rodents. Exp Clin Psychopharmacol. 2014;22(1):9–22.
176. Plescia F, Brancato A, Venniro M, Maniaci G, Cannizzaro E, Sutera FM et al. Acetaldehyde self-administration by a two-bottle choice paradigm: consequences on emotional reactivity, spatial learning, and memory. Alcohol. 2015;49(2):139–48.
177. Smith TT, Rupprecht LE, Cwalina SN, Onimus MJ, Murphy SE, Donny EC et al. Effects of monoamine oxidase inhibition on the reinforcing properties of low-dose nicotine. Neuropsychopharmacology. 2016;41(9):2335–43.
178. Biswas L, Harrison E, Gong Y, Avusula R, Lee J, Zhang M et al. Enhancing effect of menthol on nicotine self-administration in rats. Psychopharmacology. 2016;233(18):3417–27.
179. Advisory note: banning menthol in tobacco products. WHO Study Group on Tobacco Product Regulation. Geneva: World Health Organization; 2016 (http://apps.who.int/iris/bitstream/10665/205928/1/9789241510332_eng.pdf, accessed 15 May 2019).
180. Vagg R, Chapman S. Nicotine analogues: a review of tobacco industry research interests. Addiction. 2015;100:701–12.
181. Desai RI, Doyle MR, Withey SL, Bergman J. Nicotinic effects of tobacco smoke constituents in nonhuman primates. Psychopharmacology. 2016;233(10):1779–89.
182. Wong LP, Alias H, Agha Mohammadi N, Ghadimi A, Hoe VCW. E-cigarette users' attitudes on the banning of sales of nicotine e-liquid, its implication on e-cigarette use behaviours and alternative sources of nicotine e-liquid. J Community Health. 2017;42(6):1225–32.
183. O'Brien EK, Nguyen AB, Persoskie A, Hoffman AC. US adults' addiction and harm beliefs about

nicotine and low nicotine cigarettes. Prev Med. 2017;96:94–100.
184. Denlinger-Apte RL, Joel DL, Strasser AA, Donny EC. Low nicotine content descriptors reduce perceived health risks and positive cigarette ratings in participants using very low nicotine content cigarettes. Nicotine Tob Res. 2017;19(10):1149–54.
185. Mercincavage M, Saddleson ML, Gup E, Halstead A, Mays D, Strasser AA. Reduced nicotine content cigarette advertising: how false beliefs and subjective ratings affect smoking behavior. Drug Alcohol Depend. 2017;173:99–106.
186. Bolcic-Jankovic D, Biener L. Public opinion about FDA regulation of menthol and nicotine. Tob Control. 2015;24(e4):e241–5.
187. Kiviniemi MT, Kozlowski LT. Deficiencies in public understanding about tobacco harm reduction: results from a United States national survey. Harm Reduct J. 2015;12:21.
188. Czoli CD, Fong GT, Mays D, Hammond D. How do consumers perceive differences in risk across nicotine products? A review of relative risk perceptions across smokeless tobacco, e-cigarettes, nicotine replacement therapy and combustible cigarettes. Tob Control. 2017;26(e1):e49–58.
189. Darredeau C, Stewart SH, Barrett SP. The effects of nicotine content information on subjective and behavioural responses to nicotine-containing and denicotinized cigarettes. Behav Pharmacol. 2013;24(4):291–7.
190. Mercincavage M, Smyth JM, Strasser AA, Branstetter SA. Reduced nicotine content expectancies affect initial responses to smoking. Tob Regul Sci. 2016;2(4):309–16.
191. Harrell PT, Juliano LM. A direct test of the influence of nicotine response expectancies on the subjective and cognitive effects of smoking. Exp Clin Psychopharmacol. 2012;20(4):278–86.
192. Gu X, Lohrenz T, Salas R, Baldwin PR, Soltani A, Kirk U et al. Belief about nicotine modulates subjective craving and insula activity in deprived smokers. Front Psychiatry. 2016;7:126.
193. Harrell PT, Simmons VN, Piñeiro B, Correa JB, Menzie NS, Meltzer LR et al. E-cigarettes and expectancies: why do some users keep smoking? Addiction. 2015;110(11):1833–43.
194. Pearson JL, Abrams DB, Niaura RS, Richardson A, Vallone DM. Public support for mandated nicotine reduction in cigarettes. Am J Public Health. 2013;103(3):562–7.
195. Li J, Newcombe R, Walton D. Responses towards additional tobacco control measures: data from a population-based survey of New Zealand adults. N Z Med J 2016;129:87–92.
196. Fraser T, Kira A. Perspectives of key stakeholders and smokers on a very low nicotine content cigarette-only policy: qualitative study. N Z Med J. 2017;130:36–45.
197. Denlinger-Apte RL, Tidey JW, Koopmeiners JS, Hatsukami DK, Smith TT, Pacek LR et al. Correlates of support for a nicotine-reduction policy in smokers with 6-week exposure to very low nicotine cigarettes. Tob Control. 2018. pii: tobaccocontrol-2018-054622.
198. Donny EC, Walker N, Hatsukami D, Bullen C. Reducing the nicotine content of combusted tobacco products sold in New Zealand. Tob Control. 2017;26(e1):e37–42.
199. Laugesen M. Modelling a two-tier tobacco excise tax policy to reduce smoking by focusing on the addictive component (nicotine) more than the tobacco weight. N Z Med J. 2012;125:35–48.
200. Yuke K, Ford P, Foley W, Mutch A, Fitzgerald L, Gartner C. Australian urban Indigenous smokers' perspectives on nicotine products and tobacco harm reduction. Drug Alcohol Rev. 2018;42(6):1225–32.
201. Danielle LJ, Hatsukami DK, Hertsgaard L, Dermody SS, Donny EC. Gender differences in the perceptions of very low nicotine content cigarettes (abstract). Annual meeting of the College on Problems of Drug Dependence. Drug Alcohol Depend. 2014;140:e97.
202. Mercincavage M, Wileyto EP, Saddleson ML, Lochbuehler K, Donny EC, Strasser AA. Attrition during a randomized controlled trial of reduced nicotine content cigarettes as a proxy for understanding acceptability of nicotine product standards. Addiction. 2017;112(6):1095–103.

5. A regulatory strategy for reducing exposure to toxicants in cigarette smoke

Stephen S. Hecht and Dorothy K. Hatsukami, Masonic Cancer Center, University of Minnesota, Minneapolis (MN), USA

Contents

5.1 Introduction
5.2 Regulation of cigarette smoke constituents described in WHO Technical Report Series, No. 951
5.3 Industry response to WHO Technical Report Series, No. 951
5.4 Relation between smoke constituent levels and biomarkers
5.5 Use of mandated lowering of toxicant levels in a product regulatory strategy
5.6 Toxicants recommended for mandated lowering and recommended limits
5.7 Implementation of mandated lower levels of toxicants
5.8 Conclusions and recommendations
5.9 References

5.1 Introduction

The report *Scientific Basis of Tobacco Product Regulation* (WHO Technical Report Series No. 951) *(1)*, published in 2008, specifies an approach to lowering the levels of selected toxicants in mainstream cigarette smoke. We review these recommendations and summarize several papers published by the tobacco industry that discuss aspects of the proposed regulatory strategy. We also discuss the significant progress in evaluating biomarkers of tobacco smoke toxicants and carcinogens that has been made since the 2008 report and the relation of these biomarkers to smoke constituent levels and, in some cases, to disease incidence. We consider the use of mandated toxicant levels in a product regulatory strategy and present recommendations for an updated regulatory strategy based on nicotine levels in cigarettes.

5.2 Regulation of cigarette smoke constituents described in WHO Technical Report Series No. 951

The report *Scientific Basis of Tobacco Product Regulation* (WHO Technical Report Series No. 951) recognizes that regulatory strategies based on machine-measured tar, nicotine and CO yields **per cigarette** under the ISO smoking machine regimen were causing harm by misleading smokers to believe that so-called "low-yield cigarettes" were less harmful. Thus, the report recommended establishing limits for certain smoke constituents expressed **per mg nicotine** rather than per cigarette, as people smoke cigarettes to obtain adequate nicotine to satisfy their physiological and behavioural needs, and most toxicants in smoke are delivered in proportion to nicotine. This shifts the emphasis towards the potential toxicity of smoke generated under standardized conditions. The report also recommended prohibiting any communications to consumers that were based on machine measurements, as these can be easily misunderstood.

The report concluded that it was premature to consider biomarkers in a regulatory strategy, because, while biomarkers of exposure existed at the time, there were no validated biomarkers of harm. As discussed below, significant progress has been made in the application of biomarkers to evaluating the potential toxicity and carcinogenicity of cigarette smoke. There is presently no doubt that the average levels of certain biomarkers of exposure reflect the measured amounts of their parent compounds in cigarette smoke.

In WHO Technical Report Series No. 951, TobReg selected toxicants for regulation on the basis of a number of factors, including established cardiovascular and pulmonary toxicity and carcinogenicity, the feasibility of lowering their concentrations with available technology and variation in chemical classes and brands. Consideration was also given to the inclusion of compounds in both the gas and the particulate phases of cigarette smoke. Of these characteristics, the most important was evidence of toxicity.

Data from Health Canada and a paper on constituent levels in Philip Morris brands in both the USA and elsewhere *(2)* were available to assess differences in the levels of the selected toxicants in cigarettes on the market in Canada and the USA. These data were used to select the initial levels of the regulated constituents as a first step in an overall strategy to eliminate brands with higher levels from the market, thus lowering the overall mean values of the toxicants in the remaining brands. A rolling mean was envisioned, which would eventually drive down levels of the selected toxicants in all brands left on the market and prevent introduction onto the market of brands with toxicant levels higher than the mean.

The report recommended that the regulatory strategy be implemented in phases, beginning with required annual reporting of selected toxicant levels, followed by promulgation of the levels for toxicants above which brands could not be sold and, finally, enforcement of the established levels. Some progress has been made towards achieving these goals, but it is unclear whether there has been sufficient momentum.

The toxicants for which mandatory lowering is recommended and the initially recommended levels are summarized in columns 1–3 of Table 5.1. The modified intense machine smoking regimen used by Health Canada (HCI) was selected for measuring toxicants. This regimen is modified from the ISO method by increasing the puff volume from 35 mL to 55 mL and decreasing the puff interval from 60 s to 30 s. In addition, all ventilation holes are blocked. This method is referred to in the report as the "modified intense smoking" regimen. The levels in Table 5.1 for "international brands" (i.e. all those other than United States brands) are derived from a sample of brands of United-States-style cigarettes with a blend of tobaccos *(2)*. The levels in "Canadian brands" are derived from reports to Health Canada and are for cigarettes made mainly with Bright (flue-cured) tobacco. These were selected as examples because of the availability of data and the different types of tobacco and were expected roughly to represent the markets to be regulated, depending on the type of product prevalent in that market. In addition to the compounds listed in Table 5.1, priority toxicants recommended for reporting were acrylonitrile, 4-aminobiphenyl, 2-aminonapthalene, cadmium, catechol, crotonaldehyde, hydrogen cyanide, hydroquinone and nitrogen oxides *(1)*. An additional group of 20 compounds to be considered for reporting was added in 2013 (see section 8).

Table 5.1. Levels of toxicants recommended for mandated lowering (μg/mg nicotine under the Canadian intense smoking regimen)

Toxicant	Levels recommended in WHO Technical Report Series No. 951 *(1)*		Levels in some published surveys			Criterion for selecting recommended levels
	International brands *(2)*	Canadian brands[a]	*(3)*[b]	*(4)*[c]	*(5)*[d]	
NNK	0.072	0.047	0.080	0.050–0.072	0.0273	Median value
NNN	0.114	0.027	0.134	0.077–0.142	0.0916	Median value
Acetaldehyde	860	670	676	468–626	684.1	125% of median value
Acrolein	83	97	85.4	48.8–64.3	64.5	125% of median value
Benzene	48	50	43.1	40.8–53.7	66.1	125% of median value
Benzo[a]pyrene	0.011	0.011	0.0096	0.0074–0.0084	0.0085	125% of median value
1,3-Butadiene	67	53	50.5	38.1–52.1	71.9	125% of median value
Carbon monoxide	18 400	15 400	15 700	11 500–14 800	16 100	125% of median value
Formaldehyde	47	97	47.3	28.2–43.7	81.4	125% of median value
Nicotine (per cigarette)	–	–	2.13 mg	1.56–2.33 mg	1.6 mg	–

NNK: 4-(methylnitrosamino)-1-(3-pyridyl)-1-butanone; NNN: *N*´-nitrosonornicotine. [a] Data from Health Canada. [b] 61 United States brand styles, mean. [c] Range of means of three international brands. [d] 20 brands on the Chinese market, mean.

The list of compounds in Table 5.1 is still highly relevant to regulation, as all have well documented toxic and/or carcinogenic activity. The FDA has published a list of "harmful and potentially harmful constituents (HPHCs) in tobacco products and tobacco smoke" that includes all the compounds in Table 5.1 as well as those recommended for reporting *(6)*. The FDA list of HPHCs, which focuses on chemicals linked to cancer, cardiovascular disease, reproductive problems, addiction and respiratory effects, is longer than that presented here, with 93 constituents, because it includes every identified toxic or carcinogenic compound in each class rather than representatives (e.g. NNK and NNN for *N*-nitrosamines and benzo[a]pyrene for PAHs in Table 5.1).

Table 5.1 also presents the levels of the selected toxicants per mg of nicotine as reported in three industry studies published subsequent to WHO Technical Report Series No. 951 *(3–5)*. Generally, the means and ranges of constituent concentrations were similar to those in the WHO report. One industry study, not shown in Table 5.1, reported levels of NNN and NNK in smoke and concluded that they were decreasing *(7)*.

WHO Technical Report Series No. 951 also addressed questions on variations in measurement of constituent levels and among brands and reported the coefficient of variation (standard deviation of a series of measurements divided by the mean of that series) for the levels of constituents in various brands. This was divided by the coefficient of variation of multiple measurements of a given constituent (e.g. variation in the analytical method) to produce a ratio. Constituents with high ratios, e.g. with wide variation in levels, as determined by analytical methods with little variation, would be the most appropriate targets for regulation. The report concluded that the ratios for individual toxicants in the Philip Morris International brands and the Canadian brands indicated that there was sufficient variation in the levels of most toxicants among brands that mandated reductions would have a substantial effect on the levels in brands remaining on the market.

5.3 Industry response to WHO Technical Report Series No. 951

Bodnar et al. *(3)* conducted a survey of selected mainstream smoke constituents from commercially marketed United States cigarettes, segmenting the market into 13 strata according to "tar" category and cigarette design parameters. The HCI smoking regimen was used, and the yields of mainstream smoke constituent per mg nicotine were estimated. Normalization per mg nicotine gave an inverse ranking of constituent yields from those expressed per cigarette according to the ISO/FTC Cambridge filter method "tar" categories, i.e. "Full flavour" > "Lights" > "Ultra-lights" – designations that are prohibited by the WHO FCTC.

Purkis et al. *(8)* reviewed measurements of smoke constituents and aspects of variability that can affect regulatory standards. They discussed method development, harmonization and standardization and reviewed such issues as variation among products, machine-smoking regimens and laboratory measurements and their potential consequences, such as misinterpretation of data. They called for internationally agreed testing standards and measurement tolerances.

Haussmann *(9)* provided a detailed discussion of the use of hazard indices for evaluating cigarette smoke composition, including the advantages and limitations of the approach described in WHO Technical Report Series No. 951. Haussmann concluded that application of the suggested approach would result in banning 39 of the 50 brands in the data set of Counts et al. (referred to as "International brands" in Table 5.1). He demonstrated that excluding one brand because the yield of one constituent exceeded the recommended ceiling could mean that the yields of the other eight constituents that were present below their respective ceilings were not accounted for. The result would be to increase rather than decrease the average yield of those constituents in the remaining brands. Thus, the recommended levels of various constituents should be assessed carefully.

Piade et al. *(10)* determined the levels of the 18 priority constituents (expressed per mg nicotine) identified in WHO Technical Report Series No. 951 in 262 commercial brands, including various blends from 13 countries. Principal component analysis was used to identify inverse correlations and other patterns. Three principal components explained about 75% of the data variation. The first was sensitive to the relative levels of gas- and particle-phase compounds, while the other two components grouped American and Virginia blends, revealing inverse correlations. For example, the levels of nitrogen oxides and amino- or nitroso-aromatic compounds were inversely correlated with either formaldehyde, acrolein, benzo[a]pyrene or dihydroxybenzenes. They concluded that regulatory approaches might be confounded by such inverse correlations.

Belushkin et al. *(11)* studied the variation in yields of smoke constituents for a Kentucky reference cigarette and one commercial brand analysed on several occasions over 7 years. They showed that statistically significant differences in the yields of some smoke constituents did not necessarily correspond to differences between products and that tolerances should be defined. They concluded that use of two approaches – minimal detectable differences and statistical equivalence – was more meaningful for comparisons than the statistical Student t test.

Eldridge et al. *(4)* measured toxicant levels in the tobacco and smoke of reference cigarettes and three commercial products, monthly for 10 months. The monthly variation was < 15% for most analytes but increased somewhat when reported as a ratio to nicotine level. They concluded that measurement of emissions from commercial cigarettes was subject to considerable variation, particularly for toxicants present at low levels.

Deng et al. *(5)* studied the influence of measurement uncertainty on the recommended regulation of the nine constituents listed in Table 5.1. Uncertainty was evaluated in collaborative studies conducted in 2012–2016 of the compliance of 20 representative cigarette samples. They concluded that measurement uncertainty would strongly influence implementation of the proposed regulations.

Collectively, these studies indicate some cautionary and practical notes for application of the proposed regulation of nine constituents per mg nicotine. The uncertainty in analytical chemistry measurements is well known but in most cases is relatively minor, particularly for constituents that occur at high levels. Validated methods are available for measuring all the compounds proposed for measurement (*12–18*; see also http://www.who.int/tobacco/industry/product_regulation/toblabnet/en/). Thus, further studies on standardization of methods, including time-consuming ring trials that may last years, are probably not required. Furthermore, the inverse correlations among constituent levels in certain brands may be problematic, as lowering the level of one constituent may increase those of others. This has been known for decades *(19, 20)*, and it is industry's responsibility to address these levels in products that they wish to maintain on the market. In conclusion, the main goal of these industry studies was to provide reasons to challenge and delay implementation of product standards for cigarette smoke emissions, and there are no valid scientific reasons that implementation cannot proceed.

5.4 Relation between smoke constituent levels and biomarkers

Biomarkers of carcinogens and toxicants in tobacco, most of which are urinary metabolites of tobacco smoke constituents, are robust, validated, reliable indicators of human exposure to tobacco product constituents *(21)*. Several of these biomarkers are directly relevant to the toxicants recommended for mandated lowering listed in Table 5.1. For NNK, total NNAL (the sum of free NNAL and its glucuronides) is an accepted biomarker of exposure *(22)*. For acrolein, the urinary metabolite 3-hydroxypropyl mercapturic acid has been widely used *(23, 24)*. Similarly, for benzene and 1,3-butadiene, the urinary mercapturic acids, S-phenyl mercapturic acid and monohydroxybutyl mercapturic acid, respectively, have been quantified *(25–27)*. The urinary metabolite of pyrene, 1-hydroxypyrene, has been used extensively as a biomarker of uptake of PAH, the class of compounds that includes benzo[*a*]pyrene *(21, 28)*. Exhaled CO is an accepted biomarker of CO exposure in cigarette smokers *(21)*. Total nicotine equivalents – the sum of nicotine, cotinine and 3´-hydroxycotinine and their glucuronides in urine – represent a high percentage of the nicotine dose received by a tobacco user and are an excellent biomarker of nicotine dose *(29)*.

Most of these biomarkers have been quantified in large studies, such as the Health Measures Study in Canada, the National Health and Nutrition Examination Survey in the USA and the industry-sponsored Total Exposure Study, as well as in epidemiological and clinical studies *(26, 30–42)*. The levels of all the biomarkers decrease significantly when people stop smoking, and all the urinary biomarkers correlate with total nicotine equivalents *(24, 26–28, 43, 44)*. The levels of constituents in smoke and the levels of acrolein, NNK, nicotine and pyrene in the mouth are clearly related to the respective urinary biomarkers *(33, 34, 38, 41)*. Collectively, these studies demonstrate that evaluation of biomarkers can buttress analyses of smoke constituents in regulations by providing a definitive link between the levels of smoke constituents and actual human exposure. Furthermore, nicotine biomarkers and total NNAL have been positively related to lung cancer in prospective epidemiological studies *(45–47)* and can thus be considered biomarkers of potential harm.

5.5 Use of mandated lowering of toxicant levels in a product regulatory strategy

The public health impact of mandated lowering of toxicant levels of cigarettes is virtually unknown. In order to improve public health by decreasing the disease and death due to tobacco product use, evaluation of a product standard must include population effects. According to the Family Smoking Prevention and Tobacco Control Act in the USA, which gives the FDA regulatory authority over tobacco products in the country, tobacco product standards can be adopted if the standard is "appropriate for the protection of the public health." The scientific evidence that is considered in evaluating the public health impact of a product standard includes:

- the risks and benefits to the population as a whole, including users and non-users of tobacco products, of the proposed standard;
- the increased or decreased likelihood that existing users of tobacco products will stop using such products; and
- the increased or decreased likelihood that those who do not use tobacco products will start using such products.

To date, a cross-sectional study has shown that urinary levels of NNAL are lower in countries in which cigarettes have lower levels of NNK and NNN *(48)*, but it is not known whether these differences are related to the country-specific incidence of cancer. A dose–response relation has, however, been found between biomarkers of exposure to NNK and NNN and risks for cancers of the lung *(46, 47, 49)* and the oesophagus *(50)*, respectively.

To avoid replicating the negative public health impact of the low-tar and low-nicotine-yield cigarette experience *(51)* before implementation of mandated standards, their potential effects on smoking behaviour, exposure to toxicants and potential harm, consumer perceptions and misperceptions of the regulated product and uptake and continued use of the product *(52–54)* should be determined and appropriate action taken as necessary. One constituent that was not targeted in WHO Technical Report Series No. 951 but was targeted in a subsequent advisory note *(55)* is nicotine. Regulation of nicotine, as discussed in the previous section, might result in the greatest public health benefit (see below). Reducing nicotine to minimally addictive levels would prevent the development of dependence on cigarettes and facilitate abstinence among smokers, thereby reducing the prevalence of smoking.

It is notable that, since the WHO Technical Report was issued, products such as ENDS, with significantly lower toxicant levels and toxicity than conventional cigarettes *(56–58)*, have been introduced in some markets in full force. If products that are demonstrated to be substantially less toxic replace cigarettes among smokers and are used exclusively by smokers, the public health impact might be substantial provided that there are no adverse consequences such as inappropriate promotion or use by children and/or young people or continuing dual use with conventional cigarettes or other tobacco products.

5.6 Toxicants recommended for mandated lowering and recommended limits

Donny et al. *(40)* obtained highly relevant results that could change the regulatory landscape. They conducted a double-blind, parallel, randomized clinical trial of cigarettes with nicotine contents varying from 15.8 mg/g of tobacco, which is typical of commercial brands in the USA, to 0.4 mg/g. Regular smokers were assigned to smoke either low-nicotine cigarettes or their usual brand of cigarettes for 6 weeks. At week 6, the average number of cigarettes smoked per day was significantly lower for the participants randomly assigned to cigarettes containing 2.4 mg, 1.3 mg or 0.4 mg of nicotine per gram of tobacco (16.5, 16.3 and 14.9 cigarettes/day, respectively) than for participants who were randomly assigned to their usual brand or to cigarettes containing 15.8 mg/g tobacco (22.2 and 21.3 cigarettes/day, respectively). The cigarettes with a lower nicotine content reduced exposure to and dependence on nicotine and also craving during abstinence from smoking, without increasing the expired CO level or total puff volume, which suggests that there was minimal compensation. These results are consistent with those of several smaller trials *(30, 59–63)* and of a trial with vulnerable adults *(64)*.

In the trial by Donny et al. *(40)*, participants assigned to cigarettes containing ≤ 2.4 mg of nicotine per gram of tobacco smoked 23–30% fewer cigarettes per day at week 6 than did participants assigned to cigarettes containing 15.8 mg/g of tobacco. The results showed a clear, significant breakpoint in the number of cigarettes smoked per day and total nicotine equivalents in urine between the cigarettes containing 5.2 mg/g nicotine and those with 2.4 mg/g. Furthermore, and most significantly, dependence, as assessed in the Wisconsin inventory of smoking dependence motives, was significantly lower at week 6 among participants who smoked cigarettes with 0.4 mg/g nicotine than among those who smoked cigarettes with 15.8 mg/g nicotine ($P = 0.001$). The cigarettes containing 0.4 mg/g nicotine delivered 0.04 mg nicotine in their smoke under HCI conditions *(65)*.

These results suggest that a product standard could have a significant public health benefit for countries with the appropriate resources (see reference 55 for more details). The amount of nicotine in cigarette tobacco filler should be regulated such that the level of nicotine in mainstream smoke does not exceed 0.04 mg per cigarette under HCI conditions (39–58 times lower than in typical current brands, Table 5.1). With respect to other constituents, our new recommended levels are the median values per mg of nicotine of the constituents listed in Table 5.1, according to the latest market survey information, which would be obtained from major tobacco companies and Health Canada data on the brands with compliant nicotine levels, e.g. that deliver 0.04 mg nicotine per cigarette under HCI smoking conditions. This recommendation is stronger than the "125% of the median value" previously established by TobReg. We also retain the requirement for reporting levels of acrylonitrile, 4-aminobiphenyl, 2-aminonaphthalene, cadmium, catechol, crotonaldehyde, hydrogen cyanide, hydroquinone and nitrogen oxides. These toxicants satisfy several criteria for inclusion: sufficient evidence of carcinogenicity, known respiratory or cardiotoxicity, differences in levels of toxicants in different brands in different countries, measurability and potential for decreased yields in a product *(1)*.

We note that, when expressed per mg nicotine, the mandated toxicant levels would be higher than the current mandated levels for the toxicants listed in Table 5.2, based on data for the Spectrum cigarettes used in the clinical study described above *(65)*. This is obviously a consequence of the lower recommended nicotine level of 0.04 mg per cigarette in mainstream smoke as compared with current levels, which range from 1.5 to 2 mg/cigarette. Measures of biomarkers, however, clearly show that smokers of these low-nicotine cigarettes excrete significantly lower levels of total nicotine equivalents and total NNAL in their urine and thus have less uptake of toxic and carcinogenic compounds *(40)*.

Table 5.2. Levels of selected toxicants in mainstream smoke of Spectrum cigarettes (NRC 102, 0.04 mg nicotine per cigarette; Health Canada intense smoking regimen)

Toxicant	µg/mg nicotine
Acetaldehyde	41 500
Acrolein	2070
Benzo[*a*]pyrene	0.220
Carbon monoxide	720 000
Formaldehyde	258
NNK	0.798
NNN	2.88

NNK: 4-(methylnitrosamino)-1-(3-pyridyl)-1-butanone; NNN: *N'*-nitrosonornicotine. *Source*: reference 65.

5.7 Implementation of mandated lower levels of toxicants

In order for regulators to effectively mandate lower levels of toxicants and nicotine in cigarettes, appropriate procedures and sufficient infrastructure and resources are required. These include requiring the disclosure of levels of harmful constituents in cigarettes and an independent laboratory to validate and monitor these levels routinely. In the previous WHO report *(1)*, analysis of the constituent yields of cigarettes was considered to be the first step towards a mandate to reduce toxicants. Standardized testing procedures should be instituted that include not only laboratory methods (e.g. machine-determined yields that reflect a range of human exposures and validated analytical chemistry methods) but also the way in which the cigarettes to be tested are systematically procured (e.g. different geographical locations, from several retail outlets within a geographical location, directly from manufacturers, in different months). Tracking cigarettes that are licit and illicit should also be considered, as well as a plan to reduce the entry of illicit cigarettes to the market. Furthermore, education campaigns and messages should be designed to minimize any misperception of the harm associated with smoking cigarettes. Thus, the message that combusted products are hazardous to health, regardless of a mandate to lower toxicant levels, must be communicated explicitly and clearly. Finally, a surveillance system should be in place to monitor any unintended consequences, and methods for correction should be considered.

5.8 Conclusions and recommendations

The regulatory strategy recommended here retains the principle of the previous recommendations by TobReg, revising and strengthening it such that brands on the market cannot exceed the median values of the constituents listed in Table 5.1. We also retain the requirement for reporting of several other constituents. Furthermore, we recommend considering a regulatory strategy in which the nicotine level in tobacco does not exceed 0.4 mg/g of tobacco (0.04 mg nicotine per cigarette in mainstream smoke under HCI smoking conditions). The new regulation based on nicotine is the result of decades of research that has demonstrated that nicotine is the main chemical that drives people to smoke *(66, 67)* or otherwise use tobacco and the results of clinical trials that demonstrate that low-nicotine cigarettes reduce exposure to and dependence on nicotine and the number of cigarettes smoked. Thus, we recommend that the previous strategy of limiting specific toxicants and the present new recommendation of limiting nicotine be pursued aggressively in parallel. The first, critical step in the toxicant reduction strategy is collection of data on current products by methods established by WHO Tobacco Laboratory Network (TobLabNet).

We are aware that challenges may exist in implementing product standards for cigarettes and that unintended consequences are possible (e.g. illegal markets). Therefore, it will be necessary to enforce regulations such that cigarettes with levels of toxicants and nicotine above the established means are removed from the market. A reliable system of monitoring regulated constituents in tobacco and smoke will have to be established. Representative analyses of cigarettes on the international market will have to be performed, so that the mean levels of the mandated toxicants in their smoke and possibly nicotine in tobacco are reported. Vigilant monitoring for evidence that no harm results from the new regulations will be necessary.

5.9 References

1. The scientific basis of tobacco product regulation. Second report of a WHO study group (WHO Technical Report Series, No. 951). Geneva: World Health Organization; 2008: 45–277.
2. Counts ME, Morton MJ, Laffoon SW, Cox RH, Lipowicz PJ. Smoke composition and predicting relationships for international commercial cigarettes smoked with three machine-smoking conditions. Regul Toxicol Pharmacol. 2005;41(3):185–227.
3. Bodnar JA, Morgan WT, Murphy PA, Ogden MW. Mainstream smoke chemistry analysis of samples from the 2009 US cigarette market. Regul Toxicol Pharmacol. 2012;64(1):35–42.
4. Eldridge A, Betson TR, Gama MV, McAdam K. Variation in tobacco and mainstream smoke toxicant yields from selected commercial cigarette products. Regul Toxicol Pharmacol. 2015;71(3):409–27.
5. Deng H, Li Z, Bian Z, Yang F, Liu S, Fan Z et al. Influence of measurement uncertainty on the World Health Organization recommended regulation for mainstream cigarette smoke constituents. Regul Toxicol Pharmacol. 2017;86:231–40.
6. Food and Drug Administration. Harmful and potentially harmful constituents in tobacco products and tobacco smoke; established list. Fed Reg. 2012;77:20034–7.
7. Gunduz I, Kondylis A, Jaccard G, Renaud JM, Hofer R, Ruffieux L et al. Tobacco-specific N-nitrosamines NNN and NNK levels in cigarette brands between 2000 and 2014. Regul Toxicol Pharmacol. 2016;76:113–20.
8. Purkis SW, Meger M, Wuttke R. A review of current smoke constituent measurement activities and aspects of yield variability. Regul Toxicol Pharmacol. 2012;62(1):202–13.
9. Haussmann HJ. Use of hazard indices for a theoretical evaluation of cigarette smoke composition. Chem Res Toxicol. 2012;25(4):794–810.
10. Piade JJ, Wajrock S, Jaccard G, Janeke G. Formation of mainstream cigarette smoke constituents prioritized by the World Health Organization – yield patterns observed in market surveys, clustering and inverse correlations. Food Chem Toxicol. 2013;55:329–347.
11. Belushkin M, Jaccard G, Kondylis A. Considerations for comparative tobacco product assessments based on smoke constituent yields. Regul Toxicol Pharmacol. 2015;73(1):105–13.
12. Eschner MS, Selmani I, Groger TM, Zimmermann R. Online comprehensive two-dimensional characterization of puff-by-puff resolved cigarette smoke by hyphenation of fast gas chromatography to single-photon ionization time-of-flight mass spectrometry: quantification of hazardous volatile organic compounds. Anal Chem. 2011;83(17):6619–27.
13. Pang X, Lewis AC. Carbonyl compounds in gas and particle phases of mainstream cigarette smoke. Sci Total Environ. 2011;409(23):5000–9.
14. Pappas RS, Fresquez MR, Martone N, Watson CH. Toxic metal concentrations in mainstream smoke from cigarettes available in the USA. J Anal Toxicol. 2014;38(4):204–11.
15. Vu AT, Taylor KM, Holman MR, Ding YS, Hearn B, Watson CH. Polycyclic aromatic hydrocarbons in the mainstream smoke of popular US cigarettes. Chem Res Toxicol. 2015;28(8):1616–26.
16. Ding YS, Yan X, Wong J, Chan M, Watson CH. In situ derivatization and quantification of seven carbonyls in cigarette mainstream smoke. Chem Res Toxicol. 2016;29(1):125–31.
17. Cecil TL, Brewer TM, Young M, Holman MR. Acrolein yields in mainstream smoke from commercial cigarette and little cigar tobacco products. Nicotine Tob Res. 2017;19(7):865–70.
18. Edwards SH, Rossiter LM, Taylor KM, Holman MR, Zhang L, Ding YS et al. Tobacco-specific nitrosamines in the tobacco and mainstream smoke of U.S. commercial cigarettes. Chem Res Toxicol. 2017;30(2):540–51.
19. Adams JD, Lee SJ, Hoffmann D. Carcinogenic agents in cigarette smoke and the influence of nitrate on their formation. Carcinogenesis. 1984;5:221–3.
20. Ding YS, Zhang L, Jain RB, Jain N, Wang RY, Ashley DL et al. Levels of tobacco-specific nitro-

samines and polycyclic aromatic hydrocarbons in mainstream smoke from different tobacco varieties. Cancer Epidemiol Biomarkers Prev. 2008;17(12):3366–71.
21. Hecht SS, Yuan JM, Hatsukami DK. Applying tobacco carcinogen and toxicant biomarkers in product regulation and cancer prevention. Chem Res Toxicol. 2010;23:1001–8.
22. Hecht SS, Stepanov I, Carmella SG. Exposure and metabolic activation biomarkers of carcinogenic tobacco-specific nitrosamines. Acc Chem Res. 2016;49(1):106–14.
23. Yan W, Byrd GD, Brown BG, Borgerding MF. Development and validation of a direct LC-MS-MS method to determine the acrolein metabolite 3-HPMA in urine. J Chromatogr Sci. 2010;48(3):194–9.
24. Park SL, Carmella SG, Chen M, Patel Y, Stram DO, Haiman CA et al. Mercapturic acids derived from the toxicants acrolein and crotonaldehyde in the urine of cigarettes smokers from five ethnic groups with differing risks for lung cancer. PLoS One. 2015;10:e0124841.
25. Dougherty D, Garte S, Barchowsky A, Zmuda J, Taioli E. NQO1, MPO, CYP2E1, GSTT1 and GSTM1 polymorphisms and biological effects of benzene exposure – a literature review. Toxicol Lett. 2008;182(1–3):7–17.
26. Roethig HJ, Munjal S, Feng S, Liang Q, Sarkar M, Walk RA et al. Population estimates for biomarkers of exposure to cigarette smoke in adult US cigarette smokers. Nicotine Tob Res. 2009;11(10):1216–25.
27. Haiman CA, Patel YM, Stram DO, Carmella SG, Chen M, Wilkens L et al. Benzene uptake and glutathione S-transferase T1 status as determinants of S-phenylmercapturic acid in cigarette smokers in the Multiethnic Cohort. PLoS One. 2016;11(3):e0150641.
28. Suwan-ampai P, Navas-Acien A, Strickland PT, Agnew J. Involuntary tobacco smoke exposure and urinary levels of polycyclic aromatic hydrocarbons in the United States, 1999 to 2002. Cancer Epidemiol Biomarkers Prev. 2009;18(3):884–93.
29. Hukkanen J, Jacob P III, Benowitz NL. Metabolism and disposition kinetics of nicotine. Pharmacol Rev. 2005;57(1):79–115.
30. Benowitz N, Hall SM, Stewart S, Wilson M, Dempsey D, Jacob P III. Nicotine and carcinogen exposure with smoking of progressively reduced nicotine content cigarette. Cancer Epidemiol Biomarkers Prev. 2007;16(11):2479–85.
31. Le Marchand L, Derby KS, Murphy SE, Hecht SS, Hatsukami D, Carmella SG et al. Smokers with the CHRNA lung cancer-associated variants are exposed to higher levels of nicotine equivalents and a carcinogenic tobacco-specific nitrosamine. Cancer Res. 2008;68(22):9137–40.
32. Lowe FJ, Gregg EO, McEwan M. Evaluation of biomarkers of exposure and potential harm in smokers, former smokers and never-smokers. Clin Chem Lab Med. 2009;47(3):311–20.
33. Shepperd CJ, Eldridge AC, Mariner DC, McEwan M, Errington G, Dixon M. A study to estimate and correlate cigarette smoke exposure in smokers in Germany as determined by filter analysis and biomarkers of exposure. Regul Toxicol Pharmacol. 2009;55(1):97–109.
34. Morin A, Shepperd CJ, Eldridge AC, Poirier N, Voisine R. Estimation and correlation of cigarette smoke exposure in Canadian smokers as determined by filter analysis and biomarkers of exposure. Regul Toxicol Pharmacol. 2011;61(3 Suppl):S3–12.
35. Xia Y, Bernert JT, Jain RB, Ashley DL, Pirkle JL. Tobacco-specific nitrosamine 4-(methylnitrosamino)-1-(3-pyridyl)-1-butanol (NNAL) in smokers in the United States: NHANES 2007–2008. Biomarkers. 2011;16(2):112–9.
36. Appleton S, Olegario RM, Lipowicz PJ. TSNA exposure from cigarette smoking: 18 years of urinary NNAL excretion data. Regul Toxicol Pharmacol. 2014;68(2):269–74.
37. Murphy SE, Park S-SL, Thompson EF, Wilkens LR, Patel Y, Stram DO et al. Nicotine N-glucuronidation relative to N-oxidation and C-oxidation and UGT2B10 genotype in five ethnic/racial groups. Carcinogenesis. 2014;35:2526–33.
38. Sakaguchi C, Kakehi A, Minami N, Kikuchi A, Futamura Y. Exposure evaluation of adult male

Japanese smokers switched to a heated cigarette in a controlled clinical setting. Regul Toxicol Pharmacol. 2014;69(3):338–47.
39. Czoli C, Hammond D. TSNA exposure: levels of NNAL among Canadian tobacco users. Nicotine Tob Res. 2015;17(7):825–30.
40. Donny EC, Denlinger RL, Tidey JW, Koopmeiners JS, Benowitz NL, Vandrey RG et al. Randomized trial of reduced-nicotine standards for cigarettes. N Engl J Med. 2015;373(14):1340–9.
41. Theophilus EH, Coggins CR, Chen P, Schmidt E, Borgerding MF. Magnitudes of biomarker reductions in response to controlled reductions in cigarettes smoked per day: a one-week clinical confinement study. Regul Toxicol Pharmacol. 2015;71(2):225–34.
42. Wei B, Blount BC, Xia B, Wang L. Assessing exposure to tobacco-specific carcinogen NNK using its urinary metabolite NNAL measured in US population: 2011–2012. J Expo Sci Environ Epidemiol. 2016;26(3):249–56.
43. Carmella SG, Chen M, Han S, Briggs A, Jensen J, Hatsukami DK et al. Effects of smoking cessation on eight urinary tobacco carcinogen and toxicant biomarkers. Chem Res Toxicol. 2009;22(4):734–41.
44. Park SL, Carmella SG, Ming X, Stram DO, Le Marchand L, Hecht SS. Variation in levels of the lung carcinogen NNAL and its glucuronides in the urine of cigarette smokers from five ethnic groups with differing risks for lung cancer. Cancer Epidemiol Biomarkers Prev. 2015;24:561–9.
45. Church TR, Anderson KE, Caporaso NE, Geisser MS, Le C, Zhang Y et al. A prospectively measured serum biomarker for a tobacco-specific carcinogen and lung cancer in smokers. Cancer Epidemiol Biomarkers Prev. 2009;18:260–6.
46. Yuan JM, Butler LM, Stepanov I, Hecht SS. Urinary tobacco smoke-constituent biomarkers for assessing risk of lung cancer. Cancer Res. 2014;74(2):401–11.
47. Yuan JM, Nelson HH, Carmella SG, Wang R, Kuriger-Laber J, Jin A et al. CYP2A6 genetic polymorphisms and biomarkers of tobacco smoke constituents in relation to risk of lung cancer in the Singapore Chinese Health Study. Carcinogenesis. 2017;38(4):411–8.
48. Ashley DL, O'Connor RJ, Bernert JT, Watson CH, Polzin GM, Jain RB et al. Effect of differing levels of tobacco-specific nitrosamines in cigarette smoke on the levels of biomarkers in smokers. Cancer Epidemiol Biomarkers Prev. 2010;19(6):1389–98.
49. Church TR, Anderson KE, Caporaso NE, Geisser MS, Le CT, Zhang Y et al. A prospectively measured serum biomarker for a tobacco-specific carcinogen and lung cancer in smokers. Cancer Epidemiol Biomarkers Prev. 2009;18(1):260–6.
50. Yuan JM, Knezevich AD, Wang R, Gao YT, Hecht SS, Stepanov I. Urinary levels of the tobacco-specific carcinogen N′-nitrosonornicotine and its glucuronide are strongly associated with esophageal cancer risk in smokers. Carcinogenesis. 2011;32(9):1366–71.
51. Risks associated with smoking cigarettes with low machine-measured yields of tar and nicotine. Bethesda (MD): Department of Health and Human Services, National Institutes of Health, National Cancer Institute; 2001.
52. Stratton K, Shetty P, Wallace R, Bondurant S, editors. Clearing the smoke: assessing the science base for tobacco harm reduction. Washington, DC: Institute of Medicine; 2001.
53. Hatsukami DK, Biener L, Leischow SJ, Zeller MR. Tobacco and nicotine product testing. Nicotine Tob Res. 2012;14(1):7–17.
54. Berman ML, Connolly G, Cummings KM, Djordjevic MV, Hatsukami DK, Henningfield JE et al. Providing a science base for the evaluation of tobacco products. Tob Regul Sci. 2015;1(1):76–93.
55. WHO Study Group on Tobacco Product Regulation. Advisory note: global nicotine reduction strategy (WHO Technical Report Series No. 989). Geneva: World Health Organization; 2015.
56. Nicotine without smoke: tobacco harm reduction. London: Royal College of Physicians; 2016.
57. Benowitz NL, Fraiman JB. Cardiovascular effects of electronic cigarettes. Nat Rev Cardiol. 2017;14(8):447–56.

58. Shahab L, Goniewicz ML, Blount BC, Brown J, McNeill A, Alwis KU et al. Nicotine, carcinogen, and toxin exposure in long-term e-cigarette and nicotine replacement therapy users: a cross-sectional study. Ann Intern Med. 2017;166(6):390–400.
59. Donny EC, Houtsmuller E, Stitzer ML. Smoking in the absence of nicotine: behavioral, subjective and physiological effects over 11 days. Addiction. 2007;102(2):324–34.
60. Benowitz NL, Dains KM, Dempsey D, Herrera B, Yu L, Jacob P III. Urine nicotine metabolite concentrations in relation to plasma cotinine during low-level nicotine exposure. Nicotine Tob Res. 2009;11(8):954–60.
61. Hatsukami DK, Kotlyar M, Hertsgaard LA, Zhang Y, Carmella SG, Jensen JA et al. Reduced nicotine content cigarettes: effects on toxicant exposure, dependence and cessation. Addiction. 2010;105:343–55.
62. Benowitz NL, Dains KM, Hall SM, Stewart S, Wilson M, Dempsey D et al. Smoking behavior and exposure to tobacco toxicants during 6 months of smoking progressively reduced nicotine content cigarettes. Cancer Epidemiol Biomarkers Prev. 2012;21(5):761–9.
63. Hatsukami DK, Hertsgaard LA, Vogel RI, Jensen JA, Murphy SE, Hecht SS et al. Reduced nicotine content cigarettes and nicotine patch. Cancer Epidemiol Biomarkers Prev. 2013;22(6):1015–24.
64. Higgins ST, Heil SH, Sigmon SC, Tidey JW, Gaalema DE, Hughes JR et al. Addiction potential of cigarettes with reduced nicotine content in populations with psychiatric disorders and other vulnerabilities to tobacco addiction. JAMA Psychiatry. 2017;74(10):1056–64.
65. Ding YS, Richter P, Hearn B, Zhang L, Bravo R, Yan X et al. Chemical characterization of mainstream smoke from SPECTRUM variable nicotine research cigarettes. Tob Regul Sci. 2017;3(1):81–94.
66. The health consequences of smoking. Nicotine addiction. A report of the Surgeon General. Rockville (MD): Department of Health and Human Services; 1988.
67. How tobacco smoke causes disease: the biology and behavioural basis for smoking-attributable disease. A report of the Surgeon General. Atlanta (GA): Centers for Disease Control and Prevention; 2010.

6. The science of flavour in tobacco products

Suchitra Krishnan-Sarin and Stephanie S. O'Malley, Yale School of Medicine, New Haven (CT), USA

Barry G. Green, John B. Pierce Laboratory and Yale School of Medicine, New Haven (CT), USA

Sven-Eric Jordt, Duke University School of Medicine, Durham (NC), USA

Contents
- 6.1 Introduction
- 6.2 Epidemiology of use of flavoured tobacco and nicotine products
- 6.3 Flavoured products: perceptions, experimentation, uptake and regulation
- 6.4 A brief history of the development of flavoured tobacco by the tobacco and e-cigarette industry
- 6.5 Sensory systems that contribute to flavour
- 6.6 Flavour receptors: a new science of flavour sensing and coding
- 6.7 Toxicological effects of flavours
- 6.8 Conclusions
    - 6.8.1 Recommended priorities for research
    - 6.8.2 Recommended policies
- 6.9 References

6.1 Introduction

Most conventional, traditional and new or emerging tobacco products, including cigarettes, cigars, cigarillos, smokeless tobacco products (including *snus*), waterpipe tobacco and e-cigarettes with nicotine (also known as electronic nicotine delivery systems – ENDS) and without nicotine (also known as electronic non-nicotine delivery systems ENNDS) contain flavourings. Broadly defined, flavour is the sensory experience produced when something is ingested or inhaled through the mouth *(1)*. Tobacco itself imparts a flavour (e.g. "natural tobacco" flavour), which depends on the type of tobacco and the curing process *(2)*, and many, if not most, products include added flavours. Some flavourings are derived from natural products, such as cocoa, liquorice, honey and sucrose, while other are created synthetically, such as pyrazines for chocolate flavour and sucralose for sweetness. In the context of tobacco product regulation, flavours that are present at levels that impart a strong non-tobacco smell or taste are considered "characterizing". While use of the term "characterizing" is much debated and it is not used worldwide, the European Union Tobacco Product Directive *(3)* defined a characterizing flavour as a:

> clearly noticeable smell or taste other than one of tobacco, resulting from an additive or combination of additives, including, but not limited to, fruit, spice, herb, alcohol, candy, menthol or vanilla, which is noticeable before or during the consumption of the tobacco product.

The FDA does not directly define "characterizing flavours" but used the following language in guidance on the flavoured cigarette ban in 2009 *(4)*:

> "… a cigarette or any of its component parts (including the tobacco, filter, or paper) from containing as a constituent (including a smoke constituent) or additive, an artificial or natural flavor (other than tobacco or menthol) or an herb or spice, including strawberry, grape, orange, clove, cinnamon, pineapple, vanilla, coconut, liquorice, cocoa, chocolate, cherry, or coffee, that is a characterizing flavor of the tobacco product or tobacco smoke."

The tobacco industry has introduced flavours into tobacco products to increase the appeal of tobacco products. It is believed that flavours are added to these products to reduce their harshness, increase their appeal and improve their palatability. Further, the display of the names or graphic representations of flavours on packaging and advertising materials may enhance the attractiveness of a product.

We provide here a brief overview of the epidemiology of use of flavoured tobacco and nicotine products worldwide and perceptions of these products. We also discuss technological innovations that have been applied to flavoured products, including how they enhance appeal. We provide a general overview of the sensory processes underlying the perception of flavour, known pharmacological targets for flavour chemicals and evidence on the potential toxicity of flavours. Throughout, implications for research and regulation are highlighted.

6.2 Epidemiology of use of flavoured tobacco and nicotine products

Flavoured tobacco and nicotine products are used throughout the world. While there is limited systematic evidence about the availability and use of these products, preferences and use are often specific to countries and regions. For example, in Indonesia and other south Asian countries, traditional clove cigarettes (*kretek*), which contain clove pieces, oils and flavours, are highly popular. In India, spiced smokeless tobacco (*pan masala, gutka*) is used, which contains tobacco mixed with food spices and oils, flavourings, betel nut and other ingredients. Hookah smoking, which originated in India and the Middle East and is now increasingly popular among young people in Europe and north America, involves use of heavily flavoured, sweetened tobacco, known as *maassel*.

The tobacco industry has long used flavours in tobacco products as a marketing strategy *(5)*, including towards young people *(6)*. For example, in the USA in 2013, the flavoured form of non-cigarette tobacco products comprised 52.3% "cigarette-sized cigars", 81.3% "cigar wraps" ("blunt wraps", "tobacco wraps" and "wraps"), 55.1% moist snuff, 86.1% *shisha* and 81.5% dissolvable tobacco products (e.g. Ariva, Camel Orb/Sticks) *(7)*. A longitudinal study of market trends showed that the presence of flavours accounted for 59.4% of the growth in smokeless tobacco sales between 2005 and 2011 *(8)*. More recently introduced nicotine-containing e-cigarettes are now available in many flavours *(9, 10)*.

Surveys indicate high rates of flavoured tobacco use, although there is little systematic evidence on the use of these products throughout the world. The National Adult Tobacco Survey in the USA in 2013–2014 revealed that an estimated 10.2 million e-cigarette users (68.2%), 6.1 million hookah users (82.3%), 4.1 million cigar smokers (36.2%) and 4.0 million smokeless tobacco users (50.6%) had used flavoured products in the past 30 days *(11)*. The addition of multiple flavours (menthol/mint, clove/spice/herb, fruit, alcohol, candy/chocolate/other sweet flavours) was also assessed. The most prevalent flavours used by type of tobacco product were: smokeless tobacco: menthol/mint (76.9%); hookah: fruit (74.0%); cigars, cigarillos, filtered little cigars: fruit (52.4%), candy, chocolate and other sweet flavours (22.0%) and alcohol (14.5%); e-cigarettes: fruit (44.9%), menthol/mint (43.9%) and candy, chocolate and other sweet flavours (25.7%); and pipes: fruit (56.6%), candy, chocolate and other sweet flavours (26.5%) and menthol/mint (24.8%) *(11)*. In Poland, 26% of female smokers and 10.5% of male smokers reported current use of flavoured cigarettes (menthol, vanilla or other flavour) according to an analysis of data from the Global Adult Tobacco Survey conducted in 2009–2010 *(12)*.

Flavoured tobacco products are very popular among young people *(13–15)*. In a nationally representative sample of Canadian young people who used a variety of tobacco products (cigarettes, pipes, cigars, cigarillos, *bidis*, smokeless tobacco, hookah, blunts, roll-your-own cigarettes), 52% reported using flavoured products *(16)*. Similarly, evidence from a national survey conducted in Poland showed that younger smokers were more likely to use flavoured cigarettes *(12)*. Evidence from the Population Assessment of Tobacco Health study *(15)*, a longitudinal national survey conducted in the USA, indicated that use of flavoured tobacco products was highest (80%) among children aged 12–17 years, 73% among tobacco users aged 18–24 years and lowest in users aged ≥ 65 years (29%). Most of the children (81%) and young adults (86%) and only 54% of adults aged ≥ 25 years reported that their first product had been flavoured. Furthermore, first use of a flavoured tobacco product was associated with a higher prevalence of current tobacco use among young people and adults.

Summary: future regulations on flavoured products should consider the significant differences in the availability (marketing and access) and use of flavoured tobacco by country and by demographic group. Unfortunately, current surveillance tools may be inadequate for monitoring use of these products. For example, the Global Youth Tobacco Survey *(17)*, an important global tool for monitoring tobacco use by young people, does not request information specifically about flavoured tobacco use. In the Global Adult Tobacco Survey *(18)*, use of flavoured and unflavoured tobacco is investigated only for waterpipe tobacco, and information about flavoured cigarettes can be obtained only by analysing text responses about the brand of cigarette smoked. To address this information gap, additional questions about flavoured and unflavoured products could be added to iterations of these surveys. Surveillance tools should also assess different use of flavours. Given that menthol is often regulated separately from other flavours, it would be advisable that questions distinguish menthol or mint from other flavours.

6.3 Flavoured products: perceptions, experimentation, uptake and regulation

There is a general perception that flavoured tobacco products are less harmful than other tobacco products *(15, 19–21)*. Favourable perceptions of flavoured tobacco and nicotine products were recorded among both users and non-users of a wide range of products (e.g. cigars, e-cigarettes, hookah, *kreteks*, *bidis* and smokeless tobacco) *(22)*. Flavoured non-menthol tobacco products (hookahs, little cigars and cigarillos, e-cigarettes) were perceived to be less harmful than cigarettes *(20)*.

Perhaps not surprisingly, the presence of flavours in tobacco and nicotine products is associated with greater willingness to experiment with these products *(20, 23, 24)*, an important consideration for young people. The availability of flavours has been associated with initiation of use of a variety of tobacco products, including hookah, e-cigarettes and cigars *(25–27)*. The presence of flavours may also promote a move from experimentation to regular use; the presence of more flavours in e-cigarettes was associated with greater frequency of e-cigarette use among adolescents but not among adults *(28)*. Further, advertising for sweet and fruit flavours was shown to activate brain reward areas in young adults and to interfere with recall of health warnings *(29)*. For example, adolescents in the USA who had tried flavoured tobacco were almost three times more likely to be current cigarette smokers than those who had not tried these products *(30)*, and more than 80% of current tobacco or nicotine users reported that they had first used a flavoured product *(15)*.

In order to reduce appeal, experimentation and uptake, many countries have regulated flavoured tobacco products, as summarized by the Tobacco Legal Control Consortium *(31)*. Some countries ban cigarettes with characterizing flavours, other than menthol (for example in European Union countries), while others have extended the ban to flavours in other tobacco products, e.g. the USA, Canada (little cigars, blunt wraps) and Ethiopia (all products). Several countries have enacted legislation also banning menthol as a flavour additive in cigarettes and other tobacco products. For example, Turkey has banned the addition of menthol at any level and of any menthol derivatives (e.g. mint); and Canada has finalized a ban on the addition of any menthol in cigarettes, blunt wraps and most cigars sold on the Canadian market. Canada has also banned the use of any promotional materials, including packaging that depicts the e-liquid containing confectionery, desserts, cannabis, soft drinks or energy drink flavours.

Summary: both the presence of specific flavours in tobacco and nicotine products and the presence of flavour descriptors on packaging and in advertising have been linked to the appeal and uptake of tobacco and nicotine products, particularly among young people. Given these associations, many countries have regulated flavoured tobacco products. In view of differences in the timing and specificities of regulations in different countries and local ordinances, there is an opportunity for surveillance research (for example, within the International Tobacco Control project or other surveillance systems) to study the impact of regulations on the uptake of tobacco by youth, tobacco sales and use patterns and the subsequent response from the tobacco industry to the regulations.

6.4 A brief history of the development of flavoured tobacco by the tobacco and e-cigarette industry

The tobacco industry has been at the forefront in integrating flavour science and technology to increase product appeal and marketability. An early example is the addition of saccharin to chewing tobacco in the 1890s by the Reynolds Tobacco Company. This high-potency artificial sweetener replaced the more expensive sugar and increased the product's shelf life and uniformity. The industry continued to improve tobacco breeding, selection, curing and manufacturing methods to control the harshness of tobacco smoke, added fillers, casings, humectants (moisteners) and designed wrapper papers and filter systems to improve appeal and the uniformity of flavour delivery.

For decades, tobacco companies have been the major consumers of cocoa and liquorice on the world market. Both are added to cigarette tobacco as casings, not to dominate the flavour experience but to increase taste fullness and reduce the harshness of smoke. In contrast, menthol in menthol cigarettes is a characterizing flavour, an added non-tobacco flavour listed on the product label, which dominates the flavour experience *(32)*. Menthol cigarettes were first marketed in the USA in the 1920s, and they remain popular. Many cigarettes that are not labelled as mentholated may also contain low levels *(33)*, which may be sufficient to enhance the appeal of the product *(34)*.

While most early products contained natural flavours, including menthol extracted from mint, fruit extracts and natural terpenes and aldehydes, the increasing demand of the tobacco industry for standardized flavourings and the limited supply from natural sources prompted suppliers to develop and optimize chemical synthetic processes to deliver bulk amounts of key flavourings. For example, most of today's menthol supply is produced by three major chemical companies that developed procedures for stereo-selective synthesis or purification of L-menthol, the minty, cooling form of menthol *(35)*. Similarly, chemicals in non-menthol products, including vanillin (vanilla flavour), pyrazines (coffee and chocolate flavourings), aldehydes such as benzaldehyde (a berry and candy flavouring) and many other tobacco flavourings are chemically synthesized, often from petrochemical hydrocarbon precursors *(35)*. Modern flavoured tobacco products contain finely tuned mixtures of purified and recombined flavourings to create characterizing flavours, which are marketed after exhaustive testing by industry flavour panels and consumer volunteers.

Flavourings are usually added to tobacco products in the tobacco casing, sprayed on tobacco in humectants, in the filter and in the foil liner of tobacco packaging. The tobacco and flavour industries continue to introduce new flavours and flavour delivery systems to diversify products and increase their appeal. For example, the introduction of flavour capsules within the filter has resulted in a new category of cigarettes that release bursts of flavour when the filter is pressed. Cigarettes with a menthol capsule were first introduced in 2007 in Japan, then in 2008 in Europe and the USA and in 2011–2012 in Australia and Mexico *(36)*. Most capsules in products contain menthol, but they may also contain aldehydes or other flavour chemicals to create flavours like spearmint, lemon mint, apple mint and strawberry mint *(37)*. Some brands contain two flavour capsules, one containing menthol and the other a tangy, fruity flavour or more menthol. The capsules can be crushed individually, giving the user control over the dual flavour experience. The presence of flavour capsules was shown to enhance the appeal of cigarettes to young people *(37)*.

After the ban on cigarettes with characterizing flavours (except for mentholated products) in some countries, the industry quickly developed alternative flavoured products, such as small cigars, some of which share the basic design of cigarettes and have tobacco in the wrapping paper. Flavoured small cigars are highly popular among young people. Patents filed by the tobacco industry describe innovative techniques for the application and enrichment of flavourings, specifically sweeteners, to cigar wrapping papers and mouthpieces to ensure that the product is perceived as sweet. In one study, high levels of saccharin and other synthetic high-intensity sweeteners were found in many popular small cigars or cigarillos on the United States market. Saccharin levels were especially high in the mouth sections and tips of the small cigars, with a sweetness intensity exceeding that of sugar *(38)*. Small cigars of all flavour categories, regardless of whether they were labelled as sweet, contained saccharin *(39, 40)*.

Tobacco industry patents also describe the use of recombinant, sweet taste-stimulating proteins like thaumatin, which is 2000 times sweeter than table sugar, as sweeteners and of synthetic menthol derivatives as innovative cooling agents. After the ban on menthol cigarettes in Canada, the tobacco industry began marketing cigarettes without menthol in packaging with designs almost identical to those of previously marketed menthol cigarettes, signalling the same cool freshness *(41)*. It is unclear whether these cigarettes contain cooling additives to replace menthol.

Many of the smokeless tobacco products found in Asia (e.g. *zarda*, *quiwam*, *gutkha*, *khaini*) contain flavour chemicals *(42, 43)*. Novel smokeless tobacco products have also been introduced in several countries, including *snus* (previously marketed only in Sweden) and new varieties of moist snuff. *Snus* and snuff are sold in many flavours, including menthol/mint, cherry, vanilla and strawberry, and it has been suggested that the significant rise in sales of moist snuff is related to the introduction of flavours *(8, 44)*. Newly marketed *snus* products have also been found to contain very high levels of sucralose (Splenda®), a synthetic high-intensity sweetener that tastes more like sugar than saccharin, which is still used in snuff *(45)*.

Flavoured waterpipe tobacco (*maassel*), is a sweetened, flavoured form of tobacco manufactured by fermentation of tobacco with molasses, glycerol and fruit essence. It is the preferred form of tobacco used in waterpipes, especially by adolescents and young adults *(46)*.

E-cigarettes and ENDS, which appeared on the market in 2004, in Europe in 2006 and in the USA in 2007, are available in an ever-expanding range of e-liquid flavours, and the customized flavour and nicotine combinations are attractive to users. These products electrically heat and vaporize e-liquids and are available in a variety of devices, including cigarette-like products, vape or hookah pens, advanced devices known as "mods" or personal vaporizers and more discrete pod-devices like JUUL. In 2013–2014, 466 brands of e-cigarette were available online, with over 7000 unique flavour names; 242 new flavours were added each month *(9)*. By 2016–2017, the number had doubled to over 15 000 flavours and flavour combinations *(10)*. The counterbalancing effects of flavour options on promoting a switch from combustibles to e-cigarettes or ENDS against increasing the risk for uptake among young people is a hotly debated topic.

Summary: the tobacco and now the e-cigarette industry is constantly introducing innovative additives, such as flavourings and sweeteners, to increase the appeal of tobacco products, including waterpipes, smokeless tobacco, e-cigarettes and ENDS. Regulatory surveillance to detect and follow such product manipulations is critical to curtail the appeal of tobacco products.

6.5 Sensory systems that contribute to flavour

To understand the appeal of flavours in tobacco products, it is important to understand the role of the sensory components that contribute to the experience of flavour. The senses of taste and smell (olfaction) play the largest role in flavour perception; however, flavour also includes sensations of temperature, touch and chemaesthesis (i.e. sensations that are produced by chemical stimulation of the senses of temperature, touch and pain) *(47)*.

Taste: the sense of taste derives from taste buds located on the tongue and soft palate, which provide the sensations of sweetness, saltiness, sourness, bitterness and savoury. Because of the bitter taste of nicotine *(48)*, bitterness is a dominant quality of the flavour of tobacco. The addition of flavours to tobacco products may serve to mask the bitter taste and improve flavour and appeal.

Olfaction: the sense of smell provides information not only about odours in the environment (orthonasal olfaction) but also about odours emanating from the mouth and airways (retronasal olfaction), which contribute significantly to the flavour of tobacco products. An opening at the rear of the nasal cavity that is connected to the back of the oral cavity *(49, 50)* allows odours from products that are taken into or inhaled through the mouth to stimulate olfactory receptors during exhalation.

Temperature: temperature can contribute to the flavour of tobacco in at least two ways: directly, via sensations of warmth and heat produced by combustible or heated tobacco products, and indirectly, by modulating the rate of release of volatile flavour molecules that are sensed by retronasal olfaction.

Touch: in tobacco smoking, touch provides the oral "feel" of cigarettes, cigars, pipes and mouthpieces during smoking or vaping as well as the texture of smokeless tobacco products in the mouth. Touch also senses changes in oral mucosal surfaces such as dryness and astringency *(51)* that can be caused by inhaled or oral tobacco products.

Chemaesthesis: in addition to evoking tastes and smells, some chemicals can produce sensations of temperature, touch or pain by stimulating receptors in the mucous membranes that normally respond to either weak or strong (noxious) mechanical or thermal stimulation *(47)*. Chemaesthesis plays a key role in the flavour of tobacco products, including the harshness of nicotine, acrolein and other chemical irritants in tobacco smoke and the coolness and "burning" cold sensations of menthol.

Interactive effects of flavour: a defining characteristic of flavour is integration of its multisensory components into coherent perceptions in the mouth. A particularly important interaction is the referral of retronasal odours to the mouth *(52, 53)*, which leads to mislabelling of odours as tastes, most commonly for odours that have taste-like qualities, such as "sweet" odours like vanilla, cherry and strawberry *(54)*. This is important, because it means that characterizing flavours often emerge as a seamless combination of sensations evoked by multiple, interacting sensory systems.

Flavour constituents of tobacco and e-cigarette products can also be important for their masking or inhibitory effects, such as masking a bitter taste by a sweet taste *(55)*. Some studies have found evidence that sweeteners also suppress pain, particularly in infants *(56, 57)*. Sweeteners are present in smokeless tobacco products, for example, at levels that exceed those in confectionery products *(45)*. Patents filed by the tobacco industry describe procedures for adding artificial sweeteners to cigar wrappings and mouthpieces to ensure that the consumer perceives the product as sweet. Similarly, the appeal of mentholated products owes much to the appeal of their cool, minty flavour, but menthol also has analgesic effects that can reduce the sensory irritation and harshness of nicotine *(34, 58)* and other constituents of tobacco smoke *(59)* and smokeless tobacco products *(58, 60, 61)*.

Summary: the multimodal, integrative characteristics of flavour make it an important contributor to the appeal of tobacco products, ranging from essential tactile and thermal qualities in the mouth and throat to both taste and olfactory constituents in chemically derived characterizing flavours. While regulating individual well-known flavouring ingredients such as menthol and artificial sweeteners could have a straightforward effect on the appeal of tobacco products, interactions among flavour constituents designed to heighten appeal through combined sweetness or mechanisms like masking and analgesia pose more difficult problems for designing regulatory strategies. As flavour perception integrates sensory experience and is subjective, both human sensory testing and chemical analysis will be required to identify "the concentrations above which an additive will impart a characterizing flavor" *(3)* or increase the appeal of a product by masking aversive tastes or sensations. For products in which the flavour is perceptible but ambiguous, labelling may serve to boost the perception and identification of the flavour.

6.6 Flavour receptors: a new science of flavour sensing and coding

Molecular, genetic, pharmacological and behavioural approaches have revolutionized flavour research, including the discovery of flavouring receptors. Olfactory receptors, also called odorant receptors, were identified first in nerve endings in the nasal olfactory epithelium. These receptors play a dominant role in the retronasal sensing of volatile flavourings in tobacco and nicotine products. Taste receptors were identified in the taste papillae of the tongue and palate. Humans have a single sweet taste receptor made up of two protein subunits (TAS1R2 and TAS1R3) that bind sugars such as sucrose (table sugar) and, with much greater affinity, artificial high-potency sweeteners, such as saccharin and sucralose. Bitter taste receptors, the TAS2R receptors, of which humans express 38 different versions, signal the presence of potentially poisonous chemicals and are probably involved in the perception of nicotine and other tobacco alkaloids as bitter. The presence of polymorphisms of the gene encoding a human bitter receptor, TAS2R38, has been linked with menthol cigarette smoking, suggesting perhaps that smokers who are more sensitive to bitter tastes may use menthol to mask the bitter taste of nicotine or cigarette smoke *(62, 63)*.

Receptors that mediate the chemaesthetic properties of flavours and other tobacco constituents, the transient receptor potential (TRP) ion channels, are located on nerves that transmit noxious and innocuous chemical, mechanical and thermal stimuli from the oral and nasal passages and airways. TRPA1 is the receptor for noxious aldehydes in tobacco smoke, eliciting burning and irritating sensations. TRPA1 also mediates the irritating effects of nicotine and flavouring aldehydes such as cinnamaldehyde and benzaldehyde, present in many fruity flavours. TRPM8 is the receptor for menthol and mediates its cooling and soothing effects. Experiments in mice indicated that TRPM8 is essential for the suppression of the irritating and aversive effects of tobacco smoke and nicotine by menthol *(60, 64)*.

Summary: with more knowledge about flavour receptors and their pharmacology, the flavour industry can use molecular and pharmacological approaches to develop new, highly optimized flavour receptor modulators. These include novel sweet taste enhancers, bitter blockers (to reduce bitter taste), savoury (non-glutamate) taste enhancers and novel cooling agents. Several are approved as food additives, and more are in development. The tobacco industry has experimented with synthetic cooling agents that activate the menthol receptor, TRPM8, with a less minty odour, reduced irritancy and greater stability. The identification of flavouring receptors may provide an opportunity to regulate flavourings on the basis of their receptor-mediated pharmacological and behavioural effects in humans and animal model systems. For example, instead of regulating an individual flavour, such as menthol, which can be substituted quickly by alternative cooling agents, regulators could decide instead to control all TRPM8 receptor agonists in tobacco as a receptor-specific flavouring class.

6.7 Toxicological effects of flavours

While legislation and regulations vary by country, in many, inclusion of flavours as food additives requires scientific premarket review. In the European Union, flavouring additives are regulated by the European Food Safety Authority. In the USA, the FDA has designated the generally recognized as safe (GRAS) classification for food additives, in which stakeholders submit data on safety, which is reviewed for the intended use of a flavouring. The GRAS declaration for use of a flavour in confectionery products does not, however, automatically apply to its use in tobacco products, especially products that are inhaled, including cigarettes, cigars, hookah and electronic cigarettes. For example, cinnamaldehyde, the GRAS cinnamon flavouring widely used in baked goods and confectionaries, is added at very high concentrations to some electronic cigarette fluids *(65)*. In toxicological studies, vapours from these liquids damaged lung epithelial cells and caused pulmonary inflammation in mice *(66)*. Other sweet flavours, including characterizing banana and cherry flavours, were also found to damage cells and expose e-cigarette users to benzaldehyde, a key component of many berry flavour mixes, with known toxic effects on the respiratory system *(67, 68)*. Diacetyl, a flavour chemical commonly found in buttery flavours, also has a well-known toxic effect on respiration *(69, 70)*. Flavour chemicals can react with solvents and other components of e-liquids to form irritant compounds with unknown toxicological effects *(71)*.

Despite these concerns, some manufacturers of e-cigarette and ENDS have used GRAS labelling in their advertising, implying that the flavourings in their liquids are safe because they were previously approved for addition to food. Such health claims were strongly refuted by the Flavor and Extracts Manufacturing Association, a body of the United States flavour industry that submitted GRAS applications to the FDA, which stated that use of flavouring in e-cigarettes is not considered an intended use *(72)*.

The lack of a regulatory process for evaluating the safety of flavouring agents is also a limitation for tobacco products such as *snus*, snuff and flavoured cigars, cigarettes and hookah tobacco, which may result in ingestion or inhalation of toxic levels of flavourings situations by consumers. For example, analytical studies showed that wintergreen-flavoured snuff products can expose regular users to levels of the wintergreen flavouring, methyl salicylate, that exceed the acceptable daily intake determined by the Food and Agriculture Organization of the United Nations and WHO for food by 12 times *(73)*. Smokeless tobacco products (including *zarda*, *quiwam*, *gutka* and *khaini* varieties) have been found to have high levels of flavour chemicals such as eugenol, coumarin, camphor and diphenyl ether *(43)*. Some *snus* products contain such high levels of sweeteners that regular use might exceed the recommended daily consumption of sugar if users consume other sweetened products at the same time *(45)*.

There is little information about the fate of flavourings in combusted tobacco products (cigarettes, cigars, hookah) or heated nicotine solutions (as in e-cigarettes). The tobacco industry reported no major differences in the levels of toxic constituents in the smoke of menthol capsules or *kretek* (clove) cigarettes and in equivalent cigarettes without flavouring *(74, 75)*; however, any toxic effects of flavours may have been masked by the overwhelming contributions of the other toxic constituents of cigarettes. In other studies, increased levels of VOCs were found in the smoke of flavoured cigarettes, and industry documents provide evidence that the industry knew about the carcinogenicity of some types of flavoured cigarettes *(76, 77)*. In these studies, only the main smoke constituents (such as TSNAs and PAHs) were measured; the specific chemical products resulting from combusted flavourings were not examined. Analysis of the chemical fate of e-cigarette flavourings may provide information on the toxicity of flavours. When high-voltage settings are used in e-cigarettes, exposure to flavourings increases, and the levels of oxidation products such as formaldehyde exceed toxic levels. As cigarettes, cigars and hookah tobacco burn at higher temperatures than vapour-producing e-cigarettes, the possibility that larger amounts of flavouring oxidation products and other toxic chemical species may be formed is a concern. Evidence obtained for e-liquid flavours suggests that flavour chemicals can have toxic effects *(65, 78, 79)*.

Summary: regulations to specify acceptable levels of flavourings depend on the type of tobacco product, the level at which the flavouring is delivered to the consumer's oral and respiratory systems and the potential for chemical changes during storage, heating and combustion. The chemical fate of flavourings, especially when heated at high temperatures, should be investigated further. This may result in the addition of flavourings and their chemical products to the lists of HPHCs curated by national regulators such as the FDA and in TobReg reports on toxicants in tobacco products. Future regulation of tobacco products might require specific reporting of the levels and toxicity of flavour chemicals.

6.8 Conclusions

Critical questions remain about the role of flavours in the appeal of and addiction to tobacco and nicotine products. Surveillance should be conducted worldwide to assess the use of flavoured tobacco products and also perceptions about the appeal and addictive potential of flavoured and unflavoured products. Flavours and sweeteners are chemicals that can be appealing independently and can also increase the use of tobacco products by enhancing the palatability of nicotine and other bitter or harsh constituents. A flavour like menthol may do this by pharmacologically attenuating the aversive effects of exposure to nicotine (e.g. cough, harshness, heat), and the presence of menthol in cigarettes is probably associated with increased rates of initiation of and progression to regular cigarette smoking, development of addiction and difficulty in quitting smoking *(80)*. Much less is known about whether and how other flavour constituents influence preferences for and use of tobacco and nicotine products.

As flavours are complex and subjective, testing of flavoured tobacco and nicotine products for their appeal will require human behavioural and toxicological evaluations, in combination with chemical analyses. Methods for determining the presence and levels of flavourings should be based on what is delivered to the consumer (e.g. in smoke or vapour) and should also be used to determine potential changes in flavourings during storage, heating and combustion. Both the appeal and toxicity of flavourings should be evaluated. While it is important to determine the optimal doses of flavours that produce these effects, it is also important to understand the influence of the concentrations of flavours. All this information should be used to further refine the definition of a "characterizing flavour".

Surveillance and testing methods will have to keep up with the constant innovations of the tobacco industry designed to enhance the appeal of flavoured tobacco products, including use of new synthetic compounds and strategic placement of flavouring molecules in tobacco products. Declaration of flavourings and their concentrations on tobacco product labels should be considered. Additionally, regulatory work will have to contend with alternative marketing strategies introduced by the tobacco industry to deal with regulations on flavours, such as manipulation of packaging to continue to convey brand features associated with flavours *(41)*.

Future regulations should account for the possibility that some flavours and flavour constituents, like sweeteners, are also positively reinforcing in themselves and, when combined with nicotine, might enhance the rewarding effects of low-dose nicotine and promote a transition to higher levels of nicotine, leading to addiction. Experimental evidence and monitoring of these complex issues will be crucial to ensure that regulations are designed to reduce the appeal and addictive potential of tobacco products. Better understanding of the science of flavours and how they enhance the appeal and addictive potential of tobacco products will be required to regulate these molecules.

6.8.1 Recommended priorities for research

- Systematically monitor the global epidemiology of flavoured conventional, traditional, new and emerging tobacco and nicotine products.
- Identify flavour chemicals, their reaction products and their concentrations in tobacco and nicotine products and in aerosols, vapours and smoke.
- Determine how depiction of flavours on product packaging and in marketing alter public perceptions and the appeal of tobacco and nicotine products, especially among young people.
- Evaluate whether the presence of specific flavours and their concentrations in tobacco and nicotine products alters their appeal and abuse potential.
- Among smokers, investigate the role of the availability of flavoured products as a motive for switching to less harmful tobacco products (relative to the degree of harm reduction).
- Determine the toxicity and health effects of various concentrations of inhaled flavour chemicals and their metabolites and adducts.
- Monitor the use of alternative chemical moieties to replace prohibited flavours, and determine their appeal, toxicity and health effects.

6.8.2 Recommended policies

- Consider obtaining systematic global evidence on the use of flavoured tobacco and e-cigarette and ENDS products.
- Consider banning the use of flavours, including menthol, in harmful combusted products.
- Consider limiting the levels, number of and/or specific flavours allowed in tobacco and nicotine products for which there is evidence of modified or reduced risk, to reduce initiation by young people and support cessation of use of combusted tobacco products.
- Consider requiring that the types and concentrations of flavour chemicals in various products be listed among the product constituents.

6.9 References

1. Small DM, Green BG. A model of flavor perception. In: Murray MM, Wallace MT, editors. The neural bases of multisensory processes. New York: CRC Press; 2012:717–38.
2. Talhout R, van de Nobelen S, Kienhuis AS. An inventory of methods suitable to assess additive-induced characterising flavours of tobacco products. Drug Alcohol Depend. 2016;161:9–14.
3. Directive 2014/40/EU of the European Parliament and of the Council on the approximation of the laws, regulations and administrative provisions of the Member States concerning the manufacture, presentation and sale of tobacco and related products. Brussels: European Union; 2014 (http://ec.europa.eu/health/sites/health/files/tobacco/docs/dir_201440_en.pdf, accessed 15 May 2019).
4. Federal Food, Drug, and Cosmetic Act. Section 907. Tobacco product standard. General questions and answers on the ban on cigarettes that contain certain characterizing flavors (edition 2). Silver Spring (MD): Center for Tobacco Products, Food and Drug Administration; 2009.
5. Carpenter CM, Wayne GF, Pauly JL, Koh HK, Connolly GN. New cigarette brands with flavors that appeal to youth: tobacco marketing strategies. Health Affairs. 2005;24(6):1601–10.
6. Connolly GN. Sweet and spicy flavours: new brands for minorities and youth. Tob Control. 2004;13(3):211–2.
7. Morris DS, Fiala SC. Flavoured, non-cigarette tobacco for sale in the USA: an inventory analysis of Internet retailers. Tob Control. 2015;24(1):101–2.
8. Delnevo CD, Wackowski OA, Giovenco DP, Manderski MT, Hrywna M, Ling PM. Examining market trends in the United States smokeless tobacco use: 2005–2011. Tob Control. 2014;23(2):107–12.
9. Zhu SH, Sun JY, Bonnevie E, Cummins SE, Gamst A, Yin L et al. Four hundred and sixty brands of e-cigarettes and counting: implications for product regulation. Tob Control. 2014;23(Suppl 3):iii3–9.
10. Hsu G, Sun JY, Zhu SH. Evolution of electronic cigarette brands from 2013–2014 to 2016–2017: analysis of brand websites. J Med Internet Res. 2018;20(3):e80.
11. Bonhomme MG, Holder-Hayes E, Ambrose BK, Tworek C, Feirman SP, King BA et al. Flavoured non-cigarette tobacco product use among US adults: 2013–2014. Tob Control. 2016;25(Suppl 2):ii4–13.
12. Kaleta D, Usidame B, Szosland-Faltyn A, Makowiec-Dabrowska T. Use of flavoured cigarettes in Poland: data from the global adult tobacco survey (2009–2010). BMC Public Health. 2014;14:127.
13. Corey CG, Ambrose BK, Apelberg BJ, King BA. Flavored tobacco product use among middle

and high school students – United States, 2014. Morbid Mortal Wkly Rep. 2015;64(38):1066–70.
14. King BA, Dube SR, Tynan MA. Flavored cigar smoking among US adults: findings from the 2009–2010 National Adult Tobacco Survey. Nicotine Tob Res. 2013;15(2):608–14.
15. Villanti AC, Johnson AL, Ambrose BK, Cummings KM, Stanton CA, Rose SW et al. Flavored tobacco product use in youth and adults: findings from the first wave of the PATH study (2013–2014). Am J Prev Med. 2017;53(2):139–51.
16. Minaker LM, Ahmed R, Hammond D, Manske S. Flavored tobacco use among Canadian students in grades 9 through 12: prevalence and patterns from the 2010–2011 youth smoking survey. Prev Chronic Dis. 2014;11:E102.
17. Global Youth Tobacco Survey (GYTS). Core questionnaire with optional questions, version 1.0. Geneva: World Health organization; Atlanta (GA): Centers for Disease Control and Prevention; 2012.
18. Global Adult Tobacco Survey (GATS). Core questionnaire with optional questions, version 2.0. Atlanta (GA): Centers for Disease Control and Prevention; 2010.
19. Ashare RL, Hawk LW Jr, Cummings KM, O'Connor RJ, Fix BV, Schmidt WC. Smoking expectancies for flavored and non-flavored cigarettes among college students. Addict Behav. 2007;32(6):1252–61.
20. Kowitt SD, Meernik C, Baker HM, Osman A, Huang LL, Goldstein AO. Perceptions and experiences with flavored non-menthol tobacco products: a systematic review of qualitative studies. Int J Environ Res Public Health. 2017;14(4):338.
21. Sokol NA, Kennedy RD, Connolly GN. The role of cocoa as a cigarette additive: opportunities for product regulation. Nicotine Tob Res. 2014;16(7):984–91.
22. Feirman SP, Lock D, Cohen JE, Holtgrave DR, Li T. Flavored tobacco products in the United States: a systematic review assessing use and attitudes. Nicotine Tob Res. 2016;18(5):739–49.
23. Choi K, Fabian L, Mottey N, Corbett A, Forster J. Young adults' favorable perceptions of snus, dissolvable tobacco products, and electronic cigarettes: findings from a focus group study. Am J Public Health. 2012;102(11):2088–93.
24. Manning KC, Kelly KJ, Comello ML. Flavoured cigarettes, sensation seeking and adolescents' perceptions of cigarette brands. Tob Control. 2009;18(6):459–65.
25. Hammal F, Wild TC, Nykiforuk C, Abdullahi K, Mussie D, Finegan BA. Waterpipe (hookah) smoking among youth and women in Canada is new, not traditional. Nicotine Tob Res. 2016;18(5):757–62.
26. Kong G, Morean ME, Cavallo DA, Camenga DR, Krishnan-Sarin S. Reasons for electronic cigarette experimentation and discontinuation among adolescents and young adults. Nicotine Tob Res. 2015;17(7):847–54.
27. Delnevo CD, Giovenco DP, Ambrose BK, Corey CG, Conway KP. Preference for flavoured cigar brands among youth, young adults and adults in the USA. Tob Control. 2015;24(4):389–94.
28. Morean ME, Butler ER, Bold KW, Kong G, Camenga DR, Cavallo DR et al. Preferring more e-cigarette flavors is associated with e-cigarette use frequency among adolescents but not adults. PLoS One. 2018;4:13(1):e0189015.
29. Garrison KA, O'Malley SS, Gueorguieva R, Krishnan-Sarin S. A fMRI study on the impact of advertising for flavored e-cigarettes on susceptible young adults. Drug Alcohol Depend. 2018;186:233–41.
30. Farley SM, Seoh H, Sacks R, Johns M. Teen use of flavored tobacco products in New York City. Nicotine Tob Res. 2014;16(11):1518–21.
31. How other countries regulate flavored tobacco products. Saint Paul (MN): Tobacco Control Legal Consortium; 2015 (http://www.publichealthlawcenter.org/sites/default/files/resources/tclc-fs-global-flavored-regs-2015.pdf, accessed 15 May 2019).
32. Henkler F, Luch A. European Tobacco Product Directive: How to address characterizing flavors as a matter of attractiveness? Arch Toxicol. 2015;89(8):1395–8.

33. Giovino GA, Sidney S, Gfroerer JC, O'Malley PM, Allan JA, Richter PA et al. Epidemiology of menthol cigarette use. Nicotine Tob Res. 2004;6 Suppl 1:S67–81.
34. Krishnan-Sarin S, Green BG, Kong G, Cavallo DA, Jatlow P, Gueorguieva R. Studying the interactive effects of menthol and nicotine among youth: an examination using e-cigarettes. Drug Alcohol Depend. 2017;180:193–9.
35. Noyori R. Asymmetric catalysis: science and opportunities (Nobel lecture). Angew Chem Int Ed Engl. 2002;41(12):2008–22.
36. Thrasher JF, Abad-Vivero EN, Moodie C, O'Connor RJ, Hammond D, Cummings KM et al. Cigarette brands with flavour capsules in the filter: trends in use and brand perceptions among smokers in the USA, Mexico and Australia, 2012–2014. Tob Control. 2016;25(3):275–83.
37. Abad-Vivero EN, Thrasher JF, Arillo-Santillan E, Pérez-Hernández R, Barrientos-Guitierrez I, Kollath-Cattano C et al. Recall, appeal and willingness to try cigarettes with flavour capsules: assessing the impact of a tobacco product innovation among early adolescents. Tob Control. 2016;25(e2):e113–9.
38. Erythropel HC, Kong G, deWinter TM, O'Malley SS, Jordt SE, Anastas PT et al. Presence of high intensity sweeteners in popular cigarillos of varying flavor profiles. JAMA. 2018;2:320(13):1380–3
39. Sweeney WR, Marcq P, Pierotti J, Thiem D. Sweet cigar. Google Patents; 2013.
40. Cundiff RH. Letter to Dr Paul Whiter, Wilkinson-Sword Ltd, Berkeley Heights (NJ)., 24 March 1975. Winston-Salem (NC): R.J. Reynolds Tobacco Co.; 1975 (Menthol Derivatives. RJ Reynolds Records. 1975 (https://www.industrydocumentslibrary.ucsf.edu/tobacco/docs/#id=ngdp0061, accessed January 2019).
41. Brown J, DeAtley T, Welding K, Schwartz R, Chaiton M, Kittner DL et al. Tobacco industry response to menthol cigarette bans in Alberta and Nova Scotia, Canada. Tob Control. 2017;26(e1):e71–4.
42. Stanfill SB, Croucher RE, Gupta PC, Lisko JG, Lawler TS, Kuklenyik P et al. Chemical characterization of smokeless tobacco products from South Asia: nicotine, unprotonated nicotine, tobacco-specific N´-nitrosamines, and flavor compounds. Food Chem Toxicol. 2018;118:626–34.
43. Lisko JG, Stanfill SB, Watson CH. Quantitation of ten flavor compounds in unburned tobacco products. Anal Meth. 2014;6(13):4698–704.
44. Kostygina G, Ling PM. Tobacco industry use of flavourings to promote smokeless tobacco products. Tob Control. 2016;25(Suppl 2):ii40–9.
45. Miao S, Beach ES, Sommer TJ, Zimmerman JB, Jordt SE. High-intensity sweeteners in alternative tobacco products. Nicotine Tob Res. 2016;18(11):2169–73.
46. Maziak W, Taleb ZB, Bahelah R, Islam F, Jaber R, Auf R et al. The global epidemiology of waterpipe smoking. Tob Control. 2015;24(Suppl 1):i3–12.
47. Green BG. Chemesthesis: pungency as a component of flavor. Trends Food Sci Technol. 1996;7:415–20.
48. Oliveira-Maia AJ, Phan THT, Melone PD, Mummalaneni S, Nicolelis MAL, Simon SA et al. Nicotinic acetylcholine receptors (NACHRS): novel bitter taste receptors for nicotine. Chem Senses. 2008;33(8):S113.
49. Lim J, Johnson MB. Potential mechanisms of retronasal odor referral to the mouth. Chem Senses. 2011;36(3):283–9.
50. Hummel T. Retronasal perception of odors. Chem Biodivers. 2008;5(6):853–61.
51. Green BG. Oral astringency: a tactile component of flavor. Acta Psychol. 1993;84(1):119–25.
52. Rozin P. "Taste–smell confusions" and the duality of the olfactory sense. Percept Psychophys. 1982;31:397–401.
53. Lim J, Green BG. Tactile interaction with taste localization: influence of gustatory quality and intensity. Chem Senses. 2008;33(2):137–43.
54. Small DM, Prescott J. Odor/taste integration and the perception of flavor. Exp Brain Res. 2005;166(3–5):345–57.

55. Green BG, Lim J, Osterhoff F, Blacher K, Nachtigal D. Taste mixture interactions: suppression, additivity, and the predominance of sweetness. Physiol Behav. 2010;101(5):731–7.
56. Kassab M, Foster JP, Foureur M, Fowler C. Sweet-tasting solutions for needle-related procedural pain in infants one month to one year of age. Cochrane Database Syst Rev. 2012;12:CD008411.
57. Schobel N, Kyereme J, Minovi A, Dazert S, Bartoshuk L, Hatt H. Sweet taste and chorda tympani transection alter capsaicin-induced lingual pain perception in adult human subjects. Physiol Behav. 2012;107(3):368–73.
58. Rosbrook K, Green BG. Sensory effects of menthol and nicotine in an e-cigarette. Nicotine Tob Res. 2016;18(7):1588–95.
59. Ferris Wayne G, Connolly GN. Application, function, and effects of menthol in cigarettes: a survey of tobacco industry documents. Nicotine Tob Res. 2004;6(Suppl 1):S43–54.
60. Fan L, Balakrishna S, Jabba SV, Bonner PE, Taylor SR, Picciotto MR et al. Menthol decreases oral nicotine aversion in C57BL/6 mice through a TRPM8-dependent mechanism. Tob Control. 2016;25(Suppl 2):ii50–4.
61. Liu B, Fan L, Balakrishna S, Sui A, Morris JB, Jordt SE. TRPM8 is the principal mediator of menthol-induced analgesia of acute and inflammatory pain. Pain. 2013;154(10):2169–77.
62. Oncken C, Feinn R, Covault J, Duffy V, Dornela E, Kranzler HR et al. Genetic vulnerability to menthol cigarette preference in women. Nicotine Tob Res. 2015;17(12):1416–20.
63. 6Risso D, Sainz E, Gutierrez J, Kirchner T, Niaura R, Drayna D. Association of TAS2R38 haplotypes and menthol cigarette preference in an African American cohort. Nicotine Tob Res. 2017;19(4):493–4.
64. Willis DN, Liu B, Ha MA, Jordt SE, Morris JB. Menthol attenuates respiratory irritation responses to multiple cigarette smoke irritants. FASEB J. 2011;25(12):4434–44.
65. Behar RZ, Davis B, Wang Y, Bahl V, Lin S, Talbot P. Identification of toxicants in cinnamon-flavored electronic cigarette refill fluids. Toxicol In Vitro. 2014;28(2):198–208.
66. Lerner CA, Sundar IK, Yao H, Gerloff J, Ossip D, McIntosh S et al. Vapors produced by electronic cigarettes and e-juices with flavorings induce toxicity, oxidative stress, and inflammatory response in lung epithelial cells and in mouse lung. PloS One. 2015;10(2):e0116732.
67. Gerloff J, Sundar IK, Freter R, Sekera ER, Friedman AE, Robinson R et al. Inflammatory response and barrier dysfunction by different e-cigarette flavoring chemicals identified by gas chromatography–mass spectrometry in e-liquids and e-vapors on human lung epithelial cells and fibroblasts. Appl In Vitro Toxicol. 2017;3(1):28–40.
68. Kosmider L, Sobczak A, Prokopowicz A, Kurek J, Zaciera M, Knysak J et al. Cherry-flavoured electronic cigarettes expose users to the inhalation irritant, benzaldehyde. Thorax. 2016;71(4):376–7.
69. Allen JG, Flanigan SS, LeBlanc M, Vallarino J, MacNaughton P, Stewart JH et al. Flavoring chemicals in e-cigarettes: diacetyl, 2,3-pentanedione, and acetoin in a sample of 51 products, including fruit-, candy-, and cocktail-flavored e-cigarettes. Environ Health Perspect. 2016;124(6):733–9.
70. Klager S, Vallarino J, MacNaughton P, Christiani DC, Lu Q, Allen JG. Flavoring chemicals and aldehydes in e-cigarette emissions. Environ Sci Technol. 2017;51(18):10806–13.
71. Erythropel HC, Jabba SV, deWinter TM, Mendizabal M, Anastas PT, Jordt SE et al. Formation of flavorant–propylene glycol adducts with novel toxicological properties in chemically unstable e-cigarette liquids. Nicotine Tob Res. 2018. doi: 10.1093/ntr/nty192.
72. Safety assessment and regulatory authority to use flavors – focus on electronic nicotine delivery systems and flavored tobacco products. Washington (DC): Flavor and Extract Manufacturers Association; 2018 (https://www.femaflavor.org/safety-assessment-and-regulatory-authority-use-flavors-focus-electronic-nicotine-delivery-systems, accessed 15 May 2019).
73. Chen C, Isabelle LM, Pickworth WB, Pankow JF. Levels of mint and wintergreen flavorants: smokeless tobacco products vs. confectionery products. Food Chemical Toxicol. 2010;48(2):755–63.

74. Dolka C, Piade JJ, Belushkin M, Jaccard G. Menthol addition to cigarettes using breakable capsules in the filter. Impact on the mainstream smoke yields of the health Canada list constituents. Chem Res Toxicol. 2013;26(10):1430–43.
75. Roemer E, Dempsey R, Hirter J, Deger Evans A, Weber S, Ode A et al. Toxicological assessment of kretek cigarettes. Part 6: the impact of ingredients added to kretek cigarettes on smoke chemistry and in vitro toxicity. Regul Toxicol Pharmacol. 2014;70(Suppl 1):S66–80.
76. Gordon SM, Brinkman MC, Meng RQ, Anderson GM, Chuang JC, Kroeger RR et al. Effect of cigarette menthol content on mainstream smoke emissions. Chem Res Toxicol. 2011;24(10):1744–53.
77. Hurt RD, Ebbert JO, Achadi A, Croghan IT. Roadmap to a tobacco epidemic: transnational tobacco companies invade Indonesia. Tob Control. 2012;21(3):306–12.
78. Behar RZ, Luo W, McWhirter KJ, Pankow JF, Talbot P. Analytical and toxicological evaluation of flavor chemicals in electronic cigarette refill fluids. Sci Rep. 2018;8(1):8288.
79. Bengalli R, Ferri E, Labra M, Mantecca P. Lung toxicity of condensed aerosol from e-cig liquids: influence of the flavor and the in vitro model used. Int J Environ Res Public Health. 2017;14(10):1254.
80. Menthol cigarettes and public health: review of the scientific evidence and recommendations. Washington (DC): Tobacco Products Advisory Committee, Food and Drug Administration; 2011.

7. Sugar content of tobacco products

Lotte van Nierop and Reinskje Talhout, Centre for Health Protection, National Institute for Public Health and the Environment (RIVM), Bilthoven, Netherlands

Contents

7.1 Introduction
7.2 Sugars in different types of tobacco product
    7.2.1 Types of sugars and sugar-containing additives
    7.2.2 Amounts of sugars and sugar-containing additives
    7.2.3 Total levels of added and endogenous sugars
    7.2.4 Regional and cultural differences in tobacco varieties, products and use
7.3 Effects of sugars on levels of emissions from tobacco products
    7.3.1 Smokeless tobacco products
    7.3.2 Pyrolysis products of sugars and simple mixtures
    7.3.3 Cigarettes and other combusted tobacco products
7.4 Effects of sugars on the toxicity of tobacco products
    7.4.1 Smokeless tobacco
    7.4.2 Cigarettes with and without added sugar
7.5 Effects of sugars on the addictiveness of tobacco products
    7.5.1 pH and free nicotine
    7.5.2 Formation of pharmacologically active compounds in tobacco and smoke
7.6 Dependence and quitting
7.7 Effects of sugars on the attractiveness of tobacco products
    7.7.1 Perception: sensory characteristics
    7.7.2 Smoking experience and behaviour: flavour, palatability, ease of inhalation, frequency of use
    7.7.3 Initiation
7.8 Regulation of sugars according to jurisdiction
7.9 Conclusions
7.10 Recommendations
    7.10.1 Further research
    7.10.2 Policy
7.11 Acknowledgements
7.12 References

7.1 Introduction

Sugar is a component of tobacco products and is often present in large amounts. Sugars are naturally present in tobacco leaves and may also be added during manufacture *(1–3)*. Unprocessed tobacco leaves contain many types of sugar, including glucose, fructose and sucrose. Drying (curing) of the leaf can affect these levels; while air-cured tobacco contains virtually no sugars, flue-cured tobacco may contain up to 25% of its weight *(4, 5)*. In addition, during manufacture, various types of sugar and sugar-containing ingredients, such as honey and fruit syrups, may be added to the casing. The amount added depends on the tobacco blend and the product type. Added sugars can serve as binders, casing ingredients, flavours, formulation aids or humectants *(3, 4, 6)*. When sugars are burnt during combustion of tobacco products, they yield many toxic products, some of which have been hypothesized to affect the addictiveness of tobacco products *(7)*. Sugars also contribute significantly to the flavour of tobacco products *(3, 8)*. In smokeless tobacco, the sugars themselves flavour the tobacco product, whereas, in combusted tobacco products, sugars react with other components in the tobacco to produce caramel flavours and make the smoke milder to inhale *(3, 9)*. Some casing materials that contain sugars have been referred to as "ameliorants" by the tobacco industry and are claimed to "… smooth out harshness and bitterness and/or eliminate pungent aromas from tobaccos" *(10)*.

As sugar is one of the main components of tobacco products and may be harmful to consumers' health by increasing the toxicity, addictiveness and/or attractiveness of products, it may be an appropriate candidate for regulation of tobacco contents. We review the evidence on the presence and effects of sugars in different tobacco products, including smokeless and waterpipe tobacco. First, we describe the variety of sugars naturally present in tobacco leaves, those that result from curing and those that are added to processed tobacco during manufacture. We also report the percentages of products with added sugars and the amounts of sugars added. Secondly, we summarize the evidence on the effect of sugars on the levels of emissions from tobacco products. Thirdly, we review the health effects of sugars in tobacco products, as assessed by their toxicity, addictiveness and attractiveness. Finally, we highlight implications for research and regulation.

7.2 Sugars in different types of tobacco product

Tobacco is a natural plant, and its leaves contain high levels of carbohydrates such as sugars (mono- and disaccharides), starch, cellulose and pectin (polysaccharides) *(4)*. During drying (curing) of the green leaves into usable brown, dried tobacco, enzymes degrade starch into sugars. The curing conditions depend on the tobacco type. Virginia tobacco is flue-cured in humid conditions and at elevated temperatures, which allows starch to be degraded into sugar and stops enzymatic degradation of sugar, resulting in high levels (8–30%) in the cured tobacco *(11)*. Oriental tobacco is sun-cured in a process similar to flue-curing but less controlled, resulting in sugar levels of 10–20% in cured tobacco *(2)*. In contrast, Burley tobacco is air-cured at ambient temperature in a slow process that results in degradation of starch into sugars and breakdown of the sugars by enzymes, resulting in a low sugar content ($\leq 0.2\%$) in the cured tobacco *(4, 11)*. As tobacco products contain one or a blend of tobacco varieties, the sugar levels of tobacco products differ widely, depending on the blend. In general, the sugars found most abundantly in the cured tobacco leaf are glucose, fructose and sucrose *(4, 11)*.

Besides mono- and disaccharides, tobacco also contains considerable amounts of (added or naturally present) polysaccharides, such as cellulose, pectin and starch *(9, 12)*. A conventional cigarette blend contains about 10% cellulose, 10% pectin and 2% starch *(4, 12)*. Thus, carbohydrates may comprise over 40% of tobacco and therefore have a substantial impact on the chemical composition of tobacco smoke *(3)*.

7.2.1 Types of sugars and sugar-containing additives

Besides the sugars present in natural tobacco leaves, various types of sugar and sugar-containing ingredients are added to tobacco products, primarily in the casing. The types of sugars depend on the many natural sources and processing methods used. The sugars most often added to combusted and smokeless tobacco products are sucrose and invert sugar (a mixture of fructose and glucose) *(2, 13, 14)*. Other tobacco additives that contain large amounts of sugars are fruit juices, corn and maple syrups, molasses extracts, honey and caramel *(3, 4, 6, 15)*. Other substances may also have a sweet taste and characteristics similar to those of sugar but are not classified as such. These are acesulfame K, aspartame, ethyl maltol, glycerol, propylene glycol, maltitol, maltol, saccharin, sorbitol and thaumatin.

Sugars in tobacco products identified from manufacturers' lists of ingredients in the electronic database EMTOC *(16)* are: brown sugar, caramel, (corn) syrup (solids), dextrin, dextrose, (fruit) concentrate, (fruit) juice, glucose, high-fructose sugar, honey, invert sugar, lactose, malt extract, maltodextrin, (maple) syrup, molasses, (partially) inverted sugar, sucrose (syrup), sugar cane (syrup), sugar syrup and unspecified sugar. Monosaccharides are simple sugars such as fructose, galactose and glucose, with the general formula $C_6H_{12}O_6$. Disaccharides are formed by the combination of two monosaccharide molecules with the exclusion of a molecule of water; these include sucrose, lactose and maltose, with the general formula $C_{12}H_{22}O_{11}$.

7.2.2 Amounts of sugars and sugar-containing additives

Sugars are usually added to tobacco leaves as a casing ingredient, although in some cases they are added in very small amounts to non-tobacco material like paper, filter and glue *(16)*. The amount of sugar added to conventional cigarettes represents 2–18% of the weight of the tobacco *(2, 12, 17, 18)*. The levels in waterpipe tobacco (*maassel*) are much higher, representing 50–70% of the weight of the product *(19)*. Little information is available on the amounts and types of sugars in tobacco products. In the Netherlands, manufacturers are obliged to disclose the ingredients of all tobacco products on the Dutch market annually to an electronic database (EMTOC) *(16)*. Analysis of these data shows that, in 2015, sugars were added to a substantial proportion of all types of tobacco product *(20)*.

Further analysis of the data for the purposes of this paper showed that most roll-your-own products (56%), cigarettes (77%), pipe tobacco (86%) and waterpipe products (100%) contained added sugars (Fig. 7.1A), whereas few cigars (7%) or oral tobacco products (22%) did so. The sugars used most frequently in all tobacco products are invert sugar and glucose, with sugar-containing ingredients such as caramel, honey, syrups and fruit juices (Fig. 7.1B–G). The weight percentage of sugar added was 2% in cigars, 3% in roll-your-own tobacco and in cigarettes and 12% in pipe tobacco. Remarkably, an average waterpipe tobacco product contained added sugars that represented 164% of the tobacco weight, indicating more sugar than tobacco in the final product. Although the amount of sugar added to oral tobacco was only 10% of the tobacco weight in this analysis, the final sugar levels in these smokeless tobacco products may be about 35% *(21)* because of the high levels of natural sugars present in tobacco leaf. These numbers depend on the types of tobacco used in products in different countries.

Fig. 7.1. Concentrations of sugars in tobacco products

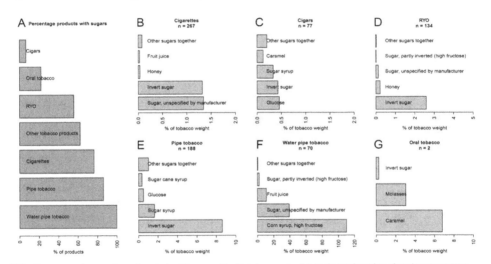

RYO, roll-your-own. A: percentage of products with sugar added to tobacco according to the total number of products containing additives, for each category of tobacco product; B–G, average amount of the sugars (percentage per mg tobacco) used most often for each category of tobacco product; n, number of products containing added sugar; "other sugars together" shows all remaining sugars in the product group. *Source*: EMTOC database *(16)*, in which manufacturers register all marketed tobacco products in the Netherlands annually and which contains information on additive composition, quantities added and function.

7.2.3 Total levels of added and endogenous sugars

Natural tobacco contains large amounts of sugar formed during growing and curing, as discussed above, besides the sugars added during manufacture. Therefore, it is more informative to measure or report the final sugar levels in marketed tobacco products rather than that on the list of added sugars, which will provide an underestimate of the final level *(22)*. Few reports of laboratory analyses of total sugar levels in tobacco products have been published. Chemical analysis of final tobacco products for the most commonly added carbohydrates (glucose, fructose and sucrose) indicated total sugar levels of ≤ 19% of the weight of tobacco in cigarettes *(23)*, 0.1% in cigars, 0.03% in snuff and 27.7% in chewing tobacco *(24)*.

7.2.4 Regional and cultural differences in tobacco varieties, products and use

The variety of tobacco products marketed and used depends on regional and cultural preferences. In a study of 2the patterns of tobacco use among adolescents in 32 countries *(25)*, the national prevalence of current cigarette smoking ranged from 1.8% in Rwanda to 32.9% in Latvia, whereas the prevalence of current smokeless tobacco use ranged from 1.1% in Montenegro to 14.4% in Lesotho. In most European countries and the USA, the prevalence of current smoking was significantly higher than that of smokeless tobacco use, in contrast to the patterns observed in low- and middle-income countries such as in south-east Asia. Eurobarometer reported that, in Europe, smokeless tobacco is particularly popular in Sweden, whereas waterpipe use is popular in other northern European countries and northern Africa *(26)*. Experience with tobacco use may vary by culture due to differences in availability, affordability and acceptability *(25)*.

More information is available on the variety and blends of tobacco used in cigarettes, although little is known about the sugar content of specific products and brands. The use of sugars differs by brand and even within a brand because of national preferences *(2)*. The variety of tobacco also differs by region; in general, however, the cigarette market appears to be dominated by two varieties: flue-cured cigarettes and American-blend cigarettes. According to PMI, in 2008, Virginia-style (flue-cured, high in sugar and few additives) cigarettes dominated the market in Canada (99%), New Zealand (95%), Australia (92%), the United Kingdom (91%), Ireland (87%) and South Africa (76%). American-blend cigarettes (a blend of Virginia, Burley and Oriental tobacco and many additives, including sugars) are preferred in continental western European countries, with a market share of 85–100%, and in the USA, with 99% *(1)*.

7.3 Effects of sugars on levels of emissions from tobacco products

7.3.1 Smokeless tobacco products

Users of smokeless tobacco products are exposed to emissions resulting from chewing or sucking. While smokeless tobacco contains high levels of sugars (see above), the amounts extracted during use and the resulting exposure to sugars are unknown.

7.3.2 Pyrolysis products of sugars and simple mixtures

Combustion in a burning cigarette has been simulated in pyrolysis studies *(27, 28)* in order to understand precursor–smoke constituent relations in cigarette mainstream smoke *(1)*. Studies with this method indicate that only small amounts of nonvolatile sugars in tobacco (approximately 0.5%) are transferred intact into cigarette mainstream smoke, whereas most of the sugar combusts, pyrolyses or is part of other pyrosynthesis processes *(1, 3)*. Combustion processes such as caramelization result in many different chemical compounds, including aldehydes (e.g. acetaldehyde, acrolein, 2-furfural), furan derivatives, VOCs, organic acids, acrylamide and PAHs (usually at high temperatures). The pyrolysis conditions used in these studies only approximate the temperature and other conditions in a burning cigarette, however, and do not account for the interaction of other tobacco and/or smoke components with sugars during combustion. For example, sugars can react with amines in tobacco (ammonium compounds, amino acids, proteins) to form Maillard products (see also section 7.2), which can break down into chemical species such as aldehydes, ketones (e.g. diacetyl), acids, acrylamide, pyrazines and pyridines *(1, 3)*.

7.3.3 Cigarettes and other combusted tobacco products

An independent review of the effects of the addition of sugar on smoke composition in 2006 concluded that adding sugars to cigarette tobacco primarily enhances the levels of aldehydes and ketones, especially formaldehyde, acetaldehyde, acetone, acrolein, 2-furfural and other furans *(3)*. Additionally, the concentration of total acids in mainstream smoke was increased, resulting in a lower pH and less free-base nicotine. Cheah et al. *(29, 30)* also showed that the addition of sugars to Burley tobacco, which contains virtually no natural sugars, increases the concentrations of the aldehydes acetaldehyde, acrolein, crotonaldehyde, propionaldehyde and butanal in mainstream tobacco smoke. The increase is specific for aldehydes, as much smaller increases were observed in tar, nicotine and CO levels.

While many of these conclusions have been discussed in tobacco industry papers and reviews published subsequently, there has been considerable debate about whether sugars increase the levels of acetaldehyde and, to a lesser extent, of acrolein. For instance, Seeman et al. *(12)* pointed out that cellulose rather than sugar is the main precursor of acetaldehyde in smoke but that this by itself does not imply that sugars do not also contribute. In a review of cigarette tobacco additives, Klus et al. *(1)* concluded (from the results of many studies) that adding sugars to tobacco increases the concentrations of phenol, furans, organic acids, 2-furfural and formaldehyde and results in more acidic smoke. He did not report an association between sugars and acetaldehyde. Baker *(31, 32)* reported an increase in formaldehyde and found that sugars increase the levels of 2-furfural in mainstream cigarette smoke but not those of acetaldehyde or acrolein. Coggins et al. *(33)* found that the addition of carbohydrates and sugar-containing natural products to tobacco resulted in only minimal changes in smoke chemistry but consistently resulted in a small increase in formaldehyde over that in the smoke of additive-free cigarettes. Roemer et al. *(2)* reported that, while most of the smoke constituents determined did not show any change in yield (per mg nicotine), increasing added sugar levels either increased (formaldehyde, acrolein, 2-butanone, isoprene, benzene, toluene, benzo[*k*]fluoranthene) or decreased (4-aminobiphenyl, *N*-nitrosodimethylamine, NNN) the levels of constituents. Hahn and Schaub *(34)* reported an increase in the concentration of formaldehyde but not that of acetaldehyde after addition of 5% sucrose as compared with a reference tobacco blend consisting of 50% Virginia tobacco (flue-cured), 20% Burley tobacco (air-cured), 20% tobacco stems and 10% Oriental tobacco (sun-cured). As Virginia tobacco contains high levels of natural sugars, relative increases in sugar and resulting aldehyde levels after addition of sugar are less evident.

O'Connor and Hurley *(35)* found that the relation between sugar and acetaldehyde levels is obscured by normalizing acetaldehyde yields to tar or total particulate matter, both of which are directly related to design features such as ventilation and mass of tobacco. They re-analysed a study by Zilkey et al. *(36)* and found that sugar accounted for over 50% of the variation in acetaldehyde levels in smoke. Re-analysis of the data of Phillpotts et al. *(37)* showed that sugars in tobacco blends accounted for an additional 11% variation in aldehyde levels, while total particulate matter accounted for 23% of the variation.

Waterpipe tobacco contains the highest levels of sugar of all tobacco products. Although this may result in higher levels of volatile aldehydes in smoke *(38, 39)*, no data were available on the relation between sugar content and the composition of the smoke of waterpipe tobacco or of any other combusted tobacco products. Furthermore, correlations have been observed between sugar levels in e-cigarette liquids and aldehyde levels in their emissions *(40)*.

7.4 Effects of sugars on the toxicity of tobacco products

Sugars are GRAS additives when used in food products but not when used in tobacco *(3)*. When tobacco products are smoked, sugars are combusted to yield many toxic and carcinogenic reaction products. Furthermore, compounds are generally more toxic when inhaled than when ingested, because the respiratory system largely lacks the detoxifying metabolic pathways of the digestive system *(3)*.

7.4.1 Smokeless tobacco

Commercially available smokeless tobacco contains 0.5–2 g of sugars per chew of about 10 g *(41)*. The product is kept in the oral cavity for an average of 30 min and used repeatedly each day *(42, 43)*. Smokeless tobacco causes several types of cancer, cardiovascular disease, adverse reproductive outcomes and adverse effects on oral health, including oral mucosal lesions, leukoplakia and periodontal disease *(44)*, but little is known about the contribution of the sugar content in smokeless tobacco to these diseases. Use of smokeless tobacco has been associated with dental caries, and a plausible cause is the high level of sugars, although other causes related to tobacco use cannot be excluded. The sugar in smokeless tobacco forms an acid that may eat away the tooth enamel, cause cavities and result in mouth sores *(42)*. Furthermore, the high levels of fermentable sugar can stimulate the growth of cariogenic bacteria *(14, 43)*.

Sugars in smokeless tobacco may also influence blood glucose. A case study showed that blood glucose levels in a patient with diabetes who stopped swallowing the juice resulting from chewing tobacco dropped by 50% (from 300–400 mg/dL to 160–200 mg/dL) *(41)*.

7.4.2 Cigarettes with and without added sugar

Many pyrolysis products of sugar have been reported to have toxic effects *(45, 46)*. These include formaldehyde (irritant, carcinogenic), acrolein (irritant) and PAHs (carcinogenic). The extent to which they contribute to the total toxicity of combusted tobacco products is unknown.

In a review of tobacco additives, Klus et al. *(1)* concluded that "sugars and the kind of sugar used as additive have – if any – only small and unimpressive effects on cigarette mainstream smoke toxicity". Mainstream cigarette smoke resulting from the combustion of cigarette tobacco with various levels of sugars was investigated in assays for cytotoxicity, mutagenicity and genotoxicity in vitro, in subchronic studies of inhalation toxicity in vivo and in studies of dermal tumorigenicity. None of the studies showed significant differences in markers of toxicity *(2, 33)*. According to the Scientific Committee on Emerging and Newly Identified Health Risks of the European Union *(47)*, however:

comparative toxicity testing strategies, …, are not considered suitable to address the properties outlined in the terms of reference with the currently available methodology. Indeed, at present, these studies lack discriminative power due to the high background toxicity of tobacco products and their results cannot be generalized to all products and brands, having a different composition with respect to tobacco type, blend and additives.

7.5 Effects of sugar on the addictiveness of tobacco products

No information was available on the role of sugar in the addictiveness of smokeless tobacco products. Sugar may, however, increase craving and reinforcement of use of combusted tobacco products, in addition to the well documented addictive effects of nicotine *(7)*. The addictive potency of sugar in combusted tobacco products may be increased by several direct and indirect pathways, including the pH of smoke, free nicotine levels and the formation of pharmacologically active compounds.

7.5.1 pH and free nicotine

The amount of nicotine that reaches the brain depends on the availability of free-base nicotine (the uncharged, volatile form), which formed at specific pH. High pH results in more free-base nicotine *(48, 49)*, which readily crosses the cell membranes of the oral cavity and lung epithelium, resulting in higher levels of nicotine that reach the brain. Tobacco products and brands with high carbohydrate or sugar contents generate more acidic smoke, resulting in lower concentrations of free-base nicotine *(50)*. To maintain a satisfying level of nicotine absorption from cigarettes with lower levels of free-base nicotine, smokers inhale more deeply and/or more frequently or increase their cigarette smoking frequency *(51, 52)*. This leads to higher exposure to carcinogenic and toxic compounds. (For other effects of changed smoke pH see section 7.7.) Other studies suggest, however, that the pH-buffering capacity of the lung epithelium diminishes the effect of changed nicotine bioavailability when the pH of tobacco or smoke is changed *(1)*.

7.5.2 Formation of pharmacologically active compounds in tobacco and smoke

Although sugars may have no addictive potential per se, combustion of sugar in a tobacco product results in several compounds in tobacco smoke that may have addictive potential. Of these, acetaldehyde is the most important *(12, 53, 54)*, as it increases the firing rate of dopaminergic neurons in the ventral tegmental area *(54)*. In experimental animals, intravenous acetaldehyde was addictive and synergistically enhanced the reinforcing effects of nicotine *(53, 56–60)*. Although acetaldehyde clearly reinforces the effects in rodents, it is not known whether these effects also occur in humans and at the concentrations found in cigarette smoke.

Acetaldehyde exerts its addictive potential by reacting with the amino acids tryptamine and tryptophan, present in tobacco or in the body, resulting in the condensation product harman (and norharman when amino acids reacts with formaldehyde) *(61, 62)*. Harman inhibits the enzyme monoamine oxidase, which degrades neurotransmitters involved in drug addiction, like dopamine, noradrenaline and serotonin *(56, 61, 63)*. Exposure to harman therefore increases the amounts of dopamine and serotonin in the nucleus accumbens and potentiates the action of nicotine *(64–66)*. Interestingly, an L-cysteine lozenge that reacts with acetaldehyde in saliva lowers the levels available for the formation of harman and showed potential for use as smoking cessation therapy in an initial double-blind, randomized, placebo-controlled trial *(67)*.

Several studies have shown that the levels of harman and norharman in smokers are related to the number of cigarettes smoked *(68–70)*, and lower activity of monoamine oxidase in tobacco smokers increased nicotine self-administration *(57)* and maintenance of behavioural sensitization to nicotine *(56, 71)* as compared with nonsmokers.

7.6 Dependence and quitting

Sugar in tobacco products can contribute to maintenance of tobacco use by enhancing dependence and making it difficult to quit. The tobacco industry performed several intercountry comparisons to investigate the influence of additives on the addictiveness of tobacco. A meta-analysis of clinical studies on smoking cessation rates did not show a significant difference in quit rates between tobacco markets dominated by Virginia flue-cured cigarettes (high in natural sugar but few added ingredients and no added sugars) and by American-blended cigarettes (many added ingredients including sugars) *(72, 73)*. An important assumption in the study, however, was that the difficulty of quitting smoking is a valid measure of tobacco addictiveness, and no correction is needed for country-specific factors such as product availability and cessation programmes *(74)*. Also, no relevant differences were found between the two markets in smoking prevalence, intensity, some markers of dependence, nicotine uptake or mortality from smoking-related lung cancer or chronic obstructive pulmonary disease *(73)*. According to Klus et al. *(1)*, no effect of sugars was to be expected, given that the total sugar level (naturally present and added) in the tobacco of American-blend and Virginia cigarettes is comparable.

Little is known about the dependence and quitting rates of smokers of specific types of tobacco products with different sugar contents *(72)*. Much of the research on the abuse potential associated with acetaldehyde is based on studies of ethanol, as acetaldehyde is also a metabolite of ethanol *(75)*, and no proper experiments have been conducted to evaluate withdrawal from acetaldehyde vapour *(76)*. Nevertheless, the synergistic action of nicotine and acetaldehyde was observed only in young and not in adult rats *(58)*, which may support the observation that, in humans, adolescents appear to be more prone to tobacco addiction than adults *(58, 65, 77)*.

7.7 Effects of sugars on the attractiveness of tobacco products

7.7.1 Perception: sensory characteristics

The sensory characteristics of tobacco products, such as taste and smell, significantly influence their attractiveness *(7)*. For instance, initial attraction may be established by the presence of flavourings like sugar and other sweeteners *(3, 9, 78)*. Manufacturers select tobacco for certain chemical criteria, such as sugar and nicotine content, and the annual variation in tobacco leaf composition is equilibrated by the use of sugars and other additives to ensure the consistent taste of a product over time *(7)*. Sugar in smokeless tobacco products is perceived by the taste receptors in the oral cavity with olfactory stimulation. In contrast, users of combusted tobacco products perceive only the pyrolysis products of sugar. Overall, manufacturers of tobacco products adjust the taste and characteristics of products to the desires of consumers and to create a "pleasant experience" *(79)*.

7.7.2 Smoking experience and behaviour: flavour, palatability, ease of inhalation, frequency of use

Sugars can improve several characteristics of tobacco products, such as masking the harshness of the smoke and improving the taste. Volatile basic components like ammonia, nicotine and alkaloids give tobacco smoke a harsh taste, which prevents smokers from inhaling *(3, 80)*. Combustion of sugar during smoking results in acids, which reduce the pH of inhaled smoke *(50)* and thus decrease its harshness and irritability *(4, 9)*, increase the palatability of the product and facilitate inhalation. More frequent, deeper inhalation, which is eased by the addition of sugar, increases exposure to nicotine and other smoke chemicals. Sugars also play an important role in tobacco flavour *(4, 6, 81)*. In smokeless tobacco, the sugar itself is consumed. During curing, storing, processing and smoking of tobacco, amino acids or ammonia react with sugars (Maillard reaction) to produce chemicals with highly diverse structures and flavouring potential. Many of these products are heterocyclic compounds, which include aromatic pyrazines, an important class responsible for the characteristic taste of certain cigarette brands. In addition, caramelization of the sugar improves the taste and smell of the tobacco smoke for both users and bystanders *(79)*. This is, unlike the Maillard reactions, a non-enzymatic browning reaction of sugar (no amines involved).

7.7.3 Initiation

The acceptance of tobacco smoke by smokers is partly proportional to the sugar level in the tobacco *(3, 8)*. The extent to which sugar in tobacco products and its level influence initiation of tobacco consumption has not been investigated. As sugar reduces pH, masking the bitter taste of cigarette smoke, new users experience less harshness. Furthermore, the sweet taste of the caramel flavours generated by the combustion of sugars is particularly attractive to adolescents *(53, 82)*. Thus, the presence of sugar in tobacco products may encourage adolescents to start smoking earlier, continue smoking for longer and increase their tobacco use *(3)*. The addition of sugars to cigarettes to stimulate or enhance the sensory attributes of cigarette smoke and encourage smoking initiation and maintenance is not only scientifically plausible but has been discussed by industry as part of their marketing strategy *(74, 83)*.

7.8 Regulation of sugars according to jurisdiction

Several international and national authorities regulate additives in tobacco products. The most important of them and the main regulations and implementations are listed below.

- **WHO FCTC**: Article 9 addresses the testing and measurement of the contents and emissions of tobacco products and their regulation, and

Article 10 addresses the regulation of tobacco product disclosures, by allowing Parties to the WHO FCTC to adopt and implement effective legislative, executive, administrative or other measures requiring manufacturers and importers of tobacco products to disclose to government authorities information about the contents and emissions of tobacco products. The Framework facilitates adoption of comprehensive tobacco control measures by States Parties, which may include reporting and disclosure, such as the requirement to submit information on sugar levels (and the emission of sugar-related compounds) in tobacco products *(84)*.

- **Health Canada**: the use of many ingredients, including sugars, in cigarettes, little cigars and blunt wraps has been prohibited since 2009 in an amendment to Bill C-32 *(85)*. Moreover, Canada requires manufacturers and importers to provide information on tobacco product ingredients, as set out in the Tobacco Reporting Regulations *(86)*.
- **The European Tobacco Product Directive:** in accordance with Directive 2014/40/EU, a ban on characterizing flavours in cigarette and roll-your-own products (but not in other tobacco products) entered into force in 2014 *(87)*. Furthermore, manufacturers in each European Union Member State are required to disclose all marketed tobacco products and their composition (e.g. amount and type of added sugars, tobacco variety or blend, curing method) annually to an electronic database (EU-CEG). The Directive states that the sugars lost during curing of leaves may be replaced during manufacture; however, the initial sugar content of the tobacco leaf and loss during processing are not defined, which makes it difficult to regulate the amounts of sugar that can be added.
- In the **United Kingdom** in 2003, a voluntary agreement with the tobacco industry set limits on the use of several sugars (sucrose, invert sugar, syrups and molasses), with a maximum of 10% weight of the product in cigarettes, roll-your-own tobacco and cigars and 15% weight in pipe tobacco *(88)*. The rationale for determining these limits was not clear. The agreement has been superseded since 2014 by the revised European Tobacco Product Directive *(89)*.
- **Brazilian Health Regulatory Agency (Anvisa)**: the use of some additives in tobacco products has been restricted in a regulatory resolution with legal power (RDC 14/2012) *(90)*. The resolution prohibits the use of sweeteners, honey, molasses, (any product originating from) fruits and substances that can impart a sweet flavour, apart from sugars. An exception was made for sugar, allowing manufac-

turers to restore the quantity lost during tobacco leaf curing. In September 2013, this resolution was suspended by a lawsuit filed by the National Confederation of Industries on behalf of tobacco product manufacturers *(90)*. A working group of independent experts recommended that RDC 14/2012 be amended such that sugars would no longer be excluded from the ban on additives *(91)*. As of June 2015, the Supreme Court injunction allowing the use of additives remained in force, with a final decision pending.

- **FDA:** the Family Smoking Prevention and Tobacco Control Act in 2009 included a ban on cigarettes containing certain characterizing flavours *(92)*. Restrictions are considered in the use and sale of certain flavored e-liquids *(93)*. No specific requirements or limits are set for the sugar content of cigarettes or other tobacco products.

7.9 Conclusions

Added sugar is one of the main ingredients in many types of tobacco product. The amount may be substantial, making up 2–164% of the weight of tobacco, depending on the type of product. As tobacco leaves themselves may contain high levels of natural sugar, the total amount of sugar (natural and added) in a tobacco product may be quite high. Limited data were available on the total sugar levels in different types of tobacco products.

There is little evidence that sugar contributes to the toxicity and addictiveness of smokeless tobacco products, apart from effects on dental health and possibly diabetes; however, sugars contribute to the attractiveness of the products by improving their taste.

Burnt and unburnt sugars contribute to the appeal of all tobacco products, and the combustion of sugar in tobacco products may contribute to the formation of toxic, carcinogenic and addictive compounds.

Some jurisdictions, such as Canada and Brazil, regulate the addition of sugars to tobacco products, and limits have been set in the United Kingdom. Stricter control measures are needed to prevent the harmful effects on health of the use of tobacco products containing both naturally occurring and added sugars.

7.10 Recommendations

7.10.1 Further research

There are still large gaps in understanding the health effects of sugar in tobacco and tobacco products, as stressed by the editor of *Nicotine and Tobacco Research (94)*. In particular, more information is required on:

- the relation between sugars in waterpipe tobacco and toxicants in smoke;

- biomarkers of exposure relevant to the effects of sugars on health and the impact of changing sugar levels on those biomarkers;
- the role of sugars in the attractiveness of tobacco products to children and young adults, e.g. by monitoring market shares according to sugar content and by marketing studies;
- the amount of sugars that alters the overall sweetness of emissions and imparts a noticeable characterizing flavour;
- the effect of sugars on overall rates of initiation, addictiveness, dependence and quitting according to the type of tobacco product;
- the role of acetaldehyde, in the amounts measured in tobacco smoke, on the addictive potential of (nicotine in) tobacco through monoamine inhibitors such as harman and formation of acetaldehyde–biogenic amine condensation products in vivo; and
- the effects of sugars on the overall toxicity of tobacco products, including:

  - for smokeless tobacco products, more information on the toxic effects of the sugars themselves, e.g. whether they are associated with diabetes and dental caries;
  - for combusted tobacco products, more information on the role of sugars in inhalation toxicity; and
  - interactions of sugars with other smoke components that may increase the toxicity of smoke.

7.10.2 Policy

Several policy measures may be considered for reducing the negative effects of sugar (and its pyrolysis products) on product toxicity, addictiveness and attractiveness.

Mandate disclosure of sugar levels

Manufacturers of tobacco products should disclose the levels of sugar (and other ingredients) in all types of tobacco product. This would provide valuable information for future regulations and regulatory decisions. Simple, fast methods are available for verifying reported levels, e.g. extraction of tobacco, followed by high-performance liquid chromatography–evaporative light-scattering detection *(23)*.

Mandate lowering of sugars in final tobacco products

Defining and regulating the maximum sugar levels in tobacco products might best protect consumers, as the actual source of sugars present naturally in tobacco or added during manufacture does not define its health effects. If desired and supported by sufficient scientific evidence, the maximum sugar level could be lowered to virtually 0, the sugar level present in Burley tobacco leaves.

Mandate lowering of sugar in all tobacco products

Canada has banned added sugar in cigarettes, little cigars and blunt wraps but not in other products, such as waterpipe and smokeless tobacco. Regulation of sugar in only some tobacco products is likely to encourage users to shift to other products. Thus, the ban on flavouring additives in cigarettes in Canada and the USA increased the market and consumption of flavoured small cigars *(86, 95)*.

Mandate disclosure and lowering of the most harmful components of tobacco smoke

To decrease the negative effects of smoking, upper limits could be set for the most harmful components resulting from the combustion of sugar in tobacco smoke *(3)*. An analytical method for aldehydes, the most harmful emission compounds resulting from the combustion of sugar, is validated by WHO TobLabNet, and TobReg has included aldehydes on its list of smoke components proposed for mandated lowering *(45)*. The COP identified these as priority toxicants for which methods should be validated. Cunningham et al., of British American Tobacco, supported the inclusion of aldehydes as a priority. In their paper on segregation of tobacco smoke toxicants for risk assessment and management purposes, acetaldehyde, acrolein and formaldehyde, with a margin of exposure < 10 000, are considered high priorities for research on reducing exposure *(96)*.

Support research on the toxicity, addictiveness and attractiveness of sugar

Regulators could support research and reporting on compounds that result from the combustion of sugar in tobacco products and their toxicity, carcinogenicity, mutagenicity, reproductive toxicity, addictiveness and attractiveness. Scientific guidelines and recommendations have been issued for assessing the impact of tobacco additives on toxicity *(97)*, addictiveness *(98)* and attractiveness *(99)*.

Change the status and definition of "sugar"

Another important means of increasing attention to and scientific research on sugar would be to include it on national and international lists of target ingredients and constituents, e.g. the European priority list *(100)*, in the next intended round of revision. The list was established in line with Article 13 of the European Union Tobacco Products Directive *(87)* and includes additives for which more scientific experimental reporting is required on cigarettes and roll-your-own tobacco, including their toxicity, addictiveness and carcinogenic, mutagenic or reproductive toxicity in unburnt and burnt forms. Sugar is not yet included, although it meets all the selection criteria (Articles 6.1 and 6.2 of the Directive).

Require disclosure, and monitor levels of sweeteners in tobacco products, in addition to sugar

Besides sugar, the (high-intensity) sweetener content of tobacco products is used efficiently to control product palatability and to increase initiation among adolescents. High-intensity sweeteners are several hundred times sweeter than sucrose *(101)* and are present in many alternative tobacco products. Regulation of sweetener content might therefore be a means of controlling the palatability of a wide range of products and reducing initiation of tobacco product use. For instance, Canada has already banned sweeteners and several flavouring additives, in addition to sugar.

7.11   Acknowledgements

We thank Rob van Spronsen for a systematic search of the literature and Jeroen Pennings for analysing the EMTOC database for sugars in tobacco products.

7.12   References

1. Klus H, Scherer G, Muller L. Influence of additives on cigarette related health risks. Contrib Tob Res. 2012;25(3):412–93.
2. Roemer E, Schorp MK, Piadé JJ, Seeman JI, Leyden DE, Haussmann HJ. Scientific assessment of the use of sugars as cigarette tobacco ingredients: a review of published and other publicly available studies. Crit Rev Toxicol. 2012;42(3):244–78.
3. Talhout R, Opperhuizen A, van Amsterdam JG. Sugars as tobacco ingredient: effects on mainstream smoke composition. Food Chem Toxicol. 2006;44(11):1789–98.
4. Leffingwell JC. Leaf chemistry: basic chemical constituents of tobacco leaf and differences among tobacco types. In: Davis DL, Nielsen MT, editors. Tobacco: production, chemistry and technology. Oxford: Blackwell Science; 1999:265–84.
5. Cahours X, Verron T, Purkis S. Effect of sugar content on acetaldehyde yield in cigarette smoke. Beitr Tabakforsch Int. 2014;25(2):381–95.
6. Seeman JI, Laffoon SW, Kassman AJ. Evaluation of relationships between mainstream smoke acetaldehyde and "tar" and carbon monoxide yields in tobacco smoke and reducing sugars in tobacco blends of US commercial cigarettes. Inhal Toxicol. 2003;15(4):373–95.
7. Addictiveness and attractiveness of tobacco additives. Brussels: European Commission Scientific Committee on Emerging and Newly Identified Health Risks; 2010 (https://ec.europa.eu/health/scientific_committees/emerging/docs/scenihr_o_029.pdf, accessed October 2018).
8. Bernasek PF, Furin OP, Shelar GR. Sugar/nicotine study. Industry documents library, 1992 (ATP 92-210:22) (https://www.industrydocumentslibrary.ucsf.edu/tobacco/docs/#id=fxbl0037, accessed 1 January 2019).
9. Rodgman A. Some studies of the effects of additives on cigarette mainstream smoke properties. II. Casing materials and humectants. Beitr Tabakforsch Int. 2002;20(4):83–103.
10. Jenkins CR et al. British American Tobacco Company limited product seminar. Tobacco documents 760076123–408, 1997 (https://www.industrydocumentslibrary.ucsf.edu/tobacco/docs/#id=rjgf0205, accessed 1 January 2019).
11. Fisher P. Tobacco blending. In: Davis DL, Nielsen MT, editors. Tobacco: production, chemistry and technology. Oxford: Blackwell Science; 1999:346–52.
12. Seeman JI, Dixon M, Haussmann HJ. Acetaldehyde in mainstream tobacco smoke: formation and

occurrence in smoke and bioavailability in the smoker. Chem Res Toxicol. 2002;15(11):1331–50.
13. Hsu SC, Pollack RL, Hsu AF, Going RE. Sugars present in tobacco extracts. J Am Dent Assoc. 1980;101(6):915–8.
14. Going RE, Hsu SC, Pollack RL, Haugh LD. Sugar and fluoride content of various forms of tobacco. J Am Dent Assoc. 1980;100(1):27–33.
15. Rustemeier K, Stabbert R, Haussmann HJ, Roemer E, Carmines EL. Evaluation of the potential effects of ingredients added to cigarettes. Part 2: chemical composition of mainstream smoke. Food Chem Toxicol. 2002;40(1):93–104.
16. Electronic Model Tobacco Control (EMTOC). Amsterdam: Rijks Instituut voor Volksgezondheiden Milieu and Ministerium für Gesundheit und Frauen; 2015.
17. Newson JR. Evaluation of competitive cigarettes. Liggett & Myers; 1975 (https://industrydocuments.library.ucsf.edu/tobacco/docs/#id=zslf0014, accessed 10 February 2016).
18. Gordon DL. PM's global strategy: Marlboro product technology. A summary developed with input from Batco, Batcf, Souza Cruz & B&W. Winston Salem (NC): Brown & Williamson; 1992.
19. Soussy S, El-Hellani A, Baalbaki R, Salman R, Shihadeh A, Saliba NA. Detection of 5-hydroxymethylfurfural and furfural in the aerosol of electronic cigarettes. Tob Control. 2016;25(Suppl 2):ii88–93.
20. van Nierop LE, Pennings JLA, Schenk E, Kienhuis AS, Talhout R. Analysis of manufacturer's information on tobacco product additive use. Tob Regul Sci. (accepted).
21. Lawler TS, Stanfill SB, Zhang L, Ashley DL, Watson CH. Chemical characterization of domestic oral tobacco products: total nicotine, pH, unprotonated nicotine and tobacco-specific N-nitrosamines. Food Chem Toxicol. 2013;57:380–6.
22. van Nierop LE, Talhout R. Sugar as tobacco additive tastes "bitter". J Addict Res Ther. 2016;7(4):2.
23. Jansen EHJM, Cremers J, Borst S, Talhout R. Simple determination of sugars in cigarettes. Anal Bioanal Tech. 2014;5(6):219.
24. Clarke MB, Bezabeh DZ, Howard CT. Determination of carbohydrates in tobacco products by liquid chromatography–mass spectrometry/mass spectrometry: a comparison with ion chromatography and application to product discrimination. J Agric Food Chem. 2006;54(6):1975–81.
25. Agaku IT, Ayo-Yusuf OA, Vardavas CI, Connolly G. Predictors and patterns of cigarette and smokeless tobacco use among adolescents in 32 countries, 2007–2011. J Adolesc Health. 2014;54(1):47–53.
26. Attitudes of Europeans towards tobacco and electronic cigarettes (Special Eurobarometer 458). Brussels: European Commission; 2017.
27. Busch C, Streibel T, Liu C, McAdam KG, Zimmermann R. Pyrolysis and combustion of tobacco in a cigarette smoking simulator under air and nitrogen atmosphere. Anal Bioanal Chem. 2012;403(2):419–30.
28. Baker RR, Bishop LJ. The pyrolysis of tobacco ingredients. J Anal Appl Pyrolysis. 2004;71(1):223–311.
29. Cheah NP. Volatile aldehydes in tobacco smoke: source, fate and risk. Maastricht: Maastricht University; 2016:2016846.
30. Cheah NP, Borst S, Hendrickx L, Cremers H, Jansen E, Opperhuizen A et al., Effect of adding sugar to Burley tobacco on the emission of aldehydes in mainstream tobacco smoke. Tob Regul Sci. 2018;4(2):61–72.
31. Baker RR. Sugars, carbonyls and smoke. Food Chem Toxicol. 2007;45(9):1783–6.
32. Baker RR. The generation of formaldehyde in cigarettes – overview and recent experiments. Food Chem Toxicol. 2006;44(11):1799–822.
33. Coggins CR, Wagner KA, Weley MS, Oldham MJ. A comprehensive evaluation of the toxicology of cigarette ingredients: carbohydrates and natural products. Inhal Toxicol. 2011;23(Suppl 1):13–40.
34. Hahn J, Schaub J. Influence of tobacco additives on the chemical composition of mainstream

smoke. Beitr Tabakforsch Int. 2014;24(3):101–16.
35. O'Connor RJ, Hurley PJ. Existing technologies to reduce specific toxicant emissions in cigarette smoke. Tob Control. 2008;17(Suppl 1):i39–48.
36. Zilkey B, Court WA, Binns MR, Walker EK, Dirks VA, Basrur PK. Chemical studies on Canadian tobacco and tobacco smoke. 1. Tobacco, tobacco sheet and cigarette smoke. Chemical analyses on various treatments of Bright and Burley. Tob Int. 1982;184:83–9.
37. Phillpotts DF, Spincer D, Westcott DT. The effect of the natural sugar content of tobacco upon the acetaldehyde concentration found in cigarette smoke. London: British American Tobacco; 1974 (Bates Number: 505242960).
38. Shihadeh A, Salman R, Jaroudi E, Saliba N, Sepetdjian E, Blank MD et al. Does switching to a tobacco-free waterpipe product reduce toxicant intake? A crossover study comparing CO, NO, PAH, volatile aldehydes, tar and nicotine yields. Food Chem Toxicol. 2012;50(5):1494–8.
39. Al Rashidia M, Shihadeh A, Saliba NA. Volatile aldehydes in the mainstream smoke of the narghile waterpipe. Food Chem Toxicol. 2008;46(11):3546–9.
40. Fagan P, Pokhrel P, Herzog TA, Moolchan ET, Cassel KD, Franke AA et al. Sugar and aldehyde content in flavored electronic cigarette liquids. Nicotine Tob Res. 2018;20(8):985–92.
41. Pyles ST, Van Voris LP, Lotspeich FJ, McCarty SA. Sugar in chewing tobacco. N Engl J Med. 1981;304(6):365.
42. Vellappally, S, Fiala Z, Smejkalová J, Jacob V, Shriharsha P. Influence of tobacco use in dental caries development. Centr Eur J Public Health. 2007;15(3):116–21.
43. Tomar SL, Winn DM. Chewing tobacco use and dental caries among US men. J Am Dent Assoc. 1999;130(11):1601–10.
44. WHO Study Group on Tobacco Product Regulation. Report on the scientific basis of tobacco product regulation: fifth report of a WHO study group (WHO Technical Report Series, No. 989). Geneva: World Health Organization; 2015.
45. Burns DM, Dybing E, Gray N, Hecht S, Anderson C, Sanner T et al. Mandated lowering of toxicants in cigarette smoke: a description of the World Health Organization TobReg proposal. Tob Control. 2008;17(2):132–41.
46. Talhout, R, Schulz T, Florek E, van Benthem J, Wester P, Opperhuizen A. Hazardous compounds in tobacco smoke. Int J Environ Res Public Health. 2011;8(2):613–28.
47. Opinion on additives used in tobacco products (opinion 2). Tobacco additives II. Brussels: European Commission Scientific Committee on Emerging and Newly Identified Health Risks; 2016.
48. Willems EW, Rambali B, Vleeming W, Opperhuizen A, van Amsterdam JG. Significance of ammonium compounds on nicotine exposure to cigarette smokers. Food Chem Toxicol. 2006; 44(5):678–88.
49. Elson LA, Betts TE. Sugar content of the tobacco and pH of the smoke in relation to lung cancer risks of cigarette smoking. J Natl Cancer Inst. 1972;48(6):1885–90.
50. Wayne GF, Carpenter CM. Tobacco industry manipulation of nicotine dosing. Handb Exp Pharmacol. 2009;192:457–85.
51. Hammond, D, Fong GT, Cummings KM, O'Connor RJ, Giovino GA, McNeill A. Cigarette yields and human exposure: a comparison of alternative testing regimens. Cancer Epidemiol Biomarkers Prev. 2006;15(8):1495–501.
52. Jarvis MJ, Boreham R, Primatesta P, Feyerabend C, Bryant A. Nicotine yield from machine-smoked cigarettes and nicotine intakes in smokers: evidence from a representative population survey. J Natl Cancer Inst. 2001;93(2):134–8.
53. Bates C, Jarvis M, Connolly G. Tobacco additives: cigarette engineering and nicotine addiction. A survey of the additive technology used by cigarette manufacturers to enhance the appeal and addictive nature of their product. London: Action on Smoking and Health, Imperial Cancer Research Fund; 1999.

54. Paschke T, Scherer G, Heller W. Effects of ingredients on cigarette smoke composition and biological activity: a literature overview. Beitr Tabakforsch Int. 2002;20(3):107–47.
55. Foddai M, Dosia G, Spiga S, Diana M. Acetaldehyde increases dopaminergic neuronal activity in the VTA. Neuropsychopharmacology. 2004;29(3):530–6.
56. Villegier AS, Blanc G, Glowinski J, Tassin JP. Transient behavioral sensitization to nicotine becomes long-lasting with monoamine oxidases inhibitors. Pharmacol Biochem Behav. 2003;76(2):267–74.
57. Guillem K, Vouillac C, Azar MR, Parsons LH, Koob GF, Cador M et al. Monoamine oxidase inhibition dramatically increases the motivation to self-administer nicotine in rats. J Neurosci. 2005;25(38):8593–600.
58. Belluzzi JD, Wang R, Leslie FM. Acetaldehyde enhances acquisition of nicotine self-administration in adolescent rats. Neuropsychopharmacology. 2005;30(4):705–12.
59. Gray N. Reflections on the saga of tar content: why did we measure the wrong thing? Tob Control. 2000;9(1):90–4.
60. Henningfield JE, Benowitz NL, Connolly GN, Davis RM, Gray N, Myers ML et al., Reducing tobacco addiction through tobacco product regulation. Tob Control. 2004;13(2):132–5.
61. Herraiz T, Chaparro C. Human monoamine oxidase is inhibited by tobacco smoke: beta-carboline alkaloids act as potent and reversible inhibitors. Biochem Biophys Res Commun. 2005;326(2): 378–86.
62. Pfau W, Skog K. Exposure to beta-carbolines norharman and harman. J Chromatogr B. 2004;802(1):115–26.
63. Koob GF. Drugs of abuse: anatomy, pharmacology and function of reward pathways. Trends Pharmacol Sci. 1992;13(5):177–84.
64. Fowler JS, Logan J, Wang GJ, Volkow ND. Monoamine oxidase and cigarette smoking. Neurotoxicology. 2003;24(1):75–82.
65. Talhout R, Opperhuizen A, van Amsterdam JG. Role of acetaldehyde in tobacco smoke addiction. Eur Neuropsychopharmacol. 2007;17(10):627–36.
66. Berlin I, Anthenelli RM. Monoamine oxidases and tobacco smoking. Int J Neuropsychopharmacol. 2001;4(1):33–42.
67. Syrjanen K, Salminen J, Aresvuo U, Hendolin P, Paloheimo L, Eklund C et al. Elimination of cigarette smoke-derived acetaldehyde in saliva by slow-release L-cysteine lozenge is a potential new method to assist smoking cessation. A randomised, double-blind, placebo-controlled intervention. Anticancer Res. 2016;36(5):2297–306.
68. Spijkerman R, van den Eijnden R, van de Mheen D, Bongers I, Fekkes D. The impact of smoking and drinking on plasma levels of norharman. Eur Neuropsychopharmacol. 2002;12(1):61–71.
69. Rommelspacher H, Meier-Henco M, Smolka M, Kloft C. The levels of norharman are high enough after smoking to affect monoamine oxidase B in platelets. Eur J Pharmacol. 2002;441(1–2):115–25.
70. Breyer-Pfaff U, Wiatr G, Stevens I, Gaertner HJ, Mundle G, Mann K. Elevated norharman plasma levels in alcoholic patients and controls resulting from tobacco smoking. Life Sci. 1996;58(17):1425–32.
71. Rose JE, Mukhin AG, Lokitz SJ, Turkington TG, Herskovic J, Behm FM et al. Kinetics of brain nicotine accumulation in dependent and nondependent smokers assessed with PET and cigarettes containing ^{11}C-nicotine. Proc Natl Acad Sci U S A. 2010;107(11):5190–5.
72. Sanders E, Weitkunat R, Utan A, Dempsey R. Does the use of ingredients added to tobacco increase cigarette addictiveness? A detailed analysis. Inhal Toxicol. 2012;24(4):227–45.
73. Lee PN, Forey BA, Fry JS, Hamling JS, Hamling JF, Sanders EB et al. Does use of flue-cured rather than blended cigarettes affect international variation in mortality from lung cancer and COPD? Inhal Toxicol. 2009;21(5):404–30.
74. Ferreira CG, Silveira D, Hatsukami DK, Paumgartten FJ, Fong GT, Glória MB et al. The effect of to-

bacco additives on smoking initiation and maintenance. Cad Saude Publica. 2015;31(2):223–5.
75. Correa M, Salamone JD, Segovia KN, Pardo M, Longoni R, Spina L et al. Piecing together the puzzle of acetaldehyde as a neuroactive agent. Neurosci Biobehav Rev. 2012;36(1):404–30.
76. Hoffman AC, Evans SE. Abuse potential of non-nicotine tobacco smoke components: acetaldehyde, nornicotine, cotinine, and anabasine. Nicotine Tob Res. 2013;15(3):622–32.
77. Adriani W, Spijker S, Deroche-Gamonet V, Laviola G, Le Moal M, Smit AB et al. Evidence for enhanced neurobehavioral vulnerability to nicotine during periadolescence in rats. J Neurosci. 2003;23(11):4712–6.
78. Leffingwell JC. Nitrogen components of leaf and their relationship to smoking quality and aroma. Recent Adv Tob Sci. 1976;2:1–31.
79. Opinion on additives used in tobacco products (opinion 1). Brussels: European Commission Scientific Committee on Emerging and Newly Identified Health Risks; 2016.
80. Hoffmann D, Hoffmann I. The changing cigarette, 1950–1995. J Toxicol Environ Health. 1997;50(4):307–64.
81. Weeks WW. Relationship between leaf chemistry and organoleptic properties of tobacco smoke. In: Davis DL, Nielsen MT, editors. Tobacco: production, chemistry and technology. Oxford: Blackwell Science; 1999:304–12.
82. Fowles J. Chemical factors influencing the addictiveness and attractiveness of cigarettes in New Zealand. Wellington: Institute of Environmental Science and Research; 2001 (http://www.moh.govt.nz/notebook/nbbooks.nsf/0/D0B68B2D9CB811ABCC257B81000D9959/$file/chemicalfactorsaddictivenesscigarettes.pdf, accessed 15 May 2019).
83. Truth Tobacco Industry Documents. San Francisco (CA): University of California at San Francisco Library and Center for Knowledge Management (https://www.industrydocumentslibrary.ucsf.edu/tobacco/, accessed 15 May 2019).
84. WHO Framework Convention on Tobacco Control. 2013 Guidelines for implementation. Article 5.3, Article 8, Articles 9 and 10, Article 11, Article 12, Article 13, Article 14. Geneva: World Health Organization; 2013 (http://apps.who.int/iris/bitstream/10665/80510/1/9789241505185_eng.pdf, accessed 1 October 2018).
85. An act to amend the Tobacco Act, Bill C-32. Ottawa: Government of Canada; 2009.
86. Tobacco Reporting Regulations (SOR/2000-273). Tobacco and Vaping Products Act, Tobacco Act. Ottawa: Government of Canada; 2006.
87. Tobacco Product Directive (2014/40/EU) of the European Parliament and of the Council. Brussels: European Commission; 2014 (http://ec.europa.eu/health/tobacco/docs/dir_201440_en.pdf, accessed 15 May 2019).
88. Permitted additives to tobacco products in the United Kingdom. London: Department of Health; 2003.
89. Consultation on implementation of the revised Tobacco Products Directive (2014/40/EU), section 5.11. London: Department of Health; 2015 (https://assets.publishing.service.gov.uk/government/uploads/system/uploads/attachment_data/file/440991/TPD_Consultation_Doc.pdf, accessed 16 January 2019).
90. Richter AP. Best practices in implementation of Article 9 of the WHO FCTC Case study: Brazil and Canada (National Best Practices Series, Paper No. 5). Geneva: World Health Organization; 2015.
91. Report of the Working Group on Tobacco Additives. Brasilia: Agência Nacional de Vigilância Sanitária; 2014 (http://portal.anvisa.gov.br/documents/106510/106594/Report+Working+Group+Tobacco+Additives/b99ad2e7-23d9-4e88-81cd-d82c28512199, accessed 15 May 2019).
92. Family Smoking Prevention and Tobacco Control Act. Washington (DC): Food and Drug Administration; 2009 (http://www.fda.gov/tobaccoproducts/labeling/rulesregulationsguidance/ucm246129.htm#ingredients, accessed 1 January 2019).

93. Gottlieb S. FDA statement from FDA Commissioner Scott Gottlieb, MD, on proposed new steps to protect youth by preventing access to flavored tobacco products and banning menthol in cigarettes. Washington (DC): Food and Agriculture Administration; 2018 (https://www.fda.gov/NewsEvents/Newsroom/PressAnnouncements/UCM625884.htm, accessed 10 January 2019).
94. Munafo M. Understanding the role of additives in tobacco products. Nicotine Tob Res. 2016;18(7):1545.
95. Corey CG, Ambrose BK, Apelberg BJ, King BA. Flavored tobacco product use among middle and high school students – United States, 2014. Morb Mortal Wkly Rep. 2015;64(38):1066–70.
96. Cunningham FH Fiebelkorn S, Johnson M, Meredith C. A novel application of the Margin of Exposure approach: segregation of tobacco smoke toxicants. Food Chem Toxicol. 2011;49(11):2921–33.
97. Kienhuis AS, Staal YC, Soeteman-Hernández LG, van de Nobelen S, Talhout R. A test strategy for the assessment of additive attributed toxicity of tobacco products. Food Chem Toxicol. 2016;94:93–102.
98. van de Nobelen S, Kienhuis AS, Talhout R. An inventory of methods for the assessment of additive increased addictiveness of tobacco products. Nicotine Tob Res, 2016;18(7):1546–55.
99. Talhout R, van de Nobelen S, Kienhuis AS. An inventory of methods suitable to assess additive-induced characterising flavours of tobacco products. Drug Alcohol Depend. 2016;161:9–14.
100. European Commission. Priority list of additives contained in cigarettes and roll-your-own. Off J Eur Union. 2016:C/2016/2923.
101. Miao S, Beach ES, Sommer TJ, Zimmerman JB, Jordt SE. High-intensity sweeteners in alternative tobacco products. Nicotine Tob Res. 2016;18(11):2169–73

8. Updated priority list of toxicants in combusted tobacco products

Irina Stepanov, Associate Professor, Division of Environmental Health Sciences and Masonic Cancer Center, University of Minnesota, Minneapolis (MN), USA

Marielle Brinkman, Senior Research Scientist, College of Public Health, Ohio State University, Columbus, OH, USA

Contents
8.1 Introduction
8.2 Background of preparation of the priority list
    8.2.1 Criteria for selection of toxicants for the priority list
    8.2.2 Key decisions of the Conference of the Parties on priority contents and emissions of combustible tobacco
8.3 Overview of new scientific knowledge on toxicity
    8.3.1 Aldehydes (acetaldehyde, acrolein, formaldehyde, crotonaldehyde, propionaldehyde, butyraldehyde)
    8.3.2 Aromatic amines (3-aminobiphenyl, 4-aminobiphenyl, 1-aminonaphthalene, 2-aminonaphthalene)
    8.3.3 Hydrocarbons (benzene, 1,3-butadiene, isoprene, toluene)
    8.3.4 Polycyclic aromatic hydrocarbons (benzo[a]pyrene)
    8.3.5 Tobacco-specific N-nitrosamines
    8.3.6 Alkaloids (nicotine)
    8.3.7 Phenols (catechol, m-, p-, and o-cresols, phenol, hydroquinone, resorcinol)
    8.3.8 Other organic compounds (acetone, acrylonitrile, pyridine, and quinolone)
    8.3.9 Metals and metalloids (arsenic, cadmium, lead, mercury)
    8.3.10 Other constituents (ammonia, carbon monoxide, hydrogen cyanide, nitrogen oxides)
8.4 Availability of analytical methods
    8.4.1 Standardized WHO TobLabNet Methods for the analysis of priority toxicants
    8.4.2 Overview of methods for the remaining priority toxicants
8.5 Update on toxicant variations among brands
8.6 Criteria for future re-evaluation of toxicants on the list
8.7 Criteria for selection of new toxicants
8.8 Research needs and regulatory recommendations
    8.8.1 Research needs, data gaps and future work
    8.8.2 Regulatory recommendations and support to Parties
8.9 References

8.1 Introduction

At the fifth meeting of the COP of the WHO FCTC, WHO was requested to compile, make available to Parties and update a non-exhaustive list of toxic contents and emissions of tobacco products and provide advice about how such information could best be used by Parties. TobReg at its meeting in Rio de Janeiro, Brazil, on 4–6 December 2013 provided an updated list of priority toxicants for reporting and regulation.

In this paper, we re-evaluate the list, as it has been four years since the updated list was issued, and new knowledge on the subject matter has become available. In particular, this report includes:

- background information on the current TobReg priority list of toxicants, including the criteria used to select specific contents and emissions (8.2);
- an overview of new data related to the priority list, including new publications on the toxicity of specific constituents, new or modified analytical methods and new data on variations in levels of toxicant in combustible tobacco brands (8.3–8.5);
- a discussion of the criteria for future re-evaluation of toxicants on the list (8.6);
- a discussion of criteria for selecting new toxicants for the list (8.7); and
- research needs and regulatory recommendations for testing selected contents and emissions in combusted tobacco products (8.8).

Literature was searched primarily in the PubMed database and SciFinder, which retrieves citations from the Medline and CAplus databases. Relevant articles cited in publications obtained in the database search were also included. In addition, the websites of the United States Centers for Disease Control and Prevention, the United States Environmental Protection Agency, the Cooperation Centre for Scientific Research Relative to Tobacco (CORESTA) and other relevant websites that provide information on the toxicology of the constituents of interest and on methods were used.

8.2 Background of preparation of the priority list

Regulation of tobacco products requires establishment of a metric or set of metrics by which tobacco products can be assessed. The measurements that have been made most commonly for cigarettes have been machine-measured tar, nicotine and CO yields per cigarette, based on the ISO regimen and the United States Federal Trade Commission (FTC) method. It is well established, however, that such measures do not provide valid estimates of human exposure or toxicity and thereby cause harm by misleading smokers and most regulators (1, 2). New product assessment approaches for setting regulatory measures were therefore considered necessary.

The new approach proposed by TobReg requires quantification of the levels of known harmful toxicants that accompany a specified amount of nicotine, the principal addictive substance in smoke sought by smokers, as measured in a standardized machine testing regimen. Normalization of specific toxicants to the nicotine yield allows measurement of the toxicity of smoke generated in a standardized regimen rather than the quantity of smoke generated. Standardized measures of toxicant yields will allow regulators to reduce the levels of identified priority toxicants in tobacco smoke, consistent with other regulatory approaches that mandate reductions of known toxicants in products used by humans. Selection of high-priority smoke toxicants for product assessment is a critical step in this strategy.

The current non-exhaustive priority list of toxic contents and emissions of combustible tobacco products was drawn up by TobReg from among more than 7000 chemicals found in cigarette smoke, after an evaluation of lists of harmful and toxic chemicals associated with cancer, cardiovascular and pulmonary diseases published by several regulatory bodies. Consideration was also given to constituents in both the particulate and the gas phases of smoke and in different chemical classes. As the list represents only a small fraction of the total complex mixture of chemicals present in tobacco smoke, the overall toxicity of the emissions of tobacco products is not necessarily characterized by the toxicity of these chemicals. Regulation of a very large number of toxicants would, however, lead to significant distortions in the existing market and increase the complexity of regulatory oversight. Limiting the number of toxicants to a carefully selected priority list recognizes the practical reality of a regulatory structure.

8.2.1 Criteria for selection of toxicants for the priority list

The following criteria were used to select priority tobacco contents and emissions of cigarette smoke for testing, reporting and future regulation.

The presence of specific chemicals in cigarette smoke at levels that are toxic for smokers, as determined by well established scientific toxicity indices

Evidence of toxicity was the most important criterion. The toxicologically important constituents considered were those on the list reported by law to Health Canada for the year 2004 *(3)* and measured by Counts et al. *(4)* by comparing several international brands manufactured by Philip Morris. The list was selected by Health Canada to represent those constituents of tobacco smoke that contribute most to its toxicity.

Quantitative data for characterizing the hazards of the reviewed toxicants were generated by calculating "toxicant animal carcinogenicity indices" and non-cancer response indices with a modification of a simplified system presented by Fowles and Dybing *(5)*. For these calculations, published toxicant yields (obtained with the modified intense smoking regimen) were normalized per milligram of nicotine and multiplied by cancer and non-cancer potency factors. "Cancer potency factors" were defined as T_{25} per milligram ($1/T_{25}$), where T_{25} is the long-term daily dose that will produce tumours at a specific tissue site above the background rate in 25% of animals *(6)*. For non-cancer potency factors, the long-term reference exposure levels published by the California Office of Environmental Health Hazard Assessment (USA) in February 2005 (http://www.oehha.ca.gov/air/chronic_rels/AllChrels.html) were used. Toxicant yields measured in a consistent manner with the modified intense smoking regimen are available from three sources: Counts et al. *(4)*, a set of Canadian brands *(3)* and a set of Australian brands *(7)*.

Variations in concentrations among cigarette brands that are substantially greater than the variation in repeated measurements of the toxicant in a single brand

Mandated reductions are likely to have the greatest impact on lowering the mean levels per milligram of nicotine for those toxicants with the broadest variation in levels around the midpoint. The variation in levels of different toxicants can be expressed as the coefficient of variation, which is the standard deviation of the measurement across brands divided by the mean value for all the brands. As the reproducibility of testing can vary substantially for different toxicants and with different testing methods, repeated measurements are used to estimate the mean value for a brand, and the variation of the repeated measurements defines the CI around that mean value. A second approach is direct determination of the range of toxicants in brands by determining the maximum and minimum values for each toxicant in the brand data set, expressed as the ratio of the median value for that toxicant. This does not include adjustment for variation in replicate measurements of the toxicant. Another useful approach to assessing variations in toxicant levels is to compare the mean levels of toxicants in different data sets. This analysis can identify levels of toxicants that differ within the same brand sold in different markets or by different manufacturers, which indicates that it is clearly possible to manufacture cigarettes that yield lower levels of that particular toxicant.

The availability of technology to reduce the concentration of a given toxicant in smoke, should an upper limit be mandated

The ability of the tobacco industry to modify their products to comply with lower toxicant levels is another factor to be considered in selecting toxicants for tobacco product regulation. For instance, the levels of the carcinogenic TSNAs NNK and NNN in smoke can be reduced by changing agricultural practices, curing and tobacco blending *(8)*. The levels of volatile toxicants, such as acetaldehyde, acrolein and formaldehyde, can be reduced by reducing the concentration of sugars added to tobacco or using charcoal filters or other filter modifications *(9)*. Reductions in benzo[*a*]pyrene yields can be achieved by treating, extracting or modifying the tobacco blend *(10)*.

8.2.2 Key decisions of the Conference of the Parties on priority contents and emissions of combustible tobacco

Table 8.1 summarizes the key COP decisions and TobReg's progress in regulating the contents and emissions of combustible tobacco products under WHO FCTC Articles 9 and 10. The initial list of nine priority emissions identified by TobReg at the third COP and for which validation of testing methods in cigarette smoke was recommended are listed in Table 8.2. These toxicants were selected among 43 toxicologically relevant compounds for which the hazard indices were calculated and other criteria were reviewed as described above. Nicotine was not included in the list of emissions but was recommended for testing in tobacco filler (contents). Following the COP mandate, analytical methods for these selected priority emissions were developed and validated by the WHO Tobacco Laboratory Network (TobLabNet).

Table 8.1. History of COP decisions and TobReg progress relevant to the testing and reporting of emissions

COP session	COP decisions	TobReg progress and reports on contents and emissions
COP 1 (Geneva, 2006)	COP/1/INF.DOC./3: set up a working group to prepare guidelines pursuant to Articles 9 and 10 of the WHO FCTC. First phase to comprise testing and measuring tobacco product contents and emissions.	
COP 2 (Bangkok, 2007)	COP/2/DIV/9: continue the work of TobReg, including on product characteristics, such as design features, to the extent that they affect the objectives of the WHO FCTC.	COP/2/8: TobReg presented a progress report that included the proposal that the provisional list of cigarette emissions consist of 44 "Hoffmann analytes".
COP 3 (Durban, 2008)	COP/3/DIV/3: decisions relevant to the priority list included: • validate, within 5 years, analytical chemical methods for testing and measuring cigarette contents and emissions identified as priorities in the progress report of the working group, with two smoking regimens (ISO and modified intense method with blocked ventilation holes). • when appropriate, design and validate methods for testing and measuring product characteristics identified in the progress report of the working group.	COP/3/6: the new progress report presented by TobReg identified: • three contents for which methods for testing and measuring should be validated as a priority (nicotine, ammonia, humectants); • nine emissions for which methods for testing and measuring should be validated as a priority (NNK, NNN, acetaldehyde, acrolein, benzene, benzo[a]pyrene, 1,3-butadiene, CO, formaldehyde); • the smoking regimens for validation of the test methods: (i) ISO 3308:2000 and (ii) a modified intense method with blocked ventilation holes; and • a provisional list of product characteristics for testing and disclosure.
COP 4 (Punta del Este, 2010)	COP/4/DIV/6: continue validation of analytical chemical methods for testing and measuring cigarette contents and emissions.	COP/4/INF.DOC./2: three analytical methods have been validated: • CO in emissions; • NNN and NNK in emissions; and • nicotine content of tobacco.
COP 5 (Seoul, 2012)	COP/5/DIV/5: continue validation of the analytical chemical methods for testing and measuring cigarette contents and emissions; compile a non-exhaustive list of toxic contents and emissions of tobacco products.	COP/5/INF.DOC./1: work in progress: • validation of methods for humectants and ammonia in cigarette tobacco filler; • validation of the method for benzo[a]pyrene in mainstream cigarette smoke.

COP 6 (Moscow, 2014)	COP6(12): finalize validation of the analytical chemical methods for testing and measuring cigarette contents and emissions.	COP/6/14: TobReg proposed: • a non-exhaustive list of 39 toxicants in tobacco products for monitoring and eventual regulation; • a shorter list of nine toxicants for mandated lowering; • development of the following standardized methods by TobLabNet: cadmium and lead content; nicotine in the smoke of waterpipes (shisha); and nicotine, NNN, NNK and benzo[a]pyrene in smokeless tobacco products.
COP 7 (Delhi, 2016)	COP7(14): finalize validation of the analytical chemical methods for aldehydes and VOCs in cigarette emissions; assess the availability of validated analytical methods for the extended list of toxicants in contents and emissions of tobacco products, as reported in FCTC/COP/6/14.	COP/7/INF.DOC./1: method validation is completed for all mandated contents and emissions, except for aldehydes and VOCs.
COP 8 (Geneva, 2018)	COP8(21): encourage Parties to acknowledge and implement TobLabNet methods.	COP/8/8: Method validation is completed for aldehydes and VOCs. All mandated contents and emissions are now validated. TobReg identified the following opportunities for extending the list of toxicants in contents and emissions: • extend SOPs to include the remaining aldehydes and VOCs; • prepare a SOP for the metal content of cigarette tobacco filler; • devise methods for the analysis of waterpipe tobacco and charcoal.

CO: carbon monoxide; COP: Conference of the Parties; ISO: International Organization for Standardization; NNK: 4-(methylnitrosamino)-1-(3-pyridyl)-1-butanone; NNN: N'-nitrosonornicotine; SOP: standard operating procedure; VOC: volatile organic compound.

Table 8.2. Emissions of combusted tobacco products considered and evaluated for inclusion in the lists of priorities for testing, reporting and regulation

	Carcinogenicity or toxicity data		Inclusion on priority lists		
			COP/3/6	COP/6/14	
Toxicants evaluated by TobReg	TACI	TNCRI	Toxicants selected for testing and measuring	Expanded priority list for monitoring and regulation	Proposed for mandatory reduction
Alkaloids					
Nicotine[a]	–	–	X	X	–
Aldehydes					
Acetaldehyde	6.1	67.1	X	X	X
Acrolein	–	1099	X	X	X
Formaldehyde	–	19.8	X	X	X
Crotonaldehyde[a]	–	–	–	X	–
Propionaldehyde	–	–	–	X	–
Butyraldehyde	–	–	–	X	–
Aromatic amines					
3-Aminobiphenyl	–	–	–	X	–
4-Aminobiphenyl[a]	–	–	–	X	–
1-Aminonaphthalene	0.00036	–	–	X	–

2-Aminonaphthalene	0.00068	–	–	x	–	
Hydrocarbons						
Benzene	2.6	0.64	X	x	x	
1,3-Butadiene	9.9	2.4	X	x	x	
Isoprene	3.7	–	–	x	–	
Styrene[b]	–	0.01	–	–	–	
Toluene	–	0.22	–	x	–	
PAHs						
Benzo[a]pyrene[a]	0.0086	–	X	x	x	
TSNAs[a]						
NNK	3.4	–	X	x	x	
NNN	0.29	–	X	x	x	
NAB	–	–	–	x	–	
NAT	–	–	–	x	–	
Phenols						
Catechol	0.58	–	–	x	–	
m- and p-Cresol	–	0.01	–	x	–	
o-Cresol	–	0.01	–	x	–	
Phenol	–	0.07	–	x	–	
Hydroquinone	1.2	–	–	x	–	
Resorcinol	–	–	–	x	–	
Other organic compounds						
Acetone	–	–		x		
Acrylonitrile	1.4	2.1		x		
Quinoline	–	–		x		
Pyridine	–	–		x		
Metals and metalloids						
Arsenic[c]	–	0.16				
Cadmium	1.7	2.6		x		
Chromium[b]	–	–				
Lead	0.00	–		x		
Mercury	–	0.02		x		
Nickel[b]	–	–				
Selenium[b]	–	–				
Other constituents						
Ammonia	–	0.07		x		
CO[a]	–	1.3	X	x	x	
Hydrogen cyanide	–	17.2		x		
Nitrogen oxides	–	3.1		x		

CO: carbon monoxide; NAB: N′-nitrosoanabasine; NAT: N′-nitrosoanatabine; NNK: 4-(methylnitrosamino)-1-(3-pyridyl)-1-butanone; NNN: N′-nitrosonornicotine; PAH: polycyclic aromatic hydrocarbon; TACI: toxicant animal carcinogenicity index; TNCRI: toxicant non-cancer response index; TSNA: tobacco-specific N-nitrosamine.
[a] Justification for inclusion on the priority lists (initial and/or expanded) of some toxicants: nicotine was not on the initial priority list of emissions but was recommended for testing in tobacco (contents) and was subsequently listed with other toxicants in the extended list. Benzo[a]pyrene was included despite its low TACI because it is a proxy for the family of PAHs found in smoke and because there is a wealth of evidence for the carcinogenicity of many of these PAHs. The toxicant 4-aminobiphenyl was added because it is a human carcinogen, although experimental data did not allow proper calculation of T_{25}. NNK and NNN were included as they had already been identified in the first report on mandating reductions in toxicant yields (12). Crotonaldehyde was included because of its reactive α,β-unsaturated aldehyde structure, although a tolerable level value was lacking. CO was also included even though it has a relatively low toxicant non-cancer response index, as it is thought to be mechanistically related to cardiovascular disease.
[b] Originally considered (11) but not included in the final recommended list because it occurs at low levels and is not considered to contribute appreciably to hazard indices.
[c] Added after the expanded list was reported in document COP/6/14.

In response to the subsequent COP request to compile a non-exhaustive list of toxic contents and emissions of tobacco products, TobReg at its meeting in Rio de Janeiro, Brazil, on 4–6 December 2013 re-evaluated the priority toxicant list by examining the lists of harmful and toxic chemicals published by several regulatory bodies, including Health Canada, the National Institute for Public Health and the Environment in the Netherlands (RIVM) and the FDA, and reviewing the list of toxicants assessed in an earlier WHO technical report on the scientific basis of tobacco product regulation *(11)*. TobReg subsequently drew up a non-exhaustive list by adding toxicants that satisfied the criteria for sufficient evidence of carcinogenicity or with known respiratory or cardiac toxicity, variations in levels in different brands in different countries that were readily measurable and possibilities for lowering yields in a product (Table 8.2). The list, composed of 37 such toxicants plus nicotine, was presented to the COP for discussion at its sixth session. Only a few toxicants on the originally assessed list of 43 constituents were not included in the updated list. For instance, chromium, nickel and selenium occur either at low levels or could not be quantified in the three analysed data sets; therefore, although these elements are quite toxic, they were not considered to contribute appreciably to hazard indices and were not included.

In the same report (COP/6/14), TobReg proposed that the nine compounds originally selected for method validation (emissions) should be considered for mandatory reduction. TobReg concluded that these compounds are the most hazardous toxicants in cigarette smoke that could be reduced in emissions; they represent different chemical families of toxicants and different phases of smoke, are toxic to the pulmonary and cardiovascular systems and are carcinogenic. The nine toxicants are: acetaldehyde, acrolein, benzene, benzo[*a*] pyrene, 1,3-butadiene, CO, formaldehyde, NNK and NNN (Table 8.2). The remaining toxicants on the priority list of 39 should be measured and reported.

8.3 Overview of new scientific knowledge on toxicity

8.3.1 Aldehydes (acetaldehyde, acrolein, formaldehyde, crotonaldehyde, propionaldehyde, butyraldehyde)

Aldehydes are toxic to the respiratory system *(13)*. Acetaldehyde and formaldehyde are also toxic to the cardiovascular system and are respiratory tumorigens *(14, 15)*. Acrolein is an intense irritant, is toxic to lung cilia and has been proposed as a lung carcinogen *(16, 17)*. Crotonaldehyde is a potent irritant and a weak hepatocarcinogen and forms DNA adducts in the human lung *(18)*.

New studies on mechanisms of toxicity and carcinogenicity

The genotoxicity of acetaldehyde, acrolein and formaldehyde in human lung cells was demonstrated in the γH2AX assay, which detects early cellular response to DNA double-strand breaks and is used as a biomarker of DNA damage in human epidemiology (19). All three aldehydes were genotoxic in a dose- or time-dependent manner, acrolein having the strongest potential to induce DNA damage, followed by formaldehyde and acetaldehyde. A computational fluid dynamics modelling study of simulated human oral breathing, representative aldehyde yields from cigarette smoke and lifetime average daily doses indicated that the order of concern for human exposure is acrolein > formaldehyde > acetaldehyde (20). An analysis of the mode of action of the toxicity of acrolein in the lower respiratory system, reflecting the exposure of smokers who inhale tobacco smoke, suggested that the mechanisms of acrolein toxicity include oxidative stress, chronic inflammation, necrotic cell death, necrosis-induced inflammation, tissue remodelling and destruction and subsequent loss of lung elasticity and enlarged lung airspaces (21). These processes are consistent with the inflammation and necrosis in the middle and lower regions of the respiratory tract that occur in chronic obstructive pulmonary disease. The findings of another study suggested that acrolein contributes to the dysfunctional innate immune responses observed in the lung during cigarette smoking (22). In addition, acrolein may contribute to bladder carcinogenesis in smokers. Lee et al. (23) analysed acrolein–deoxyguanosine adducts in normal human urothelial mucosa and in bladder tumour tissues and measured their mutagenicity in human urothelial cells. The adduct levels in both types of tissue were higher than the levels of those due to the known bladder carcinogen 4-aminobiphenyl, and they induced mutation signatures and spectra that appeared to be much more mutagenic than those due to 4-aminobiphenyl. The levels of acrolein–deoxyguanosine were two times higher in bladder tumour tissues than in normal cells.

A study of transcription responses to aldehyde exposure in human lung carcinoma lines showed a different ranking from that in the studies described above (24). Formaldehyde gave the strongest response, with differential expression of 66 genes (mostly involved in apoptosis and DNA damage) by more than 1.5 times. Acetaldehyde dysregulated 57 genes, while acrolein caused upregulation of only one gene involved in oxidative stress.

While formation of DNA adducts is a critical factor in the mutagenicity and carcinogenesis of aldehydes, protein modifications may also play a role. For instance, in addition to N^2-ethylidene-2′-deoxyguanosine, its major DNA adduct, acetaldehyde also formed hybrid protein adducts with malondialdehyde, a product of lipid peroxidation induced by toxic constituents and reactive species in cigarette smoke. Formation of hybrid malondialdehyde–acetaldehyde protein adducts in the lung has been shown to initiate several pathological conditions, including inflammation and inhibition of wound healing (25). Antibodies (immunoglobulins) involved in the immune response to these adducts (i.e. IgM, IgG and IgA) are predictive of progression of atherosclerotic disease and of cardiovascular events such as acute myocardial infarction or coronary artery bypass grafting *(26)*.

It is also important to consider the effect of gene polymorphisms on the response of humans to toxicants. For instance, risk assessments of acetaldehyde have so far been based on thresholds determined in animal toxicology studies, which do not account for the genetic-epidemiological and biochemical evidence that *ALDH2*-deficient humans are highly vulnerable to the carcinogenic effects of this chemical. In a study on exposure to acetaldehyde through alcohol consumption, *ALDH2* inactivity was associated with odds ratios of up to 7 for head-and-neck and oesophageal cancers *(27)*. A study of a Japanese cohort of patients with myocardial infarction and stable angina and matching controls suggested that the inactive *ALDH2* genotype may increase the risk for myocardial infarction in smokers *(28)*.

Additional potential mechanisms of the toxicity and carcinogenicity of aldehydes have been investigated. Acrolein has been shown to affect the metabolism and fate of aromatic amines by interacting with enzymes responsible for their acetylation *(29)*. A study of potential interactions between acrolein and the antiretroviral drug zidovudine, which is used widely in limited-resource countries but is associated with hepatotoxicity, strongly suggest that exposure to acrolein through smoking and/or alcohol consumption can contribute to the major mechanisms by which zidovudine induces hepatotoxicity *(30)*. Another study suggested that exposure to acetaldehyde and formaldehyde may induce carcinogenesis in carriers of the *BRCA2* mutation *(31)*. Mutations in *BRCA2*, which is a tumour suppressor gene, increase the risk for breast cancer and are also associated with susceptibility to ovarian, pancreatic and other cancers. Both aldehydes selectively deplete *BRCA2* by proteasomal degradation, which may trigger spontaneous mutagenesis during DNA replication *(31)*.

New studies on health effects other than respiratory toxicity and carcinogenicity

Addictiveness: there is accumulating evidence that acetaldehyde also contributes to the addictiveness of tobacco *(32)*. Studies of self-administration in laboratory animals add to understanding of the behavioural correlates of acetaldehyde administration and possible interactions with the neurotransmitters for motivation, reward and stress-related responses, such as dopamine and endocannabinoids *(33)*. Another possible mechanism is via the formation of harman, a condensation product of acetaldehyde and amines in saliva. Harman is a monoamine oxidase inhibitor and can help to maintain behavioural sensitization to nicotine. In support of this hypothesis, it was shown that smokers given a lozenge containing L-cysteine (an amino acid that reacts with acetaldehyde) had higher rates of smoking cessation than those treated with placebo *(34)*. It should be noted, however, that harman is also present in tobacco and cigarette smoke.

Cardiovascular effects: the cardiovascular effects of acrolein have been reviewed *(35)*. In vitro and in vivo, cardiovascular tissues appear to be particularly sensitive to the toxic effects of acrolein, which can generate oxidative stress in the heart, form protein adducts with myocyte and vascular endothelial cell proteins and cause vasospasm. Therefore, chronic exposure to acrolein could contribute to cardiomyopathy and cardiac failure in humans. This conclusion is supported by the results of a study of mice exposed chronically to acrolein, which suggested that even relatively low exposure to acrolein, such as that from second-hand smoke or e-cigarettes, could increase cardiovascular risk by reducing endothelium repair, suppressing immune cells or both *(36)*. Exposure to acrolein was also assessed in 211 participants in the Louisville Healthy Heart Study who were at moderate-to-high risk for cardiovascular disease *(37)*. Exposure to acrolein was associated with platelet activation, suppression of circulating angiogenic cell levels and increased cardiovascular disease risk.

Glucose and lipid metabolism: in rats, exposure to acrolein caused metabolic impairment by inducing hyperglycaemia and glucose intolerance, accompanied by a significant increase in the level of corticosterone and modest but insignificant increases in the level of adrenaline *(38)*. The association between urinary levels of acrolein metabolites and diabetes and biomarkers of insulin resistance was investigated in 2027 adults who participated in the 2005–2006 National Health and Nutrition Examination Survey in the USA *(39)*. A positive association was found between the biomarkers analysed, suggesting a potential role of acrolein in the etiology of type-2 diabetes and insulin resistance in humans.

Other effects: acrolein may affect intestinal epithelial integrity. After oral exposure of mice to acrolein, damage was found to the intestinal epithelial barrier, resulting in increased permeability and subsequent bacterial translocation *(40)*. Song et al. *(41)* provided additional evidence that acrolein is a risk factor for otitis media.
The developmental and reproductive toxicity of aldehydes has also been studied. Amiri & Turner-Henson *(42)* reported the results of a cross-sectional study of a convenience sample of 140 healthy pregnant women, in which the relation between exposure to formaldehyde and fetal growth during the second trimester was examined. A linear regression model showed that the dichotomized level of formaldehyde exposure (< 0.03 and > 0.03 parts per million) was a significant predictor of biparietal diameter percentile after control for maternal race ($P < .006$). In a study in an animal model, exposure to 5 mg/kg acrolein in utero resulted in significantly decreased testosterone synthesis in male offspring *(43)*.

New epidemiological evidence

A study of more than 2200 smokers in the Multiethnic Cohort study in the USA showed that urinary biomarkers of acrolein and crotonaldehyde exposure were significantly different among five ethnic groups *(44)*. Native Hawaiians had the highest and Latinos the lowest geometric mean levels of urinary biomarkers of both aldehydes after adjustment for confounders. These results are consistent with the findings of an epidemiological study in this cohort, in which native Hawaiians had a higher risk for lung cancer and Latinos a lower risk as compared with Whites, for the same number of cigarettes smoked *(45)*. These results suggest that acrolein and crotonaldehyde may be involved in the etiology of lung cancer in smokers.

8.3.2 Aromatic amines (3-aminobiphenyl, 4-aminobiphenyl, 1-aminonaphthalene, 2-aminonaphthalene)

Exposure to aromatic amines is associated with bladder cancer *(46)*, and 2-aminonaphthalene and 4-aminobiphenyl are known human bladder carcinogens *(15)*. In one study, the mean level of 4-aminobiphenyl was 4.8 times higher in smokers (> 20 cigarettes/day) than nonsmokers, reaffirming that tobacco smoke is a major source of exposure to this carcinogen *(47)*.

Aromatic amines may present a risk for non-urological cancer. A cohort of 224 male workers at a single factory who were followed from 1953 to 2011 had high risks for both lung cancer and bladder cancer associated with exposure to 2-aminonaphthalene *(48)*. A systematic review and meta-analysis of studies on the risk for lung cancer among workers exposed to 2-aminonaphthalene showed a significantly increased lung cancer risk, the effect estimates being similar in studies with and without concomitant occupational exposure to other lung toxicants and carcinogens *(49)*. Exposure to benzidine and 2-aminonaphthalene was monitored in both studies.

8.3.3 Hydrocarbons (benzene, 1,3-butadiene, isoprene, toluene)

Benzene and butadiene cause cancers of the haematolymphatic organs and are classified as known human carcinogens *(15)*. Isoprene causes tumours at various sites in laboratory animals *(50)*. 1,3-Butadiene and toluene are respiratory toxicants, and toluene is also toxic to the central nervous system and is a reproductive toxicant. These compounds are present in high amounts in cigarette smoke and probably play a role in lung cancer in smokers *(14, 15, 51)*.

The tobacco smoke-related health effects of 1,3-butadiene and the possible impacts of risk reduction strategies were evaluated from the ratio (margin of exposure) between the most sensitive toxicity end-point and appropriate estimates of exposure to 1,3-butadiene in mainstream and second-hand tobacco smoke *(52)*. The authors concluded that the risks for cancer (leukaemia) and non-cancer (ovarian atrophy) could be significantly reduced by lowering the levels of 1,3-butadiene in smoke. They proposed that analysis of the margin of exposure is a practical means for assessing the impact of risk reduction strategies on human health. In a review of non-cancer health effects of benzene, it was concluded that exposure to benzene can have numerous outcomes in the reproductive, immune, nervous, endocrine, cardiovascular and respiratory systems *(53)*.

Cardiovascular effects

A potential association between exposure to benzene and an increased risk for cardiovascular disease has been investigated in mice and humans *(54)*. The effects of benzene in mice were assessed by direct inhalation, while the effects in humans were assessed in 210 people with mild-to-high risks for cardiovascular disease risk by measuring urinary biomarkers. Mice had significantly reduced levels of circulating angiogenic cells and higher plasma levels of low-density lipoprotein than control mice that breathed filtered air. In humans, smokers and people with dyslipidaemia had higher exposure to benzene, which was negatively correlated with populations of circulating angiogenic cells and associated with the risk for cardiovascular disease assessed on the Framingham risk score.

New epidemiological evidence

The results of a study with a sample of adult participants in the Gulf Long-term Follow-up Study in the USA during 2012 and 2013 suggested that ambient exposure to benzene and toluene is associated with haematological effects, including decreased haemoglobin and mean corpuscular haemoglobin concentration and increased red cell distribution width *(55)*. The evidence was particularly strong for benzene.

In the Multiethnic Cohort study, benzene uptake was compared in smokers in five different ethnic groups by analysing urinary *S*-phenylmercapturic acid, a specific biomarker of exposure to benzene *(56)*. African Americans had significantly higher and Japanese Americans significantly lower levels of *S*-phenylmercapturic acid than Whites. While benzene is not generally considered to cause lung cancer, these differences are consistent with those for lung cancer risk in this cohort.

8.3.4 Polycyclic aromatic hydrocarbons (benzo[a]pyrene)

PAHs are formed during incomplete combustion of organic matter and always occur as mixtures. Many PAHs are potent carcinogens or toxicants in laboratory animals *(57)*, and many are present in cigarette smoke, including the prototypic PAH benzo[*a*]pyrene, classified as a human carcinogen by a working group convened by the International Agency for Research on Cancer (IARC) *(57, 58)*.

PAHs are widely accepted to be major contributors to lung cancer in smokers *(57, 59–61)*. A study of benzo[*a*]pyrene and the TSNA NNK in A/J mice showed dose-dependent tumorigenesis at lower doses than previously reported *(62)*.

Zaccaria and McClure *(63)* analysed published studies on benzo[*a*]pyrene and other PAHs and found a correlation between the derived relative potency factors for immune suppression for nine PAHs and their potency factors for cancer, confirming previous observations of an association between the carcinogenicity of PAHs and immunosuppression.

Environmental exposure to benzo[*a*]pyrene is correlated with impaired learning and memory in adults and poor neurodevelopment in children. A comprehensive literature review was conducted to determine the potential mechanism of neurotoxicity by benzo[*a*]pyrene *(64)*. The results suggest that neurotoxic effects are observed at lower exposure than those associated with cancer. It was proposed that benzo[*a*]pyrene binding to the aryl hydrocarbon receptor results in loss of neuronal activity and decreased long-term potentiation, compromising learning and memory.

8.3.5 Tobacco-specific N-nitrosamines

TSNAs are important constituents of tobacco products, and two, NNK and NNN, are probably responsible for cancers of the lung, pancreas, oral cavity and oesophagus in tobacco users *(65, 66)*. Both have been classified as human carcinogens by working groups at IARC *(66, 67)*. These nitrosamines are formed from tobacco alkaloids during tobacco processing. The amounts that are formed depend on tobacco type, nitrate content and tobacco processing techniques, resulting in wide variation in the amounts in various cigarette brands *(65, 68–70)*. In smokers, increases in smoke TSNA yields due to changes in cigarette design, including filter ventilation, were accompanied by an increase in the incidence of lung adenocarcinoma, the type of lung cancer that is induced by NNK in laboratory animals *(71, 72)*.

New evidence in laboratory animals

The carcinogenicity of NNK and its metabolite NNAL was studied in male F-344 rats treated for 70 weeks. Both compounds induced a high incidence of lung tumours, and metastases were observed from primary pulmonary carcinomas to the pancreas. The results clearly demonstrate the potent pulmonary carcinogenicity and DNA damaging activity of NNK and NNAL in rats *(73)*. In another study by the same group *(74)*, NNN induced 96 oral cavity tumours and 153 oesophageal tumours in 20 male F-344 rats treated chronically with this carcinogen in their drinking water. This study showed for the first time the carcinogenic potency of NNN in the oesophagus and identified NNN as a strong oral cavity carcinogen present in tobacco.

New evidence in humans

Consistent with the data on carcinogenicity in animals, a positive association was found between prospectively measured exposure to NNN and NNK and risks for oesophageal and lung cancer, respectively, in smokers in the Shanghai Cohort Study. Additional analyses of the same cohort indicated that exposure to NNK was not associated with oesophageal cancer, and exposure to NNN was not associated with the risk of smokers for lung cancer *(75)*. Together, these results reaffirm the organ specificity of NNN and NNK towards the oesophagus and the lung, respectively, in smokers, consistent with the findings in F-344 rats. The uptake of NNK was also measured in 2252 smokers in the Multiethnic Cohort study *(76)*. After adjustment for age at urine collection, sex, creatinine and total nicotine equivalents, a marker of total nicotine uptake, the highest exposure to NNK was found for African Americans and the lowest for Japanese Americans. These findings are consistent with the findings on lung cancer risk of smokers in these groups.

Mechanistic studies and studies in laboratory animals indicate that DNA adduct formation is a critical step in NNN- and NNK-induced carcinogenesis. Higher levels of these adducts were found in the oral cells of smokers with head-and-neck squamous cell carcinoma than in cancer-free smokers *(77)*.

8.3.6 Alkaloids (nicotine)

Nicotine is the major known addictive agent in tobacco and cigarette smoke *(78)* and is a key driver of tobacco use. Nicotine content can greatly influence the extent and pattern of product use and can also define the category of users to whom a product will appeal. Further, it affects exposure to other toxicants and carcinogens in the product.

New evidence on the effects of nicotine reduction in cigarette smoke

Reduction of the nicotine content of cigarettes to a minimal or non-addictive levels has been recommended for consideration by TobReg and proposed by the FDA as an approach for reducing or eliminating the use of combusted tobacco products *(79, 80)*. The feasibility of this approach is confirmed by the results of a number of studies. In a double-blind, parallel, randomized clinical trial conducted between June 2013 and July 2014 at 10 sites in the USA, 840 participants were randomly assigned to smoke either their usual brand of cigarettes or one of six types of investigational cigarettes with a nicotine content ranging from 0.4 mg/g to 15.8 mg/g of tobacco (comparable to the nicotine content of commercial brands) for 6 weeks *(81)*. At the end of the study, participants assigned to cigarettes containing 2.4, 1.3 or 0.4 mg of nicotine per gram of tobacco had smoked a smaller average number of cigarettes per day than those assigned to their usual brand or to cigarettes containing 15.8 mg/g (P <.001). Cigarettes with a lower nicotine content than control cigarettes reduced exposure to and dependence on nicotine and also reduced craving during abstinence from smoking, without significantly increasing the expired CO level or total puff volume, suggesting minimal compensation.

In a follow-up randomized, parallel-arm 8-week study by Hatsukami et al. *(82)*, smokers who were unwilling to quit were randomly assigned to normal or VLNC cigarettes, and use of alternative nicotine products, smoking behaviour and biomarkers of tobacco exposure were assessed. The offer of and instructions for use of reduced-nicotine cigarettes led to reduced smoking rates, reduced biomarkers of exposure to smoke toxicants and greater use of alternative tobacco or nicotine products than continued use of cigarettes with normal nicotine.

Health effects other than addiction

Nicotine can contribute to acute cardiovascular events and accelerated atherogenesis in tobacco users, probably due to stimulation of the sympathetic nervous system, decreasing coronary blood flow, impairment of endothelial function and other pharmacological effects *(83)*. A systematic review of studies in humans and animals on the health effects of exposure to nicotine during pregnancy and adolescence indicated that nicotine contributes critically to adverse effects, including reduced pulmonary function, auditory processing defects and impaired infant cardiorespiratory function, and it may contribute to cognitive and behavioural deficits in later life *(84)*. The study also found that exposure to nicotine during adolescence is associated with deficits in working memory, attention and auditory processing, as well as increased impulsivity and anxiety, and studies in animals suggest that nicotine increases the liability for addiction to other drugs.

8.3.7 Phenols (catechol, m-, p- and o-cresols, phenol, hydroquinone, resorcinol)

Catechol is a co-carcinogen *(85)* that is present in high amounts in cigarette smoke *(66)*. The United States Environmental Protection Agency has classified *m*-, *o*- and *p*-cresols as possible human carcinogens on the basis of genetic toxicity and increased incidences of skin and nasal tumours in rodents. Cresols are also respiratory toxicants. Hydroquinone is mutagenic in vitro and in vivo, including in a study in which it had significant genotoxicity in vitro in the γH2AX assay *(19)*. It has been shown reproducibly to induce benign neoplasms in the kidneys of male F-344 rats dosed orally, but the data on humans are inadequate. Phenol is a respiratory toxicant, elicits cardiovascular effects and is a tumour promoter. Resorcinol was reported to have a range of toxic effects in various studies and is considered to be a respiratory toxicant.

8.3.8 Other organic compounds (acetone, acrylonitrile, pyridine, quinolone)

Acetone is a respiratory toxicant and can irritate the respiratory tract. Acrylonitrile is a respiratory toxicant and is classified by the IARC as possibly carcinogenic to humans (Group 2B) *(14)*. It readily forms adducts with proteins, and the levels of such adducts are higher in smokers than in nonsmokers *(14, 86)*. Acrylonitrile is also mutagenic in some assays *(14)*. Few studies have been conducted on the toxicity of pyridine; some suggest effects on the respiratory tract, the central nervous system and the liver. Quinoline is an irritant after acute exposure and showed liver toxicity and carcinogenicity in animals. No significant new toxicological findings on these organic compounds have been identified since 2013.

8.3.9 Metals and metalloids (arsenic, cadmium, lead, mercury)

Arsenic and cadmium are human carcinogens *(87)*. These metals are lung carcinogens and could also play a role in bladder (arsenic) and kidney (cadmium) cancers. Arsenic also has cardiovascular and reproductive effects, and cadmium is a neurological and respiratory toxicant. Lead is a neurological, reproductive and cardiovascular toxicant and a probable human carcinogen *(88)*. Mercury is classified by IARC in Group 2B and is also a reproductive toxicant *(89)*. These elements are present in varying amounts in cigarette smoke *(66)* and smokeless tobacco *(90)*; the levels are probably affected by their concentrations in the soils in which the tobacco is grown. Pinto et al. *(91)* detected significantly higher levels of arsenic, lead and cadmium in the lung tissue of smokers than nonsmokers. The exposure of young children to second-hand tobacco smoke can result in blood lead levels that are associated with decreased IQ and cognition *(92)*.

Carcinogenicity

The association between long-term exposure to cadmium, measured in urine, and mortality from cancer was investigated in 3792 American Indians in Arizona, Oklahoma and North and South Dakota (USA) who participated in the Strong Heart Study during 1989–1991. Exposure to cadmium was associated with mortality from all cancers and with that from cancers of the lung and pancreas *(93)*. In a review of the literature, Feki-Tounsi & Hamza-Chaffai *(94)* concluded from the available in vitro and epidemiological studies that exposure to cadmium is associated with an increased risk of bladder cancer and may be involved in urothelial toxicity and carcinogenesis.

Neurotoxicity

Cadmium and lead are neurotoxicant components of tobacco smoke and could contribute to depression associated with smoking. The association between blood cadmium and lead levels and current depressive symptoms was investigated in a cross-sectional study of adult participants in the National Health and Nutrition Examination Survey 2011–2012 in the USA (N = 3905). Blood cadmium was associated with higher odds for depressive symptoms in male participants aged 20–47 years, and blood lead, cigarette smoking and obesity were associated with depressive symptoms in female participants in this age range *(95)*. The finding of effects on spatial and nonspatial working memory, anxiety-related behaviour and motor activities in female adolescent mice exposed to cadmium and/or nicotine supports these conclusions *(96, 97)*. Nicotine and cadmium increased the metabolism, food intake and weight of treated mice as compared with controls. Nicotine administration increased motor function, while cadmium decreased motor activity. Both compounds induced a reduction in the memory index. Combined treatment with nicotine and cadmium induced decreases in weight and motor activity, increased anxiety and a significant decrease in nonspatial working memory.

Cardiovascular disease

It has been hypothesized that cadmium contributes to the cardiovascular risk associated with smoking by injuring vascular endothelial cells *(98)*. In a systematic review of epidemiological studies of the association between exposure to cadmium and cardiovascular disease, the pooled relative risks (95% CIs) for cardiovascular disease, coronary heart disease, stroke and peripheral arterial disease were: 1.36 (1.11, 1.66), 1.30 (1.12, 1.52), 1.18 (0.86, 1.59) and 1.49 (1.15, 1.92), respectively *(99)*. With the experimental evidence, the review supports an association between exposure to cadmium and cardiovascular disease, especially coronary heart disease.

Other health effects

Mice exposed to sodium arsenite for 90 days showed increased micronucleated polychromatic erythrocytes and an increase in genotoxic and germ-cell toxic effects in liver, kidney and intestinal tissues as compared with a control group. Combined treatment with smokeless tobacco extract induced a significant increase in sperm head abnormality as compared with either material alone *(100)*.

The effect of exposure to lead in cigarette smoke on fetal growth was studied by measuring blood lead concentrations in 150 healthy pregnant women *(101)*. The birth weight of the infants of mothers who smoked was significantly lower than that of infants born to non-smoking mothers ($P < .001$) and was negatively correlated with lead levels in plasma ($r = -0.38$; $P < .001$) and in whole blood ($r = -0.27$; $P < .001$).

Exposure to mercury has been hypothesized to lead to metabolic syndrome and diabetes mellitus. Reviews of the literature indicated that, while epidemiological data suggest a possible association between total mercury concentrations in biological matrices and the incidence of these health outcomes, the relation is not consistent *(102, 103)*. A more comprehensive review of the health effects associated with exposure to mercury suggests that chronic exposure, even to low concentrations of mercury, can cause cardiovascular, reproductive and developmental toxicity, neurotoxicity, nephrotoxicity, immunotoxicity and carcinogenicity *(104)*.

8.3.10 Other constituents (ammonia, carbon monoxide, hydrogen cyanide, nitrogen oxides)

Ammonia is a respiratory irritant and toxicant, which increases the prevalence of respiratory symptoms, asthma and impaired pulmonary function in various industrial and agricultural settings *(105–108)*. Limited studies suggest that people with asthma are more sensitive to the respiratory effects of ammonia *(108, 109)*. CO is a well established cardiovascular toxicant, which competes with oxygen for binding to haemoglobin. In smokers, it is considered to reduce oxygen delivery, cause endothelial dysfunction and promote the progression of atherosclerosis and other cardiovascular diseases *(110–112)*.

Hydrogen cyanide is a well-known toxic agent, its primary targets being the cardiovascular, respiratory and central nervous systems. It acts by inhibiting cytochrome oxidase in the respiratory chain. Cigarette smoke can reduce detoxification of hydrogen cyanide, leading to chronic exposure of smokers to this toxin and consequent amblyopia, retrobulbar neuritis, sterility and a potential contribution to impaired wound healing *(112, 113)*.

Nitrogen oxides are respiratory and cardiovascular toxicants. Nitric oxide, the primary form in fresh cigarette smoke, induces vasodilation and causes DNA strand breaks and lipid peroxidation, possibly contributing to carcinogenesis *(114)*. It may also contribute to nicotine addiction by increasing nicotine absorption, reducing symptoms of stress and increasing post-synaptic dopamine levels *(115)*. Nitrogen dioxide is a pulmonary irritant.

8.4 Availability of analytical methods

8.4.1 Standardized WHO TobLabNet methods for the analysis of priority toxicants

To ensure implementation of Articles 9 and 10 of the WHO FCTC, laboratory capacity must be available that meets the highest standards of excellence, transparency, reliability and credibility *(116)*. Standardized, reliable, accurate analytical methods are required by laboratories to conduct the scientifically rigorous testing required for tobacco products globally *(117)*. Consensus on a set of methods may partly depend on successful transfer of such methods to other laboratories. Established, validated WHO TobLabNet standard operating procedures (SOPs) as of October 2018 are summarized in Table 8.3.

Table 8.3. Standardized WHO TobLabNet standard operating procedures for the analysis of priority toxicants

SOP	Title	Priority toxicants	Year of publication
SOP 01	Intense smoking of cigarettes	Necessary for mainstream cigarette smoke generation	2012
SOP 02	Validation of analytical methods of tobacco product contents and emissions	Describes method validation	2017
SOP 03	Determination of tobacco-specific nitrosamines in mainstream cigarette smoke under ISO and intense smoking conditions	Mainstream NNK, NNN (emissions)	2014
SOP 04	Determination of nicotine in cigarette tobacco filler	Nicotine (content)	2014
SOP 05	Determination of benzo[*a*]pyrene in mainstream cigarette smoke	Mainstream benzo[*a*]pyrene (emissions)	2015
SOP 06	Determination of humectants in cigarette tobacco filler	Propylene glycol, glycerol, triethylene glycol (content)	2016
SOP 07	Determination of ammonia in cigarette tobacco filler	Ammonia (content)	2016
SOP 08	Determination of aldehydes in mainstream cigarette smoke under ISO and intense smoking conditions	Acetaldehyde, acrolein, formaldehyde (emissions)	2018
SOP 09	Determination of volatile organics in mainstream cigarette smoke under ISO and intense smoking conditions	1,3-Butadiene, benzene (emissions)	2018
SOP 10	Determination of nicotine and carbon monoxide in mainstream cigarette smoke under intense smoking conditions	Mainstream nicotine and CO (emissions)	2016

CO: carbon monoxide; ISO: International Organization for Standardization; NNK: 4-(methylnitrosamino)-1-(3-pyridyl)-1-butanone; NNN: *N*′-nitrosonornicotine; SOP: standard operating procedure.

8.4.2 Overview of methods for the remaining priority toxicants

The priority toxicants for which standardized WHO TobLabNet SOPs are not available are described below, with published methods for their quantification in tobacco products and mainstream smoke.

Aldehydes (butyraldehyde, crotonaldehyde, propionaldehyde).

Ding et al. *(118)* reported a significant improvement in terms of ease of use, efficiency and environmental friendliness to previous methods for the analysis of acetaldehyde, acrolein, acetone, crotonaldehyde, formaldehyde, propionaldehyde and methyl ethyl ketone. The method involves use of a double filter (Cambridge pads, one pre-treated with dinitrophenylhydrazine and one dry) to trap and derivatize carbonyls in mainstream smoke simultaneously. The hydrazones are then quantified by ultra-high-pressure liquid chromatography (LC) coupled with tandem mass spectrometry (MS/MS) with a 4-min run time. The accuracy of the method, determined from spiking at low, medium and high levels, was 83–106%, and its precision, determined from 30 replicate measurements of reference cigarette smoke (3R4F), was < 20% relative standard deviation for all target analytes. The limits of detection of the WHO priority toxicants were 2.7 μg for acetaldehyde, 0.1 μg for acrolein, 0.2 μg for crotonaldehyde, 2.4 μg for formaldehyde and 0.6 µg for propionaldehyde. This method is ideal for regulatory analyses, as it can be used with linear smoking machines, which greatly increases sample throughput.

Researchers at the China National Tobacco Corporation established a simple method for rapid determination of acrolein, acetone, propionaldehyde, crotonaldehyde, butanone and butyraldehyde in mainstream smoke *(119)*. Although the analysis time is short (4 min), data mining of the full and daughter scans must be conducted to overcome difficulties in separating and quantifying isomers of acetone–propionaldehyde and butanone–butyraldehyde with this atmospheric pressure chemical ionization MS/MS technique. The results obtained with a reference cigarette (3R4F, 35-mL puff volume, eight puffs) are consistent with those reported in the literature. The limits of detection of the WHO priority toxicants were 0.007 µg/L for acrolein, 0.021 µg/L for acetone, 0.008 µg/L for propionaldehyde, 0.004 µg/L for crotonaldehyde, 0.012 µg/L for butanone and 0.006 µg/L for butyraldehyde. Modifications were made to the chemical ionization source in order to introduce the gas sample directly into the ionization region.

In another reported method, mainstream smoke from cigarettes was collected in sulfuric acid (20%) and ascorbic acid (25 mmol/L) impingers *(120)*, and a dispersive liquid–liquid microextraction method was used to simultaneously extract the solution and convert benzaldehyde, butyraldehyde and furfural into their hydrazone derivatives, which were then quantified by high-performance LC. The matrix spike recovery was 88.0–109%, and the relative standard deviation for inter- and intra-day assays were < 8.50%. The limits of detection of the WHO priority toxicants were 14.2 µg/L for benzaldehyde, 21.3 µg/L for butyraldehyde and 7.92 for furfural. This method is quicker, simpler and less expensive than previously published liquid phase microextraction methods for the analytes tested.

Aromatic amines (1-aminonaphthalene, 2-aminonaphthalene, 3-aminobiphenyl, 4-aminobiphenyl)

In another method, the gas (0.6 M hydrochloric acid impinger) and particulate (glass-fibre filter) phases of mainstream smoke were collected, and the resulting acid solution was used to extract the filter in an ultrasonic bath *(121)*. The extract was cleaned by passage through two solid-phase extraction cartridges of different polarity and separated on a phenyl-hexyl column. Nine aromatic amines were quantified by LC–MS/MS. The method was validated by spiking the target analytes into the extract from a reference cigarette (3R4F) at three levels (low, medium, high). Recovery of the WHO priority toxicants was 84.8–97.8%, and intra- and inter-day precision was < 9% and < 14%, respectively. The limits of detection for the WHO priority toxicants were 0.08 ng/cigarette for 1-aminonaphthalene, 0.09 ng/cigarette for 2-aminonaphthalene, 0.05 ng/cigarette for 3-aminobiphenyl and

0.03 ng/cigarette for 4-aminobiphenyl. This method is similar to that reported by Xie et al. *(122)*, but gave better separation and baseline resolution of the target aromatic amines.

Deng et al. *(123)* used a novel dispersive liquid–liquid microextraction clean-up method and faster ultraperformance convergence chromatography MS/MS to quantify nine aromatic amines in mainstream smoke. This approach resulted in an instrument run time of 5 min, a substantial improvement over the 30 min required for the high-performance LC method of Zhang et al. *(121)*. In this method, supercritical CO_2 is used as the primary mobile phase to ensure higher flow rates with a lower pressure drop across the column and faster run times. The method was validated with a reference cigarette (3R4F). The recovery from spiked (low, medium, high) mainstream smoke extracts of the 3R4F was 69.4–120% for the WHO priority toxicants. The limits of detection of the WHO priority toxicants were 0.29 ng/cigarette for 1-aminonaphthalene, 0.21 ng/cigarette for 2-aminonaphthalene, 0.02 ng/cigarette for 3-aminobiphenyl and 0.03 ng/cigarette for 4-aminobiphenyl.

Hydrocarbons (isoprene, toluene)

Sampson et al. *(124)* reported an automated, high-throughput method for accurate quantification of a broad range of hazardous VOCs, including acrylonitrile, benzene, butyraldehyde, 1,3-butadiene and toluene, in mainstream smoke. The gas phase was collected into a polyvinyl fluoride gas sampling bag, and isotopically labelled analogues were added as internal standards to account for any losses due to handling and ageing. The particle phase was collected onto a glass-fibre filter and spiked with internal standards. After heating, the bags and filter headspace were sampled by automated solid-phase microextraction and quantified by gas chromatography (GC) with mass selective detection. The method was validated with reference cigarettes (1R5F and 3R4F) smoked under ISO and HCI puffing regimens. The results were comparable to those in other reports, except for toluene, which was found at a level ~30% lower than in previous reports. The inter- and intra-run precision was ≤a20%. There was a high correlation (0.97) between toluene and *m*- and *p*-xylene levels in mainstream smoke. This method was later used by the same group to quantify the target analytes in 50 US commercial brands *(125)*. The limits of detection of the WHO priority toxicants were not reported in either article; however, the lowest calibration standards used were 0.14 parts per billion by volume for all WHO priority toxicants except benzene, which was 0.1 parts per billion by volume.

Phenols (catechol, m-, p-, o-cresols, phenol, hydroquinone, resorcinol)

A simple, precise method for quantifying WHO priority phenols in mainstream smoke was reported by Saha et al. *(126)*. The method involves single-drop microextraction and LC–MS/MS to reduce organic solvent consumption, shorten sample preparation time and eliminate the derivatization steps required in previously published GC–MS methods. The limits of detection of the WHO priority toxicants were 0.30 ng/mL for catechol, 0.05 ng/mL for o- and p-cresol, 0.15 ng/mL for phenol, 0.30 ng/mL for hydroquinone and 0.20 ng/mL for resorcinol. The method reported by Wu et al. at Labstat International *(127)* with LC–MS/MS improved on the Health Canada methods (T-114, T-211) by use of a shorter analytical column with smaller particle size to resolve the three cresol positional isomers fully.

Other organic compounds (acetone, acrylonitrile, pyridine, quinoline)

A multi-analyte method that includes acrylonitrile is discussed above in the section on hydrocarbons, and a multi-analyte method that includes acetone is discussed in the section on aldehydes. Qualitative two-dimensional chromatography time-of-flight MS methods for the analysis of organic compounds in the vapour *(128)* and particle phases *(129)* of mainstream tobacco smoke were reported by tobacco industry researchers. These methods are not suitable for testing emissions but could be useful for regulatory purposes to differentiate mainstream tobacco smoke emissions resulting from different product designs or to identify adulteration.

Metals and metalloids (arsenic, cadmium, lead, mercury)

Traditional sample collection methods for quantifying cadmium in mainstream smoke were critically evaluated *(130)*. A platinum trap was used to determine breakthrough in two different sample collection methods: electrostatic precipitation and Cambridge filter pad. The detection limit of cadmium that had passed through the primary traps to the platinum traps was 0.30 ng/cigarette. Cadmium breakthrough from the Cambridge filter was significant (4–23% of the sample) but was negligible with electrostatic precipitation (< 1%). This technique was used by Fresquez et al. *(131)* in a high-throughput method for the analysis of mercury in the gas phase of cigarette and little cigar smoke emitted by ISO and HCI puffing regimens. The method is much quicker, simpler and more environmentally friendly than previously reported methods, because it eliminates the need for impingers, strong oxidizing agents (e.g. permanganate) and strong acids (e.g. sulfuric acid). The limit of detection was 0.27 ng cadmium/g in tobacco and 0.097 ng mercury/cigarette. The same team used high-throughput methods to establish the amounts of all WHO priority toxicant metals in the tobacco filler (content) of 50 brands of cigarettes on the United States market *(132)*. Microwave digestion was used for sample preparation, except for mercury, which was introduced directly into the analyser. The analytical instrumentation included a direct combustion analyser for mercury and an inductively coupled plasma MS for all other elements except arsenic and selenium, which were run separately with a different inductively coupled plasma MS method. The accuracy of the method was determined with standard tobacco reference materials, and the results were within the target ranges and the results for WHO priority toxicant metals, except for lead, which was slightly lower (4%) than the lower range of the target value for the Oriental tobacco reference material. The limits of detection of the WHO priority toxicants in tobacco were 0.082 µg/g (magnetic sector) and 0.25 µg/g (quadrupole) for arsenic, 0.23 µg/g for cadmium, 0.16 µg/g for lead and 0.00063 µg/g for mercury.

Other constituents (hydrogen cyanide, nitric oxides)

Hydrogen cyanide was quantified in mainstream smoke by collecting smoke samples on a sodium hydroxide-treated Cambridge filter pad and quantified by ion chromatography with pulsed amperometric detection *(133)*. In comparison with the traditional continuous flow analyser method, the filter-based method offers convenience, greater accuracy and a wider linear quantification range. The method performance was evaluated by spiking hydrogen cyanide onto filters containing particulate from commercial cigarettes made of blended and flue-cured tobaccos. The mean recovery was 97%, and the intra- and inter-day relative standard deviation was < 6%. The limit of detection of hydrogen cyanide was 3 µg/L for a 25-µL injection loop. Mahernia et al. *(134)* used polarography to measure hydrogen cyanide concentrations in 50 large cigars and cigarettes purchased in the Islamic Republic of Iran. Mainstream smoke was drawn from each tobacco product with a pump and passed through a glass tube containing sodium hydroxide (0.1 M). The concentration of the cyanide was determined by fortifying the solution with known quantities of cyanide and using the standard addition method. The hydrogen cyanide level in the tobacco products was 17.6–1550 µg per rod. Limits of detection for hydrogen cyanide were not provided. No recent methods for the analysis of nitric oxides were found.

8.5 Update on variations in toxicants among brands

The variations in toxicants reported in mainstream tobacco smoke generated by standardized machine smoking methods, the HCI puffing regimen for cigarettes and little cigars and the Beirut puffing regimen for waterpipe tobacco are summarized in Table 8.4. Variations in toxicants in the tobacco filler (content) of cigarettes, little cigars, waterpipe tobacco and waterpipe charcoal are summarized in Table 8.5. More research is needed on emissions of WHO priority toxicants from the increasingly popular, newer tobacco products (in comparison with cigarettes) in order to determine the 10th, 25th, 75th and 90th percentiles, so that global comparisons can be made as a first step towards establishing limits for the WHO priority toxicants.

Table 8.4. Variations in WHO priority toxicants, expressed as mass per rod or session (waterpipe), in mainstream smoke from combustible products

Toxicant	Cigarettes[a]	Little cigars[a]	Waterpipes
Aldehydes			
Acetaldehyde (µg)	1198–1947 *(118)* (USA)	NR	120–2520 *(135)*
Acetone (µg)	385–724 *(118)* (USA)	NR	20.2–118 *(135)*
Acrolein (µg)	107–169 *(118)* (USA)	105–185 *(136)* (USA)	10.1–892 *(135)*
Crotonaldehyde (µg)	25–72 *(118)* (USA)	NR	NR
Formaldehyde (µg)	55–108 *(118)* (USA)	NR	36–630 *(135)*
Propionaldehyde (µg)	116–232 *(118)* (USA)	NR	5.71–403 *(135)*
Aromatic amines			
1-Aminonaphthalene (ng)	4.31–33.37 *(121)* (China)	NR	6.20 *(135)*
2-Aminonaphthalene (ng)	0.89–4.60 *(121)* (China)	NR	2.84 *(135)*
3-Aminobiphenyl (ng)	1.46–4.68 *(121)* (China)	NR	< 3.30 *(137)* (Germany)
4-Aminobiphenyl (ng)	0.38–5.95 *(121)* (China)	NR	
Hydrocarbons			
Benzene	58.7–128.5 *(125)* (USA)	NR	271 *(135)*
1,3-Butadiene	67.8–118.3 *(125)* (USA)	NR	Present *(138)*[a]
Isoprene	NR	NR	4.00 *(135)*
Toluene	81–178 *(125)* (USA)	NR	9.92 *(135)*
Polycyclic aromatic hydrocarbons			
Benzo[*a*]pyrene (ng)	NR	17–32 *(139)* (USA)	LOD–307 *(135)*
Pyrene[b] (ng)	NR	NR	30–12 950 *(135)*
Tobacco-specific *N*-nitrosamines			
NAB (ng)	NR	NR	8.45 *(135)*
NAT (ng)	NR	NR	103 (135)
NNK (ng)	NR	438–995 *(139)* (USA)	LOD–46.4 *(135)*
NNN (ng)	NR	434–1550 *(139)* (USA)	34.3 *(135)*
Alkaloids			
Nicotine, mg	NR	1.85–6.15 *(140)* (USA)	> 0.01–9.29 *(135)*
Phenols			
Catechol (µg)	49.6–118 *(127)* (Canada)	NR	166–316 *(135)*
m-Cresol (µg)	1.93–6.92 *(127)* (Canada)	NR	NR
p-Cresol (µg)	5.28–17.7 *(127)* (Canada)	NR	NR
m+*p*-Cresol (µg)	7.24–24.6 *(127)* (Canada)	NR	2.37–4.66 *(135)*
o-Cresol (µg)	2.21–7.93 *(127)* (Canada)	NR	2.93–4.41 *(135)*
Phenol (µg)	8.34–32.7 *(127)* (Canada)	NR	3.21–58.0 *(135)*
Hydroquinone (µg)	60.1–140 *(127)* (Canada)	NR	21.7–110.7 *(135)*
Resorcinol (µg)	1.25–2.46 *(127)* (Canada)	NR	1.69–1.87 *(135)*
Other organic compounds			
Acetone	NR	NR	20.2–118 *(135)*
Acrylonitrile (µg)	19.7–37.7 *(125)* (USA)	NR	Present *(138)*[a]
Pyridine	NR	NR	4.76 *(135)*
Quinoline	NR	NR	BLQ
Metals and metalloids			
Arsenic (ng)	NR	NR	165 *(135)*
Cadmium (ng)	NR	NR	< 100 *(141)* (UK)

Chromium[c] (ng)	0.60–1.03 *(142)* (USA)	NR	250–1340 *(135)*
Lead (ng)	NR	NR	200–6870 *(135)*
Mercury (ng)	NR	5.2–9.6 *(131)* (USA)	< 100 *(141)* (UK)
Other priority toxicants			
Carbon monoxide, mg	NR	NR	5.7 - 367 *(135)*
Hydrogen cyanide	NR	NR	NR
Nitric oxide, mg	140.9–266.8 *(143)* (Russian Federation)	NR	0.325–0.440 *(135)*

BLQ: below limit of quantification; LOD: below the limit of detection; NAB: N'-nitrosoanabasine; NAT: N'-nitrosoanatabine; NNK: 4-(methylnitrosamino)-1-(3-pyridyl)-1-butanone; NNN: N'-nitrosonornicotine; NR: not reported; UK: United Kingdom.
[a] Health Canada intense puffing regimen: 55 mL puff volume, 2-s puff duration every 30 s, 100% vent blocking).
[b] Present, presence in mainstream smoke extrapolated from urinary biomarker.
[c] Currently not a TobReg priority toxicant.

Table 8.5. Variations in WHO priority toxicant content in tobacco filler or charcoal (waterpipe) of combustible tobacco products, expressed as mass of chemical per mass of tobacco (g)

Toxicant	Cigarettes	Little cigars	Waterpipe tobacco	Waterpipe charcoal
Nicotine (mg)	16.2–26.3 *(144)* (USA) 10.5–17.8 *(145)* (Pakistan)	10.3–19.1 *(144)* (USA)	0.48–2.28 *(146)* (USA) 1.8–41.3 *(147)* (Jordan)	NA
Benzo[*a*]pyrene (ng)	NR	NR	NR	1–26 *(148, 149)* (Lebanon, USA)
Pyrene (ng)	NR	NR	NR	6–170 *(148, 149)* (Lebanon, USA)
Arsenic (μg)	0.22–0.36 *(132)* (USA)	NA	0.062 (0.023)[b] *(150)* (Egypt)	0.018 (0.013) *(150)* (Egypt)
Cadmium (μg)	1.0–1.7 *(132)* (USA)	NA	0.34 (0.007) *(150)* (Egypt)	0.005 (0.017)[b] *(150)* (Egypt)
Chromium[a] (μg)	1.3–3.1 *(132)* (USA)	NR	0.15–0.37 *(151)* (USA, Middle East)	0.161–8.32 *(152)* (Worldwide)
Lead (μg)	0.60–1.16 *(132)* (USA)	NR	0.15 (0.008)[b] *(150)* (Egypt)	0.97 (0.01)[b] *(150)* (Egypt)
Mercury (μg)	0.013–0.020 *(132)* (USA)	0.017–0.024 *(132)* (USA)	NR	NR

NR: not reported. [a] Currently not a TobReg priority toxicant. [b] Mean (relative standard deviation).

8.6 Criteria for future re-evaluation of toxicants on the list

The criteria for selecting priority toxicants and the overall recommendations of TobReg indicate four areas that should guide periodic re-evaluation of the priority list.

Toxicological profile of specific contents and emissions: the priority list of toxicants was identified to help WHO FCTC Parties and Member States to fulfil the requirements of Articles 9 and 10 of the WHO FCTC. Monitoring of the priority list is viewed as an initial step towards regulating the contents and emissions of combustible tobacco products. Up-to-date scientific knowledge on the toxicity of the selected emissions is critical to ensure the continuous relevance of the priority list and to inform future regulatory measures.

Validated methods: the contents and emissions of priority toxicants must be measured according to WHO TobLabNet SOPs to ensure accurate, reproducible results that can serve as a basis for regulatory measures. Development of standardized methods by TobLabNet is guided by robust, well-characterized, published protocols, which are reviewed, tested and further validated by TobLabNet. Advances in analytical chemistry and toxicological data are reviewed periodically by TobLabNet to identify priorities for testing methods.

Brand variations: according to the partial guidelines on Articles 9 and 10 to the WHO FCTC, the tobacco industry is ultimately responsible for generating and reporting data on priority contents and emissions for each brand. Such data can be generated according to the WHO TobLabNet SOPs, as noted in the decision of the eighth COP of the WHO FCTC in October 2018 (FCTC/COP8(21)). Emissions of many priority chemicals have not been measured systematically in all commercially available products. Thus, periodic review of new reports on the chemical composition of combustible tobacco products establishes a frame of reference for understanding variations in emissions among product types and the lowest practicably achievable levels of priority toxicants.

Correlations among constituents within a brand: a positive correlation among several toxicants would suggest that the levels of one could serve as a proxy for several other toxicants in a regulatory strategy. A negative correlation would suggest that mandatory lowering of one toxicant could result in increases in the levels of negatively correlated toxicants. While caution should be exercised in relying on these assumptions, close monitoring of the relations among constituents could provide insight for future selection of toxicants for mandatory reduction.

8.7 Criteria for selection of new toxicants

Addition of new constituents to the non-exhaustive priority list should be considered in the future. Several criteria for the selection of new toxicants are listed below.

Substantial evidence of risk to human health: for instance, Talhout et al. *(153)* listed 98 hazardous smoke constituents on the basis of the risk for human inhalation. In addition, the FDA identified 93 HPHCs for potential reporting and regulation *(154)*. The HPHC list comprises chemicals and chemical compounds in tobacco or tobacco smoke that are taken into the body (inhaled, ingested or absorbed) and cause or have the potential to cause direct or indirect harm to users and non-users of tobacco products. The FDA selected constituents that had been identified as known, probable or possible human carcinogens by either the IARC, the United States Environmental Protection Agency or the United States National Toxicology Program, as well as those identified by the United States National Institute for Occupational Safety and Health as potential occupational carcinogens. Constituents identified by the United States Environmental Protection Agency or the Agency for Toxic Substances and Disease Registry as having adverse respiratory or cardiac effects, constituents identified by the California Environmental Protection Agency as reproductive or developmental toxicants and constituents reported in the literature as contributing to abuse liability were also included. Forty toxic and carcinogenic constituents are listed on both the FDA HPHC list and by Talhout et al. *(153)* but are not currently on the TobReg list (Table 8.6). These constituents could be prioritized for future consideration by TobReg, for instance by applying hazard indices and generating data on variations in their levels among brands.

Table 8.6. Constituents to be considered for future inclusion on the TobReg priority toxicant list

Constituent	Health effect *(154)*	Risk associated with inhalation (mg/m³) *(153)*
Acetamide	CA	5.0×10^{-4} (cancer)
Acrylamide	CA	8.0×10^{-3} (cancer)
3-Amino-1,4-dimethyl-5H-pyrido[4,3-b]indole	CA	1.4×10^{-6} (cancer)
2-Amino-3-methyl)-9H-pyrido[2,3-b]indole	CA	2.9×10^{-5} (cancer)
2-Amino-6-methyldipyrido[1,2-a:3',2'-d]imidazole	CA	7.1×10^{-6} (cancer)
2-Amino-3-methylimidazo[4,5-f]quinoline	CA	2.5×10^{-5} (cancer)
2-Amino-9H-pyrido[2,3-b]indole	CA	8.8×10^{-5} (cancer)
2-Aminodipyrido[1,2-a:3',2'-d]imidazole	CA	2.5×10^{-5} (cancer)
Benz[a]anthracene	CA, CT	9.1×10^{-5} (cancer)
Beryllium	CA	4.2×10^{-6} (cancer)
Chromium	CA, RT, RTD	8.3×10^{-7} (cancer)
Chrysene	CA, CT	9.1×10^{-4} (cancer)
Cobalt	CA, CT	5.0×10^{-4} (respiratory)
Dibenz[a,h]anthracene	CA	8.3×10^{-6} (cancer)
Dibenzo[a,e]pyrene	CA	9.1×10^{-6} (cancer)
Dibenzo[a,h]pyrene	CA	9.1×10^{-7} (cancer)
Dibenzo[a,i]pyrene	CA	9.1×10^{-7} (cancer)
Dibenzo[a,l]pyrene	CA	9.1×10^{-7} (cancer)
Ethyl benzene	CA	7.7×10^{-1} (liver and kidney)
Ethyl carbamate	CA, RDT	3.5×10^{-5} (cancer)
Ethylene oxide	CA, RT, RDT	1.1×10^{-4} (cancer)
Hydrazine	CA, RT	2.0×10^{-6} (cancer)
Indeno[1,2,3-cd]pyrene	CA	9.1×10^{-5} (cancer)
Methyl ethyl ketone	RT	5.0 (developmental)
1-Methyl-3-amino-5H-pyrido[4,3-b]indole	CA	1.1×10^{-5} (cancer)
5-Methylchrysene	CA	9.1×10^{-6} (cancer)
Naphthalene	CA, RT	3.0×10^{-3} (nasal)
Nickel	CA, RT	9.0×10^{-5} (lung fibrosis)
N-Nitrosodiethanolamine	CA	1.3×10^{-5} (cancer)
N-Nitrosodiethylamine	CA	2.3×10^{-7} (cancer)
N-Nitrosodimethylamine	CA	7.1×10^{-7} (cancer)
N-Nitrosomethylethylamine	CA	1.6×10^{-6} (cancer)
N-Nitrosopiperidine	CA	3.7×10^{-6} (cancer)

N-Nitrosopyrrolidine	CA	1.6×10^{-5} (cancer)
Polonium-210	CA	925.9 (cancer)
Propylene oxide	CA, RT	2.7×10^{-3} (cancer)
Selenium	RT	8.0×10^{-4} (respiratory)
Styrene	CA	9.2×10^{-2} (neurotoxicity)
Vinyl acetate	CA, RT	2.0×10^{-1} (nasal)
Vinyl chloride	CA	1.1×10^{-3} (cancer)

CA: carcinogen; CT: cardiovascular toxicant; RDT: reproductive or developmental toxicant; RT: respiratory toxicant.

Constituents in the same chemical class as current priority toxicants: the rationale for including such constituents is that several compounds can be analysed simultaneously with the same analytical technique. Examples include metals and other trace elements, as well as PAHs. The elements beryllium, chromium, cobalt, nickel, polonium-210 and selenium, listed in Table 8.6, can be analysed by a multi-element method with cadmium and other metals already on the TobReg priority list. At least 23 different PAHs could be analysed with the same analytical technique *(155)*. Eight PAHs listed in Table 8.4 – benz[*a*]anthracene, dibenz[*a,h*]anthracene, dibenzo[*a,e*]pyrene, dibenzo[*a,h*]pyrene, dibenzo[*a,i*]pyrene, dibenzo[*a,l*]pyrene, indeno[*1,2,3-cd*]pyrene and naphthalene – could be analysed with benzo[*a*]pyrene. It has been demonstrated that the levels of these PAHs correlate negatively with those of nitrate and TSNAs; however, the relations depend on tobacco type *(156)*. In addition, in many studies of human exposure, urinary levels of 1-hydroxypyrene or phenanthrene tetraols, which are biomarkers of exposure to the non-carcinogenic PAHs pyrene and phenanthrene, respectively, are analysed to assess exposure to PAHs *(157, 158)*. As these two compounds are present in tobacco smoke at much higher levels than benzo[*a*]pyrene and other carcinogenic PAHs, simultaneous analysis of pyrene and phenanthrene with benzo[*a*]pyrene could provide information for future monitoring of changes in human exposure due to reductions in the levels of these constituents in smoke.

Chemicals or chemical compounds that are precursors to toxic emissions in tobacco smoke: for instance, nitrite and nitrate in tobacco leaf are precursors to the carcinogenic nitrosamines NNK and NNN in tobacco and smoke and to nitrogen oxides in smoke. Well established data from both academic and industry researchers show that nitrate and nitrite in tobacco significantly affect the composition of tobacco smoke *(159, 160)*. In addition, nitrate levels in tobacco affect the levels of ammonia in cigarette smoke, which in turn influences smoke pH and the bioavailability of nicotine. Therefore, nitrite and nitrate levels in tobacco filler of cigarettes are important predictors of the toxicological properties of tobacco smoke. The current TobReg priority list includes nitric oxides in the gas phase of tobacco smoke; however, analysis of such constituents of tobacco smoke may be complicated, with inconsistent results among laboratories. Therefore, analysis of nitrites and nitrates in tobacco filler would be a robust, informative alternative.

Constituents that contribute to the palatability and/or addictiveness of tobacco products: the impact of additives on the attractiveness and abuse liability of tobacco products has been reviewed *(161)*. For instance, sugars in tobacco filler of combustible products are a group of additives that could be considered for future monitoring and regulation. Sugars in tobacco contribute to the formation of acetaldehyde in tobacco smoke *(162, 163)*, potentially contributing to the addictive potential of tobacco smoke, either directly or through the formation of harman *(32–34)*.

8.8 Research needs and regulatory recommendations

8.8.1 Research needs, data gaps and future work

This review of the toxicity, analytical methods and reports on the levels of the priority toxicants in combustible tobacco products indicates that the following areas require research.

Despite toxicological evidence of the importance of the 39 priority toxicants and the availability of analytical methods, **brand- and product-specific information on the levels of emissions of toxicants in diverse combustible tobacco products is lacking**. Information on variations in the levels of toxicants in different brands and types of combustible tobacco products (e.g. cigarettes, cigarillos, waterpipes) could justify regulation of specific emissions. In view of the widespread global use of combustible products manufactured locally (e.g. *bidis*) or made by consumers (e.g. roll-you-own cigarettes), better characterization of these products is essential.

Evaluation of standardized methods for the analysis of certain constituents in tobacco filler (contents) is a more robust alternative to measurements in cigarette smoke in some cases. For instance, nitrate and nitrite levels in tobacco strongly influence the levels of nitric oxides in tobacco smoke. Analysis of these constituents in tobacco filler will require less time and resources and could provide more consistent data from laboratories than analysis of nitric oxide emissions. Furthermore, the validated methods can be used directly for the analysis of smokeless tobacco products. Carcinogenic TSNAs are formed in tobacco during processing, and the contribution of pyrolysis to their levels in smoke, if any, is minimal. Therefore, the levels in tobacco are strong predictors of the levels in cigarette smoke *(68, 86, 164, 165)*.

While TobReg previously recommended that upper limits for emissions of toxicants in tobacco products should be set on the basis of established toxicological principles *(166)*, **the extent to which toxicant levels must be reduced, as a complex mixture or singly, in order to minimize the harm caused by tobacco products remains unknown.** Laboratory in vitro and in vivo studies are required to better understand the effect of the complex chemistry of tobacco products on the carcinogenic and toxic potency and the toxicity thresholds of individual constituents. Studies of human exposure and molecular epidemiology, including studies of prospective cohorts, would indicate the optimal reductions in tobacco toxicant levels necessary to achieve the maximum public health benefit.

Research is needed to **better understand which ingredients or constituents contribute to the addictiveness and palatability of tobacco products,** either independently or by increasing the bioavailability of nicotine. This information will have an impact on future regulatory approaches. For instance, it has been suggested that measures to reduce the attractiveness and palatability of tobacco products might have a greater impact on public health than reducing toxic emissions *(167)*.

Current work to prioritize toxic emissions and the development of methods for product testing should be **extended to human exposure and health outcomes.** The smoker–cigarette interaction is more complex than any single machine-based regimen *(1)*, and it is not clear whether reductions in per-milligram nicotine emissions will lead to corresponding reductions in human exposure. Laboratory capacity should be built for analysing biomarkers of the priority toxicants in human biological samples. Analyses of spent cigarette filters might be considered as a less expensive measure of human toxicant intake *(69, 168–170)*. In addition, research should be conducted to identify suitable biomarkers of potential harm that could be used to evaluate the long-term health impact of future product standards.

8.8.2 Regulatory recommendations and support to Parties

TobLabNet has completed validation of methods for measuring selected emissions in cigarette tobacco filler and in mainstream cigarette smoke, and the SOPs for these methods are available on the WHO Tobacco Product Regulation website. It is recommended that Parties consider requiring cigarette manufacturers to conduct emission testing in accordance with the TobLabNet SOPs and to report the results to national authorities. This recommendation is part of a decision by the eighth COP (FCTC/COP8(21)), which encourages Parties to acknowledge and implement the WHO TobLabNet methods as appropriate and emphasizes the need for further support of TobLabNet's capacity-building activities by WHO.

Assessment of studies on the toxicity of the extended list of priority toxicants has demonstrated their continued role in the toxic and addictive potential of combustible tobacco products. The methods validated by TobLabNet can be used for other toxicants on the extended priority list, as reviewed in this report. Therefore, it is recommended that Parties consider requesting, as applicable and appropriate, manufacturers to report on the emissions of the additional priority toxicants with methods based on the TobLabNet SOPs.

SOPs for the analysis of priority contents and emissions for which methods validated by TobLabNet are not yet available should be developed next, as recommended by TobReg *(166)*: cadmium and lead in tobacco; nicotine in waterpipe smoke; and nicotine, TSNAs and benzo[*a*]pyrene in smokeless tobacco.

Of the constituents not currently on the priority list, it is recommended that Parties consider requiring that manufacturers report on the levels of nitrate and nitrite in tobacco filler (content), as a potential proxy for nitric oxide emissions. An easy, cost–effective method for the analysis of nitric oxide in tobacco smoke is not yet available. Nitrate and nitrite in aqueous tobacco extracts can be analysed by adapting the TobLabNet method for ammonia in tobacco filler (SOP 07). In addition, given their importance to human biomonitoring research, pyrene and phenanthrene (and potentially other carcinogenic PAHs listed in Table 8.6) could be monitored simultaneously with benzo[*a*]pyrene by the TobLabNet method.

It is recommended that reference tobacco product materials be made available to TobLabNet laboratories participating in WHO method validation to aid in determining the success of method transfer, in terms of the accuracy, repeatability and reproducibility of each method. Suitable sample matrix-matched certified and standard reference materials can be analysed at the same time as test materials as a form of quality control for the evaluation of data generated in interlaboratory testing. Commercially available certified and standard reference materials are listed in Table 8.7.

Table 8.7. Certified and standard reference materials suitable for quality control of samples

Name	Description	Issuer	Certified values
SRM 3222	Cigarette tobacco filler	NIST	Nicotine, NNN, NNK, VOCs
1R6F	Reference cigarette	KTRDC	Mainstream smoke emissions, puff count for ISO and HCI puffing, filler content, physical properties
RT5	High-TSNA ground tobacco	KTRDC	Nicotine, nornicotine, anabasine, anatabine, NNN, NAT, NAB, NNK, moisture
RT4	Burley tobacco		
RTDAC	Dark, air-cured, ground tobacco		
RT3	Turkish Oriental ground tobacco		
RT2	Flue-cured ground tobacco		
RTDFC	Dark fire-cured ground tobacco		
1R6F (RT1)	Ground filler		
1R5F (RT7)	Ground filler		
STRP 1S1	Loose leaf chewing tobacco	NCSU	Content values are not certified, but may be reported in the scientific literature
STRP 2S1	Loose leaf chewing tobacco		
STRP 1S2	Dry snuff		
STRP 1S3	Moist snuff		
STRP 2S3	Moist snuff		

CRP1	Snus
CRP2	Moist snuff
CRP3	Dry snuff
CRP4	Loose leaf chewing tobacco

HCI: Health Canada intense; ISO: International Organization for Standardization; KTRDC: University of Kentucky Center for Tobacco Reference Products; NAB: N'-nitrosoanabasine; NAT: N'-nitrosoanatabine; NCSU: North Carolina State University Smokeless Tobacco Reference Products; NIST: National Institute of Standards and Technology; NNK: 4-(methylnitrosamino)-1-(3-pyridyl)-1-butanone; NNN: N'-nitrosonornicotine; VOCs: volatile organic compounds.

It is recommended that a research agenda be prepared on which constituents contribute to the addictiveness and palatability of tobacco products, with identification or development of methods for their quantification in tobacco and smoke and establishment of ranges of emissions in various cigarette brands.

8.9 References

1. Djordjevic MV, Stellman SD, Zang E. Doses of nicotine and lung carcinogens delivered to cigarette smokers. J Natl Cancer Inst. 2000;92:106–11.
2. Jarvis MJ, Boreham R, Primatesta P, Feyerabend C, Bryant A. Nicotine yield from machine-smoked cigarettes and nicotine intakes in smokers: evidence from a representative population survey. J Natl Cancer Inst. 2001;93:134–8.
3. Health Canada data for 2004 using the intense smoking protocol. Ottawa: Health Canada Tobacco – Reports and Publications; 2004.
4. Counts ME, Morton MJ, Laffoon SW, Cox RH, Lipowicz PJ. Smoke composition and predicting relationships for international commercial cigarettes smoked with three machine-smoking conditions. Regul Toxicol Pharmacol. 2005;41:185–227.
5. Fowles J, Dybing E. Application of toxicological risk assessment principles to the chemical constituents of cigarette smoke. Tob Control. 2003;12:424–30.
6. Dybing E, Sanner T, Roelfzema H, Kroese D, Tennant RW. T25: a simplified carcinogenic potency index. Description of the system and study of correlations between carcinogenic potency and species/site specificity and mutagenicity. Pharmacol Toxicol. 1997;80:272–9.
7. Cigarette emissions data, 2001. Canberra: Australian Department of Health and Aging; 2001 (http://www.health.gov.au/internet/main/publishing.nsf/content/tobacco-emis, accessed 15 May 2019).
8. Stepanov I, Hatsukami D. Call to establish constituent standards for smokeless tobacco products. Tob Regul Sci. 2016;2:9–30.
9. Talhout R, Richter PA, Stepanov I, Watson CV, Watson CH. Cigarette design features: effects on emission levels, user perception, and behavior. Tob Regul Sci. 2018;4:592–604.
10. Rodgman A. Studies of polycyclic aromatic hydrocarbons in cigarette mainstream smoke: identification, tobacco precursors, control of levels: a review. Beitr Tabakforsch Int. 2001;19:361–79.
11. WHO Study Group on Tobacco Product Regulation. The scientific basis of tobacco product regulation. Second report of a WHO study group (WHO Technical Report Series, No. 951). Geneva: World Health Organization; 2008:45–125.
12. WHO Study Group on Tobacco Product Regulation. Setting maximal limits for toxic constituents in cigarette smoke. Geneva: World Health Organization; 2001 (http://www.who.int/tobacco/global_interaction/tobreg/tsr/en/index.html, accessed 15 May 2019).
13. O'Brien PJ, Siraki AG, Shangari N. Aldehyde sources, metabolism, molecular toxicity mechanisms, and possible effects on human health. Crit Rev Toxicol. 2005;35:609–62.
14. Re-evaluation of some organic chemicals, hydrazine and hydrogen peroxide (IARC Mono-

graphs on the Evaluation of the Carcinogenic Risk of Chemicals to Humans, Vol. 71). Lyon: International Agency for Research on Cancer; 1999.
15. A review of human carcinogens: chemical agents and related occupations (IARC Monographs on the Evaluation of Carcinogenic Risks to Humans, Vol. 100F), Lyon: International Agency for Research on Cancer; 2012.
16. Dry cleaning, some chlorinated solvents and other industrial chemicals (IARC Monographs on the Evaluation of Carcinogenic Risks to Humans, Vol. 63). Lyon: International Agency for Research on Cancer; 1995:337–72.
17. Feng Z, Hu W, Hu Y, Tang MS. Acrolein is a major cigarette-related lung cancer agent. Preferential binding at p53 mutational hotspots and inhibition of DNA repair. Proc Natl Acad Sci USA. 2006;103:15404–9.
18. Zhang S, Villalta PW, Wang M, Hecht SS. Analysis of crotonaldehyde- and acetaldehyde-derived 1,N^2-propanodeoxyguanosine adducts in DNA from human tissues using liquid chromatography-electrospray ionization–tandem mass spectrometry. Chem Res Toxicol. 2006;19:1386–92.
19. Zhang S, Chen H, Wang A, Liu Y, Hou H, Hu Q. Assessment of genotoxicity of four volatile pollutants from cigarette smoke based on the in vitro gH2AX assay using high content screening. Environ Toxicol Pharmacol. 2017;55:30–6.
20. Corley RA, Kabilan S, Kuprat AP, Carson JP, Jacob RE, Minard KR et al. Comparative risks of aldehyde constituents in cigarette smoke using transient computational fluid dynamics/physiologically based pharmacokinetic models of the rat and human respiratory tracts. Toxicol Sci. 2015;146:65–88.
21. Yeager RP, Kushman M, Chemerynski S, Weil R, Fu X, White M et al. Proposed mode of action for acrolein respiratory toxicity associated with inhaled tobacco smoke. Toxicol Sci. 2016;151:347–64.
22. Takamiya R, Uchida K, Shibata T, Maeno T, Kato M, Yamaguchi Y et al. Disruption of the structural and functional features of surfactant protein A by acrolein in cigarette smoke. Sci Rep. 2017;7:8304.
23. Lee HW, Wang HT, Weng MW, Hu Y, Chen WS, Chou D et al. Acrolein- and 4-aminobiphenyl-DNA adducts in human bladder mucosa and tumor tissue and their mutagenicity in human urothelial cells. Oncotarget. 2014;5:3526–40.
24. Cheah NP, Pennings JL, Vermeulen JP, van Schooten FJ, Opperhuizen A. In vitro effects of aldehydes present in tobacco smoke on gene expression in human lung alveolar epithelial cells. Toxicol In Vitro. 2013;27:1072–81.
25. Sapkota M, Wyatt TA. Alcohol, aldehydes, adducts and airways. Biomolecules. 2015;5:2987–3008.
26. Antoniak DT, Duryee MJ, Mikuls TR, Thiele GM, Anderson DR. Aldehyde-modified proteins as mediators of early inflammation in atherosclerotic disease. Free Radical Biol Med. 2015;89:409–18.
27. Lachenmeier DW, Salaspuro M. ALDH2-deficiency as genetic epidemiologic and biochemical model for the carcinogenicity of acetaldehyde. Regul Toxicol Pharmacol. 2017;86:128–36.
28. Morita K, Miyazaki H, Saruwatari J, Oniki K, Kumagae N, Tanaka T et al. Combined effects of current-smoking and the aldehyde dehydrogenase 2*2 allele on the risk of myocardial infarction in Japanese patients. Toxicol Lett. 2015;232:221–5.
29. Bui LC, Manaa A, Xu X, Duval R, Busi F, Dupret JM et al. Acrolein, an a,ß-unsaturated aldehyde, irreversibly inhibits the acetylation of aromatic amine xenobiotics by human arylamine N-acetyltransferase 1. Drug Metab Dispos. 2013;41:1300–5.
30. Chare SS, Donde H, Chen WY, Barker DF, Gobejishvilli L, McClain CJ et al. Acrolein enhances epigenetic modifications, FasL expression and hepatocyte toxicity induced by anti-HIV drug zidovudine. Toxicol In Vitro. 2016;35:76.
31. Tan SLW, Chadha S, Liu Y, Gabasova E, Perera D, Ahmed K et al. A class of environmental and endogenous toxins induces BRCA2 haploinsufficiency and genome instability. Cell. 2017;169:1105–18.

32. Hoffman AC, Evans SE. Abuse potential of non-nicotine tobacco smoke components: acetaldehyde, nornicotine, cotinine, and anabasine. Nicotine Tob Res. 2013;15:622–32.
33. Brancato A, Lavanco G, Cavallaro A, Plescia F, Cannizzaro C. Acetaldehyde, motivation and stress: behavioral evidence of an addictive ménage à trois. Front Behav Neurosci. 2017;11:23.
34. Syrjänen K, Salminen J, Aresvuo U, Hendolin P, Paloheimo L, Eklund C et al. Elimination of cigarette smoke-derived acetaldehyde in saliva by slow-release L-cysteine lozenge is a potential new method to assist smoking cessation. A randomised, double-blind, placebo-controlled intervention. Anticancer Res. 2016;36:2297–306.
35. Henning RJ, Johnson GT, Coyle JP, Harbison RD. Acrolein can cause cardiovascular disease: a review. Cardiovasc Toxicol. 2017;17:227–36.
36. Conklin DJ, Malovichko MV, Zeller I, Das TP, Krivokhizhina TV, Lynch BH et al. Biomarkers of chronic acrolein inhalation exposure in mice: implications for tobacco product-induced toxicity. Toxicol Sci. 2017;158:263–74.
37. DeJarnett N, Conklin DJ, Riggs DW, Myers JA, O'Toole TE, Hamzeh I et al. Acrolein exposure is associated with increased cardiovascular disease risk. J Am Heart Assoc. 2014;3:e000934.
38. Snow SJ, McGee MA, Henriquez A, Richards JE, Schladweiler MC, Ledbetter AD et al. Respiratory effects and systemic stress response following acute acrolein inhalation in rats. Toxicol Sci. 2017;158:454–64.
39. Feroe AG, Attanasio R, Scinicariello F. Acrolein metabolites, diabetes and insulin resistance. Environ Res. 2016;148:1–6.
40. Chen WY, Wang M, Zhang J, Barve SS, McClain CJ, Joshi-Barve S. Acrolein disrupts tight junction proteins and causes endoplasmic reticulum stress-mediated epithelial cell death leading to intestinal barrier dysfunction and permeability. Am J Pathol. 2017;187(12):2686–97.
41. Song JJ, Lee JD, Lee BD, Chae SW, Park MK. Effect of acrolein, a hazardous air pollutant in smoke, on human middle ear epithelial cells. Int J Pediatr Otorhinolaryngol. 2013;77:1659–64.
42. Amiri A, Turner-Henson A. The roles of formaldehyde exposure and oxidative stress in fetal growth in the second trimester. J Obstet Gynecol Neonatal Nurs. 2017;46:51–62.
43. Yang Y, Zhang Z, Zhang H, Hong K, Tang W, Zhao L et al. Effects of maternal acrolein exposure during pregnancy on testicular testosterone production in fetal rats. Mol Med Rep. 2017;16:491–8.
44. Park SL, Carmella SG, Chen M, Patel Y, Stram DO, Haiman CA et al. Mercapturic acids derived from the toxicants acrolein and crotonaldehyde in the urine of cigarette smokers from five ethnic groups with differing risks for lung cancer. PLoS One. 2015;10:e0124841.
45. Haiman CA, Stram DO, Wilkens LR, Pike MC, Kolonel LN, Henderson BE et al. Ethnic and racial differences in the smoking-related risk of lung cancer. N Engl J Med. 2006;354:333–42.
46. Some aromatic amines, organic dyes, and related exposures (IARC Monographs on the Evaluation of Carcinogenic Risks to Humans, Vol. 99). Lyon: International Agency for Research on Cancer; 2010.
47. Cai T, Bellamri M, Ming X, Koh WP, Yu MC, Turesky RJ. Quantification of hemoglobin and white blood cell DNA adducts of the tobacco carcinogens 2-amino-9H-pyrido[2,3-b]indole and 4-aminobiphenyl formed in humans by nanoflow liquid chromatography/ion trap multistage mass spectrometry. Chem Res Toxicol. 2017;30:1333–43.
48. Tomioka K, Obayashi K, Saeki K, Okamoto N, Kurumatani N. Increased risk of lung cancer associated with occupational exposure to benzidine and/or beta-naphthylamine. Int Arch Occup Environ Health. 2015;88:455–65.
49. Tomioka K, Saeki K, Obayashi K, Kurumatani N. Risk of lung cancer in workers exposed to benzidine and/or beta-naphthylamine: a systematic review and meta-analysis. J Epidemiol. 2016;26:447–58.
50. Some industrial chemicals (IARC Monographs on the Evaluation of the Carcinogenic Risk of Chemicals to Humans, Vol. 60). Lyon: International Agency for Research on Cancer; 1994:215–32.

51. Hayes RB, Yin SN, Dosemeci M, Li GL, Wacholder S, Chow WH et al. Mortality among benzene-exposed workers in China. Environ Health Perspect. 1996;104(Suppl 6):1349–52.
52. Soeteman-Hernandez LG, Bos PM, Talhout R. Tobacco smoke-related health effects induced by 1,3-butadiene and strategies for risk reduction. Toxicol Sci. 2013;136:566–80.
53. Bahadar H, Mostafalou S, Abdollahi M. Current understandings and perspectives on non-cancer health effects of benzene: a global concern. Toxicol Appl Pharmacol. 2014;276:83–94.
54. Abplanalp W, DeJarnett N, Riggs DW, Conklin DJ, McCracken JP, Srivastava S et al. Benzene exposure is associated with cardiovascular disease risk. PLoS One. 2017;12:e0183602.
55. Doherty BT, Kwok RK, Curry MD, Ekenga C, Chambers D, Sandler DP et al. Associations between blood BTEXS concentrations and hematologic parameters among adult residents of the US Gulf States. Environ Res. 2017;156:579–87.
56. Haiman CA, Patel YM, Stram DO, Carmella SG, Chen M, Wilkens LR et al. Benzene uptake and glutathione S-transferase T1 status as determinants of S-phenylmercapturic acid in cigarette smokers in the Multiethnic Cohort. PLoS One. 2016;11(3):0150641.
57. Some non-heterocyclic polycyclic aromatic hydrocarbons and some related exposures (IARC Monographs on the Evaluation of Carcinogenic Risks to Humans, Vol. 92). Lyon: International Agency for Research on Cancer; 2010.
58. Ding YS, Ashley DL, Watson CH. Determination of 10 carcinogenic polycyclic aromatic hydrocarbons in mainstream cigarette smoke. J Agric Food Chem. 2007;55:5966–73.
59. Hecht SS. Tobacco carcinogens, their biomarkers, and tobacco-induced cancer. Nature Rev Cancer. 2003;3:733–44.
60. Pfeifer GP, Denissenko MF, Olivier M, Tretyakova N, Hecht SS, Hainaut P. Tobacco smoke carcinogens, DNA damage and p53 mutations in smoking-associated cancers. Oncogene. 2002;21:7435–51.
61. Hecht SS. Tobacco smoke carcinogens and lung cancer. J Natl Cancer Inst. 1999;91:1194–210.
62. Onami S, Okubo C, Iwanaga A, Suzuki H, Iida H, Motohashi Y et al. Dosimetry for lung tumorigenesis induced by urethane, 4-(N-methyl-N-nitrosamino)-1-(3-pyridyl)-1-butanone (NNK), and benzo[a]pyrene (B[a]P) in A/JJmsSlc mice. J Toxicol Pathol. 2017;30:209–16.
63. Zaccaria KJ, McClure PR. Using immunotoxicity information to improve cancer risk assessment for polycyclic aromatic hydrocarbon mixtures. Int J Toxicol. 2013;32:236–50.
64. Chepelev NL, Moffat ID, Bowers WJ, Yauk CI. Neurotoxicity may be an overlooked consequence of benzo[a]pyrene exposure that is relevant to human health risk assessment. Mutat Res Rev Mutat Res. 2015;764:64–89.
65. Hecht SS. Biochemistry, biology, and carcinogenicity of tobacco-specific N-nitrosamines. Chem Res Toxicol. 1998;11:559–603.
66. Tobacco smoke and involuntary smoking (IARC Monographs on the Evaluation of Carcinogenic Risks to Humans, Vol. 83). Lyon: International Agency for Research on Cancer; 2004.
67. Smokeless tobacco and tobacco-specific nitrosamines (IARC Monographs on the Evaluation of Carcinogenic Risks to Humans, Vol. 89). Lyon: International Agency for Research on Cancer; 2007.
68. Wu W, Zhang L, Jain RB, Ashley DL, Watson CH. Determination of carcinogenic tobacco-specific nitrosamines in mainstream smoke from US-brand and non-US-brand cigarettes from 14 countries. Nicotine Tob Res. 2005;7:443–51.
69. Ashley DL, O'Connor RJ, Bernert JT, Watson CH, Polzin GM, Jain RB et al. Effect of differing levels of tobacco-specific nitrosamines in cigarette smoke on the levels of biomarkers in smokers. Cancer Epidemiol Biomarkers Prev. 2010;19:1389–98.
70. Stepanov I, Knezevich A, Zhang L, Watson CH, Hatsukami DK, Hecht SS. Carcinogenic tobacco-specific N-nitrosamines in US cigarettes: three decades of remarkable neglect by the tobacco industry. Tob Control. 2012;21:44–8.
71. Hoffmann D, Djordjevic MV, Hoffmann I. The changing cigarette. Prev Med. 1997;26:427–34.
72. Burns DM, Anderson CM, Gray N. Do changes in cigarette design influence the rise in adeno-

carcinoma of the lung? Cancer Causes Control. 2011;22:13–22.
73. Balbo S, Johnson CS, Kovi RC, James-Yi SA, O'Sullivan MG, Wang M et al. Carcinogenicity and DNA adduct formation of 4-(methylnitrosamino)-1-(3-pyridyl)-1-butanone and enantiomers of its metabolite 4-(methylnitrosamino)-1-(3-pyridyl)-1-butanol in F-344 rats. Carcinogenesis. 2014;35:2798–806.
74. Balbo S, James-Yi S, Johnson CS, O'Sullivan MG, Stepanov I, Wang M et al. (S)-N´-Nitrosonornicotine, a constituent of smokeless tobacco, is a powerful oral cavity carcinogen in rats. Carcinogenesis. 2013;34:2178–83.
75. Stepanov I, Sebero E, Wang R, Gao YT, Hecht SS, Yuan JM. Tobacco-specific N-nitrosamine exposures and cancer risk in the Shanghai Cohort Study: remarkable coherence with rat tumor sites. Int J Cancer. 2014;134:2278–83.
76. Park SL, Carmella SG, Ming X, Stram DO, LeMarchand L, Hecht SS. Variation in levels of the lung carcinogen NNAL and its glucuronides in the urine of cigarette smokers from five ethnic groups with differing risks for lung cancer. Cancer Epidemiol Biomarkers. 2015;24:561–9.
77. Khariwala SS, Ma B, Ruszczak C, Carmella SG, Lindgren B, Hatsukami DK et al. High level of tobacco carcinogen-derived DNA damage in oral cells is an independent predictor of oral/head and neck cancer risk in smokers. Cancer Prev Res. 2017;10:507–13.
78. Hukkanen J, Jacob P III, Benowitz NL. Metabolism and disposition kinetics of nicotine. Pharmacol Rev. 2005;57:79–115.
79. WHO Study Group on Tobacco Product Regulation. Global nicotine reduction strategy. Advisory note. Geneva: World Health Organization; 2015.
80. Food and Drug Administration. Tobacco product standard for nicotine level of combusted cigarette. Fed Regist. 2018;83.
81. Donny EC, Denlinger RL, Tidey JW, Koopmeiners JS, Benowitz NL, Vandrey RG et al. Randomized clinical trial of reduced-nicotine standards for cigarettes. N Engl J Med. 2015;373:1340–9.
82. Hatsukami DK, Luo X, Dick L, Kangkum M, Allen SS, Murphy SE et al. Reduced nicotine content cigarettes and use of alternative nicotine products: exploratory trial. Addiction. 2017;112:156–67.
83. Benowitz NL, Burbank AD. Cardiovascular toxicity of nicotine: implications for electronic cigarette use. Trends Cardiovasc Med. 2016;26:515–23.
84. England LJ, Aagaard K, Bloch M, Conway K, Cosgrove K, Grana R et al. Developmental toxicity of nicotine: a transdisciplinary synthesis and implications for emerging tobacco products. Neurosci Biobehav Rev. 2017;72:176–89.
85. Van Duuren BL, Goldschmidt BM. Cocarcinogenic and tumor-promoting agents in tobacco carcinogenesis. J. Natl Cancer Inst. 1976;56:1237–42.
86. Fennell TR, MacNeela JP, Morris RW, Watson M, Thompson CL, Bell DA. Hemoglobin adducts from acrylonitrile and ethylene oxide in cigarette smokers: effects of glutathione S-transferase T1-null and M1-null genotypes. Cancer Epidemiol Biomarkers Prev. 2000;9:705–12.
87. A review of human carcinogens: arsenic, metals, fibres, and dusts (IARC Monographs on the Evaluation of Carcinogenic Risks to Humans, Vol. 100C). Lyon: International Agency for Research on Cancer; 2012.
88. Inorganic and organic lead compounds (IARC Monographs on the Evaluation of Carcinogenic Risks to Humans, Vol. 87). Lyon: International Agency for Research on Cancer; 2006.
89. Beryllium, cadmium, mercury, and exposures in the glass manufacturing industry (IARC Monographs on the Evaluation of Carcinogenic Risks to Humans, Vol. 58). Lyon: International Agency for Research on Cancer; 1993.
90. Pappas RS, Stanfill SB, Watson CH, Ashley DL. Analysis of toxic metals in commercial moist snuff and Alaskan iqmik. J Anal Toxicol. 2008;32:281–91.
91. Pinto E, Cruz M, Ramos P, Santos A, Almeida A. Metals transfer from tobacco to cigarette smoke: evidences in smokers' lung tissue. J Hazard Mater. 2017;5:31–5.
92. Richter PA, Bishop EE, Wang J, Kaufmann R. Trends in tobacco smoke exposure and blood

lead levels among youths and adults in the United States: the National Health and Nutrition Examination Survey, 1999–2008. Prev Chronic Dis. 2013;10:E213.
93. Garcia-Esquinas E, Pollan M, Tellez-Plaza M, Francesconi KA, Goessler W, Guallar E et al. Cadmium exposure and cancer mortality in a prospective cohort: the Strong Heart Study. Environ Health Perspect. 2014;122:363–70.
94. Feki-Tounsi M, Hamza-Chaffai A. Cadmium as a possible cause of bladder cancer: a review of accumulated evidence. Environ Sci Pollut Res Int. 2014;21:10561–73.
95. Buser MC, Scinicariello F. Cadmium, lead, and depressive symptoms: analysis of National Health and Nutrition Examination Survey 2011–2012. J Clin Psychiatry. 2017;78:e515–21.
96. Adeniyi PA, Olatunji BP, Ishola AO, Ajonijebu DC, Ogundele OM. Cadmium increases the sensitivity of adolescent female mice to nicotine-related behavioral deficits. Behav Neurol. 2014;2014:36098.
97. Ajonijebu C, Adekeye AO, Olatunji BP, Ishola AO, Ogundele OM. Nicotine–cadmium interaction alters exploratory motor function and increased anxiety in adult male mice. Neurodegener Dis. 2014;359436. doi: 10.1155/2014/359436.
98. Hecht EM, Landy DC, Ahn S, Hlaing WM, Hennekens CH. Hypothesis: cadmium explains, in part, why smoking increases the risk of cardiovascular disease. J Cardiovasc Pharmacol Ther. 2013;18:550–4.
99. Tellez-Plaza M, Jones MR, Dominguez-Lucas A, Guallar E, Navas-Acien A. Cadmium exposure and clinical cardiovascular disease: a systematic review. Curr Atheroscler Rep. 2013;15:356.
100. Das S, Upadhaya P, Giri S. Arsenic and smokeless tobacco induce genotoxicity, sperm abnormality as well as oxidative stress in mice in vivo. Genes Environ. 2016;38:4.
101. Chelchowska M, Ambroszkiewicz J, Jablonka-Salach K, Gajewska J, Maciejewski TM, Bulska E et al. Tobacco smoke exposure during pregnancy increases maternal blood lead levels affecting neonate birth weight. Biol Trace Elem Res. 2013;155:169–75.
102. Tinkov AA, Ajsuvakova OP, Skalnaya MG, Popova EV, Sinitskii AI, Nemereshina ON et al. Mercury and metabolic syndrome: a review of experimental and clinical observations. Biometals. 2015;28:231–54.
103. Roy C, Tremblay PY, Ayotte P. Is mercury exposure causing diabetes, metabolic syndrome and insulin resistance? A systematic review of the literature. Environ Res. 2017;156:747–60.
104. Genchi G, Sinicropi MS, Carocci A, Lauria G, Catalano A. Mercury exposure and heart diseases. Int J Environ Res Public Health, 2017;14:pii: E74.
105. Rahman MH, Bratveit M, Moen BE. Exposure to ammonia and acute respiratory effects in a urea fertilizer factory. Int J Occup Environ Health. 2007;13:153–9.
106. Casas L, Zock JP, Torrent M, Garcia-Esteban R, Garcia-Lavedan E, Hyvarinen A et al. Use of household cleaning products, exhaled nitric oxide and lung function in children. Eur Respir J. 2013;42:1415–8.
107. Arif AA, Delclos GL. Association between cleaning-related chemicals and work-related asthma and asthma symptoms among healthcare professionals. Occup Environ Med. 2012;69:35–40.
108. Loftus C, Yost M, Sampson P, Torres E, Arias G, Breckwich Vasquez V et al. Ambient ammonia exposures in an agricultural community and pediatric asthma morbidity. Epidemiology. 2015;26:794–801.
109. Petrova M, Diamond J, Schuster B, Dalton P. Evaluation of trigeminal sensitivity to ammonia in asthmatics and healthy human volunteers. Inhal Toxicol. 2008;20:1085–92.
110. The health consequences of smoking: a report of the Surgeon General. Rockville (MD): Department of Health and Human Services, Public Health Service, Centers for Disease Control, Center for Health Promotion and Education, Office on Smoking and Health; 2004.
111. Ludvig J, Miner B, Eisenberg MJ. Smoking cessation in patients with coronary artery disease. Am Heart J. 2005;149:565–72.
112. Scherer G. Carboxyhemoglobin and thiocyanate as biomarkers of exposure to carbon monox-

ide and hydrogen cyanide in tobacco smoke. Exp Toxicol Pathol. 2006;58:101–24.
113. Silverstein P. Smoking and wound healing. Am J Med. 1992;93:22S–24S.
114. Weinberger B, Laskin DL, Heck DE, Laskin JD. The toxicology of inhaled nitric oxide. Toxicol Sci. 2001;59:5–16.
115. Vleeming W, Rambali B, Opperhuizen A. The role of nitric oxide in cigarette smoking and nicotine addiction. Nicotine Tob Res. 2002;4:341–8.
116. WHO Study Group on Tobacco Product Regulation. Guiding principles for the development of tobacco product research and testing capacity and proposed protocols for the initiation of tobacco product testing: Recommendation 6. Geneva: World Health Organization; 2004.
117. WHO Tobacco Laboratory Network standard operating procedure for validation of analytical methods of tobacco product contents and emissions. WHO TobLabNet Official Method SOP02. Geneva: World Health Organization; 2017.
118. Ding YS, Yan X, Wong J, Chan M, Watson CH. In situ derivatization and quantification of seven carbonyls in cigarette mainstream smoke. Chem Res Toxicol. 2016;29:125–31.
119. Zhao W, Zhang Q, Lu B, Sun S, Zhang S, Zhang J. Rapid determination of six low molecular carbonyl compounds in tobacco smoke by the APCI-MS/MS coupled to data mining. J Anal Meth Chem. 2017:8260860. doi: 10.1155/2017/8260860.
120. Ahad BT, Abdollahi A. Dispersive liquid–liquid microextraction for the high performance liquid chromatographic determination of aldehydes in cigarette smoke and injectable formulations. J Hazard Mater. 2013;254–255:390–6.
121. Zhang J, Bai R, Zhou Z, Liu X, Zhou J. Simultaneous analysis of nine aromatic amines in mainstream cigarette smoke using online solid-phase extraction combined with liquid chromatography-tandem mass spectrometry. Anal Bioanal Chem. 2017;409:2993–3005.
122. Xie F, Yu J, Wang S, Zhao G, Xia G, Zhang X et al. Rapid and simultaneous analysis of ten aromatic amines in mainstream cigarette smoke by liquid chromatography/electrospray ionization tandem mass spectrometry under ISO and "Health Canada intensive" machine smoking regimens. Talanta. 2013;115:435–41.
123. Deng H, Yang F, Li Z, Bian Z, Fan Z, Wang Y et al. Rapid determination of 9 aromatic amines in mainstream cigarette smoke by modified dispersive liquid liquid microextraction and ultraperformance convergence chromatography tandem mass spectrometry. J Chromatogr A. 2017;1507:37–44.
124. Sampson MM, Chambers DM, Pazo DY, Moliere F, Blount BC, Watson CH. Simultaneous analysis of 22 volatile organic compounds in cigarette smoke using gas sampling bags for high-throughput solid-phase microextraction. Anal Chem. 2014;86:7088–95.
125. Pazo DY, Moliere F, Sampson MM, Reese CM, Agnew-Heard KA, Walters MJ et al. Mainstream smoke levels of volatile organic compounds in 50 US domestic cigarette brands smoked with the ISO and Canadian intense protocols. Nicotine Tob Res. 2016;18:1886–94.
126. Saha S, Mistri R, Ray BC. A rapid and selective method for simultaneous determination of six toxic phenolic compounds in mainstream cigarette smoke using single-drop microextraction followed by liquid chromatography-tandem mass spectrometry. Anal Bioanal Chem. 2013;405:9265–72.
127. Wu J, Rickert WS, Masters A. An improved high performance liquid chromatography-fluorescence detection method for the analysis of major phenolic compounds in cigarette smoke and smokeless tobacco products. J Chromatogr A. 2012;1264:40–7.
128. Savareear B, Brokl M, Wright C, Focant JF. Thermal desorption comprehensive two-dimensional gas chromatography coupled to time of flight mass spectrometry for vapour phase mainstream tobacco smoke analysis. J Chromatogr A. 2017;1525:126–37.
129. Brokl M, Bishop L, Wright CG, Liu C, McAdam K, Focant JF. Analysis of mainstream tobacco smoke particulate phase using comprehensive two-dimensional gas chromatography time-of-flight mass spectrometry. J Separation Sci. 2013;36:1037–44.

130. Pappas RS, Fresquez MR, Watson CH. Cigarette smoke cadmium breakthrough from traditional filters: implications for exposure. J Anal Toxicol. 2014;39:45–51.
131. Fresquez MR, Gonzalez-Jimenez N, Gray N, Watson CH, Pappas RS. High-throughput determination of mercury in tobacco and mainstream smoke from little cigars. J Anal Toxicol. 2015;39:545–50.
132. Fresquez MR, Pappas RS, Watson CH. Establishment of toxic metal reference range in tobacco from US cigarettes. J Anal Toxicol. 2013;37:298–304.
133. Zhang ZW, Xu YB, Wang CH, Chen KB, Tong HW, Liu SM. Direct determination of hydrogen cyanide in cigarette mainstream smoke by ion chromatography with pulsed amperometric detection. J Chromatogr A. 2011;1218:1016–9.
134. Mahernia S, Amanlou A, Kiaee G, Amanlou M. Determination of hydrogen cyanide concentration in mainstream smoke of tobacco products by polarography. J Environ Health Sci Eng. 2015;13:57.
135. Shihadeh A, Schubert J, Klaiany J, El Sabban M, Luch A, Saliba NA. Toxicant content, physical properties and biological activity of waterpipe tobacco smoke and its tobacco-free alternatives. Tob Control. 2015;24:i22–30.
136. Cecil TL, Brewer TM, Young M, Holman MR. Acrolein yields in mainstream smoke from commercial cigarette and little cigar tobacco products. Nicotine Tob Res. 2017;19:865–70.
137. Schubert J, Kappenstein O, Luch A, Schulz TG. Analysis of primary aromatic amines in the mainstream waterpipe smoke using liquid chromatography–electrospray ionization tandem mass spectrometry. J Chromatogr A. 2011;1218:5628–37.
138. Jacob P III, Abu Raddaha AH, Dempsey D, Havel C, Peng M, Yu I et al. Comparison of nicotine and carcinogen exposure with water pipe and cigarette smoking. Cancer Epidemiol Biomarkers Prev. 2013;22:765–72.
139. Hamad SH, Johnson NM, Tefft ME, Brinkman MC, Gordon SM, Clark PI et al. Little cigars vs 3R4F cigarette: physical properties and HPHC yields. Tob Regul Sci. 2017;3:459–78.
140. Goel R, Trushin N, Reilly SM, Bitzer Z, Muscat J, Foulds J et al. A survey of nicotine yields in small cigar smoke: influence of cigar design and smoking regimens. Nicotine Tob Res. 2018;20(10):1250–7.
141. Apsley A, Galea KS, Sanchez-Jimenez A, Semple S, Wareing H, van Tongeren M. Assessment of polycyclic aromatic hydrocarbons, carbon monoxide, nicotine, metal contents and particle size distribution of mainstream shisha smoke. J Environ Health Res. 2011;11:93–103.
142. Fresquez MR, Gonzalez-Jimenez N, Gray N, Valentin-Blasini L, Watson CH, Pappas RS. Electrothermal vaporization-QQQ-ICP-MS for determination of chromium in mainstream cigarette smoke particulate. J Anal Toxicol. 2017;41:307–12.
143. Ashley M, Dixon M, Prasad K. Relationship between cigarette format and mouth-level exposure to tar and nicotine in smokers of Russian king-size cigarettes. Regul Toxicol Pharmacol. 2014;70:430–7.
144. Lawler TS, Stanfill SB, de Castro BR, Lisko JG, Duncan BW, Richter P et al. Surveillance of nicotine and pH in cigarette and cigar filler. Tob Regul Sci. 2017;3:101–16.
145. Mahmood T, Zaman M. Comparative assessment of total nicotine content of the cigarette brands available in Peshawar, Pakistan. Pak J Chest Med. 2015;16.
146. Kulak JA, Goniewicz ML, Giovino GA, Travers MJ. Nicotine and pH in waterpipe tobacco. Tob Regul Sci. 2017;3:102–7.
147. Hadidi KA, Mohammed FI. Nicotine content in tobacco used in hubble-bubble smoking. Saudi Med J. 2004;25:912–7.
148. Sepetdjian E, Saliba N, Shihadeh A. Carcinogenic PAH in waterpipe charcoal products. Food Chem Toxicol. 2010;48:3242–5.
149. Nguyen T, Hlangothi D, Matrinez RA, Jacob D, Anthony K, Nance H et al. Charcoal burning as a source of polyaromatic hydrocarbons in waterpipe smoking. J Environ Sci Health B. 2013;48:1097–102.

150. Schubert J, Muller FD, Schmidt R, Luch A, Schultz TG. Waterpipe smoke: source of toxic and carcinogenic VOCs, phenols and heavy metals? Arch Toxicol. 2015;89:2129–39.
151. Saadawi R, Figueroa JAL, Hanley T, Caruso J. The hookah series part 1: total metal analysis in hookah tobacco (narghile, shisha) – an initial study. Anal Meth. 2014;4:3604–11.
152. Saadawi R, Hachmoeller O, Winfough M, Hanley T, Caruso JA, Figueroa JAL. The hookah series part 2: elemental analysis and arsenic speciation in hookah charcoals. J Anal Atom Spectrom. 2014;29:2146–58.
153. Talhout R, Schulz T, Florek E, Van Benthem J, Wester P, Opperhuizen A. Hazardous compounds in tobacco smoke. Int J Environ Res Public Health. 2011;8:613–28.
154. Harmful and potentially harmful constituents in tobacco products and tobacco smoke; established list. Fed Regist. 2012;77, No. 64:20034–7.
155. Stepanov I, Villalta PW, Knezevich A, Jensen J, Hatsukami DK, Hecht SS. Analysis of 23 polycyclic aromatic hydrocarbons in smokeless tobacco by gas chromatography-mass spectrometry. Chem Res Toxicol. 2010;23:66–73.
156. Ding YS, Zhang L, Jain RB, Jain N, Wang RY, Ashley DL et al. Levels of tobacco-specific nitrosamines and polycyclic aromatic hydrocarbons in mainstream smoke from different tobacco varieties. Cancer Epidemiol Biomarkers Prev. 2008;17:3366–71.
157. Hecht SS, Chen M, Yoder A, Jensen J, Hatsukami D, Le C et al. Longitudinal study of urinary phenanthrene metabolite ratios: effect of smoking on the diol epoxide pathway. Cancer Epidemiol Biomarkers Prev. 2005;14:2969–74.
158. Hochalter JB, Zhong Y, Han S, Carmella SG, Hecht SS. Quantitation of a minor enantiomer of phenanthrene tetraol in human urine: correlations with levels of overall phenanthrene tetraol, benzo[a]pyrene tetraol, and 1-hydroxypyrene. Chem Res Toxicol. 2011;24:262–8.
159. Hoffmann D, Adams JD, Brunnemann KD, Hecht SS. Assessment of tobacco-specific N-nitrosamines in tobacco products. Cancer Res. 1979;39:2505–9.
160. DeRoton C, Wiernik A, Wahlberg I, Vidal B. Factors influencing the formation of tobacco-specific nitrosamines in French air-cured tobaccos in trials and at the farm level. Beitr Tabakforsch Int. 2005;21:305–20.
161. van de Nobelen S, Kienhuis AS, Talhout R. An inventory of methods for the assessment of additive increased addictiveness of tobacco products. Nicotine Tob Res. 2016;18:1546–55.
162. Talhout R, Opperhuizen A, van Amsterdam JG. Sugars as tobacco ingredient: effects on mainstream smoke composition. Food Chem Toxicol. 2006;44:1789–98.
163. O'Connor RJ, Hurley PJ. Existing technologies to reduce specific toxicant emissions in cigarette smoke. Tob Control. 2008;18:139–48.
164. Fischer S, Spiegelhalder B, Eisenbarth J, Preussmann R. Investigations on the origin of tobacco-specific nitrosamines in mainstream smoke of cigarettes. Carcinogenesis. 1990;11:723–30.
165. Djordjevic MV, Sigountos CW, Brunnemann KD, Hoffmann D. Formation of 4-(methylnitrosamino)-4-(3-pyridyl)butyric acid in vitro and in mainstream cigarette smoke. J Agric Food Chem. 1991;39:209–13.
166. WHO Study Group on Tobacco Product Regulation. Report on the scientific basis of tobacco product regulation. Fifth report of a WHO study group (WHO Technical Report Series, No. 989). Geneva: World Health Organization; 2015.
167. Gray N, Borland R. Research required for the effective implementation of the Framework Convention on Tobacco Control, Articles 9 and 10. Nicotine Tob Res. 2013;15:777–88.
168. Pauly JL, O'Connor RJ, Paszkiewicz GM, Cummings KM, Djordjevic MV, Shields PG. Cigarette filter-based assays as proxies for toxicant exposure and smoking behavior – a literature review. Cancer Epidemiol Biomarkers Prev. 2009;18:3321–33.

9. Approaches to measuring and reducing toxicant concentrations in smokeless tobacco products

Stephen B. Stanfill, Division of Laboratory Sciences, National Center for Environmental Health, Centers for Disease Control and Prevention, Atlanta (GA), USA

Patricia Richter, Office of Noncommunicable Diseases, Injury and Environmental Health, Centers for Disease Control and Prevention, Atlanta (GA), USA

Sumitra Arora, National Centre for Integrated Pest Management, Indian Council of Agricultural Research, New Delhi, India

Contents
9.1 Introduction
9.2 Product composition
9.3 Agricultural practices and manufacturing processes that result in the formation and accumulation of harmful compounds
9.4 Product additives
9.5 Harmful agents in smokeless tobacco products and methods to reduce their effects
9.6 Detection of microorganisms
    9.6.1 Rapid detection of live microorganisms by cell viability
9.7 Overview of analytical methods used to measure toxicants in smokeless tobacco
9.8 Regulatory approaches and response
9.9 Policy recommendations and summary
    9.9.1 Disclaimer
9.10 References

9.1 Introduction

This section presents possible means of measuring and decreasing the levels of certain harmful agents in smokeless tobacco products, in conformity with the decision at the seventh session of the Conference of the Parties to the WHO FCTC in 2016 (COP7) *(1)*. Global estimates suggest that 367 million people in all six WHO regions use smokeless tobacco *(2)*, comprising almost 90% of adult tobacco consumption in south-east Asia (mainly Bangladesh and India), where its use exceeds that of combustible tobacco products *(3)*. This section builds on several reports on smokeless tobacco that provide more detail on product chemistry *(4, 5)*.

    Smokeless tobacco products are complex and contain both inorganic and organic chemicals that contribute to their addictive, toxic or carcinogenic effects *(6)*. Of the harmful chemicals in smokeless tobacco, TSNAs are the most abundant and include potent compounds, such as NNN and NNK, which are known human carcinogens; NNN is known to cause oral cancer *(6)*. During cultivation and production, the levels of these agents or their precursors can be increased by soil uptake, biosynthesis of alkaloids and microbial activity, including the formation of mycotoxins, nitrite and TSNAs *(4, 7)*. In certain product types, fire-curing can increase the levels of harmful agents, such as VOCs and PAHs, while the inclusion of certain additives (areca nut, tonka bean, *khat*, diphenyl ether, caffeine, pH-boosting agents) and chemicals, can increase addiction and carcinogenicity *(4, 5)*. Microorganisms play an important role throughout production, resulting in chemical changes during the production of tobacco products. Harmful and carcinogenic agents are generated throughout production *(8, 9, 10)*.

The wide variety of smokeless tobacco products ranges from simple to elaborate hand-made and industrial preparations consisting of tobacco mixed with a wide spectrum of non-tobacco plant materials and chemical additives. Manufactured and cottage-industry products include tobacco leaves, loose flaked tobacco, finely minced tobacco, pulverized tobacco, pressed cakes, tars, gel-like pastes, tobacco-containing toothpaste and pressed pellets. Smokeless tobacco products can be generally subdivided into four main categories on the basis of key ingredients (a similar scheme was presented previously) *(4, 6, 11)*:

- category 1, which contains tobacco with few or no alkaline modifiers;
- category 2, which contains tobacco and substantial amounts of alkaline agents;
- category 3, which contains tobacco, one or more alkaline agents and areca nut; and
- category 4, which contains tobacco mixed with other chemical or plant ingredients with
- additional bioactivity, such as stimulants.

Because of the constituents it contains, each product category harms health. The tobacco contained in these products contributes to their toxicity, carcinogenicity and addictiveness. Approximately 70 *Nicotiana* species are known; however, only a few are used to make smokeless tobacco products. In general, *Nicotiana* species have different levels of alkaloids *(12, 13)*; the most commonly used species, *N. tabacum* (cultivated tobacco), has a moderate amount of nicotine, whereas *N. rustica* has extremely high levels *(14)*. Some smokeless tobacco products contain *N. glauca* (tree tobacco), which has high concentrations of the tobacco alkaloid anabasine; accidental poisoning and fatalities have been associated with use of this species in a few cases *(15, 16)*. The presence of different species of tobacco can result in exposure to different proportions of tobacco alkaloids, which can contribute to addictiveness, toxicity and carcinogenicity, as tobacco alkaloids are necessary precursors of TSNAs *(10, 12, 14, 17, 18)*. The presence of nicotine, common to all smokeless tobacco products, promotes continued use and can result in repeated, often daily, exposure to carcinogens and toxicants *(4)*.

Although tobacco contains thousands of chemicals *(19)*, many of the carcinogenic agents in smokeless tobacco products are not present or are present at very low concentrations in newly transplanted tobacco *(17)* but form and accumulate between cultivation and the finished product. Certain chemical constituents of tobacco are synthesized from metals and nitrate, which are absorbed by tobacco during growth *(18)*. According to the soil characteristics and the environment in which tobacco is grown, certain microorganisms may occur naturally inside the plant (endophytes) or on its surface (epiphytes) *(20, 21)*. During cultivation, harvesting and processing, other microorganisms may be deposited on tobacco from air, water, soil or manure (if used) or introduced by human handling or as additives. The microbial communities present during production and in the final product can affect the product constituents. Smokeless tobacco production also includes steps such as fire-curing of green leaves, which can introduce additional chemical agents, such as VOCs, phenolic compounds and PAHs *(22–24)*. During processes such as fermentation, ageing and storage, microorganisms are viable and metabolically active in tobacco *(8–10, 25)*, and their presence can result in the generation of reactive agents such as nitrite and other harmful by-products (aflatoxins, endotoxins, TSNAs, other nitrosamines, ethyl carbamate) *(9, 26, 27)* (Table 9.1). Studies published by tobacco industry scientists have shown that toxicant levels can be lowered by changes in growing practices, manufacturing processes and continuous monitoring *(19, 28)*.

Table 9.1. Potential sources of carcinogens, toxicants and biologically active compounds in smokeless tobacco products, originating mainly from soil, microbial action, fire-curing and additives

Agent class	IARC-classified carcinogens (groups 1, 2A, 2B), toxicants or biologically active compounds	Potential source or cause
Metals and metalloids	Group 1: arsenic, beryllium, cadmium, nickel compounds, polonium-210 Group 2A: inorganic lead compounds Group 2B: cobalt sensitization: aluminium, chromium, cobalt, nickel Dermal irritants: barium, mercury Copper in areca nut may contribute to oral submucosal fibrosis	Absorption from the soil or by deposition of soil particles on tobacco leaf surfaces; potentially present in other ingredients (betel leaf, slaked lime) used with tobacco
Nitrosation agents	Group 2B: nitrite	Generated by microorganisms
Mycotoxins	Group 1: aflatoxins (mixtures of) Group 2B: aflatoxin M1, ochratoxin A, sterigmatocystin	Formed by fungi (*Aspergillus*)
Nitrosamines Tobacco-specific *N'*-nitrosoamines	Group 1: *N'*-nitrosonornicotine, 4-(methylnitrosamino)-1-(3-pyridyl)-1-butanone and 4-(methylnitrosamino)-1-(3-pyridyl)-1-butanol (NNAL)	Formed by microbial-generated nitrite followed by nitrosation during curing, fermentation and ageing (nitrite reacts with alkaloids)
Volatile *N'*-nitrosoamines	Group 2A: *N*-nitrosodimethylamine Group 2B: *N*-nitrosopyrrolidine, *N*-nitrosopiperidine, *N*-nitrosomorpholine, *N*-nitrosodiethanolamine	Formed by microbial-generated nitrite followed by nitrosation during curing, fermentation and ageing (nitrite reacts with certain secondary and tertiary amines)
Nitrosoacids	Group 2B: *N*-nitrososarcosine	
Carbamates	Group 2A: ethyl carbamate	Formed during fermentation (reaction of urea and ethanol)

Polycyclic aromatic hydrocarbons	Group 1: benzo[a]pyrene Group 2A: dibenz[a,h]anthracene Group 2B: benz[a]anthracene, benzo[b]fluoranthene, benzo[j]fluoranthene, benzo[k]fluoranthene, dibenzo[a,i]pyrene, dibenzo[a,l]pyrene, indeno[1,2,3-cd]pyrene, 5-methylchrysene, naphthalene	Deposited on tobacco during fire-curing
Volatile aldehydes	Group 1: formaldehyde Group 2B: acetaldehyde	Deposited on tobacco during fire-curing
Non-tobacco plant materials	Group 1: areca nut Liver toxicant: tonka bean Stimulant: *khat*, caffeine	Additives

Source: reference 29.

Smokeless tobacco users are exposed to inorganic, organic and biological agents and their interactions. The areas of concern depend on product types and the region of the world because of different soil constituents (metals, nitrate, microbial communities), tobacco species grown, agricultural practices, product processing steps and permissible inclusion of certain additives.

9.2 Product composition

Tobacco products are highly complex mixtures that contain nicotine, Group 1 carcinogens (as evaluated by IARC working groups) *(6))* and toxic metals, such as arsenic, cadmium and lead *(30)*. These products also contain nitrate, which can be metabolized by microbes to nitrite; nitrite initiates the formation of TSNAs and other nitrosamines *(9, 31)*. Certain NO_x gases present during fire-curing can also cause nitrosation *(32)*. In particular, nitrosation of nornicotine yields N'-nitrosonornicotine, a known human carcinogen (IARC Group 1) which is known to cause oral cancer *(6)*. The concentrations of these agents in various smokeless tobacco products have been surveyed extensively *(4–6, 33)*.

The concentrations of total and free nicotine vary widely among product types. Detectable concentrations of total nicotine in smokeless tobacco products range from < 0.2 mg/g to 95 mg/g *(11, 14)*. High nicotine concentrations are found in products such as dry snuff and Sudanese *toombak (14, 34)*. The high concentrations in certain products, such as *toombak*, *gul*, tobacco leaf and *zarda*, are attributed to use of the *N. rustica* variety of tobacco, a nicotine-enriched tobacco species *(11, 14)*. For a given product, the percentage of nicotine that is unprotonated is calculated from the pH of the product and the appropriate pK_a of nicotine (8.02) with the Henderson–Hasselbach equation *(35)*. The pH of smokeless tobacco products has been reported to range from pH 4.6 to pH 11.8; in this range, < 0.1–99.9% of the nicotine present would be free nicotine *(3)*. The highest pH values were observed for *nass (36)* and also *iqmik* which is used in Alaska *(23)*. As nicotine is converted to free nicotine, it is more readily released from tobacco during use and passes across biological membranes *(37–39)*.

Limiting both total nicotine concentrations and the permissible pH could maintain lower free nicotine concentrations in smokeless tobacco products.

Metals and metalloids are naturally present in soil *(40)*, and a number of these and other inorganic compounds are absorbed into tobacco plant tissue *(41)* and are also present in smokeless tobacco products *(4, 42–45)*. In a study by Golia et al. *(41)*, certain metals, including cadmium, lead, zinc and copper, were in higher concentrations on tobacco leaf surfaces near the ground (primers) than on leaves higher up the tobacco stalk. This was the case for Oriental, flue and Burley tobaccos. It is possible that soils containing metals may be splashed on to tobacco leaves during rain or may be deposited on tobacco leaves that are spread out on the ground.

The levels of metals in smokeless tobacco products vary among countries. The amounts in products may be influenced by the soil content, pH *(46)* and industrial contamination *(43, 47–48)*. The concentration of metals varies by product type and country of origin. High concentrations have been reported in products from Ghana, India and Pakistan, with higher levels of lead, nickel and chromium in products from Pakistan than in those from Ghana and India. Products from India had high iron and copper levels, while those from Ghana had higher levels of copper, iron and aluminium than those from the other countries *(43)*. Swedish *snus* has a very low metal content *(49)* due to deliberate selection of tobacco *(28)* or other processes, such as washing tobacco leaves *(28, 50)*.

The concentrations of total TSNAs range from < 0.5 to 12 630 µg/g, with the highest concentration in Sudanese *toombak* products. The highest values are consistently seen in products that undergo fermentation (e.g. *toombak, khaini*, dry snuff, moist snuff) *(11, 14, 27, 51–53)*. Fermentation is used in tobacco processing to enhance taste but is characterized by microbial proliferation and active metabolism to nitrite, which reacts with natural tobacco alkaloids to form TSNAs *(8, 9)*. The lowest TSNA concentrations are found in products such as *snus* and dissolvable oral tobacco that do not contain fermented tobacco. ("Dissolvables" are products that essentially include tobacco and other ingredients compressed into chewable tablets, wafer-like strips or elongated sticks.) *Snus* products that are not fermented but are pasteurized eliminate microorganisms and result in low TSNA concentrations such as that in moist snuff. Some products in India that are labelled "*snus*" have high TSNA levels *(53)*.

A wide variety of microbes, including bacteria (*Staphylococcus, Corynebacterium, Lactobacillus, Enterobacter, Clostridium, Bacillus, Serratia* and *Escherichia*) and fungi (*Aspergillus, Fusarium, Cladosporium, Candida, Alternaria* and *Acremonium*), can generate nitrite. Some of these organisms are potentially harmful or pathogenic *(26, 54, 55)*. Microbial production of nitrite is a major determinant of the concentrations of TSNAs and other nitrosamines in tobacco *(9, 36, 56)*. A number of species, including *Staphylococcus, Corynebacterium,*

Lactobacillus and genera in the *Enterobacteriaceae* family, present in smokeless tobacco products contain a respiratory nitrate reductase and associated transporters that export nitrite (encoded in the *nar* gene operon) *(57)*. Dry snuff also contains a periplasmic nitrate reductase (*nap* gene), mainly in species in the *Enterobacteriaceae* family, which may also result in the release of nitrite *(58)*. In some species, *nar* genes are activated when oxygen levels decrease *(59)*.

As tobacco is cured, tobacco cells rupture and release nitrate *(31)*. Nitrate-reducing microbes can convert nitrate to nitrite under low-oxygen conditions. Certain bacteria can convert nitroalkanes (if present) to nitrite in the presence of oxygen *(60)*. Nitrite that is not further metabolized is generally expelled due to its toxicity. Fermentation, ageing, storage and tightly sealed packaging, with low oxygen, provide conditions in which respiration by nitrate-reducing bacteria involves generating and exporting nitrite *(8–10)*. Regardless of how nitrite is formed, once formed, it can combine with alkaloids and other secondary and tertiary amines to form nitrosamines *(31)*. Certain secondary or tertiary amines can also react with nitrite to form volatile nitrosamines, such as *N*-nitrosodimethylamine, and nitrosoacids *(31)*.

9.3 Agricultural practices and manufacturing processes that result in the formation and accumulation of harmful compounds

Many of the harmful agents in tobacco products are at lower levels or almost entirely absent from freshly transplanted tobacco *(61)* but begin accumulating in the early stages of cultivation and curing. Growing tobacco involves a number of agronomic decisions – type or species of tobacco to be grown, harvest timing and procedures, types of fertilizers and agrichemicals and application rates – and environmental factors such as soil composition, climate and rainfall, which collectively determine the chemistry of a product *(4)*.

The constituents of a finished smokeless tobacco product, including nicotine (in its various ionic forms), toxicants and carcinogens, result from the presence of inorganic, organic and biological agents and their interactions *(4, 19, 31, 40, 42, 43)*. Naturally occurring organic constituents of tobacco used in smokeless tobacco products include lignins, fatty acids, sugars, alkaloids, terpenoids, polyphenols, cembranoids and carotenoid pigments. The breakdown products of some of these constituents contribute to volatile flavour chemicals and the colouration of cured tobacco *(18, 19, 61)*.

Methods of tobacco growing, harvesting and curing vary geographically, which may affect product chemistry *(19)*. During the growing season, tobacco can absorb metals *(41, 43)*, as noted above. Addition of nitrate fertilizer boosts plant mass and also increases the concentrations of nitrate, nicotine and other alkaloids in leaves – all precursors of TSNAs *(18, 31, 61, 62, 63)*. Agricultural plant protection chemicals applied to tobacco may also persist on the tobacco at harvest *(64)*. Exposure to the soil and atmosphere during cultivation and harvesting can result in the introduction of microbes and also fungi that generate mycotoxins. At harvesting, tobacco may be laid on the ground *(18)* or piled in a field for extended periods (≤ 45 days) *(65)*. Contact with the soil offers an opportunity for the introduction of microorganisms and other organisms (insects) into the tobacco.

After harvest, tobacco leaves are cured by sun, air, flue or fire, the four primary means for traditional products (except products such as dissolvables). Sun-curing involves drying tobacco in the sun, whereas air-curing involves hanging tobacco in a well-ventilated barn. Flue-curing is done by exposing the leaves to elevated temperatures fuelled by wood, coal, oil or liquid petroleum gas in a tightly constructed building equipped with ventilators and flues *(18)*. During fire-curing, leaves are dried in smoke from the burning of wood or sawdust, when smoke-derived chemicals such as PAHs and VOCs can accumulate on the tobacco leaves *(22, 23, 24)*. The levels of PAHs in fire-cured tobacco exceed those in air-cured tobacco *(23)* (see Fig. 9.1).

Fig. 9.1. Effects of type of curing on levels of polycyclic aromatic hydrocarbons in tobacco used to make smokeless products

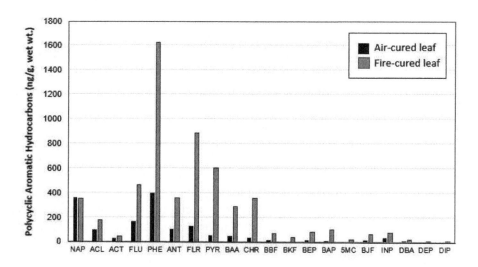

5MC: 5-methylchrysene; ACL: acenaphthylene; ACT: acenaphthene; ANT: anthracene; BAA: benz[*a*]anthracene; BAP: benzo[*a*]pyrene; BBF: benzo[*b*]fluoranthene; BEP: benzo[*e*]pyrene; BJF: Benzo[*j*]fluoranthene; BKF: benzo[*k*]fluoranthene; CHR: chrysene; DBA: Dibenz[*ah*]anthracene; DEP: Dibenzo[*ae*]pyrene; DIP: Dibenzo[*ai*]pyrene; FLR: fluoranthene; FLU: fluorene; INP: Indeno[1,2,3-*cd*]pyrene; NAP: naphthalene; PHE: phenanthrene; PYR: pyrene. BAA, BAP, BBF and BKF are on the FDA list of HPHCs *(60)*. *Source*: reference *23*.

After curing, tobacco may undergo further ageing, fermentation or long-term storage, and these production stages may increase microbial proliferation in anaerobic conditions. During these phases, dramatic chemical changes occur that include rapid catalysis of sugars, increases in pH and temperature and increases in the concentrations of nitrite and TSNAs *(8)*. Microbes may be deliberately added during fermentation *(9)*. Also, viable microorganisms are present in purchased products *(25, 55)*. Aflatoxins, carcinogens produced by certain *Aspergillus* mould species, may accumulate during cultivation or subsequent production steps and have been found in certain smokeless tobacco products *(7, 66)*. Microbial action is clearly a driver of the evolving product chemistry.

9.4 Product additives

During manufacture, almost all products are augmented with some level of additives, including flavouring compounds that are not necessarily toxic but may add to the attractiveness of products, thus promoting initiation or fostering use. Flavourings used in these products are drawn from extracts, oleoresins, spice powders, individual compounds (e.g. menthol, vanillin) and more than 60 essential oils *(67)*. Some substances used as additives that have known adverse health effects include *khat* (an addictive plant), tonka bean (a liver toxin) and areca nut (an addictive psychoactive substance); areca nut is known to cause cancer and oral malformations, such as oral submucosal fibrosis *(68, 69)*. In some cases, areca nut and other ingredients are added at substantial levels to tobacco-containing products (*zarda, rapé*) or hand-made preparations (e.g. betel quid, *tombol, dohra, moawa* and *mainpuri*) *(3, 7, 70, 71)*.

9.5 Harmful agents in smokeless tobacco products and methods to reduce their effects

Damaging agents in smokeless tobacco products can be reduced in several ways, including by changes to smokeless tobacco production *(19)*. Some of these measures may help decrease TSNA levels *(9, 28, 72)*. Means of altering the levels of harmful agents may include the following.

- **Fertilizers**: decreasing the use of nitrate-containing fertilizers or using other fertilizers (e.g. urea or other non-nitrate fertilizers late in the growing season) could limit the formation of nitrosamines by decreasing the accumulation of nitrate at harvest *(18, 61)*.
- **Surface disinfection of harvested tobacco**: washing harvested leaf material with a dilute bleach solution (hypochlorite:water solution) can disinfect leaf surfaces. This procedure removes not only microorganisms but also soil *(50,56)*. Protocols effective in disinfecting food

products, such as leafy greens, may be useful for eliminating surface contamination. Disinfection of tobacco leaves have been shown reduce the levels of surface microbes *(28, 50, 56)*.

- Practices used effectively in the food industry can **reduce contamination of tobacco by microorganisms and soil**. In Sweden, *snus* must meet food regulatory standards, which has led to a product with reproducibly lower toxicant levels *(28)*.

- **Electron beam technology:** high-energy electron beam irradiation is a non-thermal, chemical-free technology, in which compact linear accelerators generate highly energetic (10 MeV) electrons, that has been used to pasteurize foods and sterilize medical devices. The technology is often called "cold pasteurization", because it irradiates products without generating excess heat, which might cause undesirable product changes *(73)*. Although the appropriate dose of irradiation may depend on the product, electron beam technology could eliminate viable organisms present in tobacco that generate mycotoxins or nitrite and remove viable and potentially harmful organisms in smokeless tobacco. Because electron beam technology does not generate heat, it may be possible to use it to sterilize tobacco, ingredients, packaging material and the final product. Irradiation of tobacco early in production, especially before fermentation and ageing, may eliminate some microorganisms that generate nitrite resulting in lower TSNA as compared to non-irradiated tobacco.

- **Changes related to fire-curing and air-curing:** the levels of PAHs and volatile aldehydes are higher in fire-cured tobacco *(22, 23, 30, 74)*; the process of fire-curing should be omitted, if possible. During air-curing, microbial treatment with *Bacillus amyloliquefaciens* DA9, an organism that efficiently accumulates nitrite, may decrease nitrite and TSNA concentrations *(75)*. During curing, conditions of high humidity (70%) and temperatures ranging from 10°C to 32°C that may be found in poorly ventilated curing facilities are conducive to mould growth and the potential formation of mycotoxins, such as aflatoxins and ochratoxins *(11, 76)*. Efforts should be made to prevent these conditions and monitor to ensure mould growth is prevented.

- **Pasteurization:** the Swedish Match Company uses pasteurization in the preparation of Swedish *snus*, a smokeless tobacco product with a low TSNA concentration *(28, 77)*. Ground, blended leaf tobacco is mixed with water and sodium chloride in closed process blenders, then heat-treated with hot water and steam injection to achieve temperatures up to 80–100 °C for several hours. The mixture is cooled, and other ingredients, such as flavours and humectants, are added before the product is packaged. This process changes taste, reduces microbial activity and yields a product with a shelf life of 14–30 weeks when refrigerated, according to a manufacturer *(28, 76)*. *Snus* products continue to have low TSNA levels *(28)*, which have been reduced and maintained over several decades. In 2017, the average level of NNN plus NNK in the products of the Swedish Match Company was 0.47 µg/g product, dry weight *(28, 77)*.

- **Modified fermentation:** fermentation is the main process in the production of tobacco products that increases levels of nitrite and TSNAs *(8)*, and it should be modified *(9)* or avoided entirely. During fermentation, microbial populations can rapidly proliferate or decrease and can increase pH, oxalate, nitrite and TSNA levels *(8, 9)*. The steps in the formation of TSNAs are shown in Fig. 9.2. Cleaning of fermentation equipment before use and addition of non-nitrite-producing microbes at that stage can reduce TSNA levels *(9)*. Use of oxygen-rich endothermic fermentation *(78)* may also decrease the anaerobic respiratory nitrate reduction that occurs during tobacco fermentation *(8, 9)*. Addition of suitable fermentation organisms that do not produce nitrite *(9)* may be required after irradiation. A product free of microbes (as is the case for *snus*) would generally be seen as beneficial *(8, 9, 28)* as compared with products harbouring unknown microorganisms, including those that generate nitrite or are otherwise harmful to the user.

- **Microwave technology:** this technology has been used safely over the past 50 years for cooking, drying, pasteurizing, sterilizing, bacterial destruction and enzyme deactivation in food products, but has only recently been used in continuous in-line processing in the United States. The principle of microwave technology is simple: when a substance is subjected to microwave radiation, water molecules in the material absorb the energy and internal heat is generated volumetrically by their molecular vibration. Continuous microwave heating equipment and technology was first used for aseptic processing of food products in a North Carolina processing facility in 2008. Microwave technology has been used in producing pharmaceutical and nutraceutical products. In India, commercial ready-to-eat meals were processed in-pack using proprietary microwave technology. Microwave energy likely increases temperature inside of the microbial cell, denaturing critical biomolecules and resulting in reduced cell efficacy and often death *(79–83)*. Although not used for tobacco presently, microwave technology that has been used successfully with food products may be assessed for its applicability in reducing or eliminating microorganisms in tobacco or tobacco products.

- **Nitrite scavengers:** although the most efficient approach to minimizing nitrite in smokeless tobacco products is complete elimination of nitrite-generating organisms, the use of agents that act as nitrite scavengers has been investigated *(28)*. Certain polyphenols, vitamin C, tocopherol, green tea extract, a green tea component (epigallocatechin gallate) and morpholine could be investigated as potential nitrite scavengers to neutralize nitrite generated during smokeless tobacco processing *(28, 84, 85)*.
- **Product refrigeration:** continuing growth of microbial populations and potential formation of nitrosamine compounds in products can be controlled by refrigeration. One manufacturer encourages refrigeration of its products, including at points of sale *(28)*.

Fig. 9.2. Processes that can increase the concentrations of nitrosamines in smokeless tobacco products

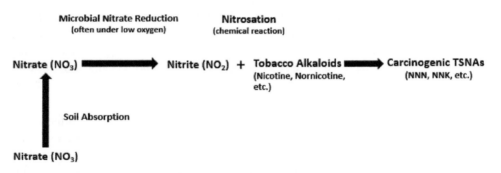

NNK: 4-(methylnitrosamino)-1-(3-pyridyl)-1-butanone; NNN: *N*´-nitrosonornicotine.
Source: references *9, 10, 18, 31, 61*

Smokeless tobacco production can result in the presence of nitrosamines and microbes, including bacteria and fungi. Tobacco contains nitrate, tobacco alkaloids and certain secondary and tertiary amines, which play a role in the formation of nitrosamines. The most abundant carcinogens in tobacco products are TSNAs, but they also include other nitrosamines such as *N*-nitrosamino acids and volatile *N*-nitrosamines. Nitrosamines can be formed at various stages in smokeless tobacco production. One TSNA, NNN, is a human carcinogen known to cause oral cancer *(6, 54)*.

9.6 Detection of microorganisms

Culture counts in standard culture media: microbial contamination can be assessed on culture media plates, but this method is time-consuming, as it involves pouring media plates, streaking and interpreting growth; it may be difficult to find the appropriate media for testing specific organisms in tobacco *(87)*. Many different microorganisms grow in tobacco, and they can be cultured on various media, including tryptic soy agar *(55)*, sheep blood agar, mannitol salt agar and MacConkey agar plates *(25)*. According to the "great plate anomaly" *(87)*, certain taxa may be overrepresented, such that less abundant yet important taxa remain unidentified, and no single medium will capture the microbial diversity of tobacco products. The expertise, volume of supplies and requirement for sterile facilities to set up, inoculate, incubate and correctly interpret media plates may be too costly in certain localities, and less expensive, less complicated methods may have practical advantages.

Culture counts on disposable microbiological films: an alternative to culture media is a small two-layer film. The upper film is lifted open so that the inoculum can be spread on the lower film surface, which is then incubated and read for the extent of microbial growth. Although other products may exist, media for testing a wide variety of bacteria are made by 3M Corp.

9.6.1 Rapid detection of live microorganisms by cell viability

Plate reader format: it would be helpful to know the extent of microbial contamination on tobacco at various points during cultivation and production and in final tobacco products. An assay of microbial viability based on luciferase (e.g. BacTiter-Glo) indicates the number of viable microorganisms by quantification of ATP, a molecule present in all living cells *(88)*. A single reagent causes cell lysis and generation of a luminescent signal, which is proportional to the ATP concentration and is directly proportional to the number of viable cells. This assay detects a variety of bacteria and fungi. Data can be recorded 5 min after addition and mixing of the reagent. The assay can detect as few as 10 bacterial cells on a luminometer or a CCD camera. The application can be used with a 96-well plate reader *(88)*.

Handheld luminometer: ATP measurement has been miniaturized in handheld readers made by at least six manufacturers. Handheld luminometers to test for ATP allow determination of the extent of microbial contamination in 5–20 seconds. Several manufacturers make ATP luminometers, and their capability has been compared. One manufacturer makes devices to enumerate specific bacterial groups, total viable count, *Enterobacteriaceae*, coliform, *E. coli* and *Listeria* spp. These devices are used in various industries (e.g. health care, pharmaceuticals, water treatment, dairy, meat, produce). A portable unit could be used to test tobacco products, preparations or related ingredients on site within minutes *(90, 91)*.

Detection of organisms by qPCR: organisms could also be detected by quantitative polymerase chain reaction (qPCR), a technique in which probes against specific gene regions bind, replicate and can be quantified. These and other organisms can be detected with the use of properly designed molecular probes. Molecular probes can be designed for qPCR that are specific for nitrate-reducing organisms and genes (e.g. *narG*, *napA*, etc.) and used quantification by qPCR *(89)*.

Detection of nitrate-reducing organisms: in addition to enumeration of microorganisms to assess the extent of contamination, actual nitrate-reducing organisms in tobacco products should be identified. One means may involve growing microbes on nitrate agar with an added nitrite indicator. Another approach is use of Greise reagent (produced from sulfanilamide with NO_2, followed by addition of naphthylethylenediamine) *(92)*.

9.7 Overview of analytical methods used to measure toxicants in smokeless tobacco

Analytical methods for measuring tobacco constituents (Table 9.2) depend mainly on GC and LC linked to various types of mass spectrometers (single and triple quad; MS, MS/MS) and other detectors, such as inductively coupled plasma and flame ionization detectors. This is not an exhaustive list, but it shows some recent papers describing approaches utilized by various laboratories for analysing tobacco products.

Table 9.2. Methods for measuring tobacco constituents

Analyte	Measurement method
Nicotine *(35, 93)*	GC-MS, GC/FID
Minor alkaloids *(13)*	GC-MS/MS
Flavours and non-tobacco plant constituents, including coumarin and camphor *(94)*	GC-MS
PAHs *(22, 23, 24)*	GC-MS
VOCs *(22)*	GC-MS
Toxic metals *(42)*	ICP-MS
TSNA *(11, 52, 53)*	LC–MS/MS
Aflatoxins *(7)*	**UHPLC-MS/MS**
Areca-nut-related compounds *(95)*	LC/MS /MS
pH *(34, 35)*	Ion probe
Nitrate and nitrite *(96)*	Ion chromatography
Fungi *(76)*	Culture media specific to fungi
Bacteria *(25)*	Culture media specific to bacteria
Identification of metals and alkaline agents *(97)*	X-ray fluorescence (XRF)
Identification of tobacco species (*Nicotiana tabacum, N. rustica, N. glauca*), non-tobacco plant materials (e.g. areca nut, tonka bean) and alkaline agents (magnesium carbonate, slake lime) *(11, 52, 71)*	Fourier transform/infrared spectroscopy (FT/IR)

GC: gas chromatography; LC: liquid chromatography; MS: mass spectrometry; PAHs: polycyclic aromatic hydrocarbons; TSNAs: tobacco-specific nitrosamines; VOCs: volatile organic compounds.

Other techniques that have been used to measure these analytes could be substituted, with appropriate validation, if necessary. For regulatory agencies with limited funding or space, compact or portable instrumentation is available, including GC-MS systems that could allow the detection and quantification of nicotine, minor alkaloids, arecoline (areca nut), flavours and non-tobacco plant constituents (coumarin, diphenyl ether, camphor). Some GC-MS systems are miniaturized, mobile and cost less than conventional bench systems.

9.8 Regulatory approaches and responses

Various regulatory and other responses have been used. While the European Union regulates smokeless tobacco under directive 2001/37/EC *(98)*, Sweden is exempted from the prohibition on marketing of certain types of tobacco for oral use. In the United States, smokeless tobacco has been regulated by the FDA since the passage of the Family Smoking Prevention and Tobacco Control Act in 2009 *(99)*. Elsewhere, there is limited regulation of smokeless tobacco use, such as prohibiting products or regulating the contents. Despite attempts by the states of Andhra Pradesh, Bihar, Goa, Maharashtra and Tamil Nadu in India to prohibit sales of *gutkha* and *pan masala* as food products *(3)*, manufacturers have circumvented the ban *(100)*, in the case of *gutkha* by dividing the product into attached packages with tobacco flakes in one and the other constituents in the other. However, Goa has reportedly sustained a ban on smokeless tobacco products through the Goa Public Health (Amendment) Act, 2005 *(101)*.

An example of a targeted approach to reducing toxicants in smokeless tobacco is the rule proposed by the FDA in January 2017 for NNN in finished smokeless tobacco products *(72)*, which was based on the carcinogenicity of NNN, a major contributor to elevated oral cancer risks associated with smokeless tobacco use. Elements of the standard include:

- required expiration date and, if applicable, storage conditions (e.g. refrigeration at point of sale) for finished smokeless tobacco products;
- a mean level of NNN in any batch of finished smokeless tobacco products of ≤ 1 µg/g of tobacco (dry weight) at any time up to the expiration date;
- testing to assess the stability of NNN levels in finished smokeless tobacco products and to establish and verify the product's expiration date and storage conditions; and
- batch-testing to determine whether the products conform to the proposed NNN level.

The FDA also proposed various evidence-based, broadly stated considerations with regard to this regulation *(72)*, which might be helpful for other regulatory bodies pursuing toxicant reduction in smokeless tobacco products.

- Regulation of smokeless tobacco toxicants may affect consumer perceptions of the harm of smokeless tobacco use.
- Users of smokeless tobacco products might interpret a reduction in one or more toxicants as resulting in less harm, which could reduce their motivation to quit. FDA, however, does not expect the proposed product standard to appreciably discourage cessation of smokeless tobacco products in such a way as to offset the beneficial public health impact from reduced cancer risk.
- Health messages must continue to educate consumers with evidence-based advice about all products.

Another approach is the GothiaTek standard instituted by Swedish Match, a *snus* manufacturer, in the late 1990s. It set maximum allowable concentrations of nitrite, TSNAs (NNN and NNK), *N*-nitrosodimethylamine, benzo[*a*]pyrene, aflatoxins, agrochemicals and various metals, and these have been met *(28, 102)*. *Snus* is made with air-cured tobacco that is pasteurized, fire-curing and fermentation are omitted, and products are refrigerated at points of sale and have very low TSNA concentrations *(22)*.

In contrast to the FDA product standard for NNN, GothiaTek limits are set on an "as is" (wet weight) basis *(22)*. Other components include:

- standards for raw materials, including guidance levels for agricultural residues
- quality control and quality assurance measures during manufacture
- heat treatment of tobacco
- reduced water content
- flavourings consistent with the Swedish Food Act for additives and flavourings.

The GothiaTek standard does not regulate product pH or nicotine content *(102)*, but, for certain harmful and carcinogenic agents, it serves as a reference for what is possible if tobacco product makers or regulators wish to decrease toxicant levels. The levels of most GothiaTek toxicants decreased or remained stable between 2009 and 2017. The agents included are related to soil content, curing, microbial activity, and agrochemical use and are shown in Table 9.3.

Table 9.3. Systematic decreases in toxicant levels regulated under GothiaTek in 2009, 2015 and 2017

Potential source and chemical component	2009 level[a]	2015 level[b]	2017 level[b]	Limit (2009)	Limit (2015, 2017)
Soil metal content					
Arsenic (µg/g)	0.1	< 0.06	0.06	0.5	0.25
Cadmium (µg/g)	0.6	0.28	0.27	1	0.5
Chromium (µg/g)	0.8	0.46	0.46	3	1.5
Lead (µg/g)	0.3	0.15	0.15	2	1
Mercury (µg/g)	NA	< 0.02	< 0.02	NA	0.02
Nickel (µg/g)	1.3	0.87	0.82	4.5	2.25
Curing[c]					
Acetaldehyde[d] (µg/g)	NA	6.5	6.3	NA	25
Crotonaldehyde[d] (µg/g)	NA	< 0.10	< 0.10	NA	0.75
Formaldehyde[d] (µg/g)	NA	2.3	2.3	NA	7.5
Benzo[a]pyrene[e] (ng/g)	1.1	< 0.6	< 0.6	20	1.25
Microorganisms (fungi)					
Aflatoxin (ng/g)	NA	< 2.1	< 2.1	NA	2.5
Ochratoxin A (ng/g)	NA	2.3	2	NA	10
Microorganisms (bacteria)					
Nitrite (µg/g)	2.0	1.7	1.5	7	3.5
NNN+NNK (µg/g)	1.6	0.39	0.47	10	0.95
NDMA (ng/g)	0.7	< 0.6	< 0.6	10	2.5
Agrochemicals (µg/g)	NA	b	b	NA	a

NA: not available; NDMA: *N*-nitrosodimethylamine; NNK: 4-(methylnitrosamino)-1-(3-pyridyl)-1-butanone; NNN: *N´*-nitrosonornicotine. These analytes are derived primarily from soil uptake (toxic metals), curing and the presence and activity of fungi and bacteria. Units are converted from per kilogram to per gram.
[a] In accordance with the Swedish Match Agrochemical Management Programme *(28)*.
[b] Below Swedish Match internal limits *(102)*.
[c] *Snus* is not fire-cured; these analytes are often formed during fire-curing in the manufacture of other products (dry and moist snuff).
[d] These aldehydes are volatile organic compounds.
[e] This compound is a polycyclic aromatic hydrocarbon.

9.9 Policy recommendations and summary

In this section, we have described research to identify the major characteristics of concern and approaches to measuring and decreasing the levels of certain harmful agents in smokeless tobacco products. We concur with the recommendations of TobReg in 2015 *(29)* for the establishment of regulatory limits for carcinogens in smokeless tobacco products, which are still valid and are reiterated below.

- **Reduce toxicity**: decrease or eliminate use of *N. rustica*; limit bacterial contamination, which can promote nitrosation and carcinogen formation; require that tobacco be flue- or sun-cured rather than fire- or air-cured to avoid bacterial growth; kill bacteria by pasteurization; improve storage conditions, such as by refrigerating products before sale; affix the date of manufacture; eliminate ingredients such as areca nut and tonka bean that are known to be carcinogenic.

- **Impose product standards**: set an upper limit of 2 μg/g (dry weight) for NNN plus NNK and an upper limit of 5 ng/g (dry weight) for benzo[a]pyrene; and monitor the levels of arsenic, cadmium and lead in tobacco.
- **Reduce appeal and addictiveness**: take steps to reduce the appeal of and addiction to tobacco products, including by prohibiting addition of sweeteners and flavourings (including herbs, spices and flowers) and setting limits on free nicotine and pH.
- **Apply uniform standards for transnational products**: hold exported smokeless tobacco products to the same (or higher) standards as in the country of manufacture.

WHO TobReg also recommended that communications about smokeless tobacco products be regulated to prevent unsubstantiated claims about exposure or disease reduction *(103)*.

We have summarized below some concerns associated with the carcinogenic, toxic and addictive agents in smokeless tobacco products and propose ways of lowering the concentrations of these agents. Decreasing the levels of nicotine, free nicotine, arsenic, cadmium, lead, benzo[a]pyrene and NNN plus NNK will not make smokeless tobacco products safe; however, it is prudent to decrease the levels of known addictive, toxic or carcinogenic agents.

The concerns associated with smokeless tobacco products include:

- use of high-nicotine tobacco, including *N. rustica*;
- inclusion of alkaline agents that substantially raise the pH, increase free nicotine concentrations and contribute to higher blood nicotine levels, cardiovascular risk and addiction;
- presence of toxic metals in tobacco due to soil uptake or deposited from contaminated soil or the environment;
- use of fire-curing, which can introduce chemicals from smoke, such as PAHs (including benzo[a]pyrene), phenols and volatile aldehydes;
- presence of microbial contamination (bacteria and fungi) on tobacco leaves, especially those organisms that promote the formation of aflatoxins, ochratoxin A, nitrite or nitrosamines (particularly TSNAs), and the conditions most conducive to their formation;
- presence of microbes (bacteria and fungi) that can cause infectious disease, become resistant to antibiotics, alter oral biofilms or displace the healthy microflora of the gastrointestinal tract;
- soil fertilization practices that result in elevated levels of nitrate, leading to increased levels of nitrite and nitrosamines in tobacco at harvest;

- presence of harmful agricultural chemical residues on tobacco at harvest;
- presence of viable microorganisms (including pathogens) in purchased products that can be transferred to users and may become established in the oral cavity or the gastrointestinal tract;
- use of fermentation, ageing, product storage or other processes that provide anaerobic conditions that can contribute to rapid formation of nitrite and TSNA;
- presence of high residual levels of nitrate in finished tobacco products;
- presence of areca nut (IARC Group 1 human carcinogen) and other additives of recognized toxicity or carcinogenicity;
- presence of flavouring additives that have toxic effects; and
- presence of flavouring additives that enhance initiation of tobacco use by adding appeal, impeding cessation or masking the recognition or severity of disease symptoms.

Attention should also be paid to the contribution of non-tobacco ingredients and plant materials in smokeless tobacco products. Steps that might decrease the level of TSNAs, the most abundant carcinogens in smokeless tobacco products, are shown in Fig. 9.3.

Fig. 9.3. Countermeasures that might contribute to reducing concentrations of TSNAs and other harmful agents in tobacco products

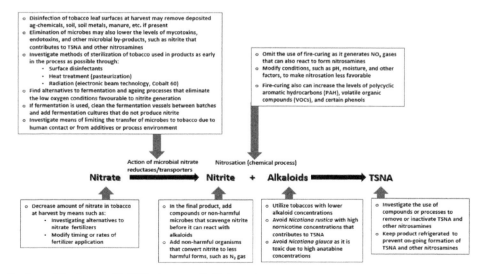

N_2: nitrogen; NO_x: nitrogen oxide; TSNA: tobacco-specific N'-nitrosamines.
To address the concerns listed above, in addition to the steps recommended by TobReg and specific product standards, product manufacturers could take a number of steps to decrease the concentrations of harmful agents in tobacco products.

- Because nornicotine is the precursor of NNN, genetically screen tobacco cultivars for lower nornicotine content as a means of reducing the formation of NNN.
- Screen soils for metal contamination and nitrate levels.
- Investigate agronomic means for minimizing the concentrations of nitrate and agricultural chemicals in or on plants at harvest.
- Use gloves when harvesting or handling tobacco to prevent the transfer of organisms on the skin (e.g. *Staphylococcus*) to the tobacco and also to protect workers from "green tobacco sickness".
- Wash harvested tobacco with dilute bleach solution (1:250 hyprochlorite:water) to remove soil, soil metals, manure, agricultural chemicals, insects and microorganisms.
- Ensure that curing facilities are clean so that other organisms (especially nitrate-reducing organisms) are not introduced.
- Identify harmless chemical or biological agents that could be added to prevent microbial growth.
- Screen cured tobacco for elevated nitrite or the presence of nitrite-generating organisms (e.g. *Staphylococcus*, *Corynebacterium*) before further processing.
- Pasteurize or heat-treat products.
- Clean fermentation vats before use.
- Add non-nitrate-reducing fermentation organisms before fermentation.
- Investigate means for lowering nitrate levels during fermentation (selective filtration, denitrification, cycle liquid from fermentation through a cell that reacts nitrite with chemicals scavengers or denitrifying organisms).
- Add nitrite scavengers (vitamin C, tocopherol, green tea extract, Kunlun tea extract, cysteine) before fermentation.
- Assay final products for microbial growth.
- Keep products in refrigerated storage before sale.

These observations provide additional support for the feasibility of monitoring and controlling the levels of some carcinogenic and other harmful agents in smokeless tobacco products.

9.9.1 Disclaimer

The findings and conclusions in this chapter are those of the author(s) and do not necessarily represent the views of the Centers for Disease Control and Prevention.

Use of trade names is for identification only and does not imply endorsement by the Centers for Disease Control and Prevention, the Public Health Service, or the United States Department of Health and Human Services.

9.10 References

1. Decision FCTC/COP7(14). Further development of the partial guidelines for implementation of Articles 9 and 10 of the WHO FCTC (Regulation of the contents of tobacco products and Regulation of tobacco product disclosures). In: Report of the seventh session of the Conference of the Parties to the WHO Framework Convention on Tobacco Control. Geneva: World Health Organization; 2016) (http://www.who.int/fctc/cop/cop7/FINAL_COP7_REPORT_EN.pdf, accessed 15 May 2019).
2. WHO global report on trends in prevalence of tobacco smoking 2000–2025, 2nd edition. Geneva: World Health Organization; 2018.
3. Eriksen M, Mackay J, Schluger N, Gomeshtapeh FI, Drope J, editors. Tobacco atlas. Fifth edition. Atlanta (GA): American Cancer Society and World Lung Foundation; 2015 (https://tobaccoatlas.org/, accessed 16 July 2019).
4. Smokeless tobacco and public health: a global perspective. Chapter 3. Global view of smokeless tobacco products: constituents and toxicity (NIH/CDC Monograph). Bethesda (MD): Department of Health and Human Services; 2014:75–114 (NIH Pub. 14-7983) (https://cancercontrol.cancer.gov/brp/tcrb/global-perspective/SmokelessTobaccoAndPublicHealth.pdf, accessed 16 July 2019).
5. Stanfill SB. Chapter 7: Toxic contents and emissions in smokeless tobacco products. In: WHO Study Group on Tobacco Product Regulation. Report on the scientific basis of tobacco product regulation. Sixth report of a WHO study group (WHO Technical Report Series, No. 1001). Geneva: World Health Organization; 2017:23–39. (http://www.who.int/tobacco/publications/prod_regulation/trs1001/en/, accessed 16 July 2019).
6. Smokeless tobacco and some tobacco-specific N-nitrosamines. Lyon: International Agency for Research on Cancer; 2007 (IARC Monographs on the Evaluation of Carcinogenic Risks to Humans, Vol. 89) (http://monographs.iarc.fr/ENG/Monographs/vol89/index.php, accessed 20 September 2018).
7. Zitomer N, Rybak ME, Li Z, Walters MJ, Holman MR. Determination of aflatoxin B1 in smokeless tobacco products by use of UHPLC-MS/MS. J Agric Food Chem. 2015;63(41):9131–8.
8. Di Giacomo M, Paolino M, Silvestro D, Vigliotta G, Imperi F, Visca P et al. Microbial community structure and dynamics of dark fire-cured tobacco fermentation. Appl Environ Microbiol. 2007;73:825–37.
9. Fisher MT, Bennett CB, Hayes A, Kargalioglu Y, Knox BL, Xu D et al. Sources of and technical approaches for the abatement of tobacco specific nitrosamine formation in moist smokeless tobacco products. Food Chem Toxicol. 2012;50:942–8.
10. Andersen RA, Burton HR, Fleming PD, Hamilton-Kemp TR. Effect of storage conditions on nitrosated, acylated, and oxidized pyridine alkaloid derivatives in smokeless tobacco products. Cancer Res. 1989;49:5895–900.
11. Stanfill SB, Connolly GN, Zhang L, Jia TL, Henningfield J, Richter P et al. Surveillance of international oral tobacco products: total nicotine, unionized nicotine and tobacco-specific nitrosamines. Tob Control. 2011;20:e2.
12. Sisson VA, Severson RF. Alkaloid composition of Nicotiana species. Beitr Tabakforsch Int. 1990;14:327–39.
13. Lisko JG, Stanfill SB, Duncan BW, Watson CH. Application of GC-MS/MS for the analysis of to-

14. Idris AM, Nair J, Ohshima H, Friesen M, Brouet I, Faustman EM et al. Unusually high levels of carcinogenic tobacco-specific nitrosamines in Sudan snuff (toombak). Carcinogenesis. 1991;12:1115–8.
15. Steenkamp PA, van Heerden FR, van Wyk B. Accidental fatal poisoning by Nicotiana glauca: identification of anabasine by high performance liquid chromatography/photodiode array/mass spectrometry. Forensic Sci Int. 2002;128:208–17.
16. Furer V, Hersch M, Silvetzki N, Breuer GS, Zevin S. Nicotiana glauca (tree tobacco) intoxication – two cases in one family. J Med Toxicol. 2011;7(1):47–51.
17. Bhide SV, Nair J, Maru GB, Nair UJ, Kameshwar Rao BV, Chakraborty MK et al. Tobacco-specific N-nitrosamines [TSNA] in green mature and processed tobacco leaves from India. Beitr Tabakforsch Int. 1987;14:29–32.
18. Burton HR, Dye NK, Bush LP. Distribution of tobacco constituents in tobacco leaf tissue. 1. Tobacco-specific nitrosamines, nitrate, nitrite, and alkaloids. J Agric Food Chem. 1992;40(6):1050–5.
19. Rodgman A, Perfetti TA, editors. The chemical components of tobacco and tobacco smoke, 2nd edition. Boca Raton (FL): CRC Press; 2013.
20. Chen Z, Xia Z, Lei L, Xu S, Wang M, Cun L et al. Characteristic analysis of endophytic bacteria population in tobacco. Acta Tabacaria Sinica. 2014;20:102–7.
21. List of tobacco pests and diseases. In: Ephytia [website]. Paris: French National Institute for Agricultural Research; 2019 (http://ephytia.inra.fr/en/C/10749/Tobacco-List-of-tobacco-pests-and-diseases, accessed 16 July 2019).
22. Stepanov I, Jensen J, Hatsukami D, Hecht SS. New and traditional smokeless tobacco: comparison of toxicant and carcinogen levels. Nicotine Tob. Res. 2008; 10:1773–82.
23. Hearn BA, Ding YS, England L, Kim S, Vaughan C, Stanfill SB et al. Chemical analysis of Alaskan iq'mik smokeless tobacco. Nicotine Tob Res. 2013;15(7):1283–8.
24. Stepanov I, Villalta PW, Knezevich A, Jensen J, Hatsukami D, Hecht SS. Analysis of 23 polycyclic aromatic hydrocarbons in smokeless tobacco by gas chromatography–mass spectrometry. Chem Res Toxicol. 2010;23:66–73.
25. Han J, Sanad YM, Deck J, Sutherland JB, Li Z, Walters MJ et al. Bacterial populations associated with smokeless tobacco products. Appl Environ Microbiol. 2016;82:6273–83.
26. Larsson L, Szponar B, Ridha B, Pehrson C, Dutkiewicz J, Krysińska-Traczyk E et al. Identification of bacterial and fungal components in tobacco and tobacco smoke. Tob Induced Dis. 2008;4:4.
27. Pauly JL, Paszkiewicz G. Cigarette smoke, bacteria, mold, microbial toxins, and chronic lung inflammation. J Oncol. 2011:819129.
28. Rutqvist LE, Curvall M, Hassler T, Ringberger T, Wahlberg I. Swedish snus and the GothiaTek standard. Harm Reduct J. 2011;8:11.
29. WHO Study Group on Tobacco Product Regulation. Report on the scientific basis of tobacco product regulation: sixth report of a WHO study group (WHO Technical Report Series No. 1001). Geneva: World Health Organization; 2018 (https://www.who.int/tobacco/publications/prod_regulation/trs1001/en/, accessed 16 July 2019).
30. Fact sheet, smokeless tobacco health effects. Atlanta (GA): Centers for Disease Control and Prevention; 2018 (https://www.cdc.gov/tobacco/data_statistics/fact_sheets/smokeless/, accessed 16 July 2019).
31. Spiegelhalder B, Fischer S. Formation of tobacco-specific nitrosamines. Crit Rev Toxicol. 1991;21:241.
32. Wang J, Yang H, Shi H, Zhou J, Bai R, Zhang M et al. Nitrate and nitrite promote formation of tobacco-specific nitrosamines via nitrogen oxides intermediates during postcured storage under warm temperature. J Chemistry. 2017;6135215 (https://doi.org/10.1155/2017/6135215, accessed 20 July 2019).

33. Bhisey RA, Stanfill SB. Chapter 13. Chemistry and toxicology of smokeless tobacco. In: Smokeless tobacco and public health in India. New Delhi: Ministry of Health and Family Welfare; 2017 (https://www.researchgate.net/, accessed 23 September 2018).
34. Lawler TS, Stanfill SB, Zhang L, Ashley DL, Watson CH. Chemical characterization of domestic oral tobacco products: total nicotine, pH, unprotonated nicotine and tobacco-specific N-nitrosamines. Food Chem Toxicol. 2013;57:380–6.
35. Notice regarding requirement for annual submission of the quantity of nicotine contained in smokeless tobacco products manufactured, imported, or packaged in the United States. Washington (DC): Centers for Disease Control and Prevention; Department of Health and Human Services; 1999 (https://www.gpo.gov/fdsys/pkg/FR-1999-03-23/pdf/99-7022.pdf, accessed 16 July 2019).
36. Brunnemann KD, Genoble L, Hoffmann D. N-Nitrosamines in chewing tobacco: an international comparison. J Agric Food Chem. 1985;33:1178–81.
37. Tomar SL, Henningfield JE. Review of the evidence that pH is a determinant of nicotine dosage from oral use of smokeless tobacco. Tob Control. 1997;6(3):219–25.
38. Fant RV, Henningfield JE, Nelson RA, Pickworth WB. Pharmacokinetics and pharmacodynamics of moist snuff in humans. Tob Control. 1999;8:387–92.
39. Alpert HR, Koh H, Connolly GN. Free nicotine content and strategic marketing of moist snuff tobacco products in the United States: 2000–2006. Tob Control. 2008;17:332–8.
40. Alloway BJ, editor. Heavy metals in soils: trace metals and metalloids in soils and their bioavailability. Dordrecht: Springer Netherlands; 2013.
41. Golia EE, Dimirkou A, Mitsios IK. Accumulation of metals on tobacco leaves (primings) grown in an agricultural area in relation to soil. Bull Environ Contam Toxicol. 2007;79:158–62.
42. Pappas RS, Stanfill SB, Watson C, Ashley DL. Analysis of toxic metals in commercial moist snuff and Alaskan iqmik. J Anal Toxicol. 2008;32(4):281–91.
43. Pappas RS. Toxic elements in tobacco and in cigarette smoke: inflammation and sensitization. Metallomics. 2011; 3(11):1181–98.
44. Zakiullah SM, Muhammad N, Khan SA, Gul F, Khuda F, Khan H. Assessment of potential toxicity of a smokeless tobacco product (naswar) available on the Pakistani market. Tob Control. 2012;(4):396–401.
45. Gupta P, Sreevidya S. Laboratory testing of smokeless tobacco products. Final report to the India Office of the WHO (allotment No: SE IND TOB 001.RB.02). New Delhi; 2004.
46. Zaprjanova P, Ivanov K, Angelova V, Dospatliev L. Relation between soil characteristics and heavy metal content in Virginia tobacco. In: World Congress of Soil Science. Brisbane, Australia. 2010:205–8 (https://www.iuss.org/19th%20WCSS/Symposium/pdf/2027.pdf, accessed 16 July 2019).
47. Mulchi CL, Adamu CA, Bell PF, Chaney RL. Residual heavy metal concentrations in sludge amended coastal plain soils – II. Predicting metal concentrations in tobacco from soil test information. Commun Soil Sci Plant Anal.1992;23(9–10):1053–69.
48. Adamu CA, Bell PF, Mulchi CL, Chaney RL. Residual metal levels in soils and leaf accumulations in tobacco a decade following farmland application of municipal sludge. Environ Pollut. 1989;56:113–26.
49. New limit values for GothiaTek. In: Swedish Match [website]. Stockholm: Swedish Match; 2016 (https://www.swedishmatch.com/Media/Pressreleases-and-news/News/New-limit-values-for-GOTHIATEK/, accessed 16 July 2019).
50. Wahlberg I, Wiernik A, Christakopoulos A, Johansson L. Tobacco-specific nitrosamines: a multidisciplinary research area. Agro Food Industry Hi Tech. 1999;10(4):23–8.
51. Richter P, Hodge K, Stanfill S, Zhang L, Watson C. Surveillance of moist snuff: total nicotine, moisture, pH, un-ionized nicotine, and tobacco-specific nitrosamines. Nicotine Tob Res. 2008;

10(11):1645–52.
52. Stanfill SB, Croucher R, Gupta P, Lisko J, Lawler TS, Kuklenyik P. Chemical characterization of smokeless tobacco products from South Asia: nicotine, unprotonated nicotine, tobacco-specific N-nitrosamines and flavor compounds. Food Chem Toxicol. 2018;118:626–34.
53. Stepanov I, Gupta PC, Dhumal G, Yershova K, Toscano W, Hatsukami D et al. High levels of
54. tobacco-specific N-nitrosamines and nicotine in Chaini Khaini, a product marketed as snus.
55. Tob Control. 2015;24,e271–4.
56. Balbo S, James-Yi S, Johnson CS, O'Sullivan MG, Stepanov I, Wang M et al. (S)-N⊠-nitrosonornicotine, a constituent of smokeless tobacco, is a powerful oral cavity carcinogen in rats. Carcinogenesis. 2013;34:2178.
57. Smyth EM, Kulkarni P, Claye E, Stanfill SB, Tyx RE, Maddox C et al. Smokeless tobacco products harbor diverse bacterial microbiota that differ across products and brands. Appl Microbiol Biotechnol. 2017;101:5391.
58. Wiernik A, Christakopoulos A, Johansson L, Wahlberg I. Effect of air-curing on the chemical composition of tobacco. Recent Adv Tob Sci. 1995;21:39–80.
59. González PJ, Correia C, Moura I, Brondino CD, Moura JJG. Bacterial nitrate reductases: molecular and biological aspects of nitrate reduction. J Inorg Biochem. 2006;100:1015–23.
60. Tyx RE, Stanfill SB, Keong LM, Rivera AJ, Satten GA, Watson CH. Characterization of bacterial communities in selected smokeless tobacco products using 16 S rDNA analysis. PLoS One 2016;11:e0146939.
61. Nishimura T, Vertes AA, Shinoda Y, Inui M, Yukawa H. Anaerobic growth of Corynebacterium glutamicum using nitrate as a terminal electron acceptor. Appl Microbiol Biotechnol. 2007;75:889–97.
62. Gadda G, Francis K. Nitronate monooxygenase, a model for anionic flavin semiquinone intermediates in oxidative catalysis. Arch Biochem Biophysics. 2010;493:53–61.
63. Burton HR, Dye NK, Bush LP. Distribution of tobacco constituents in tobacco leaf tissue. 1. Tobacco-specific nitrosamines, nitrate, nitrite, and alkaloids. J Agric Food Chem. 1992;40(6):1050–5.
64. Burton HR, Dye NK, Bush LP. Relationship between tobacco-specific nitrosamines and nitrite from different air-cured tobacco varieties. J Agric Food Chem. 1994;42:2007–11.
65. Bush LP. Alklaloid biosynthesis. In: Davis DL, Nielsen MT, editors. Tobacco: production, chemistry and technology. Oxford: Blackwell; 1999;285–91.
66. Ghosh RK, Khan Z, Rao CVN, Banerjee K, Reddy DD, Murthy TGK et al. Assessment of organochlorine pesticide residues in Indian flue-cured tobacco with gas chromatography-single quadrupole mass spectrometer. Environ Monit Assess. 2014;186:5069–75.
67. Idris AM, Ibrahim SO, Vasstrand EN, Johannessen AC, Lillehaug JR, Magnusson B et al. The Swedish snus and the Sudanese toombak: are they different? Oral Oncol. 1998;34(6):558–66.
68. Harmful and potentially harmful constituents (HPHCs) [website].Washington (DC): Food and Drug Administration; 2018 (https://www.fda.gov/TobaccoProducts/Labeling/ProductsIngredientsComponents/ucm20035927.htm, accessed 17 January 2019).
69. Smokeless tobacco ingredient list as of April 4, 1994. Report to the Subcommittee on Health and the Environment, Committee on Energy and Commerce. Washington (DC): House of Representatives; 1994 (http://legacy.library.ucsf.edu/tid/pac33f00/pdf, accessed 16 July 2019).
70. Garg A, Chaturvedi P, Gupta PC. A review of the systemic adverse effects of areca nut or betel nut. Indian J Med Paediatr Oncol. 2014;35(1):3–9.
71. Prabhu RV, Prabhu V, Chatra L, Shenai P, Suvarna N, Dandekeri S. Areca nut and its role in oral submucous fibrosis. J Clin Exp Dent. 2014;6(5):e569–75.
72. Bao P, Huang H, Hu ZY, Haggblom MM, Zhu YG. Impact of temperature, CO2 fixation and nitrate reduction on selenium reduction, by a paddy soil clostridium strain. J Appl Microbiol. 2013;114:703–12.

73. Stanfill SB, Oliveira-Silva AL, Lisko J, Lawler TS, Kuklenyik P, Tyx R et al. Comprehensive chemical characterization of rapé tobacco products: flavor constituents, nicotine, tobacco-specific nitrosamines and polycyclic aromatic hydrocarbons. Food Chem Toxicol. 2015;82:50–8.
74. Tobacco product standard for N-nitrosonornicotine level in finished smokeless tobacco products. Washington (DC): Department of Health and Human Services, Food and Drug Administration; 2017 (https://www.fda.gov/downloads/aboutfda/reportsmanualsforms/reports/economicanalyses/ucm537872.pdf, accessed 16 July 2019).
75. Pillai SD, Shayanfar S. Electron beam technology and other irradiation technology applications in the food industry. Top Curr Chem. 2017;375:6. doi: 10.1007/s41061-016-0093-4.
76. Idris AM, Ibrahim SO, Vasstrand EN, Johannessen AC, Lillehaug JR, Magnusson B et al. The Swedish snus and the Sudanese toombak: are they different? Oral Oncol. 1998;34(6):558–66.
77. Wei W, Deng X, Cai D, Ji Z, Wang C, Li J et al. Decreased tobacco-specific nitrosamines by microbial treatment with *Bacillus amyloliquefaciens* DA9 during the air-curing process of Burley tobacco. Agric Food Chem. 2014;62:12701–6.
78. Saleem S, Naz A, Safique M, Jabeen N, Ahsan SW. Fungal contamination in smokeless tobacco products traditionally consumed in Pakistan. J Pak Med Assoc. 2018;68(10):1471–77.
79. How long is the shelf life of snus. Stockholm: Swedish Match; 2018 (https://www.swedishmatch.ch/en/what-is-snus/qa/shelf-life-of-snus/, accessed 16 July 2019).
80. Brenik W, Rudhard H. Method of endothermic fermentation of tobacco. US Patent 04343318. Washington (DC): Patent Office; 1982 (https://pubchem.ncbi.nlm.nih.gov/patent/US4343318#section=Patent-Submission-Date, accessed 16 July 2019).
81. Chandrasekaran S, Ramanathan S, Basak T. Microwave food processing—a review. Food Res Intl. 2013;52:243–61 (https://doi.org/10.1016/j.foodres.2013.02.033, accessed 16 July 2019).
82. Industrial Microwave Systems LLC [website]. New Orleans (LA): Industrial Microwave Systems LLC; 2019 (http://www.industrialmicrowave.com, accessed 16 July 2019).
83. Brinley TA, Dock CN, Truong VD, Coronel P, Kumar P, Simunovic J et al. Feasibility of utilizing bioindicators for testing microbial inactivation in sweet potato purees processed with a continuous-flow microwave system. J Food Sci. 2007;72(5):235–42.
84. David JRD, Graves RH, Szemplenski T. Handbook of aseptic processing and packaging, 2nd edition. Boca Raton (FL): CRC Press; 2013.
85. Parrott DL. Microwave technology sterilizes sweet potato puree. Food Technol. 2010:66–70 (https://ucanr.edu/datastoreFiles/608-672.pdf, accessed 20 July 2019).
86. Lee S, Kim S, Jeong S, Park J. Effect of far-infrared irradiation on catechins and nitrite scavenging activity of green tea. J Agric Food Chem. 2006;54(2):399–403.
87. Rundlöf T, Olsson E, Wiernik A, Back S, Aune M, Johansson L et al. Potential nitrite scavengers as inhibitors of the formation of N-nitrosamines in solution and tobacco matrix systems. J Agric Food Chem. 2000;48(9):4381–8.
88. Cockrell WT, Roberts JS, Kane BE, Fulghum S. Microbiology of oral smokeless tobacco products. Tob Int. 1989;191:55–7.
89. Harwani D. The great plate count anomaly and the unculturable bacteria. Int J Sci Res. 2012;2(9):350–1.
90. Technical bulletin: BacTiter-Glo™ microbial cell viability assay (document TB337). Madison (WI): Promega Corporation; 2019 (https://www.promega.com/, accessed 20 July 2019).
91. Law AD, Fisher C, Jack A, Moe LA. Tobacco, microbes, and carcinogens: correlation between tobacco cure conditions, tobacco-specific nitrosamine content, and cured leaf microbial community. Microb Ecol. 2016;72:1–10.
92. Omidbakhsh N, Ahmadpour F, Kenny N. How reliable are ATP bioluminescence meters in assessing decontamination of environmental surfaces in healthcare settings? PLoS One. 2014;9(6):e99951.

93. Performance evaluation of various ATP detecting units (Food Science Center report RPN 13922). Chicago (IL): Silliker Inc; 2010.
94. Sigma-Aldrich [website] (https://www.sigmaaldrich.com, accessed 16 July 2019).
95. Stanfill SB, Jia LT, Watson CH, Ashley DL. Rapid and chemically-selective quantification of nicotine in smokeless tobacco products using gas chromatography/mass spectrometry, J Chrom Sci. 2009;47(10):902–9.
96. Lisko JG, Stanfill SB, Watson CH. Quantitation of ten flavor compounds in unburned tobacco products. Anal Methods. 2014;6(13):4698–704. doi: 10.1039/C4AY00271G.
97. Jain V, Garg A, Parascandola M, Chaturvedi P, Khariwala SS, Stepanov I. Analysis of alkaloids in areca nut-containing products by liquid chromatography-tandem mass spectrometry. J Agric Food Chem. 2017;65, 1977–1983 (https://doi.org/10.1021/acs.jafc.6b05140, accessed 20 July 2019).
98. Stepanov I, Hecht SS, Ramakrishnan S, Gupta PC. Tobacco-specific nitrosamines in smokeless tobacco products marketed in India. Carcinogenesis. 2015;116:16–9 (https://doi.org/10.1002/ijc.20966, accessed 16 July 2019).
99. Dalipi R, Margui E, Borgese L, Depero E. Multi-element analysis of vegetal foodstuff by means of low power total reflection X-ray fluorescence (TXRF) spectrometry. Food Chem. 2017; 218:348–55.
100. Directive 2001/37/EC of the European Parliament and of the Council (L194/26–L194/34). Brussels: European Commission; 2001 (https://ec.europa.eu/health/sites/health/files/tobacco/docs/dir200137ec_tobaccoproducts_en.pdf, accessed 22 July 2019).
101. Family Smoking Prevention and Tobacco Control Act 2009. Washington (DC): United States Congress (Public Law 111-31 111th Congress) (https://www.gpo.gov/fdsys/pkg/PLAW-111publ31/pdf/PLAW-111publ31.pdf, accessed 16 July 2019).
102. Vidhubala E, Pisinger C, Basumallik B, Prabhakar DS. The ban on smokeless tobacco products is systematically violated in Chennai, India. Indian J Cancer. 2016;53:325–30.
103. Arora M, Madhu R. Banning smokeless tobacco in India: policy analysis. Indian J. Cancer. 2012;49(4):336–41.
104. GothiaTek limits for undesired components. Stockholm: Swedish Match; 2017 (https://www.swedishmatch.com/Snus-and-health/GOTHIATEK/GOTHIATEK-standard/, accessed 16 July 2019).
105. Gray N, Hecht S. Smokeless tobacco – proposals for regulation. Lancet. 2010;375:1589–91

10. Waterpipe tobacco smoking: prevalence, health effects and interventions to reduce use

Mohammed Jawad and Christopher Millett, Public Health Policy Evaluation Unit, Imperial College London, London, England

Contents
10.1 Introduction
10.2 Prevalence, health effects and effective interventions to reduce use
    10.2.1 Regional and global patterns of waterpipe tobacco use
    10.2.2 Acute and chronic health effects
    10.2.3 Cultural practices and initiation and maintenance of use
    10.2.4 Influence of flavourings
    10.2.5 Dependence liability of low-nicotine products
    10.2.6 Interventions
10.3 Future research
10.4 Policy recommendations
    10.4.1 Policies relevant to waterpipe tobacco use
10.5 Conclusions
10.6 References

10.1 Introduction

At the seventh session of the COP to the WHO FCTC, document FCTC/COP/7/10 presented policy options and best practice for the control of waterpipe tobacco product use in relation to the WHO FCTC *(1)*. This section builds on the evidence presented in document FCTC/COP/7/10 in preparation for the eighth session of the COP. The purpose is to provide a comprehensive analysis of waterpipe tobacco control and to outline challenges and recommendations to improve the prevention and control of waterpipe tobacco use. The objectives are to analyse studies on waterpipe tobacco in order to:

- summarize regional and global patterns of waterpipe tobacco use, including changes in prevalence;
- evaluate the acute and chronic health effects of waterpipe tobacco use;
- describe cultural practices in relation to initiation and maintenance of use;
- explain the influence of flavourings on initiation, maintenance of use and increasing use;
- explore dependence liability with low-nicotine products;
- evaluate the evidence base for culturally relevant, waterpipe-specific interventions to prevent uptake and promote cessation;
- outline effective policies based on conceptual frameworks; and
- make recommendations for research and regulation.

This section is limited to waterpipe tobacco and its accessories. Electronic nicotine devices with synonyms of "waterpipe" (e.g. shisha or hookah pens, electronic shisha or hookah) are not addressed. We focus on the most recent literature to minimize overlap with document FCTC/COP/7/10 and the WHO advisory note on waterpipe tobacco smoking *(2)*.

10.2 Prevalence, health effects and effective interventions to reduce use

The literature on waterpipe tobacco is relatively limited but is growing exponentially. In 2016, the number of citations in five clinical databases (Medline, Embase, Web of Science, PsychInfo, Global Health) that contained synonyms of "waterpipe" in the title or abstract was more than 10 times the number in 2006 (285 versus 25) (Fig. 10.1). Academia is continuously providing strengthened evidence for the prevention and control of waterpipe tobacco use.

Fig. 10.1. Numbers of citations in five clinical databases with synonyms of "waterpipe" in the titles or abstracts, 1990–2017

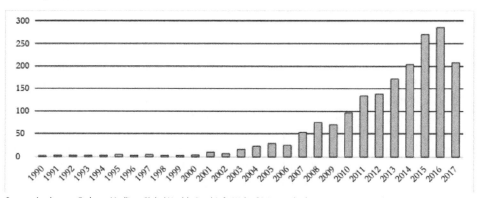

Source: databases – Embase, Medline, Global Health, PsychInfo, Web of Science; duplicate entries removed; 1 January–27 August only, in 2017.

10.2.1 Regional and global patterns of waterpipe tobacco use

Document FCTC/COP/7/10 noted that estimates of prevalence were available for some countries in all WHO regions *(1)*. Generally, studies reported high prevalence among children aged 13–15 years and university students, although these population groups were surveyed in only a few countries, and many of the studies were not nationally representative.

To the best of our knowledge, 131 studies have reported jurisdictionally representative estimates of the prevalence of waterpipe tobacco use in 84 countries *(3)*. About one third of the estimates derive from three international surveys: the Global Adult Tobacco Survey, the Global Youth Tobacco Survey and the Special

Eurobarometer Survey. Combined data is from the latest wave of the Global Adult Tobacco Survey and Special Eurobarometer Survey (wave 87.1, 2017) *(4, 5)*, both among adults, and also the results of other studies in which similar methods were used *(6–9)*. The figure reveals two important findings: an increasing prevalence in the WHO European Region and an absence of studies with comparable survey tools in the WHO Eastern Mediterranean Region, where waterpipe tobacco use is the highest globally.

The lack of data on the prevalence among adults in the WHO Eastern Mediterranean Region is in contrast to its involvement in the Global Youth Tobacco Survey. Data from the latest waves of the Global Youth Tobacco Survey *(5)* and other methodologically similar national school-based surveys *(10–13)* shows an estimated prevalence of 5.0–14.9% in many countries in all WHO regions, which is a concern and warrants further investigation and intervention. The highest estimated prevalence is in the Eastern Mediterranean Region (Lebanon, 37.0%; West Bank, 35.2%) and in Cyprus (33.2%) (Fig. 10.2).

Fig. 10.2. Use of waterpipe tobacco in the past 30 days by young people, Middle East

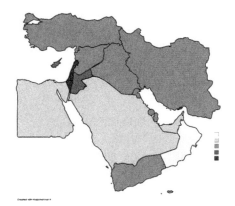

Source: references *5–10*.

Few studies have reported trends in waterpipe use. In 28 European Union Member States that were surveyed in 2009 and again in 2017, the mean prevalence of regular waterpipe tobacco use among adults increased by 11.2% *(4)*.

Waterpipe tobacco use appears to be increasingly popular among children aged 13–15 years, especially in the WHO Eastern Mediterranean Region (Fig. 10.2). In the USA, one national survey reported no or very little change in the prevalence of use in the past 30 days among young adults aged 18–24 years, which remained at 2–3% between 2011 and 2016 *(14)*. In contrast, a number of other surveys conducted between 2008 and 2015 among young people in the USA, including the National Youth Tobacco Survey, showed absolute increases in waterpipe use in the past 30 days of 0.3–1.0% each year *(15–22)*. Similar increases were seen in repeated national school surveys in Canada and Lebanon *(23–26)*, while in Jordan much larger absolute increases were seen, of about 2.9% per year between 2008 and 2011 *(27)*. In the study in Jordan, increased use over time was associated with higher maternal education, frequent physical activity, ever cigarette use and waterpipe use by peers *(28)*. In Turkey, the Global Adult Tobacco Survey indicated a decrease in waterpipe use, from 2.3% in 2008 to 0.8% in 2010 *(29)*.

While waterpipe tobacco use is best characterized as intermittent, its interactions with other tobacco products are a potential cause for concern. In a longitudinal study of schoolchildren in Jordan, the risk of cigarette initiation was significantly higher among waterpipe tobacco smokers than those who had never smoked *(30)*. A similar conclusion was made in a study of adolescents in the USA, which was of the same methodological design but additionally controlled for baseline propensity to smoke *(31)*. In a study of university students in North Carolina (USA), those whose first tobacco product was waterpipe tobacco were more likely to be dual or multiple tobacco users at the time of the survey *(32)*.

In contrast to cigarette use, there is no clear pattern of increased waterpipe use with lower socioeconomic status. Nevertheless, inequalities by socioeconomic status may develop as both the public perception of waterpipe tobacco and the regulatory environment shift.

10.2.2   Acute and chronic health effects

Waterpipe users are exposed to toxicants from both the tobacco and the charcoal used to heat it. Toxicant yield studies have consistently found substantial levels of tar, nicotine, CO and cancer-causing chemicals in waterpipe tobacco smoke *(33)*. A population-level modelling study of exposure to cigarettes and waterpipe toxicants among 13–15-year-olds in the WHO Eastern Mediterranean Region *(34)* included estimates of the population intake of tobacco toxicants while factoring in frequencies of use and waterpipe sharing. It showed that waterpipe tobacco users were exposed to about 70% of all tobacco-derived CO and benzene, while cigarette users were exposed to only 30% *(34)*. The high levels of CO and benzene probably come from the charcoal used to heat the tobacco.

In the acute phase, the absorbed components of waterpipe tobacco and charcoal combustion cause adverse cardiovascular and respiratory changes. Increased blood pressure and heart rate *(35, 36)* are expected, given the considerable but variable nicotine content of waterpipe tobacco *(37)*. Reduced lung function is also commonly reported *(38)*. Particularly with waterpipe tobacco, CO levels rise sharply secondary to charcoal combustion, and reported cases of CO poisoning among waterpipe tobacco users are widespread. It should be noted that non-tobacco, "herbal" waterpipe products are also heated with charcoal, resulting in a toxicological profile similar to that tobacco waterpipe products except for the absence of nicotine *(39, 40)*.

Growing evidence shows the risk of second-hand waterpipe tobacco smoke, particularly in waterpipe cafes. One pre–post study of 15 people passively exposed to waterpipe tobacco smoke showed an average increase in carboxyhaemoglobin from 0.8% (± 0.2) to 1.2% (± 0.8) after they had sat next to four to five active smokers for 30 min *(41)*. Another showed that the level of acrolein, a harmful respiratory toxicant, increased in nonsmokers in waterpipe cafes or at waterpipe smoking events in homes *(42)*. The effects of second-hand waterpipe tobacco smoke on other cardiorespiratory and toxicant parameters are less certain *(41, 43, 44)*. In an analysis of air quality, non-smoking rooms in waterpipe cafes had > 10 times the concentration of fine particulate matter as smoke-free rooms in other venues *(45)*. Assessments of poor air quality inside waterpipe cafes have been replicated in studies in the United Kingdom *(46)*.

Emerging evidence suggests that waterpipe tobacco has harmful long-term effects on health, similar to those of other forms of smoked tobacco. Table 10.1 summarizes the findings of several published systematic reviews *(36, 47–50)* and a further nine studies published after the reviews *(51–59)*, thus providing the latest evidence of long-term harm. The table shows that waterpipe tobacco use is associated with respiratory and cardiovascular disease, five types of cancer, adverse complications in pregnancy (low birth weight and intrauterine growth restriction) and a number of other diseases. Second-hand smoke from waterpipe tobacco use has been shown to induce wheezing in children *(60, 61)*. A further two cohort studies showed that waterpipe tobacco use was associated with increased overall mortality *(56, 62)*; one of these also showed an increase in mortality from cancer *(56)*. It is nevertheless difficult to draw clear conclusions from these studies because of the common practice of dual use of waterpipes and cigarettes. Many systematic reviews on this topic therefore rate the risk of bias as high. Consequently, the quality of the epidemiological studies on health and waterpipe tobacco use remains low, and further high-quality research, especially with enough statistical power to determine the relation between waterpipe tobacco smoking and health outcomes among never-cigarette-users, should be conducted to help to confirm these associations.

Table 10.1. Mortality and diseases associated with waterpipe tobacco use

Outcome	No. of studies	Odds ratio (95% CI)
Mortality	–	–
Overall	2	1.25 (1.03, 1.51)[a]
Cardiovascular disease	1	0.94 (0.43, 2.07)
Cancer	1	2.82 (1.30, 6.11)[a]
Cardiovascular disease	–	–
Coronary artery disease	3	1.47 (1.06, 2.04)[a]
Hypertension	1	1.95 (1.54, 2.48)[a]
Respiratory disease	–	–
Chronic obstructive pulmonary disease	5	2.33 (1.96, 2.77)[a]
Bronchitis	2	3.85 (1.92, 7.72)[a]
Asthma in children (active exposure)	2	1.32 (1.20, 1.46)[a]
Wheeze in children (passive exposure)	2	1.61 (1.25, 2.07)[a]
Cancer		
Bladder	6	1.28 (1.10, 1.48)[a]
Colorectal	3	1.20 (0.79, 1.82)
Gastric	4	2.35 (1.47, 3.76)[a]
Head and neck	7	3.00 (2.39, 3.76)[a]
Liver	1	1.13 (0.62, 2.78)
Lung	6	3.96 (2.96, 5.30)[a]
Oesophageal	5	2.61 (2.12, 3.26)[a]
Pancreatic	1	1.60 (0.91, 2.82)
Prostate	1	7.00 (0.88, 55.66)
Pregnancy-related diseases		
Low birthweight	4	1.54 (1.16, 2.04)[a]
Intrauterine growth restriction	1	3.50 (1.10, 12.60)[a]
Other diseases		
Diabetes	1	0.52 (0.27, 1.00)
Gastroesophageal reflux disease	2	1.31 (1.13, 1.53)[a]
Hepatitis C	3	0.98 (0.76, 1.25)
Infertility	1	2.50 (1.00, 6.30)[a]
Mental health	2	1.33 (1.25, 1.42)[a]
Multiple sclerosis	1	1.77 (1.36, 2.31)[a]
Periodontal disease	3	3.65 (2.22, 6.01)[a]

[a] Statistically significant positive association between waterpipe tobacco use and this disease in a random effects meta-analysis.

10.2.3 Cultural practices and initiation and maintenance of use

Cultural practices of waterpipe smoking are broadly embedded within a strong network of social acceptance, at the levels of family, peers and communities. Users and non-users alike have a positive attitude towards waterpipe tobacco use, which stems from the social environment in which it is consumed, the attraction towards the array of flavours and elaborately designed apparatuses and the portrayal of waterpipes in marketing media *(63–65)*.

One of several recommendations for research in the second WHO advisory note on waterpipe tobacco use *(2)* was on cultural practices and how they affect initiation and maintenance of use. For policy, the MPOWER package *(66)* is intended to assist countries in implementing effective interventions to reduce the demand for tobacco products. The six components of MPOWER are: Monitor tobacco use and prevention policies, Protect people from tobacco smoke, Offer help to quit tobacco use, Warn about the dangers of tobacco, Enforce bans on tobacco advertising, promotion and sponsorship, and Raise taxes on tobacco. Several Parties to the WHO FCTC, such as India, the Islamic Republic of Iran, Kenya, Malaysia, Pakistan, the United Arab Emirates and the United Republic of Tanzania, have gone beyond MPOWER and enforced (although, in some cases, then reversed), a total prohibition on public consumption of waterpipe tobacco, at either state or national levels. This policy may be based on social norms, as waterpipe tobacco use is embedded within the cultures of these Parties. The extent to which prohibitions on waterpipe tobacco use have been enforced and their impact on smoking prevalence remain uncertain, and this should be a key area of evaluation for these Parties. Evaluation within the International Tobacco Control Policy Evaluation Project is recommended *(67)*.

In Europe and north America, acculturation may determine waterpipe tobacco use. A study in the USA showed that immigrants from Arab countries who felt less acculturated into north American identity were more likely to use waterpipes *(68)*. Furthermore, qualitative research in Canada showed that immigrants from Arab countries saw themselves as expressing their cultural heritage by smoking waterpipes, and, in the United Kingdom, non-Arab users saw themselves as experiencing an alternative culture *(69)*.

Individual patterns of waterpipe use vary widely; the extent of within- and between-culture variation is, however, unclear. For example, in a study of waterpipe tobacco smoking behaviour in Lebanon, users on average drew an inhalation every 17 s *(70)*, whereas in a methodologically identical study in neighbouring Jordan, the frequency was 8 s and resulted in far greater exposure to toxicants. Gender differences in waterpipe smoking patterns were also noted in these studies, males tending to take larger puffs and spending more time smoking than females *(71)*.

Cultural differences are also apparent commercially. In commercial waterpipe premises, where about half of all waterpipe tobacco is consumed *(72, 73)*, café cultures range from a quiet coffee shop-like atmosphere to a busy, loud bar-like environment *(74)*. The café culture is likely to be associated with the patterns of use; for example, covert observation of commercial waterpipe premises in the USA indicated that some café owners insist that all clients entering the premises smoke a waterpipe *(74)*. In boisterous cafes that attract many customers, they are likely to associate waterpipe tobacco with socializing, whereas those that frequent quieter cafes may be more likely to indulge in solo, nicotine-driven use.

10.2.4 Influence of flavourings

Most of the waterpipe tobacco that is consumed is flavoured *(75, 76)*. Flavours are aggressively marketed by the waterpipe tobacco industry; in a review of 52 marketing material items from a waterpipe trade exhibition, flavours were among the most commonly elicited themes *(77)*. Alarmingly, the most common theme was that waterpipe tobacco was safe or safer than cigarettes *(77)*. Marketing practices of the waterpipe tobacco industry have indicated that tobacco and flavouring components may be sold separately to either evade flavouring bans or reduce the tobacco weight in order to lessen the excise tax *(77)*.

Waterpipe tobacco flavourings play a dominant role in perceptions of their safety and their attraction, in particular to young people. A review conducted in April 2016 of 10 qualitative studies on the role of flavours in waterpipe tobacco found that flavours were appealing or tasty *(78)*. In four studies (in Canada, the United Kingdom and the USA), young adults reported using waterpipes specifically for their flavour and because they did not wish to use other tobacco products *(78)*. There was also a perception that flavoured waterpipe tobacco was less harmful than cigarettes. In the same review, adults in Canada and Lebanon explained that young people used waterpipe tobacco because of their flavours, and another explained that flavours were responsible for initiation of waterpipe tobacco use by young adults *(78)*. In a large survey of adolescents in the USA, nearly 80% of users of waterpipes in the past 30 days reported that the flavours were a reason for use *(79)*.

These qualitative studies are supported by other research. In an experiment in the USA, 367 adult waterpipe smokers were asked to choose from menus with hypothetical combinations of different session types *(80)*. Flavoured waterpipe tobacco products were preferred significantly more than non-flavoured ones, and flavour more strongly influenced the decision to smoke waterpipes than price or nicotine content *(80)*. The association was stronger for females and for non-cigarette smokers. In another experiment in the USA, 36 adult waterpipe smokers completed two waterpipe sessions, one with their preferred flavour of waterpipe tobacco and another with a non-preferred flavour, in a randomized cross-over design *(81)*. Those who smoked their preferred flavour reported a better subjective smoking experience on a visual analogue scale, such as more interest in continued use, greater pleasure derived from smoking, increased liking and enjoyment and willingness to continue use *(81)*. The fact that the study was conducted on the premise that waterpipe tobacco users have preferred flavours is an important finding in itself. Together, these studies suggest that flavour restrictions or flavour bans might be effective in discouraging waterpipe tobacco use.

10.2.5 Dependence liability of low-nicotine products

The second WHO advisory note on waterpipe tobacco smoking clearly demarcated the role of nicotine dependence in waterpipe tobacco use *(2)*. Regular waterpipe users absorb enough nicotine to reach a dependence threshold and exhibit typical symptoms of dependence, such as craving, withdrawal symptoms and difficulty in quitting. Regular waterpipe use is, however, relatively uncommon outside the Eastern Mediterranean Region. In one study, it was estimated that users would have to smoke a waterpipe at least three times per week to become dependent *(82)*, whereas in the USA only 11.7% of young adults who had smoked a waterpipe in the past 30 days smoked at this level of frequency or more *(83)*.

There are two issues with regard to the dependence liability of low-nicotine waterpipe tobacco products. The first is that there is almost no regulation of waterpipe tobacco manufacture, which results in widely different nicotine levels among products. In an experiment in which 110 waterpipe smokers engaged in a 45-min session of smoking, brands labelled "0.05% nicotine" resulted in higher mean peak plasma levels of nicotine than brands labelled "0.5% nicotine" *(37)*. Given the lack of attention to policy on waterpipe tobacco, discussions about manufacture should prioritize flavour bans, as flavours have been shown to play a larger role in waterpipe purchase than low nicotine *(80)*.

The second issue is that waterpipe tobacco already contains less nicotine than cigarettes, and, while dependence on waterpipe tobacco is well documented, it remains relatively uncommon at a population level. When standardized to the nicotine yield per minute of smoking, waterpipe tobacco contains 2–13 ng less nicotine than cigarettes for every minute smoked (Table 10.2). Because the nicotine level is already low and dependence relatively uncommon on a population level, the justification for low-nicotine waterpipe tobacco is unclear.

Table 10.2. Nicotine yield per minute in waterpipe tobacco and cigarettes

Reference	Nicotine yield (ng/min)	No. of times less concentrated than in cigarettes[a]
34	0.11	1.9–5.6
70	0.08	2.7–8.0
84	0.05	4.4–13.2

[a] Based on data from Hoffman & Hoffman *(85)*, who estimated a nicotine yield of 1–3 ng/cigarette and an average time to consume a cigarette of 5 min.

10.2.6 Interventions

A review of interventions to reduce waterpipe tobacco use found only four individual-level and five group-level interventions *(86)*. A further search for this section found no further interventional studies, indicating lack of research in this area. Of five randomized controlled trials, only two showed statistically significant higher quit rates in the intervention group *(87, 88)*; in one trial, a cigarette-specific intervention was tested in waterpipe tobacco users *(87)*. The details of these interventions are shown in Table 10.3. Non-randomized studies had mixed results for cessation and behavioural and knowledge outcomes and were generally of much lower methodological quality.

The behavioural interventions used in these randomized studies varied widely but were broadly based on the same principles as cigarette behavioural interventions. In the study by Asfar et al. *(89)*, the intervention consisted of three 45-min, individual and in-person counselling sessions by a trained physician and five 10-min phone calls before and after the proposed quit date. Dogar et al. *(87)* provided two structured behavioural sessions, the first lasting 30 min and the second 10 min, based on the WHO "5 As" approach, and used behavioural change techniques considered to be effective for cigarette smoking cessation. Lipkus et al. *(90)* showed participants 20 slides giving factual information about waterpipe tobacco use, including its effects on health. Mohlman et al. *(88)* delivered health promotion in villages over 12 months, including deglamourizing tobacco use in primary schools and describing its health hazards in primary schools, mosques and homes; education was also given in handling peer pressure to smoke in preparatory and secondary schools. Nakkash et al. *(91)* delivered 10 sessions to students, four of which were on knowledge and six on skill-building (e.g. media literacy, decision-making and refusal skills).

Table 10.3. Randomized interventions for waterpipe tobacco cessation

Reference	Study design	Period of study	Country	Sample size, sex and mean age	Intervention	Control	Outcome	Biochemical verification	Effect size, AOR (95% CI)
89	Two-arm, parallel trial	2007–08	Syrian Arab Republic	N=50, 94% male, 30 years	Behavioural	Usual care	Prolonged abstinence	Yes	1.46 (0.69, 3.09)
89	Two-arm, parallel trial	2007–08	Syrian Arab Republic	N=50, 94% male, 30 years	Behavioural	Usual care	Seven-day point prevalence of abstinence	Yes	1.34 (0.57, 3.35)
89	Two-arm, parallel trial	2007–08	Syrian Arab Republic	N=50, 94% male, 30 years	Behavioural	Usual care	Continuous abstinence	Yes	1.07 (0.27, 4.42)
87	Three-arm cluster trial	2011–12	Pakistan	N=1955, 79% male, 52 years	Behavioural	Usual care	Continuous abstinence	Yes	2.20 (1.30, 3.80)
87	Three-arm cluster trial	2011–12	Pakistan	N=1955, 79% male, 52 years	Behavioural and bupropion	Usual care	Continuous abstinence	Yes	2.50 (1.30, 4.70)
90	Two-arm trial	2009–10	USA	N=91, 76% male, 20 years	Behavioural (health education)	Behavioural (non-health education)	No waterpipe use in past 30 days	No	1.46 (0.81, 2.62)
88	Two-arm cluster trial	2004–06	Egypt	N=7657, 45% male, 36 years	Behavioural	No intervention	No current waterpipe use	No	3.25 (1.39, 8.89)[a]
91	Two-arm cluster trial	2011–12	Lebanon and Qatar	N=1857, 6th–8th grade students	Behavioural	No intervention	No waterpipe use in past 30 days	No	1.50 (0.89, 2.53)

AOR: adjusted odds ratio; CI: confidence interval. [a] Analysis restricted to males.

10.3 Future research

Four areas of research on waterpipe tobacco should be addressed, according to the findings in this section.

1. **Surveillance of waterpipe tobacco use should continue:** nonetheless, researchers should consider using improved, standardized tools to measure prevalence, so that estimates can be compared between countries. Simple estimates of prevalence among adults in the Eastern Mediterranean Region and young people in central Europe and in the South-East Asia Region are a priority. For researchers, a toolbox of survey items is available that covers use patterns, dependence, exposure and policy *(92)*. An often overlooked but important detail is the type of waterpipe tobacco consumed. This report summarizes studies that mainly focused on *mo'assel* tobacco, a flavoured tobacco type commonly marketed in commercial venues. Other waterpipe tobacco types, often consumed in the WHO Eastern Mediterranean and South-East Asia regions, are associated with distinct patterns of use and sociodemographic correlates *(93)*.

2. **High-quality epidemiological studies on the long-term health effects of waterpipe tobacco should be conducted:** in particular, not much is known about the long-term health effects of intermittent or infrequent use or the cumulative harm of dual use of waterpipe tobacco and cigarettes *(34)*. Furthermore, understanding of the harm due to the potential of waterpipe tobacco to act as a gateway to other tobacco use will strongly inform the policy debate and action. Modelling studies could complement traditional epidemiological approaches in filling this research gap. While evidence for the long-term health effects of waterpipe tobacco use is being collected, better characterization of the product, by charcoal and tobacco type, will be an important step to better understanding its potential harm.

3. **As Parties to the WHO FCTC continue to develop and implement policies on waterpipe tobacco, the policies should be formally evaluated:** it is important to evaluate the impact of policies on the use of and attitudes towards waterpipe tobacco and any unintended consequences and to monitor industry compliance. Sharing of experience will guide efforts to reduce the health effects of waterpipe tobacco use. Key considerations are whether enforcement in waterpipe cafes will be offset by increased use at home and whether policies to reduce waterpipe tobacco use will increase use of other tobacco products.

4. **Evidence for the effectiveness of individual and group interventions is needed to support waterpipe tobacco cessation:** while research is

lacking on effective individual and group interventions, this should not delay implementation and evaluation of population-level interventions that have been shown to work. The main gaps in individual and group interventions include the effectiveness of pharmacological cessation aids *(94)* and the extent to which interventions for cigarette users can be transposed directly for waterpipe tobacco users. Researchers who are considering conducting individual or group interventions should consult an inventory of behaviour change techniques for waterpipe tobacco use *(95)*. Population-level approaches described in the MPOWER framework should be implemented immediately.

10.4 Policy recommendations

An extensive list of policy recommendations aligned with the WHO FCTC is given in documents FCTC/COP/7/10 *(1)* and FCTC/COP7(4) *(96)*. We support these recommendations as important measures for the prevention and control of waterpipe tobacco use. Below, we provide an abridged list of policy options that build on the success of the MPOWER framework and could be considered the most pertinent. This is prudent, given that countries that have not implemented MPOWER fully have a higher prevalence of waterpipe tobacco use *(97)*. While MPOWER is applicable to all tobacco products, we call for renewed implementation explicitly incorporating the particularities of waterpipe tobacco smoking, such as the predominant use of flavours, lengthy, stationary tobacco use and regular use at commercial venues. We call this revision MPOWER-W.

In implementing the policies listed below, it should be remembered that waterpipe tobacco substitutes (e.g. "herbal" substances and steam stones) still require charcoal as the heating source *(76)* and should be classified as tobacco products because of the known chemical composition of the smoke they produce, the fact that they are marketed with waterpipe tobacco products and the fact that products claimed to be tobacco-free may nevertheless contain tobacco *(39, 98)*. Similar calls have been made for products that are used as substitutes for or mimic smokeless tobacco *(99)*.

1. **Monitor tobacco use and prevention policies (Article 20 of the WHO FCTC).**

   Surveillance of waterpipe tobacco use has improved substantially during the past decade, but many countries still do not have mechanisms to estimate the prevalence of waterpipe tobacco use, exposure to second-hand smoke and waterpipe tobacco industry activities. This section shows that comparable prevalence estimates for adults are not available for the Eastern Mediterranean Region, and prevalence estimates for young people are not available for central Europe and the South-East Asia Region. High-quality surveillance is essential for the prevention and control of waterpipe tobacco use. Prevalence should be measured with standardized tools *(92)*, perhaps supplemented by routine administrative data or directories of waterpipe cafes *(100)*.

2. **Protect people from exposure to second-hand tobacco smoke (Article 8 of the WHO FCTC).**

   Commercial establishments that allow customers to smoke waterpipe tobacco on their premises should be included in comprehensive smoke-free laws. There is evidence of the harm of second-hand smoke from waterpipe smoking *(41, 43, 44)*. Studies of cigarette smoking in commercial premises have shown that comprehensive smoke-free laws reduce the harm of second-hand smoke, help smokers to quit and reduce smoking among young people *(101)*. Waterpipe tobacco sessions can be lengthy (e.g. up to several hours), and movement of smokers outdoors increases their public visibility and may be noisy and create a nuisance for nearby residents *(98)*. Turkey and several countries in the Eastern Mediterranean Region have instituted zoning laws to prohibit waterpipe cafes within a certain distance of residential areas and educational establishments *(102)*. Zoning laws should be co-implemented with smoke-free laws to maximize protection of the public.

3. **Offer help to quit tobacco use (Article 14 of the WHO FCTC).**

   Regular waterpipe tobacco smokers are nicotine dependent. Therefore, Parties to the WHO FCTC should integrate waterpipe tobacco cessation into traditional smoking cessation services. Although the literature on effective individual interventions is sparse, they show promising results and should be scaled up and evaluated for exchange among countries. It is also important to address the strong social components of waterpipe tobacco use.

4. **Warn about the dangers of tobacco (Articles 11 and 12 of the WHO FCTC).**

   Health warnings on waterpipe tobacco packs should, as for cigarettes, cover at least 50% of the principal display areas and include graphic imagery, which is more effective in reaching people with low literacy. Health warnings should also be applied to waterpipe accessories, such as the device and charcoal. Evidence strongly suggests that waterpipe tobacco use is harmful to health *(36, 47–50)*, but the industry complies poorly with WHO FCTC recommendations for health warnings *(103, 104)*. As waterpipe devices are elaborate, well decorated and a component of the positive affect of waterpipe tobacco smoking, the device should be regulated to a standardized size and pattern to reduce its attractiveness.

5. **Enforce bans on tobacco advertising, promotion and sponsorship (Article 13 of the WHO FCTC).**

    Complete bans on waterpipe tobacco advertising, promotion and sponsorship, which are commonly seen in and around waterpipe cafes and on social media, should be enforced. In contrast to the products of transnational tobacco companies, most waterpipe tobacco is advertised and promoted by retailers or at waterpipe cafes *(77, 105, 106)*. Student offers, discounts and the display of waterpipe devices in shop windows are common approaches to encouraging waterpipe tobacco use.

6. **Raise taxes on tobacco (Article 6 of the WHO FCTC).**

    Taxes on waterpipe tobacco should be raised, at least to be in line with the tax rate on cigarettes. Tobacco taxation is one of the most effective policies in tobacco control, but waterpipe tobacco is taxed at a much lower rate than cigarettes. A national survey in Lebanon indicated that a 10% increase in waterpipe tobacco taxes would result in a 14% decrease in consumption in homes *(107)*. While the impact of such a tax on use in waterpipe cafes is unknown, policy-makers should be mindful that, when waterpipe tobacco is used by a group, the effect of taxation is weakened, as the cost per user is lower. Therefore, larger increases in taxes on waterpipe tobacco will be required to achieve a similar effect to cigarette taxation.

10.4.1 Policies relevant to waterpipe tobacco use

The six policies listed above are the backbone of tobacco control policy globally as part of the comprehensive policy recommendations in the WHO FCTC; however, two further recommendations do not fall within the MPOWER framework. Introduction of MPOWER-W will guide policies to address waterpipe-specific priorities.

First, as most waterpipe tobacco is flavoured, a ban on flavours is likely to have a profound effect on the waterpipe tobacco industry *(108)*. It would reduce the attractiveness of waterpipe tobacco, promote cessation and prevent uptake, especially by young people. Policy-makers should be wary, however, that a flavour ban could be circumvented by separate sale of flavours in bottles, to be added by the user to an unflavoured tobacco mix. Such bottles are likely to be specific to waterpipe tobacco use, and any shift to this approach by the industry could be prevented by tightly regulated, enforced policy. Flavours can also be added to charcoal and water rather than to the tobacco, and parts of the apparatus can be replaced with fruit-based accessories (such as cored fruit in place of the head, where the tobacco is held). Manipulations of tobacco flavouring in retail settings should be prohibited by policy-makers to ensure that any flavour ban is upheld throughout the supply chain.

Secondly, as many of these policies target customers of waterpipe cafes, licensing might be the most effective method for reducing the burden of enforcement on local governments and ensuring that waterpipe retailers understand their legislative responsibilities *(98)*. In addition to the above policies, a licensing framework could incorporate wider protection not directly related to tobacco, such as health and safety requirements, quality control measures to promote good sanitation (e.g. disposable hoses and mouthpieces and device cleaning procedures) and age restrictions on entry. By including tobacco control policies within this licensing framework, enforcement can be cost-effective. Furthermore, depending on the country, the terms of waterpipe tobacco licenses may be under the jurisdiction of local governments, which would facilitate implementation of local policies such as flavour bans and the categorization of non-tobacco ("herbal") products as tobacco products, without the need for changes to national licensing frameworks.

10.5 Conclusions

Research on waterpipe tobacco use is increasing exponentially, providing evidence that use of this harmful tobacco product is prevalent in many countries on all continents, particularly among children aged 13–15 years. While waterpipe tobacco users may become dependent on nicotine, most are not dependent and may therefore be more susceptible to population-level policies than individual or group interventions. The prevalence of waterpipe tobacco smoking continues to rise in many countries, mainly because of lack of regulation of the industry. Public health concern about waterpipe tobacco centres on the fact that it is predominantly a flavoured product and that flavour encourages its use and may be more important than price in the decision to smoke it. A ban on flavoured waterpipe tobacco could reduce the appeal of these products, thereby reducing demand and ultimately improving public health. The ban should, however, be complementary to tobacco control policies such as higher taxation, comprehensive smoke-free laws and public education to remove misperceptions of harm. Renewed implementation of MPOWER with incorporation of the particularities of waterpipe tobacco smoking could facilitate prevention and control of waterpipe tobacco use. In the meantime, the research priorities include continued, standardized surveillance, better-quality epidemiological studies of harm, evaluation of policies and the design of prevention and cessation interventions.

10.6 References

1. Control and prevention of waterpipe tobacco products (FCTC/COP/7/10). Geneva: World Health Organization; 2016 (http://bit.ly/2xGlxSx, accessed 28 August 2017).
2. Advisory note. Waterpipe tobacco smoking: health effects, research needs and recommended actions for regulators. 2nd edition. Geneva: World Health Organization; 2015 (http://bit.ly/2i-7nAwr, accessed 10 September 2017).
3. Jawad M, Charide R, Waziry R, Darzi A, Ballout RA, Akl EA. The prevalence and trends of waterpipe tobacco smoking: A systematic review. PLoS One. 2018;13(2):e0192191.
4. Filippidis FT, Jawad M, Vardavas CI. Trends and correlates of waterpipe use in the European Union: analysis of selected Eurobarometer surveys (2009–2017). Nicotine Tob Res. 2017. doi: 10.1093/ntr/ntx255.
5. Yang Y, Jawad M, Filippidis FT, Millett C. The prevalence, correlates and trends of waterpipe tobacco smoking: a secondary analysis of the Global Youth Tobacco Survey (under review).
6. Agaku IT, King BA, Husten CG, Bunnell R, Ambrose BK, Hu SS et al. Tobacco product use among adults – United States, 2012–2013. Morb Mortal Wkly Rep. 2014;63(25):542–7.
7. Ward KD, Ahn S, Mzayek F, Al Ali R, Rastam S, Asfar T et al. The relationship between waterpipe smoking and body weight: population-based findings from Syria. Nicotine Tob Res. 2015;17(1):34–40.
8. Aden, B, Karrar S, Shafey O, Al Hosni F. Cigarette, water-pipe, and *medwakh* smoking prevalence among applicants to Abu Dhabi's pre-marital screening program, 2011. Int J Prev Med. 2013;4(11):1190–5.
9. Baron-Epel O, Shalata W, Hovell MF. Waterpipe tobacco smoking in three Israeli adult populations. Israel Med Assoc J. 2015;17(5):282–7.
10. Szklo AS, Autran Sampaio MM, Masson Fernandes E, de Almeida LM. Smoking of non-cigarette tobacco products by students in three Brazilian cities: Should we be worried? Cad Saúde Pública. 2011;27(11):2271–5.
11. Minaker LM, Shuh A, Burkhalter RJ, Manske SR. Hookah use prevalence, predictors, and perceptions among Canadian youth: findings from the 2012/2013 Youth Smoking Survey. Cancer Causes Control. 2015;26(6):831–8.
12. Kuntz B, Lampert T. Waterpipe (shisha) smoking among adolescents in Germany: results of the KiGGS study: first follow-up (KiGGS wave 1) [in German]. Bundesgesundheitsblatt Gesundheitsforsch Gesundheitsschutz. 2015;58(4–5):467–73.
13. Singh T, Arrazola RA, Corey CG, Husten CG, Neff LJ, Homa DM et al. Tobacco use among middle and high school students – United States, 2011–2015. Morbid Mortal Wkly Rep. 2016;65(14):361–7.
14. Cohn AM, Johnson AL, Rath JM, Villanti AC. Patterns of the co-use of alcohol, marijuana, and emerging tobacco products in a national sample of young adults. Am J Addict. 2016;25(8):634–40.
15. Arrazola RA, Neff LJ, Kennedy SM, Holder-Hayes E, Jones CD. Tobacco use among middle and high school students – United States, 2013. Morb Mortal Wkly Rep. 2014;63(45):1021–6.
16. Arrazola RA, Singh T, Corey CG, Husten CG, Neff LJ, Apelberg BJ et al. Tobacco use among middle and high school students – United States, 2011–2014. Morb Mortal Wkly Rep. 2015;64(14):381–5.
17. Singh T, Arrazola RA, Corey CG, Husten CG, Neff LJ, Homa DM et al. Tobacco use among middle and high school students – United States, 2011–2015. Morb Mortal Wkly Rep. 2016;65(14):361–7.
18. Centers for Disease Control and Prevention. Tobacco product use among middle and high school students –United States, 2011 and 2012. Morb Mortal Wkly Rep. 2013;62(45):893–7.
19. Bover Manderski MT, Hrywna M, Delnevo CD. Hookah use among New Jersey youth: associations and changes over time. Am J Health Behav. 2012;36(5):693–9.

20. Haider MR, Salloum RG, Islam F, Ortiz KS, Kates FR, Maziak W. Factors associated with smoking frequency among current waterpipe smokers in the United States: findings from the National College Health Assessment II. Drug Alcohol Depend. 2015;153:359–63.
21. Primack BA, Fertman CI, Rice KR, Adachi-Mejia AM, Fine MJ. Waterpipe and cigarette smoking among college athletes in the United States. J Adolesc Health. 2010;46(1):45–51.
22. Sidani JE, Shensa A, Primack BA. Substance and hookah use and living arrangement among fraternity and sorority members at US colleges and universities. J Community Health. 2013;38(2):238–45.
23. Chan WC, Leatherdale ST, Burkhalter R, Ahmed R. Bidi and hookah use among Canadian youth: an examination of data from the 2006 Canadian Youth Smoking Survey. J Adolesc Health. 2011;49(1):102–4.
24. Minaker LM, Shuh A, Burkhalter RJ, Manske SR. Hookah use prevalence, predictors, and perceptions among Canadian youth: findings from the 2012/2013 Youth Smoking Survey. Cancer Causes Control. 2015;26(6):831–8.
25. Jawad M, Lee JT, Millett C. Waterpipe tobacco smoking prevalence and correlates in 25 eastern Mediterranean and eastern European countries: cross-sectional analysis of the Global Youth Tobacco Survey. Nicotine Tob Res. 2016;18(4):395–402.
26. Saade G, Warren CW, Jones NR, Asma S, Mokdad A. Linking Global Youth Tobacco Survey (GYTS) data to the WHO Framework Convention on Tobacco Control (FCTC): the case for Lebanon. Prev Med. 2008;47(Suppl 1):S15–9.
27. McKelvey KL, Attonito J, Madhivanan P, Jaber R, Yi Q, Mzayek F et al. Time trends of cigarette and waterpipe smoking among a cohort of school children in Irbid, Jordan, 2008–11. Eur J Public Health. 2013;23(5):862–7.
28. Jaber R, Madhivanan P, Khader Y, Mzayek F, Ward KD, Maziak W. Predictors of waterpipe smoking progression among youth in Irbid, Jordan: a longitudinal study (2008–2011). Drug Alcohol Depend. 2015.153:265–70.
29. Erdol C, Ergüder T, Morton J, Palipudi K, Gupta P, Asma S. Waterpipe tobacco smoking in Turkey: policy implications and trends from the Global Adult Tobacco Survey (GATS). Int J Environ Res Public Health. 2015;12(12):15559–66.
30. Jaber R, Madhivanan P, Veledar E, Khader Y, Mzayek F, Maziak W. Waterpipe a gateway to cigarette smoking initiation among adolescents in Irbid, Jordan: a longitudinal study. Int J Tuberc Lung Dis. 2015;19(4):481–7.
31. Soneji S, Sargent JD, Tanski SE, Primack BA. Associations between initial water pipe tobacco smoking and snus use and subsequent cigarette smoking: results from a longitudinal study of us adolescents and young adults. JAMA Pediatrics. 2015;169(2):129–36.
32. Sutfin EL, Sparks A, Pockey JR, Suerken CK, Reboussin BA, Wagoner KG et al. First tobacco product tried: associations with smoking status and demographics among college students. Addict Behav 2015;51:152–7.
33. Shihadeh A, Schubert J, Klaiany J, El Sabban M, Luch A, Saliba NA. Toxicant content, physical properties and biological activity of waterpipe tobacco smoke and its tobacco-free alternatives. Tob Control. 2015;24:i22–30.
34. Jawad M, Eissenberg T, Salman R, Alzoubi KH, Khabour OF, Karaoghlianian N et al. Real-time in situ puff topography and toxicant exposure among singleton waterpipe tobacco users in Jordan. Tob Control. 2018 [Epub ahead of print]
35. Haddad L, Kelly DL, Weglicki LS, Barnett TE, Ferrell AV, Ghadban R. A systematic review of effects of waterpipe smoking on cardiovascular and respiratory health outcomes. Tob Use Insights. 2016;9:13–28.
36. El-Zaatari ZM, Chami HA, Zaatari GS. Health effects associated with waterpipe smoking. Tob Control. 2015;24:i31–43.

37. Vansickel AR, Shihadeh A, Eissenberg T. Waterpipe tobacco products: nicotine labelling versus nicotine delivery. Tob Control. 2012;21(3):377–9.
38. Raad D, Gaddam S, Schunemann HJ, Irani J, Abou Jaoude P, Honeine R et al. Effects of water-pipe smoking on lung function: a systematic review and meta-analysis. Chest. 2011;139(4):764–74.
39. Hammal F, Chappell A, Wild TC, Kindzierski W, Shihadeh A, Vanderhoek A et al. "Herbal" but potentially hazardous: an analysis of the constituents and smoke emissions of tobacco-free waterpipe products and the air quality in the cafes where they are served. Tob Control. 2015;24(3):290–7.
40. Shihadeh A, Salman R, Jaroudi E, Saliba N, Sepetdjian E, Blank MD, Cobb CO et al. Does switching to a tobacco-free waterpipe product reduce toxicant intake? A crossover study comparing CO, NO, PAH, volatile aldehydes, "tar" and nicotine yields. Food Chem Toxicol. 2012;50(5):1494–8.
41. Bentur L, Hellou E, Goldbart A, Pillar G, Monovich E, Salameh M et al. Laboratory and clinical acute effects of active and passive indoor group water-pipe (narghile) smoking. Chest. 2014;145(4):803–9.
42. Kassem NOF, Kassem NO, Liles S, Zarth AT, Jackson SR, Daffa RM et al. Acrolein exposure in hookah smokers and non-smokers exposed to hookah tobacco secondhand smoke: implications for regulating hookah tobacco products. Nicotine Tob Res. 2018;20(4):492–501.
43. Azar RR, Frangieh AH, Mroué J, Bassila L, Kasty M, Hage G et al. Acute effects of waterpipe smoking on blood pressure and heart rate: a real-life trial. Inhal Toxicol. 2016;28(8):339–42.
44. Kassem NOF, Kassem NO, Liles S, Jackson SR, Chatfield DA, Jacob P 3rd et al. Urinary NNAL in hookah smokers and non-smokers after attending a hookah social event in a hookah lounge or a private home. Regul Toxicol Pharmacol. 2017;89:74–82.
45. Cobb CO, Vansickel AR, Blank MD, Jentink K, Travers MJ, Eissenberg T. Indoor air quality in Virginia waterpipe cafes. Tob Control. 2013;22(5):338–43.
46. Gurung G, Bradley J, Delgado-Saborit JM. Effects of shisha smoking on carbon monoxide and $PM_{2.5}$ concentrations in the indoor and outdoor microenvironment of shisha premises. Sci Total Environ. 2016;548–9:340–6.
47. Akl EA, Gaddam S, Gunukula SK, Honeine R, Jaoude PA, Irani J. The effects of waterpipe tobacco smoking on health outcomes: a systematic review. Int J Epidemiol. 2010;39(3):834–57.
48. Waziry R, Jawad M, Ballout RA, Al Akel M, Akl EA. The effects of waterpipe tobacco smoking on health outcomes: an updated systematic review and meta-analysis. Int J Epidemiol. 2017;46(1):32–43.
49. Awan KH, Siddiqi K, Patil S, Hussain QA. Assessing the effect of waterpipe smoking on cancer outcome – a systematic review of current evidence. Asian Pac J Cancer Prev. 2017;18(2):495–502.
50. Mamtani, R, Cheema S, Sheikh J, Al Mulla A, Lowenfels A, Maisonneuve P. Cancer risk in waterpipe smokers: a meta-analysis. Int J Public Health. 2017;62(1):73–83.
51. Nasrallah MP, Nakhoul NF, Nasreddine L, Mouneimne Y, Abiad MG, Ismaeel H et al. Prevalence of diabetes in greater Beirut area; worsening over time. Endocr Pract. 2017;23(9):1091–1100.
52. Lotfi MH, Keyghobadi N, Javahernia N, Bazm S, Tafti AD, Tezerjani HD. The role of adverse life style factors in the cause of colorectal carcinoma in the residents of Yazd, Iran. Zahedan J Res Med Sci. 2016;18(9):e8171.
53. Larsen K, Faulkner GEJ, Boak A, Hamilton HA, Mann RE, Irving HM, To T. Looking beyond cigarettes: Are Ontario adolescents with asthma less likely to smoke e-cigarettes, marijuana, waterpipes or tobacco cigarettes? Respir Med. 2016;120:10–15.
54. Lai HT, Koriyama C, Tokudome S, Tran HH, Tran LT, Nandakumar A et al. Waterpipe tobacco smoking and gastric cancer risk among Vietnamese men. PLoS One. 2016;11(11):e0165587.
55. Fedele DA, Barnett TE, Dekevich D, Gibson-Young LM, Martinasek M, Jagger MA. Prevalence of and beliefs about electronic cigarettes and hookah among high school students with asthma. Ann Epidemiol. 2016;26(12):865–9.

56. Etemadi A, Khademi H, Kamangar F, Freedman ND, Abnet CC, Brennan P et al. Hazards of cigarettes, smokeless tobacco and waterpipe in a Middle Eastern population: a cohort study of 50 000 individuals from Iran. Tob Control. 2017;26(6):674–82.
57. Khodamoradi Z, Gandomkar A, Poustchi H, Salehi A, Imanieh MH, Etemadi A et al. Prevalence and correlates of gastroesophageal reflux disease in southern Iran: Pars Cohort Study. Middle East J Dig Dis. 2017;9(3):129–38.
58. Bandiera FC, Loukas A, Wilkinson AV, Perry CL. Associations between tobacco and nicotine product use and depressive symptoms among college students in Texas. Addict Behav. 2016;63:19–22.
59. Abdollahpour I, Nedjat S, Sahraian MA, Mansournia MA, Otahal P, van der Mei I. Waterpipe smoking associated with multiple sclerosis: a population-based incident case–control study. Mult Scler. 2017;23(10):1328–35.
60. Tamim H, Akkary G, El-Zein A, El-Roueiheb Z, El-Chemaly S. Exposure of pre-school children to passive cigarette and narghile smoke in Beirut. Eur J Public Health. 2006;16(5):509–12.
61. Mohammad Y, Shaaban R, Hassan M, Yassine F, Mohammad S, Tessier JF et al. Respiratory effects in children from passive smoking of cigarettes and narghile: ISAAC phase three in Syria. Int J Tuberc Lung Dis. 2014;18(11):1279–84.
62. Wu F, Chen Y, Parvez F, Segers S, Argos M, Islam T et al. A prospective study of tobacco smoking and mortality in Bangladesh. PLoS One. 2013;8(3):e58516.
63. Akl EA, Ward KD, Bteddini D, Khaliel R, Alexander AC, Lotfi T et al. The allure of the waterpipe: a narrative review of factors affecting the epidemic rise in waterpipe smoking among young persons globally. Tob Control. 2015;24(Suppl 1):i13–21.
64. Akl EA, Jawad M, Lam WY, Co CN, Obeid R, Irani J. Motives, beliefs and attitudes towards waterpipe tobacco smoking: a systematic review. Harm Reduction J. 2013;10(1):12.
65. Griffiths MA, Ford EW. Hookah smoking: behaviors and beliefs among young consumers in the United States. Soc Work Public Health. 2014;29(1):17–26.
66. MPOWER. Advancing the WHO Framework Convention on Tobacco Control (WHO FCTC). Geneva: World Health Organization; 2018 (http://www.who.int/cancer/prevention/tobacco_implementation/mpower/en/, accessed 1 November 2018).
67. International Tobacco Control Policy Evaluation Project. Waterloo, Ontario; 2017 (https://bit.ly/1pe5ELk, accessed 3 August 2018).
68. El Hajj DG, Cook PF, Magilvy K, Galbraith ME, Gilbert L, Corwin M. Tobacco use among Arab immigrants living in Colorado: prevalence and cultural predictors. J Transcult Nurs. 2017;28(2):179–86.
69. Roskin J, Aveyard P. Canadian and English students' beliefs about waterpipe smoking: a qualitative study. BMC Public Health. 2009;9:10.
70. Katurji M, Daher N, Sheheitli H, Saleh R, Shihadeh A. Direct measurement of toxicants inhaled by water pipe users in the natural environment using a real-time in situ sampling technique. Inhal Toxicol. 2010;22(13):1101–9.
71. Alzoubi KH, Khabour OF, Azab M, Shqair DM, Shihadeh A, Primack B et al. CO exposure and puff topography are associated with Lebanese waterpipe dependence scale score. Nicotine Tob Res. 2013;15(10):1782–6.
72. Bejjani N, I Bcheraoui C, Adib SM. The social context of tobacco products use among adolescents in Lebanon (MedSPAD-Lebanon). J Epidemiol Global Health. 2012;2(1):15–22.
73. Alzyoud S, Weglicki LS, Kheirallah KA, Haddad L, Alhawamdeh KA. Waterpipe smoking among middle and high school Jordanian students: patterns and predictors. Int J Environ Res Public Health. 2013;10(12):7068–82.
74. Carroll MV, Chang J, Sidani JE, Barnett TE, Soule E, Balbach E et al. Reigniting tobacco ritual: waterpipe tobacco smoking establishment culture in the United States. Nicotine Tob Res.

2014;16(12):1549–58.
75. Joudrey PJ, Jasie KA, Pykalo L, Singer ST, Woodin MB, Sherman S. The operation, products and promotion of waterpipe businesses in New York City, Abu Dhabi and Dubai. East Mediterr Health J. 2016;22(4):237–43.
76. Amin TT, Amr MAM, Zaza BO, Suleman W. Harm perception, attitudes and predictors of waterpipe (shisha) smoking among secondary school adolescents in Al-Hassa, Saudi Arabia. Asian Pac J Cancer Prev. 2010;11(2):293–301.
77. Jawad M, Nakkash RT, Hawkins B, Akl EA. Waterpipe industry products and marketing strategies: analysis of an industry trade exhibition. Tob Control. 2015;24(e4):e275–9.
78. Kowitt SD, Meernik C, Baker HM, Osman A, Huang LL, Goldstein AO. Perceptions and experiences with flavored non-menthol tobacco products: a systematic review of qualitative studies. Int J Environ Res Public Health. 2017;14(4):338.
79. Ambrose BK, Day HA, Rostron B, Conway KP, Borek N, Hyland A et al. Flavored tobacco product use among US youth aged 12–17 years, 2013–2014. JAMA. 2015;314(17):1871–3.
80. Salloum RG, Maziak W, Hammond D, Nakkash R, Islam F, Cheng X et al. Eliciting preferences for waterpipe tobacco smoking using a discrete choice experiment: implications for product regulation. BMJ Open. 2015;5(9):e009497.
81. Leavens EL, Driskill LM, Molina N, Eissenberg T, Shihadeh A, Brett EI, Floyd E et al. Comparison of a preferred versus non-preferred waterpipe tobacco flavour: subjective experience, smoking behaviour and toxicant exposure. Tob Control. 2018;27(3):319–24.
82. Salameh P, Waked M, Aoun Z. Waterpipe smoking: construction and validation of the Lebanon Waterpipe Dependence Scale (LWDS-11). Nicotine Tob Res. 2008.10(1):149–58.
83. Haider MR, Salloum RG, Islam F, Ortiz KS, Kates FR, Maziak W. Factors associated with smoking frequency among current waterpipe smokers in the United States: findings from the National College Health Assessment II. Drug Alcohol Depend. 2015;153:359–63.
84. Shihadeh A. Investigation of mainstream smoke aerosol of the nargileh water pipe. Food Chem Toxicol. 2003;41(1):143–52.
85. Hoffmann D, Hoffmann I. Tobacco smoke components. Beitr Tabakforsch Int. 1998;18(1):49–52.
86. Jawad M, Jawad S, Waziry RK, Ballout RA, Akl EA. Interventions for waterpipe tobacco smoking prevention and cessation: a systematic review. Sci Rep. 2016;6:25872.
87. Dogar O, Jawad M, Shah SK, Newell JN, Kanaan M, Khan MA et al. Effect of cessation interventions on hookah smoking: post-hoc analysis of a cluster-randomized controlled trial. Nicotine Tob Res. 2014;16(6):682–8.
88. Mohlman MK, Boulos DN, El Setouhy M, Radwan G, Makambi K, Jillson I et al. A randomized, controlled community-wide intervention to reduce environmental tobacco smoke exposure. Nicotine Tob Res. 2013;15(8):1372–81.
89. Asfar T, Al Ali R, Rastam S, Maziak W, Ward KD. Behavioral cessation treatment of waterpipe smoking: the first pilot randomized controlled trial. Addict Behav. 2014;39(6):1066–74.
90. Lipkus IM, Eissenberg T, Schwartz-Bloom RD, Prokhorov AV, Levy J. Affecting perceptions of harm and addiction among college waterpipe tobacco smokers. Nicotine Tob Res. 2011;13(7):599–610.
91. Nakkash RT, Al Mulla A, Torossian L, Karhily R, Shuayb L, Mahfoud ZR et al. Challenges to obtaining parental permission for child participation in a school-based waterpipe tobacco smoking prevention intervention in Qatar. BMC Med Ethics. 2014;15:70.
92. Maziak W, Ben Taleb Z, Jawad M, Afifi R, Nakkash R, Akl EA et al. Consensus statement on assessment of waterpipe smoking in epidemiological studies. Tob Control. 2017;26(3):338–43.
93. Jawad M, Lee JT, Millett C. The relationship between waterpipe and cigarette smoking in low and middle income countries: cross-sectional analysis of the global adult tobacco survey. PLoS One. 2014;9(3):e93097.

94. Maziak W, Jawad M, Jawad S, Ward KD, Eissenberg T, Asfar T. Interventions for waterpipe smoking cessation. Cochrane Database Syst Rev. 2015;7:CD005549.
95. O'Neill N, Dogar O, Jawad M, Kellar I, Kanaan M, Siddiqi K. Which behaviour change techniques may help waterpipe smokers to quit? An expert consensus using a modified delphi technique. Nicotine Tob Res. 2018;20(2):154–60.
96. Decision FCTC/COP7(4). Control and prevention of waterpipe tobacco products. Geneva: World Health Organization; 2016 (https://bit.ly/2vbWd9s, accessed 3 August 2018).
97. Heydari, G, EbnAhmady A, Lando HA, Chamyani F, Masjedi M, Shadmehr MB. Third study on WHO MPOWER tobacco control scores in Eastern Mediterranean countries 2011–2015. East Mediterr Health J. 2017;23(9):598–603.
98. Jawad M. Legislation enforcement of the waterpipe tobacco industry: a qualitative analysis of the London experience. Nicotine Tob Res. 2014;16(7):1000–8.
99. Mukherjea A, Modayil MV, Tong EK. Paan (pan) and paan (pan) masala should be considered tobacco products. Tob Control. 2015;24(e4):e280–4.
100. Cawkwell PB, Lee L, Weitzman M, Sherman SE. Tracking hookah bars in New York: utilizing Yelp as a powerful public health tool. JMIR Public Health Surveill. 2015;1(2):e19.
101. Protect people from exposure to second-hand tobacco smoke. Geneva: World Health Organization, Tobacco Free Initiative; 2018 (https://bit.ly/2vbkUCU, accessed 3 August 2018).
102. Jawad M, El Kadi L, Mugharbil S, Nakkash R. Waterpipe tobacco smoking legislation and policy enactment: a global analysis. Tob Control. 2015;24:i60–5.
103. Jawad M, Darzi A, Lotfi T, Nakkash R, Hawkins B, Akl EA. Waterpipe product packaging and labelling at the 3rd international Hookah Fair; does it comply with Article 11 of the Framework Convention on Tobacco Control? J Public Health Policy. 2017;19:19.
104. Jawad M, Choaie E, Brose L, Dogar O, Grant A, Jenkinson E et al., Waterpipe tobacco use in the United Kingdom: a cross-sectional study among university students and stop smoking practitioners. PLoS One. 2016;11 (1):e0146799.
105. Primack BA, Walsh M, Bryce C, Eissenberg T. US hookah tobacco smoking establishments advertised on the internet. Am J Prev Med. 2012;42(2):150–6.
106. Griffiths MA, Harmon TR, Gilly MC. Hubble bubble trouble: the need for education about and regulation of hookah smoking. J Public Policy Market. 2011;30(1):119–32.
107. Jawad M, Lee JT, Glantz S, Millett C. Price elasticity of demand of non-cigarette tobacco products: a systematic review and meta-analysis. Tob Control. 2018; doi: 10.1136/tobaccocontrol-2017-054056.
108. Jawad M, Millett C. Impact of EU flavoured tobacco ban on waterpipe smoking. BMJ. 2014;348:2698.

11. Overall recommendations

The WHO Study Group on Tobacco Product Regulation publishes reports to provide a scientific basis for tobacco product regulation. In line with Articles 9 and 10 of the WHO FCTC, the reports identify evidence-based approaches to the regulation of tobacco products.

At its ninth meeting, the Study Group discussed: the prevalence and health effects of waterpipe tobacco smoking and interventions to reduce use; approaches to reducing toxicant concentrations in tobacco products, including cigarettes and smokeless tobacco; the state of the science on a global nicotine reduction strategy; the clinical pharmacology of nicotine in ENDS; the sugar content of tobacco products; a regulatory strategy for reducing exposure to toxicants in cigarette smoke; an updated list of priority toxicants in tobacco products for regulatory purposes; heated tobacco products; and the science of flavours in tobacco products. The aim of the discussions was to update knowledge in these areas in order to inform policy at global level and to advance tobacco product regulation.

The report provides guidance through the Executive Board to Member States. It focuses primarily on requests of the COP to the WHO FCTC to WHO through the Convention Secretariat at its seventh session, in 2016, as articulated in decisions FCTC/COP7(4), FCTC/COP7(9) and FCTC/COP7(14). These decisions informed the content of the background papers in the above areas, for which Member States have requested technical assistance as a basis for national policies. The 10-member Study Group invited subject matter experts, who drafted background papers, contributed to discussions and provided the most up-to-date empirical data on the topics considered. Sections 2–10 of the report provide scientific information and policy recommendations to guide Member States in navigating difficult issues in tobacco product regulation. Further, the report provides guidance to Member States on the most effective evidence-based means for bridging regulatory gaps in tobacco control and for developing coordinated regulatory frameworks for tobacco products to guide international policy. Additionally, it identifies areas for further work and future research, focusing on the regulatory needs of Member States; it takes into consideration regional differences, thus providing a strategy for continued, targeted technical support to Member States.

11.1 Main recommendations

The main recommendations of this report to policy-makers and all other interested parties include the following.

- Monitor and collect reliable, independent data on heated tobacco products and alternative products in order to understand behaviour and potential risks to users and bystanders and to verify claims of reduced exposure and risk.
- Consider and examine the design features that determine nicotine flux in ENDS and the extent to which these products could promote or impede cessation of smoking, and invest in research on appropriate policies and regulations on ENDS.
- Consider a regulatory strategy for reducing exposure to toxicants in combusted tobacco product smoke that includes a nicotine level in tobacco that does not exceed 0.4 mg/g of tobacco (0.04 mg nicotine per combusted product in mainstream smoke under HCI smoking conditions). This should be accompanied by a reliable system for monitoring regulated constituents in tobacco and smoke, comprehensive tobacco control and concerted national and international efforts to prevent black markets.
- Consider a nicotine reduction policy coordinated with policies that allow adequate access to nicotine replacement therapies and other products, if and as approved by relevant authorities and with appropriate safeguards. This should be supported by population surveillance, monitoring and testing of products, enforcement of product standards, and a strong focus on protecting children and young people.
- Consider banning or restricting the use of flavours in nicotine delivery systems and tobacco products in order to reduce initiation by young people, and consider banning or restricting flavours in combusted tobacco products to promote cessation.
- Require manufacturers to disclose relevant information on sugar content, and consider lowering the level of sugars in tobacco products to reduce their effects on product toxicity, addictiveness and attractiveness.
- Require manufacturers, as applicable and appropriate, to report priority toxicants analysed with methods based on the SOPs of the WHO TobLabNet.
- Lower the levels of addictive, toxic and carcinogenic agents in tobacco products, including cigarettes and smokeless tobacco, recognizing that decreasing the levels of these agents will not make these products safe.
- Consider applying WHO FCTC provisions to waterpipe tobacco for the prevention and control of waterpipe tobacco use.

Continuing research is required to monitor product development and use, promotional strategies and other activities of the tobacco and related industry to build intelligence to protect public health. Specific recommendations on each of the topics considered in this report can be found in sections 2.9, 3.8, 4.5–4.6, 5.8, 6.8, 7.10, 8.8, 9.9 and 10.3–10.5.

11.2 Significance for public health policies

The Study Group's report provides helpful guidance for understanding the content, emissions and design features of selected products, such as smokeless tobacco, waterpipes, heated tobacco products and ENDS and describes the public health impact of these products and features. In recent years, unconventional nicotine and tobacco products have permeated several markets, for which there is no precedent, and these present unique regulatory challenges to Member States. Further, there is better understanding of the science, adverse effects, characteristics, contents and emissions of conventional products owing to advances in knowledge; therefore, the report provides updated information for Member States on novel and emerging tobacco products and nicotine delivery systems to support them in formulating effective strategies for regulating tobacco and nicotine products.

Because of the unique composition of the Study Group, with regulatory, technical and scientific experts, it can navigate and distil complex data and research and synthesize them into recommendations for policy development at country, regional and global levels. The recommendations promote international coordination of regulatory efforts and the adoption of best practices in tobacco product regulation, strengthen capacity for tobacco product regulation in all WHO regions and provide a ready, science-based resource for Member States.

11.3 Implications for the Organization's programmes

The report fulfils the mandate of the WHO Study Group on Tobacco Product Regulation to provide the Director-General with scientifically sound, evidence-based recommendations for Member States about tobacco product regulation. Tobacco product regulation is a highly technical area of tobacco control, in which Member States face complex regulatory challenges. The outcomes of the Study Group's deliberations and its main recommendations will improve Member States' understanding of tobacco and nicotine products. The report's contribution to the body of knowledge on tobacco product regulation will play a pivotal role in informing the work of the tobacco programme in the WHO Department for Prevention of Noncommunicable Diseases, especially in providing technical support to Member States. It will also contribute to further development of partial guidelines for implementation of Articles 9 and 10 of the WHO FCTC.